Fate and Effects of Anticancer Drugs in the Environment

Ester Heath • Marina Isidori • Tina Kosjek
Metka Filipič
Editors

Fate and Effects of Anticancer Drugs in the Environment

Springer

Editors
Ester Heath
Department of Environmental Sciences
Jožef Stefan Institute
Ljubljana, Slovenia

Jožef Stefan International Postgraduate School
Ljubljana, Slovenia

Tina Kosjek
Department of Environmental Sciences
Jožef Stefan Institute
Ljubljana, Slovenia

Jožef Stefan International Postgraduate School
Ljubljana, Slovenia

Marina Isidori
Department of Environmental, Biological and Pharmaceutical Sciences and Technologies
University of Campania "L.Vanvitelli"
Caserta, Italy

Metka Filipič
Department of Genetic Toxicology and Cancer Biology
National Institute of Biology
Ljubljana, Slovenia

ISBN 978-3-030-21047-2 ISBN 978-3-030-21048-9 (eBook)
https://doi.org/10.1007/978-3-030-21048-9

This Springer imprint is published by the registered company Springer Nature Switzerland AG.
The registered company address is: Gewerbestrasse 11, 6330 Cham, Switzerland

Preface

Cytostatic drugs (also known as anticancer drugs or antineoplastics) have been at the forefront of treating cancer since the 1940s. However, many of these drugs are themselves carcinogenic, mutagenic and teratogenic, triggering widespread concerns about the risks they pose to the environment. The scientific community is now turning its attention toward investigating the occurrence, effects, and fate of cytostatic pharmaceutical residues in the environment. A watershed moment was in 2011 with the funding of two major FP7 EU projects CytoThreat and Pharmas, which led to an increase in the number of scientific publications, including a special issue in *Environmental Science and Pollution Research* (Fate and effects of the residues of anticancer drugs in the environment, 2016, vol. 23, no. 15, pp. 14687–14691). The motivation for this book was the desire to bring together current knowledge and research on the presence and effects of cytostatic drug residues in the aqueous environment. This book contains 18 chapters and represents the combined work of leading scientists from different research institutions from across the globe. It covers all relevant aspects of the presence of cytostatic drug residues in wastewaters and natural aquatic systems, where numerous analogies are made between their pharmacokinetics and pharmacodynamics in humans and their effects on the environment. For example, in the case of pharmacokinetics, their levels in humans or animals can be analogous with their *occurrence* in waste and natural waters, distribution in the body with environmental cycling (*fate)*, human metabolism with their *transformation*, and their elimination from humans or animals with their removal. All these parameters are based on the *determination* (analytical method development) of cytostatic drug residues in different, often complex matrices. In terms of pharmacodynamics, their effects on target tissues along with unwanted adverse reactions are also analogous with their *effects* (toxicity) on non-target organisms, which must be known in order to perform *hazard and risk assessment*.

Scheme 1 provides the reader with an overview of how the various chapters relate to environmental pharmacokinetics and pharmacodynamics.

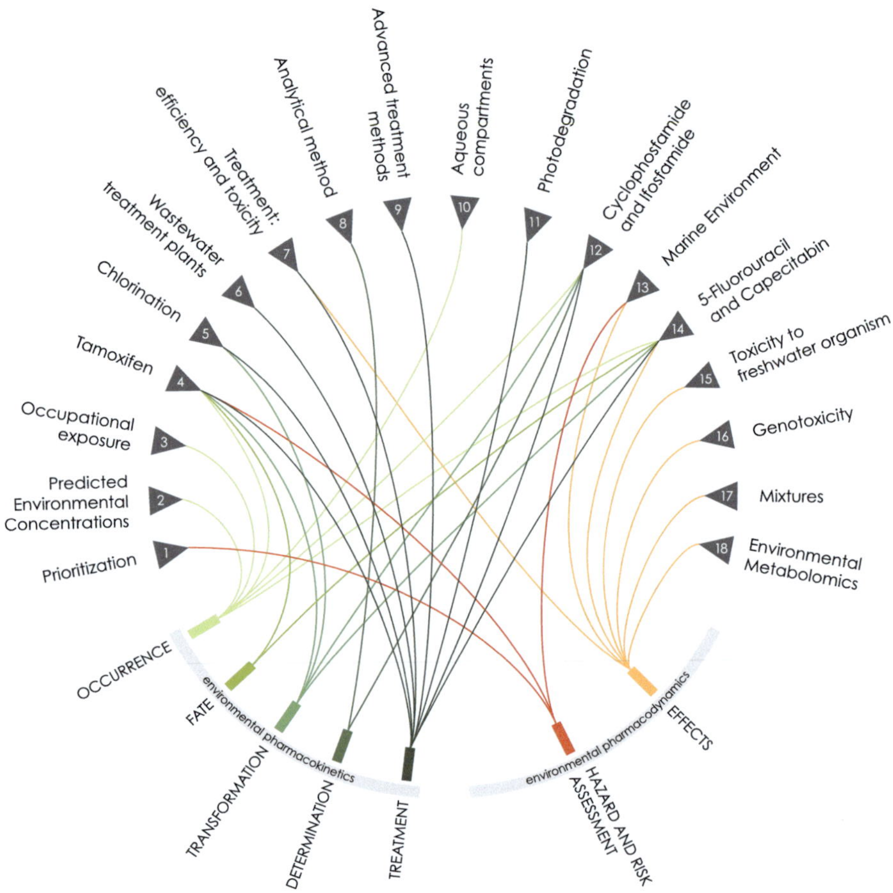

With regards to environmental pharmacokinetics, Chaps. 3, 4, 10, 12 and 14 discuss the *occurrence* of these compounds in the workplace (hospitals, pharmacies), wastewaters, surface waters and drinking water using data obtained experimentally, while Chap. 2 calculates the predicted environmental concentrations (PEC) based on consumption data, excretion, elimination in the WWTP and dilution in receiving waters. Chapter 10 reviews studies performed in 18 countries reporting the levels of compounds in different water compartments and highlights the importance of planning efficient sampling strategies to obtain representative water samples. Chapter 3 discusses occupational exposure to drugs containing 5-fluorouracil (FU), cyclophosphamide (CP), and platinum in 21 hospitals in the Czech Republic. Unwanted releases have been documented during all steps in the preparation and administration of these drugs to patients leading to contamination of both the work place and the environment. In many cases, monitoring and discussions with responsible managers resulted in the implementation of proper handling procedures and decreased contamination. Chapter 12 covers the occurrence of CP and ifosfamide (IF) residues in hospital wastewaters, municipal wastewaters and surface waters. The occurrence of tamoxifen and its metabolites in water bodies and the risk posed to aquatic organisms due to its known toxicity and its potential for bioaccumulation are subject of Chap. 4, while Chap. 14 examines the available literature on the cycling and effects of 5-FU and capecitabine (CAP) residues. Chapters 4, 5, 12 and 14 address *transformation* during various treatment processes, and Chaps. 4 and 14 cover their environmental *fate*. Method development and analysis (*determination*) is the focus of Chaps. 8 and 12. Chapter 8 provides a comprehensive overview of the different methodologies used, including extraction and clean-up techniques (solid-phase extraction for liquid samples and pressurized liquid extraction and ultrasound extraction for sludge samples) and detection methods (mostly liquid chromatography with mass spectrometry detection). Chapter 12 describes the analytical method development for determining CP and IF transformation products formed during water treatment.

Whereas the previously mentioned chapters measure the levels of cytostatic residues, Chap. 2 introduces the concept of predicted environmental concentrations (PECs), an approach first suggested by the European Medicines Agency (EMA). It describes how to calculate PECs, provides raw data for their calculation and discusses the applicability of PECs for assessing cytostatics in wastewater and river water.

Different *treatment* technologies are also extensively discussed in the book. *Biological treatment* is discussed in several of the book chapters. For instance, conventional activated sludge treatment is discussed in Chaps. 6, 12 and 14. Membrane bioreactors (MBR) are described in Chaps. 6, 9 and 12 and sequential batch reactors (SBR) in Chap. 12. Bioreactors with attached biomass are covered in Chap. 12 and biological treatment using fungi in Chaps. 4, 7 and 12. *Abiotic treatment* involves UV irradiation, chlorination, ozonation and advanced oxidation processes (AOPs). UV irradiation and direct photolysis are discussed in Chaps. 6, 7, 11, 12 and 14. Ozonation is covered in Chaps. 6, 7, 12 and 14 and chlorination in Chaps. 4, 5, 6, 7, 9. Different AOPs, including ozonation and/or UV in combination with different oxidants (H_2O_2) or catalysts (Fe^{2+}, TiO_2) and other non-conventional oxidation processes like electrochemical oxidation, are discussed in Chaps. 4, 7, 9, 11, 12 and 14. *Physical treatments* are also discussed at length, including adsorption (Chaps. 9, 12 and 14), nanofiltration (Chaps. 9 and 12) and reverse osmosis (Chaps. 9 and 12).

With regards to *environmental pharmacodynamics*, single compound *effects* are the subject of Chaps. 14, 15 and 16. Chapter 15 investigates the acute and chronic effects of cytostatic drugs in freshwater organisms, while Chap. 14 describes the effects of 5-FU and CAP residues in the aqueous environment and Chap. 16 discusses how residues of specific drugs are genotoxic for aquatic organisms. Their main conclusions are that more work into the toxicological effects of environmental mixtures of cytotoxic compounds still needs to be performed, further actions are needed to ensure more reliable environmental risk assessments, and that stricter measures are necessary to prevent contamination of the environment by cytostatic drug residues. Mixture effects are addressed in Chaps. 13 and 17. The acute and chronic effects on non-target organisms at levels typically found in the marine environment are reviewed in Chap. 13, together with their ecotoxicological potential, synergistic, additive and antagonist effects. Similarly, Chap. 17 focuses on the toxicity of mixtures of 5-FU, cisplatin, etoposide, imatinib mesylate, and CP and their transformation products on various biological models (bacteria, algae, animals, plants and human cells). Chapter 7 uses the toxicity of mixtures of cytostatic drugs, their metabolites and/or transformation products in addition to the removal of the parent compound in order to evaluate removal efficiency during wastewater treatment. In Chap. 18, environmental metabolomics is used to enrich our understanding of the response of an organism to environmental stressors, while Chaps. 1, 4 and 13 focus on *hazard and risk assessment.* Chapter 1 introduces different approaches for screening, ranking and prioritizing specific cytostatic compounds that pose the most significant risk. For example, environmental risk posed by tamoxifen is evaluated in Chap. 4, while Chap. 13 focuses on the risks posed by cytotoxic drug residues in the marine environment. It also makes recommendations about suitable biological models to assess the ecotoxicological effects on marine organisms.

We are aware that this book tackles only a small part of what is a far more extensive and complex issue, but we hope that information provided within this book will enable readers to learn about the fate and effects of cytostatic pharmaceuticals in the aqueous environment and be cognizant of the many challenges that remain. We are thankful to the authors for their contribution and their patience during the publication of the book. We thank the Springer team, namely Alexandrine Cheronet and Judith Terpos, for their continued support during the project.

Ljubljana, Slovenia — Ester Heath
Caserta, Italy — Marina Isidori
Ljubljana, Slovenia — Metka Filipič
Ljubljana, Slovenia — Tina Kosjek
January, 2019

Contents

Contributors

Mustafa Al Aukidy Department of Engineering, University of Ferrara, Ferrara, Italy

Arminda Alves LEPABE – Laboratory for Process, Environmental, Biotechnology and Energy Engineering, Faculty of Engineering, University of Porto, Porto, Portugal

Maria João Bebianno CIMA, University of Algarve, Faro, Portugal

Ludek Blaha Faculty of Science, RECETOX, Masaryk University, Brno, Czech Republic

Lucie Blahova Faculty of Science, RECETOX, Masaryk University, Brno, Czech Republic

Aina Campos Garrigós Department of Engineering, University of Ferrara, Ferrara, Italy
Faculty of Earth Science, University of Barcelona, Barcelona, Spain

Marjeta Česen Department of Environmental Sciences, Jožef Stefan Institute, Ljubljana, Slovenia
Jožef Stefan International Postgraduate School, Ljubljana, Slovenia

Chiara Russo Department of Environmental, Biological and Pharmaceutical Sciences and Technologies, University of Campania "L.Vanvitelli", Caserta, Italy

Concetta Piscitelli Department of Environmental, Biological and Pharmaceutical Sciences and Technologies, University of Campania "L.Vanvitelli", Caserta, Italy

Tainá Garcia da Fonseca CIMA, University of Algarve, Faro, Portugal

Miren López de Alda Water, Environmental and Food Chemistry Unit, Department of Environmental Chemistry, Institute of Environmental Assessment and Water Research (IDAEA-CSIC), Barcelona, Spain

Lenka Dolezalova Masaryk Memorial Cancer Institute, Brno, Czech Republic

Carlos Escudero-Oñate Norwegian Institute for Water Research (NIVA), Oslo, Norway

Despo Fatta-Kassinos Nireas-International Water Research Center, University of Cyprus, Nicosia, Cyprus
Department of Civil and Environmental Engineering, University of Cyprus, Nicosia, Cyprus

Laura Ferrando-Climent Tracer Technology Department, Fluid Flow and Environmental Technology Section, Institute for Energy Technology (IFE), Kjeller, Norway

Metka Filipič Department of Genetic Toxicology and Cancer Biology, National Institute of Biology, Ljubljana, Slovenia

Antonio Fiorentino Department of Environmental, Biological and Pharmaceutical Sciences and Technologies, University of Campania "L.Vanvitelli", Caserta, Italy

Helena Franquet-Griell Department of Environmental Chemistry, IDAEA-CSIC, Barcelona, Catalonia, Spain

Goran Gajski Mutagenesis Unit, Institute for Medical Research and Occupational Health, Zagreb, Croatia

Marko Gerić Mutagenesis Unit, Institute for Medical Research and Occupational Health, Zagreb, Croatia

Cristian Gómez-Canela Department of Environmental Chemistry, IDAEA-CSIC, Barcelona, Catalonia, Spain

Juan Carlos Guillen Water, Environmental and Food Chemistry Unit, Department of Environmental Chemistry, Institute of Environmental Assessment and Water Research (IDAEA-CSIC), Barcelona, Spain

Ester Heath Department of Environmental Sciences, Jožef Stefan Institute, Ljubljana, Slovenia
Jožef Stefan International Postgraduate School, Ljubljana, Slovenia

Lida Ioannou-Ttofa Nireas-International Water Research Center, University of Cyprus, Nicosia, Cyprus

Marina Isidori Department of Environmental, Biological and Pharmaceutical Sciences and Technologies, University of Campania "L. Vanvitelli", Caserta, Italy

Tina Kosjek Department of Environmental Sciences, Jožef Stefan Institute, Ljubljana, Slovenia
Jožef Stefan International Postgraduate School, Ljubljana, Slovenia

Sarka Kozakova Masaryk Memorial Cancer Institute, Brno, Czech Republic

Klaus Kümmerer Sustainable Chemistry and Material Resources, Institute of Sustainable and Environmental Chemistry, Faculty of Sustainability, Leuphana University of Lüneburg, Lüneburg, Germany

Jan Kuta Faculty of Science, RECETOX, Masaryk University, Brno, Czech Republic

Silvia Lacorte Department of Environmental Chemistry, IDAEA-CSIC, Barcelona, Catalonia, Spain

Margherita Lavorgna Department of Environmental, Biological and Pharmaceutical Sciences and Technologies, University of Campania "L. Vanvitelli", Caserta, Italy

Teik-Thye Lim School of Civil and Environmental Engineering, Nanyang Technological University, Singapore, Singapore

Milka Ljoncheva Department of Environmental Sciences, Jožef Stefan Institute, Ljubljana, Slovenia

Carlos Alexandre Lutterbeck Graduate Program in Environmental Technology, Universidade de Santa Cruz do Sul, Santa Cruz do Sul, Rio Grande do Sul, Brazil

Ênio Leandro Machado Graduate Program in Environmental Technology, Universidade de Santa Cruz do Sul, Santa Cruz do Sul, Rio Grande do Sul, Brazil

Giovanni Mastroianni Department of Environmental, Biological and Pharmaceutical Sciences and Technologies, University of Campania "L.Vanvitelli", Caserta, Italy

Sarah Montesdeoca-Esponda Instituto Universitario de Estudios Ambientales y Recursos Naturales (i-UNAT), Universidad de Las Palmas de Gran Canaria, Las Palmas de Gran Canaria, Spain

Matjaž Novak Department of Genetic Toxicology and Cancer Biology, National Institute of Biology, Ljubljana, Slovenia

Adrián Olalla Research Group in Environmental Toxicology and Risk Assessment (TAyER), Rey Juan Carlos University, Alcorcón, Madrid, Spain

Jose Luis Rodriguez-Gil Department of Biology, University of Ottawa, Ottawa, ON, Canada

Sara Rodríguez-Mozaz Catalan Institute for Water Research (ICRA), H2O Building, Scientific and Technological Park of the University of Girona, Girona, Spain

José Juan Santana-Rodríguez Instituto Universitario de Estudios Ambientales y Recursos Naturales (i-UNAT), Universidad de Las Palmas de Gran Canaria, Las Palmas de Gran Canaria, Spain

Sergio Santana-Viera Instituto Universitario de Estudios Ambientales y Recursos Naturales (i-UNAT), Universidad de Las Palmas de Gran Canaria, Las Palmas de Gran Canaria, Spain

Mónica S. F. Santos LEPABE – Laboratory for Process, Environmental, Biotechnology and Energy Engineering, Faculty of Engineering, University of Porto, Porto, Portugal

Monica Scognamiglio Department of Environmental, Biological and Pharmaceutical Sciences and Technologies, University of Campania "L.Vanvitelli", Caserta, Italy

Ewa Maria Siedlecka Department of Chemistry, University of Gdansk, Gdansk, Poland

Zoraida Sosa-Ferrera Instituto Universitario de Estudios Ambientales y Recursos Naturales (i-UNAT), Universidad de Las Palmas de Gran Canaria, Las Palmas de Gran Canaria, Spain

Yolanda Valcárcel Research Group in Environmental Toxicology and Risk Assessment (TAyER), Rey Juan Carlos University, Alcorcón, Madrid, Spain
Department of Medicine and Surgery, Psychology, Preventive Medicine and Public Health, Immunology and Medical Microbiology, Faculty of Health Sciences, Rey Juan Carlos University, Alcorcón, Madrid, Spain

Francesc Ventura Department of Environmental Chemistry, IDAEA-CSIC, Barcelona, Catalonia, Spain

Paola Verlicchi Department of Engineering, University of Ferrara, Ferrara, Italy
Terra&Acqua Tech, Technopole of the University of Ferrara, Ferrara, Italy

Verica Garaj Vrhovac Mutagenesis Unit, Institute for Medical Research and Occupational Health, Zagreb, Croatia

Bojana Žegura Department of Genetic Toxicology and Cancer Biology, National Institute of Biology, Ljubljana, Slovenia

Yiqing Zhang School of Civil and Environmental Engineering, Nanyang Technological University, Singapore, Singapore

Božo Žonja Water, Environmental and Food Chemistry Unit, Department of Environmental Chemistry, Institute of Environmental Assessment and Water Research (IDAEA-CSIC), Barcelona, Spain

Part I
Prioritisation and Predicted Environmental Concentrations

Chapter 1
Approaches for Hazard Assessment Screening, Ranking, and Prioritization of Cytostatic Compounds

Adrián Olalla, Jose Luis Rodriguez-Gil, and Yolanda Valcárcel

Abstract Since the 1990s, the continued input of pharmaceutical and personal care compounds (PPCPs) into the environment, mainly via wastewater treatment plants, has become one of the main areas of research for environmental chemists and toxicologists. The assessment of the risk posed by cytostatic compounds is of particular interest because of their cytotoxic, and in many cases carcinogenic, nature. The availability of data on the environmental presence and effects of cytostatic agents in the environment is currently very limited; at the same time, the amount of resources available to be dedicated to the study of these aspects is also limited. To prioritize the cytostatic agents on which we need to focus our research and monitoring, as well as to determine whether specific treatment systems should be implemented to remove these compounds before reaching natural water flows, the risk and environmental hazards of these substances must be calculated. In this context, this chapter introduces a number of screening, ranking, and prioritization approaches that could be used to highlight specific cytostatic compounds posing greater risks. Proof-of-concept application of some of these approaches is presented.

Keywords Cytostatic compounds · Priorization · Emerging contaminants · Risk and hazard

A. Olalla
Research Group in Environmental Toxicology and Risk Assessment (TAyER), Rey Juan Carlos University, Alcorcón, Madrid, Spain

J. L. Rodriguez-Gil
Department of Biology, University of Ottawa, Ottawa, ON, Canada
e-mail: jose.rodriguez@uottawa.ca

Y. Valcárcel (✉)
Research Group in Environmental Toxicology and Risk Assessment (TAyER), Rey Juan Carlos University, Alcorcón, Madrid, Spain

Department of Medicine and Surgery, Psychology, Preventive Medicine and Public Health, Immunology and Medical Microbiology, Faculty of Health Sciences, Rey Juan Carlos University, Alcorcón, Madrid, Spain
e-mail: yolanda.valcarcel@urjc.es

E. Heath et al. (eds.), *Fate and Effects of Anticancer Drugs in the Environment*,
https://doi.org/10.1007/978-3-030-21048-9_1

1.1 The Need for an Assessment of the Risk Posed by Cytostatic Agents

Since the 1990s, the continued input of pharmaceutical and personal care compounds (PPCPs) into the environment, mainly via wastewater treatment plants, and the use of reclaimed waters and biosolids, has become one of the main areas of research for environmental chemists and toxicologists (Halling-Sørensen et al. 1998; Cizmas et al. 2015). Despite these increased research efforts, many questions still remain unanswered, especially with regard to the actual risk posed by these compounds to the environment and to human health (Brooks et al. 2009; Boxall et al. 2012; Noguera-Oviedo and Aga 2016). The availability of data on the presence and environmental effects of these compounds has slowly grown over this period (with toxicity data generally increasing at a slower pace than exposure data). However, the use of these data to inform industry and regulators whether these PPCPs pose a risk to the environment and human health has been, generally, unsuccessful (Rodríguez-Gil et al. 2018). In recent years, societal interest in these compounds has expanded and is today no longer limited to the research community. In fact, the need for assessment of potential human health and environmental risks of PPCPs has now been acknowledged in a number of regulations in different jurisdictions. For example, article 8(3) of European Directive 2001/83/EC (EC 2012), as amended, lists an Environmental Risk Assessment as one of the documents required to be submitted during the registration process of medicinal products for human use. For this purpose, the European Medicines Agency (EMA) created a guideline document on environmental risk assessment (ERA) of medicinal products for human use (EMEA 2006b, 2016). Although registration is not conditional on the results of this assessment, the regulation establishes the possibility, on a case-by-case basis, for the consideration of specific arrangements to limit the impact. Similar provisions exists, as well, for veterinary pharmaceuticals (EC 2004; EMEA 2015). Furthermore, in the 2013 amendment of the Water Framework Directive (WFD) and the Environmental Quality Standards Directive (EQSD) (EC 2013), the European Union specifically flagged contamination of water and soil with pharmaceutical residues as an emerging environmental concern. Thus, the EU called for the need to study the risks of environmental effects from medicinal products and to provide an analysis of the relevance and effectiveness of current legislative framework in protecting the aquatic environment and human health via the aquatic environment. These amendments also called for the creation of a new mechanism to provide high-quality monitoring information on the concentrations of polluting substances (including PPCPs) in the aquatic environment across the EU to identify priority substances for regulation under the Water Framework Directive (WFD) (EC 2000b). To accomplish such a task, a dynamic watch list was created that included three PPCPs [diclofenac, 17-beta-estradiol (E2), and 17-alpha-ethinylestradiol (EE2)] in its first iteration (Carvalho et al. 2015).

The assessment of the risk posed by cytostatic compounds is of particular interest from their cytotoxic, and in many cases carcinogenic, nature. From a regulatory standpoint, according to the list of waste decision (LoW) 2000/532/EC (EC 2000a), discarded cytotoxic and cytostatic medicines are considered hazardous waste. This classification translates into specific obligations related to how cytostatic-containing waste streams are to be handled. However, these provisions do not apply to cytostatic compounds present in hospital or urban wastewaters. Currently, cytostatic compounds are not considered priority substances under the WFD, and none of these substances was considered as a possible candidate for the first watch list developed under the 2013 amendments to the WFD and the EQSD. Thus, no special provisions exist for their presence in wastewater or surface water.

Approximately 75–80% of cytostatic treatment takes place as outpatient treatment, with the patients leaving the hospital after treatment (Mahnik et al. 2007; Johnson et al. 2008). Once excreted by patients in hospitals or at home, cytostatic agents and their metabolites are discharged into wastewater flows before being treated in wastewater treatment plants. As these systems were not specifically designed to remove these types of compounds, removal rates for these chemicals are usually low, and thus the probability of these reaching natural water flows is high. The availability of data on the environmental presence and effects of cytostatic agents in the environment is currently very limited, and at the same time, the amount of resources available to be dedicated to the study of these aspects is also limited. To prioritize the cytostatic agents on which we need to focus our research and monitoring, as well as to determine whether specific treatment systems should be implemented to remove these compounds before reaching natural water flows, the risk and environmental hazard of these substances must be calculated.

1.2 The Risk Assessment Framework

Although the exact nomenclature of the different stages and their practical implementation might vary depending on the context or jurisdiction where the process is being applied, any risk assessment usually consist of three main stages: problem formulation, analysis, and the risk characterization itself (Suter 1997).

In the Problem Formulation stage, researchers and regulators evaluate the state of the knowledge about the compound, or compounds, to be assessed. Information about the chemical characteristics and use patterns of the compound, potential routes of exposure, environmental compartments most likely to receive/accumulate the compound, as well as the receptors that could be exposed (humans and/or the environment, and if the latter, which elements), are all considered at this stage. All existing gaps in knowledge that might limit our ability to carry out a complete risk assessment are also flagged in this step.

When the characteristics of the compound, its expected use, movement, environmental compartments and receptors, as well as all information gaps, have been evaluated, the Analysis Stage focuses on collecting all missing information and extending existing coverage. The analysis phase is usually further divided into two sub-stages that individually handle the collection of the two main types of needed information: the Characterization of Exposure Stage and the Characterization of Effects Stage. In the characterization of exposure sub-stage, information regarding the fate and transport of the compound is collected to estimate (or measure) the expected concentration that might occur in different environmental compartments. Aspects related to the toxicokinetic characteristics of the compound can be considered during this sub-stage improve the estimates of exposure to target organs or tissues. The characterization of effects sub-stage takes place simultaneously, with the aim of improving understanding of the toxicological characteristics of the compound. If not previously known, or uncertain (flagged during the problem formulation), the mode (or mechanism) of action of the compound, as well as the main target organ or tissues, can be determined. Dose–response (or concentration–response) relationships are described, whenever possible, to derive toxicological benchmark concentrations to be used in the risk characterization stage.

At the end of the analysis phase, we should have collected, at minimum, information regarding concentration levels of the compound that are to be expected in different environmental compartments, as well as a number of benchmark concentrations describing the toxicity characteristics of the compound. These data are then combined during the Risk Characterization Stage to assess whether an overlap between expected environmental concentrations and those concentrations causing toxicological effects are possible (i.e., a hazard exists), as well as (whenever sufficient information is available) what is the probability of such an overlap actually taking place (i.e., the risk). It is in this stage that the uncertainty associated with the available information is assessed and applied to the final hazard/risk estimate.

The risk assessment process often takes an iterative approach, in which the afore-described stages can be repeated in a series of linked levels, or tiers, of increasing complexity. Lower-tier risk assessments usually rely on basic estimates and modeled data, which allows for a speedy process; however, a high level of uncertainty is associated with these lower-level assessments. Large uncertainty factors are often required and applied to any hazard estimates to ensure protection of human health and the environment. These quick, lower-tier assessments allow for initial screening and prioritization of substances that could be subject to higher-tier assessments. Higher-tier assessments, on the other hand, can provide more realistic hazard/risk estimates, but they depend on reliable (often measured in the field) data, often requiring large investments of time and money.

The number of tiers applied to a particular compound depends on such factors as the nature of the compounds, their use pattern and economic value, and the jurisdiction in charge of their approval or control. For example, the REACH methodologies (EC 2006) use lower-tier assessments to allow the assessment of the hundreds of thousands of existing chemical substances. On the other hand, the EFSAs regulations, applied to the authorization of pesticide products, allow for higher-tier assessments, including the use of microcosm or mesocosm studies (EFSA 2013).

The chapters of this book present, in one way or another, different aspects of the risk assessment of cytostatic compounds. The introductory chapters present the state of the knowledge about the characteristics and use of the compounds, as well as summary information on their presence and toxicity. In this way these sections could be considered as part of the problem formulation stage for the assessment. Most of the contents of this book can be classified as belonging to the analysis stage, reporting advances in the characterization of the exposure and the effects that are expected for the compounds in this group. Thus, this chapter mainly focuses on the third stage, the risk characterization stage, or how the overlap between exposure and effects can be evaluated to determine whether the cytostatic compounds could pose a risk to human health and the environment.

1.2.1 Screening, Ranking, and Prioritization Approaches to Risk Characterization

Lower-tier hazard assessments, although limited, can be extremely useful as screening and prioritization tools when time and resources are limited. When the potential risk posed by many substances (such as cytostatic compounds) is to be evaluated, screening and prioritization methodologies can identify those individual substances expected to pose the greatest risk, so as to focus the analysis stage of the next iteration of the assessment process (laboratory testing or environmental monitoring) on those substances of greatest concern. As explored in this book, there is a limited – albeit growing – amount of data on exposure and effects for cytostatic compounds. This lack of data limits our ability to currently conduct these realistic high-tier risk assessments. Given the limited resources (financial and temporal) available for research, the screening, ranking, and prioritization techniques can orient research efforts to focus on those cytostatic compounds that might pose a larger risk.

For obvious reasons, the precautionary principle is important in the application of these screening, ranking, and prioritization tools. To be protective, these methodologies need to ensure that the number of false negatives (substances that could pose a risk, but are deemed safe in the assessment) are minimized. At the same time, if these tools are to be useful as an actual screen, the number of false positives (safe substances flagged as posing a risk) must remain manageable.

Many prioritization approaches have been used for the screening and prioritization of pharmaceutical and personal care products as related to their environmental risk. These methodologies are usually divided into two groups.

- Hazard-based screening tools: those that focus on the intrinsic characteristics of a compound as related to their potential fate, transport, and toxicity
- Risk-based screening tools: those that assess the actual potential overlap between the toxicity and environmental exposure of the compound (usually, at this stage, via models and predictions)

1.3 Hazard-Based Screening Tools

Hazard-based tools focus on only one of the elements of risk, this is, either on the exposure alone or solely on the effects (toxicity). As such, they can report only the hazard, or the possibility of a risk. Because both elements are not studied simultaneously, no assessment of the probability of an overlap can be made, and as such, no estimate of risk is possible.

As mentioned, these approaches can focus either on the potential exposure to the substance or on the potential effects resulting from that exposure. Approaches focusing on potential effects generally report the intrinsic characteristics of the assessed compound. Because the focus of these effects-oriented methodologies is the compound itself, these tools are useful as they do not depend on location-specific information, such as sales data or wastewater treatment technologies (Roos et al. 2012). Because of this location independency, however, these methodologies cannot incorporate risk-reduction factors such as reduced compound use, or advance treatment technologies that might reduce or eliminate the exposure in certain scenarios. Such questions might, then, be better addressed by exposure-based approaches.

1.3.1 Sales Data-Based Ranking Methods

This exposure-based approach aims to rank the assessed substances based on total sales over a specific geographic area or timeframe. This approach is the simplest prioritization method (when sales information is easily accessible), and thus it is, nowadays, one of the prioritization methods most commonly employed for regulatory purposes. Sales data are, in fact, one of the main triggers from further assessment under the EMA legislation (EMEA 2006b).

Some authors consider rankings based on sales data a risk-based approach as these are not based on the chemical characteristics of the substance (Roos et al. 2012). This approach, however, does not consider the potential overlap between exposure (worst-case scenario estimated from sales data) and effects (which are not considered at all). As such, we consider this as a hazard-based technique, one that is solely based on a characterization of exposure (or potential worst-case exposure).

1.3.2 Log K_{ow}-Based Ranking

Perhaps the simplest of the effects-based approaches (i.e., approaches based on the characteristics of the compound itself), this method ranks substances based on the values of their respective log octanol–water partitioning coefficients (log K_{ow}), that is, the lipophilicity. The value of the log K_{ow} for a particular substance has been correlated with its tendency to concentrate or accumulate in organisms (Meylan et al.

Table 1.1 List of common cytostatic compounds ranked by the value of the log of their octanol–water partitioning coefficient (log K_{ow})

Therapeutic action	Compound	CAS number	Log K_{ow}
Hormonal agents	Tamoxifen citrate	54,965-24-1	6.30
Hormonal agents	(*Z*)-4-Hydroxytamoxifen	68,047-06-3	5.82
Hormonal agents	Endoxifen	110,025-28-0	5.61
Taxanes	Paclitaxel	33,069-62-4	3.30
Inhibitors of kinases	Imatinib mesylate	220,127-57-1	3.01
Inhibitors of kinases	Erlotinib hydrochloride	183,319-69-9	2.78
Taxanes	6(α)-Hydroxypaclitaxel	153,212-75-0	2.62
Inhibitors of topoisomerases	Irinotecan hydrochloride trihydrate	136,572-09-3	2.33
Antitumor antibiotics	Doxorubicin hydrochloride	25,316-40-9	1.27
Alkylating agents	Ifosfamide	3778-73-2	0.86
Alkylating agents	Cyclophosphamide	50-18-0	0.63
Inhibitors of topoisomerases	Etoposide	33,419-42-0	0.60
Anti-metabolites	Capecitabine	154,361-50-9	0.55
Alkylating agents	Temozolomide	85,622-93-1	−1.30
Anti-metabolites	Hydroxy methotrexate	5939-37-7	−1.70
Anti-metabolites	Methotrexate	59-05-2	−1.85
Anti-metabolites	Gemcitabine hydrochloride	122,111-03-9	−2.01

1999; Arnot and Gobas 2006), as well as with its potential basal toxicity (narcotic effect) (US EPA 2013). When no experimental log K_{ow} values exist, available computational tools, such as the US EPA EPIWIN tool (US EPA 2013), can be used to estimate them. As an example, Table 1.1 presents a list of 17 common cytostatic compounds ranked on the basis of their estimated log K_{ow}. It is interesting to note how three hormonal agents top the ranking, with higher log K_O values.

Although easy to apply, this method presents many limitations. A correlation between log K_{ow} and bioconcentration or toxicity exists, but this relationship is not linear throughout the range of log K_{ow} values (Meylan et al. 1999), because a decrease in bioconcentration potential and toxicity is common for substances with high log K_{ow} values. A direct log K_{ow}-based ranking would, thus, likely misclassify certain substances. Of special relevance with pharmaceuticals and personal care products (such as cytostatic compounds), which are often ionizable compounds, is the fact that log K_{ow} can only correctly describe the partition coefficient of the neutral molecule. An alternative option, the log D (Schreiber et al. 2011; Fu et al. 2009), can describe the lipophilicity of the charged substance, but to calculate it, the pH of the surrounding medium must be known. This knowledge implies that information on exposure is available, which excludes this option as a strictly hazard-based methodology. Additionally, the log K_{ow} relationship to toxicity is based on the ability of a molecule to disrupt cell membranes, known as basal toxicity or narcotic effect. For substances with specific modes of action, such as the many that exist in the various cytostatic compounds, this approach would not offer a complete picture.

1.3.3 QSAR-Based Expected Effect Ranking

This method is, in part, also based on the use of log K_{ow} values, via its use (together with structural characteristics of the molecule) as the base for effect measure estimations via quantitative structure–activity relationship (QSAR) models (Sanderson et al. 2004). The QSAR estimation of effect measures (e.g., EC_{50}) solves some of the previously discussed issues with log K_{ow}-based approaches, specifically the nonlinear relationship between log K_{ow} and toxicity. Additionally, commonly used QSAR tools, such as the US EPA ECOSAR (part of the previously mentioned EPIWIN) (US EPA 2013), employ classification approaches, which generate toxicity estimations based on structural similarities to other molecules for which data are available. In this way, aspects related to mode/mechanisms of action other than basal toxicity can be taken into account, so long as they are conserved in the structure of the molecules in the evaluated class. Changes in the toxicokinetic and toxicodynamic properties of the molecule related to pH changes in the surrounding medium are still not accounted for in this method. As an example, Table 1.2 shows the lowest ECOSAR-modeled Chronic Values (ChV) for daphniids, fish, and green algae for 16 cytostatic compounds. This table is based on the values of the lowest ChV among all three groups of organisms. There are some similarities with the order

Table 1.2 Lowest ECOSAR-modeled Chronic Value (ChV) for daphniids, fish, and green algae for 16 cytostatic compounds based on the value of the lowest ChV among the three groups of organisms

Compound	*Daphnia* ChV (mg l^{-1})	Fish ChV (mg l^{-1})	Green algae ChV (mg l^{-1})	Min. ChV (mg l^{-1})
(Z)-4-Hydroxytamoxifen	0.005	0.003	0.007	0.003
Endoxifen	0.007	0.005	0.009	0.005
Cyclophosphamide	295.755	0.017	29.616	0.017
Ifosfamide	295.755	0.017	29.616	0.017
Doxorubicin hydrochloride	0.448	0.032	1.517	0.032
Irinotecan hydrochloride trihydrate	0.04	0.091	0.185	0.04
Paclitaxel	0.487	0.05	1.079	0.05
Imatinib mesylate	0.253	0.087	0.649	0.087
6(α)-Hydroxypaclitaxel	1.135	0.101	1.846	0.101
Temozolomide	1.257	2.275	2.424	1.257
Erlotinib hydrochloride	4.629	6.764	12.009	4.629
Methotrexate	5.095	73.177	45.344	5.095
Etoposide	9.129	84.22	220.62	9.129
Tamoxifen citrate	13.631	20.73	33.468	13.631
Capecitabine	52.215	58.452	55.48	52.215
Gemcitabine hydrochloride	85.624	118.813	62.252	62.252

derived from the previous method (log K_{ow}) but there are also some relevant differences. For example, tamoxifen citrate, which led the table based on log K_{ow} values, appears now as third from the bottom in this new ranking.

1.3.4 PBT/vPvB Criteria

Perhaps the most common of the hazard-based approaches is the PBT/vPvB classification used, among others, as a first step for chemical authorization under the European REACH framework (ECHA 2017; EC 2006), as well as for classification in the Toxics Release Inventory (TRI) under the United States Toxic Substances Control Act (TSCA), as amended by the Frank R. Lautenberg Chemical Safety for the Twenty-first Century Act (Toxic Substances Control Act 1976). This method is another step in improving the two previously presented approaches. The PBT approach evaluates a particular compound based on three criteria: Persistence, Bioaccumulation Potential, and Toxicity. PBT substances are substances that are Persistent, Bioaccumulative and Toxic; vPvB substances are substances that are very Persistent and very Bioaccumulative.

Persistence (P) is considered to be the ability of a substance to remain in the environment unaltered. It is usually represented using the half-life of the compound in a particular environmental compartment. The longer a compound persists, the higher the likelihood that humans or environmental receptors could be exposed to it at hazardous levels. *Bioaccumulation* (B) is the process by which a compound can build up in organisms such that the concentration of the compound in the organism exceeds the levels found in a particular environmental compartment. In addition to increased exposure and increased potential for direct toxicity, compounds with high bioaccumulation potential should trigger assessments of potential secondary poisoning to other organisms along the food web that might be indirectly exposed to the compound via consumption of organisms in which the compound has accumulated. *Toxicity* (T) is the property whereby a substance exerts a harmful effect on living organisms.

There are different approaches to the derivatization of PBT values, but they usually rely on modeling approaches for the estimation of these values from existing molecular information. A simple tool commonly used is the PBT Profiler (http://www.pbtprofiler.net), which was designed to allow for the quick hazard assessment of chemical compounds lacking experimental data. At its core, the PBT profiler tool employs many of the modules available in the EPIWIN platform (US EPA 2013).

The information required for this calculations is minimal. The user only needs to provide an identifier for the product (e.g., CAS number or name) and/or the molecular structure. The chemical structure passes through nine separate modules to estimate the physicochemical properties, and the results are used to estimate persistence, bioaccumulation, and toxicity values. The tool not only estimates the P, B, and T values but also performs a preliminary classification of the compound. The persistence, bioaccumulation, and toxicity values estimated by PBT

Table 1.3 Persistence, Bioaccumulation Potential, and Toxicity (PBT) criteria used for classification in the US EPA PBT Profiler tool

Property	PBT criteria	vPvB criteria
Persistence	A substance fulfils the persistence criterion (P) in *any* of the following situations:	A substance fulfils the "very persistent" criterion (vP) in *any* of the following situations:
	$T_{1/2} > 60$ days in water, soil, or sediment	
	$T_{1/2} > 2$ days in air	$T_{1/2} > 180$ days in water, soil, or sediment
Bioaccumulation	A substance fulfils the bioaccumulation criterion (B) at $BCF > 1000$	A substance fulfils the "very bioaccumulative" criterion (vB) when $BCF > 5000$
Toxicity	A substance fulfils the toxicity criterion (T) in *any* of the following situations:	–
	High concern: Fish chronic value ChV <0.1 mg/l	
	Moderate concern: Fish chronic value ChV 0.1 mg/l–10 mg/l	

Profiler are automatically compared with the criteria published by the US EPA (Toxic Substances Control Act 1976; US EPA 2012a). Those values that comply with or exceed the criteria are provided to the user on the results page of the PBT Profiler. The criteria used by the PBT profiler for classification of a substance as a PBT or a vPvB substance are provided in Table 1.3.

The PBT profiler is a simple tool that can be used for early assessment and research purposes, but screening and prioritization of substances based on PBT methodologies for regulatory purposes (e.g., during the initial steps of REACH) require more robust and detailed approaches that usually require expert judgment of available data before an actual decision on the classification of a substance as PBT or vPvB can be finalized. Recent amendments to the US TSCA as presented in the TSCA Work Plan Chemicals: Methods Document (US EPA 2012b) further expand on the way this information is used as a screening tool in chemical regulation in the US.

European regulation also takes advantage of the convenience of PBT/vPvB approaches for chemical authorization. The criteria used to classify a substance as a PBT/vPvB under Annex XIII of the REACH (ECHA 2017) are presented in Table 1.4. The commonly employed tool QSAR Toolbox (OECD 2017) can assist European researchers on the estimation of P, B, and T values and classification of substances as PBT or vPvB. The QSAR toolbox uses the same EPIWIN modules as the PBT profiler, but these tools consider the REACH Annex XIII criteria for the classification.

The lack of need to show toxicity for vPvB substances might come as a surprise; however, according to the European Medicines Agency Guideline on the assessment of persistent, bioaccumulative, and toxic (PBT) or very persistent and very bioaccumulative (vPvB) substances in veterinary medicinal products (EMEA 2015), the vPvB clasification was developed with the understanding that for

Table 1.4 PBT and vPvB criteria taken from REACH, Annex XIII of regulation (EC) 1907/2006, table R. 11-1 (ECHA 2017)

Property	PBT criteria	vPvB criteria
Persistence	A substance fulfils the persistence criterion (P) in *any* of the following situations:	A substance fulfils the "very persistent" criterion (vP) in *any* of the following situations:
	$T_{1/2}$ > 60 days in marine water	
		$T_{1/2}$ > 60 days in marine, freshwater, or estuarine water
	$T_{1/2}$ > 40 days in freshwater or estuarine water	$T_{1/2}$ > 180 days in marine, freshwater, or estuarine sediment
	$T_{1/2}$ > 180 days in marine sediment	$T_{1/2}$ > 180 days in soil
	$T_{1/2}$ > 120 days in freshwater or estuarine sediment	
	$T_{1/2}$ > 120 days in soil	
Bioaccumulation	A substance fulfils the bioaccumulation criterion (B) at BCF > 2000	A substance fulfils the "very bioaccumulative" criterion (vB) when BCF > 5000
Toxicity	A substance fulfils the toxicity criterion (T) in *any* of the following situations:	–
	NOEC or EC_{10} < 0.01 mg/l for marine or freshwater organisms	
	A substance is classified as carcinogenic (category 1A or 1B), germ cell mutagenic (category 1A or 1B), or toxic for reproduction (category 1A, 1B, or 2);	
	There is other evidence of chronic toxicity, as identified by the classifications: STOT (repeated exposure), category 1 (oral, dermal, inhalation of gases/vapors, inhalation of dust/mist/fume) or category 2 (oral, dermal, inhalation of gases/vapors, inhalation of dust/mist/fume) according to the CLP regulation	

substances which are very persistent and very bioaccumulative, high but unpredictable levels may be reached in wildlife or humans over extended time periods. For such substances, it is not necessary to demonstrate toxicity in laboratory testing because long-term effects can be anticipated.

As we have seen so far, PBT/vPvB approaches can be used as a direct screening tool to inform whether a compound should require a more comprehensive risk assessment (continued through higher tiers). At the same time, numerical approaches have been used as well, to rank and prioritize compounds for further research. The method developed by Wennmalm and Gunnarsson (2005), for example, involves assigning a numerical value (0 to 3) to each parameter corresponding to persistence, bioaccumulation, and toxicity. The sum of these three values results in a *PBT index*

Table 1.5 Scores for Persistence, Bioaccumulative Potential, and Toxicity for 17 commonly used cytostatic compounds: the compounds appear sorted from most to less environmental concern based on a PBT index compounding the P, B, and T scores (Olalla et al. 2018)

Cytostatic class	Compound	Log K_{ow}	P	B	T	PBT index
Hormonal agents	Tamoxifen citrate	6.30	3	3	3	9
Hormonal agents	Endoxifen	5.61	3	3	3	9
Hormonal agents	(Z)-4-Hydroxytamoxifen	5.82	3	3	3	9
Alkylating agents	Cyclophosphamide	0.63	3	0	3	6
Alkylating agents	Ifosfamide	0.86	3	0	3	6
Alkylating agents	Temozolomide	−1.30	3	0	3	6
Antitumor antibiotics	Doxorubicin	1.27	3	0	3	6
Inhibitors of topoisomerases	Irinotecan	2.33	3	0	3	6
Inhibitors of kinases	Imatinib mesilate	3.01	3	0	3	6
Taxanes	Paclitaxel	3.30	3	0	3	6
Anti-metabolites	Methotrexate	−1.85	3	0	0	3
Anti-metabolites	Hydroxymethotrexate	−1.70	3	0	0	3
Anti-metabolites	Gemcitabine	−2.01	3	0	0	3
Inhibitors of topoisomerases	Etoposide	0.60	3	0	0	3
Inhibitors of kinases	Erlotinib	2.78	3	0	0	3
Taxanes	6(a)-Hydroxypaclitaxel	2.62	3	0	0	3
Anti-metabolites	Capecitabine	0.55	0	0	0	0

for the substance evaluated, with the highest value indicating a greater potential of the substance to harm the environment. In fact, the use of this approach for the hazard assessment of cytostatic compounds was recently tested by Olalla et al. (2018). Table 1.5 presents the scores for persistence, bioaccumulative potential, and toxicity for 17 commonly used cytostatic compounds, based on data obtained from the US EPA PBT Profiler tool. The different compounds appear sorted from most to least environmental concern based on a PBT index compounding the P, B, and T scores as described in Olalla et al. (2018). It is interesting to note how, after taking chemical properties back into consideration, tamoxifen citrate moves back to the top of the list, as it appeared when ranked based on log K_{ow} alone.

A limitation of these quantitative PBT approaches for the ranking of cytostatic compounds lies in the fact that many of the substances in this group are considered genotoxic, mutagenic, or carcinogenic. If applying REACH Annex XIII criteria, substances showing any of these characteristics should be flagged with the highest toxicity scores. This categorization, in turn, would leave only the P and B criteria available for ranking these compounds, reducing the flexibility and resolving capacity of the approach. Because of this limitation some authors (Olalla et al. 2018) have chosen to temporarily ignore (within the scope of their analyses) the possible genotoxic, mutagenic, or carcinogenic nature of these substances and calculate the toxicity values (T) based on traditional approaches only [e.g., no-observed effect concentration (NOEC), EC_{50}, ChV values), thus allowing for a higher resolving capacity within the group of cytostatic compounds. The validity of this approach, nonetheless, is supported by recent experimental data suggesting that the chronic toxicity of some of these compounds to aquatic organisms could be lower than would have been expected (Parrella et al. 2014; Pichler et al. 2014).

1.3.5 Critical Environmental Concentration

This method, developed by Fick et al. (2010), proposes to rank pharmaceutical and personal care products based on their critical environmental concentrations (CECs), which can be described as the surface water concentration expected to cause a pharmacological effect in fish.

An advantage of working with pharmaceutical compounds, in general, is that considerable information on their pharmacological properties in humans is available. The method proposed by Fick et al. (2010) makes use of available human therapeutic plasma concentrations (H_TPC) and a theoretically derived fish plasma bioconcentration factor ($P_{blood:water} = 0.73 \times \log K_{ow} - 0.88$), to derive a critical water column concentration that would result in the concentration of the evaluated substance reaching therapeutic levels in the plasma of fish swimming in the receiving waters. This method assumes that target organs/tissues are conserved between humans and fish and that therapeutic doses for both humans and fish are the same.

A possible limitation of the applicability of this method to cytostatic compounds is that therapeutic doses for these compounds are commonly quite high and acutely applied, as they are intended to cause certain levels of toxicity in the patient. Information on the possible human health effects or therapeutic effects of chronic, low-dose exposures is more rarely available. As such, this method could underestimate the potential environmental risk to fish populations in ecosystems receiving wastewater effluents.

1.4 Risk-Based Screening Tools

As previously mentioned, risk-based screening tools are those that assess the actual potential overlap between the toxicity and environmental exposure of the compound (usually, at this stage, via models and predictions). Some commonly employed lower-tier risk-based tools follow.

1.4.1 Chemical Scores

As mentioned in Sect. 1.3.4, recent amendments to the US TSCA (Toxic Substances Control Act 1976) have resulted in updates to the methodologies employed during the process used by the US EPA to identify potential candidate chemicals for near-term review and assessment. As described in US EPA (2012b), these updated methodologies include a risk-based extension of the traditional PBT/vPvB approach, which assesses and scores three aspects:

- Hazard: based on human health and environmental toxicity information
- Exposure: based on a combination of chemical use, general population and environmental exposure, and release information
- Persistence/Bioaccumulation: calculated in a similar fashion to the methods described in Sect. 1.3.4.

Each of these aspects is provided a score from 1 to 3, and a total score for the compound (maximum score of 9) is calculated. This total score can then be used to support the categorization of Candidates for Inclusion as TSCA Work Plan Chemicals (US EPA 2012b).

This method is similar to the quantitative approaches to PBT assessment as described in Sect. 1.3.4 and by Olalla et al. (2018). The main difference is the inclusion of exposure data in the analysis, allowing for some initial risk estimates and making the results more representative of particular scenarios, such as national or regional use. Additionally, the toxicological properties evaluated in this method as described in US EPA (2012b) are more extensive than those considered in traditional PBT approaches, especially those considered in the PBT Profiler tool.

1.4.2 Fish Plasma Model

This method is an extension of the critical environmental concentration (CEC) method described in Sect. 1.3.5. As previously indicated, the method proposed by Fick et al. (2010) uses available human therapeutic plasma concentrations (H_TPC) and a theoretically derived fish plasma bioconcentration factor ($P_{blood:water} = 0.73 \times \log K_{ow} - 0.88$) to estimate Fish Steady-State Plasma Concentrations ($F_{ss}PC$). The main assumption of this method is that target organs/tissues are conserved between humans and fish and that therapeutic doses for both humans and fish are the same. This variant of the method described in Sect. 1.3.5 is aimed at calculating the theoretical critical environmental concentrations (CECs) at which the FssPC for a particular compound would equal its H_TPC, and, thus, an effect on the fish is to be expected.

This risk-based version incorporates data on measured or predicted exposure (i.e., water column concentration) to calculate the $F_{ss}PC$ value ($F_{ss}PC = C_{water} \times P_{blood:water}$). A simple ratio is then applied between the known H_TCP and the calculated $F_{ss}CP$ to determine "how far" the measured/predicted environmental concentration is from a concentration that can cause therapeutic effects on fish. Values greater than 1 would, thus, indicate the possibility for effects to be observed at the measured/predicted water column concentrations.

1.4.3 *Risk/Hazard Quotients*

Risk quotients (RQs), also known as hazard quotients (HQs) depending on the jurisdiction, have been broadly applied to the hazard/risk assessment of pharmaceutical and personal care products during the past decade. They are also a common feature in environmental risk assessment guidelines and regulations across many jurisdictions (ECHA 2012b; US EPA 2004). The method consists of the direct comparison of the expected exposure and expected toxicity of the evaluated compound via the ratio of two values. Commonly, in this context, the expected exposure is referred to as a predicted environmental concentration (PEC) or a measured environmental concentration (MEC). On the other hand, a predicted no-effect concentration (PNEC) is derived as an indication of a "safe" exposure concentration below which no deleterious effects are to be expected. In this way, the RQ (or HQ) could be described by the following equation:

$$\mathrm{RQ} = \frac{\mathrm{PEC}}{\mathrm{PNEC}}$$

If the risk quotient (RQ) value is less than 0.1, no adverse effect is expected and the risk is classified as insignificant. The risk for RQ values between 0.1 and 1 is low, but possible adverse effects should be taken into consideration. If the RQ value is between 1 and 10, a moderate risk and adverse effect are likely. Finally, a high risk is predicted if the calculated RQ value is greater than 10 (ECHA 2012b). HQ/RQ values should be derived for each environmental compartment/ecosystem where exposure is expected to occur. For example, under the REACH Guidelines (ECHA 2012b), this is done separately for each of the following environmental protection targets:

- Aquatic ecosystem
- Terrestrial ecosystem
- Atmosphere
- Predators (fish-eating and worm-eating)
- Microorganisms in sewage treatment plants (STPs)

The hazard/risk quotient approach clearly follows the typical risk assessment framework described in Sect. 1.2, where the risk characterization stage (i.e., the actual calculation of the HQ/RQ), requires the completion of the problem formulation and analysis stages. During the analysis stage, PEC (characterization of exposure) and PNEC (characterization of effects) values are derived. Most of the contents of this book have considered aspects related to the analysis phase, and as such, we only briefly touch on some aspects related to the derivation of PECs and PNECs in the context of the risk assessment of cytostatic compounds. Extended information on how to derive PEC/MEC and PNEC values (in the European context) can be found in the guidance on information requirements and chemical safety assessment documents in support of REACH (ECHA 2011a, b, 2012a, b, 2017; EC 2006), as well as the documents published by the European Medicines Agency in support of the proposed risk assessment process to be applied to medicine products (EMEA 2006a, b).

1.4.3.1 Exposure Characterization (PEC/MEC Derivation)

As further explored throughout this book, there are two main approaches for determining environmental exposure levels: predictive methods, which estimate the amount of compound that is to be expected in a particular environmental compartment on the basis of various parameters, such as the consumption or excretion rate of the compounds studied, and experimental methods, involving the direct measurement of the evaluated compounds. Both approaches have been explored in different chapters in this book, and both approaches present advantages and drawbacks. Although predictive methodologies require few resources and can provide good worst-case scenario estimates, their estimates also present high levels of uncertainty. On the other hand, direct measurement usually involves substantial costs, and, even though a direct measure of what is actually there, it might still fail to provide us with a complete picture. For example, temporal variability might lead to our exposure dataset missing punctual high spikes in concentration. Furthermore, analytical limitations might prevent us from reliably quantifying certain compounds that might occur above therapeutic or potentially toxic levels.

Many predictive methods are based on more or less simple "back of the envelope" calculations, involving information such as compound sales, use, excretion rates, volume of water treated by a particular WWTP, and retention/degradation rated during treatment. In other cases, more sophisticated approaches can incorporate mathematical modeling approaches, such as fugacity or activity models (Mackay 1991; Mackay and Arnot 2011).

Whether from predicted values or from direct measures, the value selected in this stage (PEC/MEC) should be a conservative, precautionary reflection of the potential environmental exposure (i.e., a worst-case scenario). In this way, it is common practice to select the highest measured or predicted value. PEC/MEC values should be generated for each environmental compartment where the evaluated compound has been predicted/measured to occur.

The aim of the study by Franquet-Griell et al. (2017) was to prioritize and evaluate their risk by calculating PECs using raw consumption data from Spain according to population and dilution factor in the eight main river basins. According to Franquet-Griell et al. (2017), for the specific case of anticancer drugs this is a suitable model because of the consumption can easily be correlated with the sales of these drugs, as all the prescribed amount is consumed by the patient.

To calculate the predicted environmental concentrations in WWTP effluents and surface waters, the equation used by Besse et al. (2008) was applied. Here, parameters such as consumption of each cytostatic (as quantity delivered by pharmacies), excreted fraction of the unchanged drug, removal fraction in WWTP, water consumption per inhabitant, and number of inhabitants in Spain were taken into account.

After calculating PECs, risk assessment was performed to determine if these predicted concentrations might pose danger to the aquatic environment as explained in Sect. 1.4.3. Franquet-Griell et al. (2017) estimated PNEC using NOEC and a security factor f_1 of 10. If NOEC was not available, PNEC was estimated by using $E(L)C_{50}$ and a security factor f_2 of 1000. As explained below in Sect. 1.4.3.2, toxic information was collected for several species, as it is important to evaluate the adverse effects ta different trophic levels.

All obtained RQ were less than 1 for drugs with PEC > 10 ng/l (in river), showing no expected risk for the aquatic environment. For the other drugs, with PEC < 10 ng/l, the highest RQ value was as much as 2.6×10^{-2}, meaning no expected risk.

The study by Franquet-Griell et al. (2017) reveals the importance of consumption data and temporal patterns for estimating the occurrence and risk of anticancer drugs in surface waters. The use of exhaustive data compilation covering several years permits the estimation of PECs.

In the work by Orias and Perrodin (2014), the highest concentrations of each PC already measured in HWW from a previous study (Orias and Perrodin 2013) were used to obtain the worst-case scenario, finding 12 anticancer PCs of the 16 sought.

1.4.3.2 Characterization of the Effect (PNEC Derivation)

This phase involves defining the sensitivity of the organism(s) potentially exposed to the evaluated compound(s) (as defined in the problem formulation stage). This step requires gathering information pertaining to the expected individual/ecological effects and the concentrations at which these are expected to take place. The ultimate goal is the derivation of the predicted no-effect concentration (PNEC).

Many approaches are used for the derivation of PNEC values. In general, at this stage the use of model-based data is the most common method: however, use of experimental toxicity data, when available, is always recommended. Common approaches involve the use of the previously mentioned ECOSAR module, available in the US EPA EPISUIT (US EPA 2013). As explained earlier, ECOSAR and similar QSAR-based models follow a categorization approach to assign the assessed molecule to one or various molecular classes based on the structure of the compound. Once the compound has been assigned to a class, or classes, predicted toxicity values are calculated from existing models relating log K_{ow} values to toxicity endpoints for compounds in those chemical categories for which experimental data are available. The output from ECOSAR provides toxicity data for organisms belonging to the main aquatic trophic levels: algae, daphniids, and fish. Recent improvements also include data for a number of saltwater organisms. Effective concentrations causing mortality to 50% of the organisms (EC_{50}) in acute tests are the most common value collected from ECOSAR; however, the software also provides effect measured for chronic exposures in the way of chronic values (ChVs). These data are calculated as the geometric mean between the lowest-observed effect concentration (LOEC) and the no-observed effect concentration (NOEC) values from a test. Data availability for chronic tests in the US EPA databases is traditionally less than for acute tests, making these ChV models marginally less reliable than those used for the estimation of acute EC_{50} data. For freshwater aquatic systems, effect measures (e.g., EC_{50}, ChV) should be collected for a minimum of organisms representing the main aquatic trophic levels: algae, daphniids, and fish.

When these data have been collected, and following the same protectionary approach as in the PEC/MEC derivation, the effect measure corresponding to the most sensitive species is selected as the benchmark concentration to be used for PNEC derivation. To derive the final PNEC value, this benchmark concentration (i.e., the effect measure for the most sensitive organism) is modified by the application of an Assessment Factor (AF), also known as an Uncertainty Factor (UF), that accounts for the uncertainty associated with the different levels of extrapolation that are to be made from the available data to the real world (e.g., move from modeled data for one organisms to real-world effects in a whole ecosystem). The following equation shows how this AF is used to correct (i.e., further reduce) the chosen effect measure

$$\text{PNEC} = \frac{\text{Effect measure (e.g.EC50)}}{\text{AF}}$$

The assessment factor (AF) to be used depends on the nature of the available data and the completeness of the ecological coverage of the data set. As an example, the AF values used for the derivation of freshwater PNEC values under the REACH guidelines (ECHA 2008) are summarized in Table 1.6.

The risk quotient approach has been employed for ranking and prioritization of chemicals on numerous occasions. Perhaps it is worth mentioning that this approach was employed for the ranking of the candidate substances to be included in the creation of the watch list to support the identification of priority substances for regulation under the Water Framework Directive (WFD) (EC 2000b; Carvalho et al. 2015).

As an example, Orias and Perrodin (2013) calculated PNECs of PCs (among those 16 anticancer compounds) according to modeled ecotoxicological data using the ECOSAR method. They also used experimental data from international

Table 1.6 Summary of criteria used for the selection of assessment factors (AF) to be used for the derivation of freshwater PNECS under REACH guidelines (ECHA 2008)

Available data	AF
At least one short-term $L(E)C_{50}$ from each of three trophic levels [fish, invertebrates (preferred, *Daphnia*), and algae]	1000
One long-term EC_{10} or NOEC (either fish or *Daphnia*	100
Two long-term results (e.g., EC_{10} or NOEC) from species representing two trophic levels (fish and/or *Daphnia* and/or algae)	50
Long-term results (e.g., EC_{10} or NOEC) from at least three species (normally fish, *Daphnia*, and algae) representing three trophic levels	10
Species sensitivity distribution (SSD) method	5-1
Field data or model ecosystems	Case by case

databases (e.g., EPA ECOTOX, Wikipharma) and from the literature. Of the anticancer compounds evaluated by Orias and Perrodin (2014), ifosfamide, etoposide, azathioprine, cyclophosphamide, methotrexate, gemcitabine, and procarbazine were grouped among those compounds with a PNEC higher than 1 μg l^{-1} (less ecotoxic). Only epirubicin was among those compounds with PNECs between 100 ng l^{-1} and 1 μg l^{-1} (intermediate ecotoxicity). The last four anticancer compounds (tamoxifen, doxorubicin, vincristine, 5-fluorouracil) had PNECs lower than 100 ng l^{-1}, which means that they were among those compounds with the greatest ecotoxicity level. After the characterization of the exposure (MEC) and the characterization of the effects (PNEC) was derived, Orias and Perrodin (2014) identified and ranked the PCs by calculating the hazard quotient for each compound, as explained previously in this section.

Cyclophosphamide, methotrexate, and etoposide were found among the compounds detected in HWW presenting a low ecotoxicological hazard for aquatic organisms (HQ lower than 1). In this study (Orias and Perrodin 2014), only one antineoplastic and immunomodulating agent (5-fluorouracil) was found among the 15 most hazardous compounds in hospital wastewaters with a hazard quotient higher than 100,000.

1.5 General Considerations and Conclusions

As presented in this chapter, many screening, ranking, and prioritization approaches are available for low-tier assessment of emerging contaminants. All the methods presented here have been applied, in one way or another, to the assessment of pharmaceutical and personal care products (PPCPs). A recent study by Roos et al. (2012) compared a set of previously proposed ranking and prioritization methods with a focus on aquatic systems that includes the majority of those methods presented in this chapter. The authors compared the results of applying the different approaches to 582 active pharmaceutical ingredients (APIs) including a number of well-studied compounds as well as 54 substances considered as antineoplastic agents. From this exercise, the authors concluded that hazard-based methods were more successful in correctly ranking the well-studied APIs, but the fish plasma model, which includes human pharmacological data, also showed a high success rate. In general, the authors reported that hazard-based approaches based on compound properties produced consistent results and that for the "reference" well-studied APIs, achieved a higher success rate in correctly identifying APIs of high, moderate, and low environmental risk, respectively, compared to the risk-based methods that rely more heavily on sales statistics. The risk-based fish-plasma model, which combines exposure (sales-based water column concentration estimates) and compound-specific data (e.g., log K_{ow}) showed the best performance for the risk-based approaches. Roos et al. (2012) also point to limitations imposed by the quantity and quality of currently available exposure data for PPCPs, which might be one of the reasons for the higher performance of hazard-based systems. Even for

PPCPs for which extensive monitoring data are available, the heterogeneity of formats in which these data are published and the lack of an easily searchable central repository limits its use to risk assessors. In a recent study, Rodríguez-Gil et al. (2018) showed some of these limitations using caffeine (one of the most data-rich PPCPs) as an example. Perhaps the early stages of research into the field of cytostatic compounds in the environment is the perfect time to promote the value of data completeness and openness.

As noted throughout this chapter, many of these screening, ranking, and prioritization approaches are already being applied in a regulatory context, where these screening tools are commonly used as triggers for further assessment. For example, the European Medicines Agency (EMEA) guidelines on the Environmental Risk Assessment of Medicinal Products for Human Use (EMEA 2006b, 2016) recommends the following steps.

Phase I:

- Estimate of predicted environmental concentration (PEC) for surface water and
- Measure or estimate of the log K_{ow}.
  - If the PEC value is equal to or above 0.01 μg/l, a Phase II assessment is performed.
  - If log $K_{ow} > 4.5$, the compound is screened for PBT.
  - If the compound is known to affect reproduction of vertebrates or invertebrates at concentrations.

Phase II:

- Phase II consists of two tiers. In the first tier (Phase II, Tier A), data on physicochemical, fate, and effect studies (experimental data are recommended) are reviewed and the predicted no effect concentration (PNEC) for water, groundwater, and microorganisms are calculated.
  - If the ratio $PEC_{surface\ water}/PNEC_{surface\ water}$ is above 1, Tier B, in Phase II, is required.

In general the selection of the most appropriate method must be a balancing act between readiness and convenience (we are conducting a screening process, after all) and the reliability and relevance of the results. As already mentioned, minimizing false negatives should be the main priority, and the number of false positives should not overburden the process.

Some authors (Wennmalm and Gunnarsson 2005) have envisioned the use of ranking and prioritization methodologies as a tool to inform healthcare practitioners about the environmental risk associated with medicinal products with the aim of including this information in their decision-making process. These ideas later translated into the Swedish Environmental Classification and Information System for Pharmaceuticals (Ågerstrand et al. 2009). The success of such classification systems has been low, especially for influencing the choices made by healthcare practitioners (Ågerstrand et al. 2009). In our opinion, this particular use of ranking and prioritization methodologies for cytostatic compounds would be even less effective because of the particular circumstances under which these compounds are used. The use of these techniques, then, should be focused as a tool to prioritize compounds for which further research is needed. We have seen how cytostatic compounds are considered as hazardous waste, according to decision 2000/532/EC (EC 2000a). In this way, the results of a hazard/risk assessment process (specially based on a ranking and prioritization approach) could be used to inform us not only about which compounds represent a larger concern but also on whether there is a need for cytostatic compound-containing waste (such as hospital wastewater) to be treated in a separate manner from other wastewaters.

Last, it is important to note that some of the specific concerns related to cytostatic compounds are not often considered in many of these screening techniques. For example, many of the presented ERA methodologies rely on acute and subacute toxicity tests, as recommended in most guidelines (EMEA 2006b, 2016; ECHA 2011b; US EPA 2012b). The mutagenic/genotoxic nature of these compounds calls for long-term–low-dose studies, data on which are not in a readily available form for most cytostatic compounds. Additionally, as compounds that interact with DNA directly, it could be proposed that no safe limit can be assumed for the presence of cytostatic compounds in the environment. As such, some authors propose (Kümmerer et al. 2016) that screening tools used as triggers during regulatory process should not be employed and that DNA-damaging drugs should be exempt from the action limit set, for example, by the EMEA guidelines for performing an environmental risk assessment. In these cases, a case-by-case evaluation of the risk associated with their presence in the environment is recommended.

References

Ågerstrand M, Wester M, Rudén C (2009) The Swedish environmental classification and information system for pharmaceuticals—an empirical investigation of the motivations, intentions and expectations underlying its development and implementation. Environ Int 35(5):778–786. https://doi.org/10.1016/j.envint.2008.12.001

Arnot JA, Gobas FAPC (2006) A review of bioconcentration factor (BCF) and bioaccumulation factor (BAF) assessments for organic chemicals in aquatic organisms. Environ Rev 14 (4):257–297. https://doi.org/10.1139/a06-005

Besse J-P, Kausch-Barreto C, Garric J (2008) Exposure assessment of pharmaceuticals and their metabolites in the aquatic environment: application to the French situation and preliminary priorization. Hum Ecol Risk Assess Int J 14:665–695

Boxall AB, Rudd MA, Brooks BW, Caldwell DJ, Choi K, Hickmann S, Innes E, Ostapyk K, Staveley JP, Verslycke T, Ankley GT, Beazley KF, Belanger SE, Berninger JP, Carriquiriborde P, Coors A, Deleo PC, Dyer SD, Ericson JF, Gagne F, Giesy JP, Gouin T, Hallstrom L, Karlsson MV, Larsson DG, Lazorchak JM, Mastrocco F, McLaughlin A, McMaster ME, Meyerhoff RD, Moore R, Parrott JL, Snape JR, Murray-Smith R, Servos MR, Sibley PK, Straub JO, Szabo ND, Topp E, Tetreault GR, Trudeau VL, Van Der Kraak G (2012) Pharmaceuticals and personal care products in the environment: what are the big questions? Environ Health Perspect 120(9):1221–1229. https://doi.org/10.1289/ehp.1104477

Brooks BW, Huggett DB, Boxall ABA (2009) Pharmaceuticals and personal care products: research needs for the next decade. Environ Toxicol Chem 28(12):2469–2472. https://doi.org/10.1897/09-325.1

Carvalho RN, Ceriani L, Ippolito A, Lettieri T (2015) Development of the first watch list under the environmental quality standards directive. European Commission Joint Research Center, Ispra

Cizmas L, Sharma V, Gray C, McDonald T (2015) Pharmaceuticals and personal care products in waters: occurrence, toxicity, and risk. Environ Chem Lett 13(4):381–394. https://doi.org/10.1007/s10311-015-0524-4

EC (2000a) 2000/532/EC: Commission Decision of 3 May 2000 replacing Decision 94/3/EC establishing a list of wastes pursuant to Article 1(a) of Council Directive 75/442/EEC on waste and Council Decision 94/904/EC establishing a list of hazardous waste pursuant to Article 1(4) of Council Directive 91/689/EEC on hazardous waste (notified under document number C(2000) 1147) (Text with EEA relevance). 6.9.2000. Official Journal of the European Union, Brussels

EC (2000b) Directive 2000/60/EC of the European Parliament and of the Council of 23 October 2000 establishing a framework for Community action in the field of water policy. 22.12.2000. Official Journal of the European Union, Brussels

EC (2004) Directive 2004/28/EC of the European Parliament and of the Council of 31 March 2004 amending Directive 2001/82/EC on the Community code relating to veterinary medicinal products (Text with EEA relevance). 30.4.2004. Official Journal of the European Union, Brussels

EC (2006) Regulation (EC) No 1907/2006 of the European Parliament and of the Council of 18 December 2006 concerning the Registration, Evaluation, Authorisation and Restriction of Chemicals (REACH), establishing a European Chemicals Agency OJ L 396/1. (EC No. 1907/2006)

EC (2012) Directive 2001/83/EC of the European Parliament and of the Council of 6 November 2001 on the Community code relating to medicinal products for human use (Consolidated version: 16/11/2012). Official Journal of the European Union, Brussels

EC (2013) Directive 2013/39/EU of the European Parliament and of the Council of 12 August 2013 amending Directives 2000/60/EC and 2008/105/EC as regards priority substances in the field of water policy. Text with EEA relevance. 24.8.2013. Official Journal of the European Union, Brussels

ECHA (2008) Guidance on information requirements and chemical safety assessment. Chapter R.10: characterisation of dose (concentration)-response for environment. European Chemicals Agency, Helsinki

ECHA (2011a) Guidance on information requirements and chemical safety assessment. Part A: introduction to the guidance document. V1.1 (ECHA-2011-G-15-EN). European Chemicals Agency, Helsinki

ECHA (2011b) Guidance on information requirements and chemical safety assessment. Part B: hazard assessment V2.1 (ECHA-11-G-16-EN). European Chemicals Agency, Helsinki

ECHA (2012a) Guidance on information requirements and chemical safety assessment. Part D: exposure scenario building. V1.2 (ECHA-12-G-17-EN). European Chemicals Agency, Helsinki

ECHA (2012b) Guidance on information requirements and chemical safety assessment. Part E: Risk Characterisation. V2.0 (ECHA-12-G-16-EN). European Chemicals Agency, Helsinki

ECHA (2017) Guidance on information requirements and chemical safety assessment. Part C: PBT/vPvB assessment. V3.0 (ECHA-17-G-09-EN). European Chemicals Agency, Helsinki

EFSA (2013) Guidance on tiered risk assessment for plant protection products for aquatic organisms in edge-of-field surface waters. European Food Safety Authority Panel on Plant Protection Products and their residues (PPR). EFSA J 11(7):3290. https://doi.org/10.2903/j.efsa.2013.3290

EMEA (2006a) The environmental risk assessment of medicinal products for human use (EMEA/CHMP/SWP/4447/00). European Medicines Agency, London

EMEA (2006b) Guideline on the environmental risk assessment of medicinal products for human use (EMEA/CHMP/SWP/4447/00). European Medicine Agency – Committee for Medicinal Products for Human Use, London

EMEA (2015) Guideline on the assessment of persistent, bioaccumulative and toxic (PBT) or very persistent and very bioaccumulative (vPvB) substances in veterinary medicinal products. Committee for Medicinal Products for Veterinary Use, European Medicine Agency, London

EMEA (2016) Questions and answers on 'Guideline on the environmental risk assessment of medicinal products for human use' (EMA/CHMP/SWP/44609/2010 Rev. 1). Committee for Medicinal Products for Human Use, European Medicine Agency, London

Fick J, Lindberg RH, Tysklind M, Larsson DGJ (2010) Predicted critical environmental concentrations for 500 pharmaceuticals. Regul Toxicol Pharmacol 58(3):516–523. https://doi.org/10.1016/j.yrtph.2010.08.025

Franquet-Griell H, Gómez-Canela C, Ventura F, Lacorte S (2017) Anticancer drugs: consumption trends in Spain, prediction of environmental concentrations and potential risks. Environ Pollut 229:505–515. https://doi.org/10.1016/j.envpol.2017.06.011

Fu W, Franco A, Trapp S (2009) Methods for estimating the bioconcentration factor of ionizable organic chemicals. Environ Toxicol Chem 28(7):1372–1379. https://doi.org/10.1897/08-233.1

Halling-Sørensen B, Nors Nielsen S, Lanzky PF, Ingerslev F, Holten Lützhøft HC, Jørgensen SE (1998) Occurrence, fate and effects of pharmaceutical substances in the environment: a review. Chemosphere 36(2):357–393. https://doi.org/10.1016/S0045-6535(97)00354-8

Johnson AC, Jürgens MD, Williams RJ, Kümmerer K, Kortenkamp A, Sumpter JP (2008) Do cytotoxic chemotherapy drugs discharged into rivers pose a risk to the environment and human health? An overview and UK case study. J Hydrol 348(1):167–175. https://doi.org/10.1016/j.jhydrol.2007.09.054

Kümmerer K, Haiß A, Schuster A, Hein A, Ebert I (2016) Antineoplastic compounds in the environment—substances of special concern. Environ Sci Pollut Res 23(15):14791–14804. https://doi.org/10.1007/s11356-014-3902-8

Mackay D (1991) Multimedia environmental models: the fugacity approach. CRC Press, Boca Raton

Mackay D, Arnot JA (2011) The application of fugacity and activity to simulating the environmental fate of organic contaminants. J Chem Eng Data 56(4):1348–1355. https://doi.org/10.1021/je101158y

Mahnik SN, Lenz K, Weissenbacher N, Mader RM, Fuerhacker M (2007) Fate of 5-fluorouracil, doxorubicin, epirubicin, and daunorubicin in hospital wastewater and their elimination by activated sludge and treatment in a membrane-bio-reactor system. Chemosphere 66(1):30–37. https://doi.org/10.1016/j.chemosphere.2006.05.051

Meylan WM, Howard PH, Boethling RS, Aronson D, Printup H, Gouchie S (1999) Improved method for estimating bioconcentration/bioaccumulation factor from octanol/water partition coefficient. Environ Toxicol Chem 18(4):664–672. https://doi.org/10.1002/etc.5620180412

Noguera-Oviedo K, Aga DS (2016) Lessons learned from more than two decades of research on emerging contaminants in the environment. J Hazard Mat 316:242–251. https://doi.org/10.1016/j.jhazmat.2016.04.058

Olalla A, Negreira N, López de Alda M, Barceló D, Valcárcel Y (2018) A case study to identify priority cytostatic contaminants in hospital effluents. Chemosphere 190(suppl C):417–430. https://doi.org/10.1016/j.chemosphere.2017.09.129

Orias F, Perrodin Y (2013) Characterization of the ecotoxicity of hospital effluents: a review. Sci Total Environ 454–455:250–276

Orias F, Perrodin Y (2014) Pharmaceuticals in hospital wastewater: their ecotoxicity and contribution to the environmental hazard of the effluent. Chemosphere 115:31–39. https://doi.org/10.1016/j.chemosphere.2014.01.016

Parrella A, Lavorgna M, Criscuolo E, Russo C, Isidori M (2014) Estrogenic activity and cytotoxicity of six anticancer drugs detected in water systems. Sci Total Environ 485-486(suppl C):216–222. https://doi.org/10.1016/j.scitotenv.2014.03.050

Pichler C, Filipič M, Kundi M, Rainer B, Knasmueller S, Mišík M (2014) Assessment of genotoxicity and acute toxic effect of the imatinib mesylate in plant bioassays. Chemosphere 115(suppl C):54–58. https://doi.org/10.1016/j.chemosphere.2014.01.010

QSAR. (2017) Toolbox V3.4. https://www.qsartoolbox.org/. Organisation for Economic Co-operation and Development. Accessed Feb 2017

Rodríguez-Gil JL, Cáceres N, Dafouz R, Valcárcel Y (2018) Caffeine and paraxanthine in aquatic systems: global exposure distributions and probabilistic risk assessment. Sci Total Environ 612:1058–1071. https://doi.org/10.1016/j.scitotenv.2017.08.066

Roos V, Gunnarsson L, Fick J, Larsson DGJ, Rudén C (2012) Prioritising pharmaceuticals for environmental risk assessment: towards adequate and feasible first-tier selection. Sci Total Environ 421-422(suppl C):102–110. https://doi.org/10.1016/j.scitotenv.2012.01.039

Sanderson H, Johnson DJ, Reitsma T, Brain RA, Wilson CJ, Solomon KR (2004) Ranking and prioritization of environmental risks of pharmaceuticals in surface waters. Regul Toxicol Pharmacol 39(2):158–183. https://doi.org/10.1016/j.yrtph.2003.12.006

Schreiber R, Gündel U, Franz S, Küster A, Rechenberg B, Altenburger R (2011) Using the fish plasma model for comparative hazard identification for pharmaceuticals in the environment by extrapolation from human therapeutic data. Regul Toxicol Pharmacol 61(3):261–275. https://doi.org/10.1016/j.yrtph.2011.08.006

Suter GW II (1997) Overview of the ecological risk assessment framework. In: Ingersoll CG, Dillon T, Biddinger GR (eds) Ecological risk assessment of contaminated sediments. SETAC Pellston workshop on sediment ecological risk assessment (1995 Apr 23–28; Pacific Grove, CA). SETAC Press, Pensacola, pp 1–11

Toxic Substances Control Act, 15 USC Ch. 53, Pub. L. No. 94-469 § 2601-2697 (1976 November 10)

US EPA (2004) Overview of the ecological risk assessment process in the office of pesticide programs. U.S. Environmental Protection Agency, Office of Prevention, Pesticides, and Toxic Substances. Office of Pesticide Programs, Washington, DC.

US EPA (2012a) Sustainable futures/P2 framework manual (EPA-748-B12-001). US Environmental Protection Agency

US EPA (2012b) TSCA work plan chemicals: methods document. US Environmental Protection Agency

US EPA (2013) Estimation programs interface suite™ for Microsoft® Windows v 4.11. US Environmental Protection Agency

Wennmalm Å, Gunnarsson B (2005) Public health care management of water pollution with pharmaceuticals: environmental classification and analysis of pharmaceutical residues in sewage water. Drug Inf J 39(3):291–297. https://doi.org/10.1177/009286150503900307

Chapter 2
Predicted Environmental Concentrations: A Useful Tool to Evaluate the Presence of Cytostatics in Surface Waters

Cristian Gómez-Canela, Mónica S. F. Santos, Helena Franquet-Griell, Arminda Alves, Francesc Ventura, and Silvia Lacorte

Abstract Cytostatic or anticancer drugs used in chemotherapy are administered in the order of tons per year in European countries. After administration, these compounds are excreted and reach the sewage system. Because of their poor elimination in wastewater treatment plants (WWTPs), a fraction of these compounds are finally released to surface waters and can produce genotoxic and mutagenic effects on aquatic organisms. Given that cancer incidence has increased over the last years and is foreseen to increase, it is anticipated that the consumption of cytostatic drugs will equally increase. Thus, their control in the environment is of utmost importance.

Consumption or prescription data have demonstrated to be very valuable to estimate the presence of pharmaceuticals in environmental waters. This approach was first suggested by the European Medicines Agency (EMA) that proposed the calculation of predicted environmental concentration (PEC) based on consumption data, excretion, elimination in the WWTP, and dilution in receiving waters. Further, EMA recommended the evaluation of risk when PEC values in surface water were equal or above the threshold value of 0.01 μg/L. The calculation of PEC results is an extremely useful information to prioritize compounds for further monitoring, to establish the potential incidence of pharmaceuticals in a specific area, and even to assess their risk according to toxicological data. Cytostatic compounds account for an exemplary family to calculate the PECs, as differing from other pharmaceuticals, all the prescribed amounts will be consumed, and thus, PECs are very accurate.

Cristian Gómez-Canela and Mónica S. F. Santos contributed equally in the writing of this chapter.

C. Gómez-Canela · H. Franquet-Griell · F. Ventura · S. Lacorte (✉)
Department of Environmental Chemistry, IDAEA-CSIC, Barcelona, Catalonia, Spain
e-mail: slbqam@cid.csic.es

M. S. F. Santos · A. Alves
LEPABE – Laboratory for Process, Environmental, Biotechnology and Energy Engineering, Faculty of Engineering, University of Porto, Porto, Portugal

E. Heath et al. (eds.), *Fate and Effects of Anticancer Drugs in the Environment*,
https://doi.org/10.1007/978-3-030-21048-9_2

The objectives of this chapter are to define the concept of PECs, show how the calculations are undertaken in all its adaptations, provide the raw data for their calculation, and demonstrate its applicability for the assessment of cytostatics in wastewater and river water.

Keywords Cytostatics · Prioritization approaches · Predicted environmental concentrations · Uncertainties · Measured environmental concentrations

2.1 Introduction

Cancer is a tumor growth of tissues, of malign character which disturb biological functions. According to the World Health Organization (WHO), in 2012, 14.1 million new cancer cases and 8.2 million cancer-related deaths were recorded worldwide (WHO 2017). Of these global values, 57% of the cases and 65% of the deaths occurred in the less developed regions. The most common causes of cancer-related deaths worldwide include cancers of the lung (1.69 million deaths), liver (788,000 deaths), colorectal (774,000 deaths), stomach (754,000 deaths), and breast (571,000 deaths). The incidence of cancer varies among countries, with levels from 138×10^5 to 368×10^5 persons/year in Europe (Generalitat de Catalunya (GENCAT) 2016). The cancer incidence in the global population is gradually increasing due to the growth of the population and its aging, which causes a greater exposure to risk factors and a decrease in cell repair mechanisms. The early detection programs allow the discovery of new cancer cases in initial stages, contributing to a decrease in the mortality rate due to improved medication.

To fight against cancer, different treatments can be applied depending on the type of tumor and its extension, the physical condition of the person to be treated, together with other medical considerations. The majority of these treatments imply the administration of pharmaceuticals, named antineoplastic, cytostatics, or anticancer drugs, which are a broad group of chemotherapy compounds with different chemical structures and modes of action. These drugs are classified by the WHO (www.whocc.no/atcddd) under class L of the Anatomical Therapeutic Chemical (ATC) classification, which belongs to antineoplastic and immunomodulating agents. Four main groups are currently used: antineoplastic agents (L01), endocrine therapy (L02), immunostimulants (L03), and immunosuppressants (L04). There are more than 300 cytostatic drugs registered by the ATC classification. Once administered, cytostatic drugs are excreted by urine or feces as parent compounds and metabolites and are directly discharged into the sewerage system (Lenz et al. 2007; Zhang et al. 2013). As cytostatic drugs have low biodegradability (Kosjek and Heath 2011) and poor removal by conventional activated sludge treatment (Martín et al. 2011), effluents from wastewater treatment plants (WWTP) receiving urban wastewaters and hospital effluents are the main sources of cytostatic compounds in the aquatic system (Besse et al. 2012; Olalla et al. 2018; Weissbrodt et al. 2009; Zhang et al. 2013). The concentrations detected in river water are generally in the ng/L level

(Garcia-Ac et al. 2009; Johnson et al. 2008; Martín et al. 2011), but despite this low concentration, cytostatic compounds are cytotoxic and can affect aquatic organisms. Short-term toxicity studies showed high lethal concentrations (Brystol-Myers Squibb Company 2010; Parrella et al. 2014; Roche 2012) and interactions with DNA, which may be a sign of long-term effects on aquatic organisms (Besse et al. 2012).

Given the high number of cytostatics currently used in chemotherapy, the assessment of the environmental occurrence of these drugs is not feasible through monitoring plans because (i) there are no analytical methods to identify and quantify all cytostatic compounds and (ii) the monitoring would become an extremely time-consuming process and would involve high costs of operation. Therefore, methods for estimating the environmental occurrence have been developed to allow the identification and prioritization of main cytostatics in water. These theoretical approaches aim at reducing the number of compounds to be analyzed and, consequently, optimize the time and costs associated with monitoring plans. Some of the prioritization systems include the use of computer models that combine geographic and environmental information, such as GREAT-ER (Geo-referenced Regional environmental Exposure Assessment Tool for European Rivers) or Low Flows 2000 software package (Johnson et al. 2008; Rowney et al. 2009; Young et al. 2003), but maps of the area under study are required, as well as programming knowledge. An alternative and more accessible method is the prediction of environmental concentrations (PECs) as proposed by EMA (European Medicines Agency (EMA) 2006), who recommends a threshold level of 0.01 μg/L for further environmental risk assessment. This prediction model has played an important role in assessing the theoretical concentration of cytostatic drugs in waters based on their consumption patterns (Franquet-Griell et al. 2015, 2017b). Unlike other pharmaceuticals, all cytostatic drugs prescribed are consumed, and thus this model is especially accurate for this class of pharmaceutical compounds.

2.2 Description of PEC in Surface Waters

PECs are the estimated concentrations of a pharmaceutical in influents and effluents of WWTPs and in rivers, calculated from available information on their consumption, excretion rates, elimination in WWTPs, and dilution into receiving waters. According to the EMA guidelines, the assessment of the potential environmental risks of human medicines is a step-wise procedure comprising two phases. In Phases I and II (Tier A), the PEC calculation is restricted to the aquatic compartment and is expressed by the following equation:

$$PEC_{surfacewater} = \frac{DOSE_{ai} \times F_{pen}}{WW \times DF}. \quad (2.1)$$

The parameter $DOSE_{ai}$ is the daily dose of a specific drug consumed per inhabitant (μg/inhab/day), and it is recommended to use the maximum dose, corresponding to the worst-case scenario. F_{pen} is a factor related to market penetration, whose default value is 0.01. This value indicates that 1% of the population is treated daily with a specific drug substance, and it resulted from the analysis of German market in 2001, taking into account 800 drug substances (European Medicines Agency (EMA) 2006). The parameter *WW* is the amount of water consumed per inhabitant per day (L/inhab/day). The default value is 200 L/inhab/day (European Medicines Agency (EMA) 2006), but the use of data specific for the region under analysis is always recommended. Finally, the dilution factor (*DF*) from wastewater treatment plant effluents to surface waters is an important parameter, which could strongly affect the final result. The default value recommended by EMA for this parameter is 10, although the European Chemicals Agency has recognized that it could vary between 1 and 1000 (European Chemicals Agency (ECHA) 2003). The calculation of an accurate DF is described in Sect. 2.4.5.

Since the administration of cytostatics is highly influenced by new chemotherapies directed to reassure the specific needs of each patient, the calculation of consumptions based on defined daily doses values ($DOSE_{ai}$) and the incidence of cancer (F_{pen}) is not accurate. Alternatively, data of cytostatics consumption in hospitals and sold in pharmacies should be considered (Eq. 2.2):

$$PEC_{surface-water} = \frac{Consumption}{WW \times inhab \times DF \times 365} \tag{2.2}$$

where *Consumption* is the total amount of a drug which is consumed in a defined region per year (μg/yr) and *inhab* the number of habitants of the same region. 365 are the days of the year and permit to convert consumption into μg/day.

The PECs resulting from Eqs. 2.1 and 2.2 are conservative, since neither metabolism in the body (i.e., 100% of the cytostatic administered is excreted unchanged) nor removal in WWTPs is assumed.

The first refinement of the PEC calculation is regarded with the inclusion of the percentage of excretion of the parent cytostatic F_{exc}:

$$PEC_{surfacewater} = \frac{Consumption \times F_{exc}}{WW \times inhab \times DF \times 365}. \tag{2.3}$$

Another refinement corresponds to the incorporation of the removal efficiency in WWTPs (F_{WWTP}) – Eq. 2.4:

$$PEC_{surfacewater} = \frac{Consumption \times F_{exc} \times (1 - F_{WWTP})}{WW \times inhab \times DF \times 365}. \tag{2.4}$$

Using the units indicated for each parameter, PECs calculated from Eq. 2.1 to Eq. 2.4 are given in μg/L and correspond to the estimated concentrations in surface waters, such as rivers. The estimated concentrations in WWTP effluents can be determined from Eqs. 2.1 to 2.4, without considering the dilution factor.

The calculation of PECs allows prioritizing those compounds which will be primarily detected in WWTP effluents or river waters according to consumption data. However, accurate consumption data and sociodemographic information are needed to provide reliable PECs for a specific area.

2.3 Compilation of Consumption Data for Cytostatic Drugs

The accurate knowledge of the consumption trends allows the identification of those compounds more widely used and that can potentially trigger a high environmental impact. In addition, this permits to focus the monitoring programs to those compounds which have a high probability to be found in environmental waters. Therefore, the ability to obtain consumption data at a country level or more regionally is key for estimating the PECs. Figure 2.1 displays a workflow of the different stages to calculate PECs in surface waters.

However, obtaining the consumption data of anticancer drugs in a given area/ country is a very demanding task for several reasons:

- Difficulty to access to consumption data.
- Information may only be available for specific drugs, e.g., individual compounds or specific ATC groups.
- Information is only available for pharmacies or hospitals. Rarely the complete information on cytostatic consumption is provided at a country base.
- The consumption data is only available for specific years.
- The data provided is very disorganized in terms of presentations of each drug, concentration of active ingredient, and number of prescriptions.

In this section, clues on how and what data has to be requested to obtain reliable information are provided so that the PECs can be calculated with precision. In addition, an explanation on how this data must be processed and normalized to be used in the PEC calculations is given.

2.3.1 Which Compounds Should Be Considered?

The consumption data for cytostatic drugs should be requested through their ATC code. The ATC classification was proposed by WHO, and it allows to distinguish different pharmaceutical compounds through a code of five levels which is attributed according to the anatomical group (1st level), therapeutic group (2nd level), and the pharmacological and chemical properties (3rd and 4th level). The 5th-level designs the substance itself (World Health Organization (WHO) 2017). According to this classification, cytostatic drugs are included in the group of antineoplastic and immunomodulating agents, designed by letter L. Nowadays, the L group contains 300 active ingredients, but the list is periodically revised, and every year new pharmaceuticals are included. Specifically, it is expected that new substances will

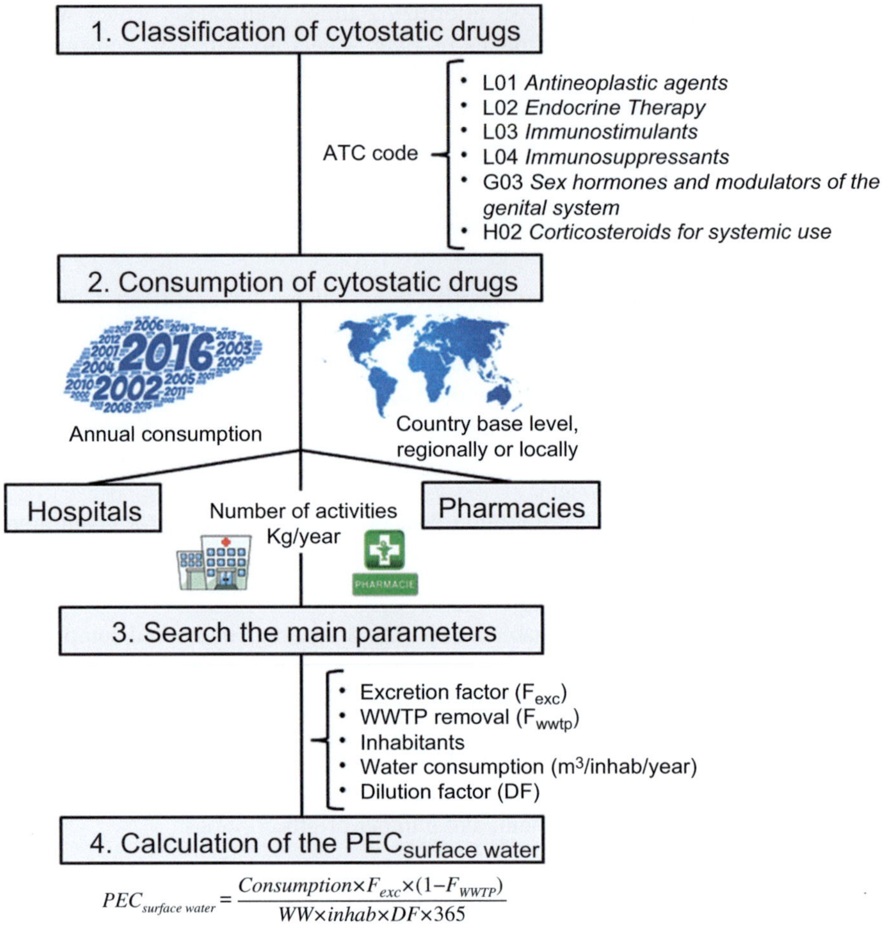

Fig. 2.1 Workflow of the calculation of PEC in surface water

be included in the future. The L group is subdivided in L01 (antineoplastic agents), L02 (endocrine therapy), L03 (immunostimulants), and L04 (immunosuppressants), which are briefly described as:

L01: Antineoplastic agents: The L01 group is subdivided into alkylating agents (L01A), antimetabolites (L01B), plant alkaloids and natural products (L01C), cytotoxic antibiotics and related substances (L01D), and other antineoplastic agents (L01X). The drugs included in this group act by inhibiting or altering the transcription of the DNA or by interacting with proteins that regulate the biological processes of the cells in order to destroy or control the growth of cancer cells (Besse et al. 2012). L01 type of anticancer drugs, such as cyclophosphamide, ifosfamide, capecitabine, fluorouracil, gemcitabine, vinblastine, vincristine, epirubicin, carboplatin, is among the most widely studied environmental contaminants.

L02: Endocrine therapy: As the name indicates, these drugs (subgroups L02A and L02B) are used in endocrine therapy, also called hormone therapy, and are responsible for inhibiting the synthesis of hormones or their receptors (Besse et al. 2012). Compounds such as megestrol, diethylstilbestrol, and tamoxifen are used in the treatment of prostate cancer or breast cancer and hormone-sensitive tumors, which develop due to the influence of estrogen or testosterone.

L03: Immunostimulants: L03/L03A groups are substances that increase the capacity of the immune system to fight against diseases or infections and are used in biological therapies, also called directed therapies. These treatments have less side effects since they attack healthy cells to a lesser extent (National Cancer Institute (NCI) 2017a). Examples of this type of drugs are the different types of interferons, among others.

L04: Immunosuppressants: Commonly, immunosuppressants (L04A subgroup) are used to suppress the immune system after a transplant (National Cancer Institute (NCI) 2017b). For example, after transplantation of the bone marrow against leukemia, drugs of the L04 group are commonly used in combination with different drugs from other L subgroups. For some of these L04 drugs, in addition to their immunosuppressive effect, it has been shown that they can also have an anticancer effect. Some examples are mycophenolic acid (Carter et al. 1969; Dun et al. 2013; Majd et al. 2014), sirolimus (Law 2005), or lenalidomide (National Cancer Institute (NCI) 2017a).

Besides the drugs of L group, other compounds are used in the treatment of cancers, such as drugs of the G03 group (sex hormones and modulators of the genital system) and H02 (corticosteroids for systemic use). Specifically, some antiandrogens (G03H), such as the cyproterone, are administered in the treatment against prostate cancer, whereas glucocorticoids (H02AB), such as prednisone, are used for the treatment of leukemia, among other types of cancer (National Cancer Institute (NCI) 2017a).

Overall, for an adequate estimation of the PECs, it is recommended to obtain consumption data of all L subclasses (and some G and H) at a country or region without previous preselection. This way the concentrations and risks associated with the presence of all cytostatics in the environment can be globally determined.

2.3.2 *From Whom the Data Should Be Requested?*

The consumption data can be obtained at a country base level, regionally or locally (e.g., hospital-based consumptions). The more detailed the consumption patterns, the more accurate the PEC calculation (Fig. 2.1). Most meaningful data to calculate river based PECs are obtained at a country level, as it provides systematic data on all the compounds prescribed annually. The consumption data on a country base is generally managed by the Ministries of Health or health institutions, who will provide data upon request, as this data is public. For example, Besse et al. obtained the consumption of 48 cytostatics in France from the French Health Products Safety Agency (Besse et al. 2012), whereas Coetsier et al. contacted the French social health care system to get the consumption of medicines reimbursed by the social system, excluding the over-the-counter products (Coetsier et al. 2009). In England, the consumption data was obtained from the Department of Health (DoH) and only represent the cytostatics consumption via pharmacies (Johnson et al. 2008; Rowney et al. 2009). In Spain, data was requested to the Ministry of Health, Social Services and Equality and correspond to the billing of prescriptions of 72 compounds from the Spanish National Health Service through pharmacies (Franquet-Griell et al. 2017b, Ortiz de García et al. 2013). The consumption data for all 132 cytostatics consumed over the period 2010–2012 in pharmacies and hospitals of Catalonia (NE Spain) were obtained from the Catalan Health Service (CatSalut) (Franquet-Griell et al. 2015). In Portugal, the consumption data of 171 different anticancer drugs over 9 years (2007–2015) was provided by the Instituto Nacional da Farmácia e do Medicamento, I.P. (Infarmed, I.P) (Santos et al. 2017). As much as possible, it is advised to request for consumption data of all cytostatics, not only preselected compounds, as this will expand the potential of PECs.

2.3.3 Which Consumption Data Should Be Requested?

Cytostatics are administered in hospitals and in pharmacies. In parallel, they are administered in public or private health institutions (Fig. 2.1). In general, administrative bodies compile data of the public health system, so the consumption through mutual insurance companies is not accounted for. Depending on the administrative body in each country, the compilation of consumption patterns may vary and can only record hospital consumption, pharmacies, and very seldom both. Regarding hospitals or pharmacies, the quantity dispensed in pharmacies is much higher than the amount of cytostatics administered in hospitals (in units of the active ingredient) (Franquet-Griell et al. 2015). This is because many patients have their treatment at home for a given period of time (outward patients). In Germany for example, the sales in pharmacies represent 78.9% of the total cytostatic consumption; the rest is administered in hospitals (Kümmerer et al. 2016). In France, the hospital administration decreased from 82% to 35%, while an increase of cytostatic sales in pharmacies was observed during the period 2004–2008 (Besse et al. 2012). Another example is Catalonia, where the pharmacies also proved to be the main route (70–80%) of anticancer drug consumption (Franquet-Griell et al. 2015). Therefore, both data from hospitals and pharmacies should be requested whenever possible, but if the administrative bodies do not have all information, data from pharmacies is the most representative of the consumption patterns. For avoiding confusions, the source and type of consumption data have to be provided.

2.3.4 What Time Period Should Be Representative?

As indicated in the introduction, the number of pharmaceuticals used in the treatment of cancer can vary among years, and there is also a continuous shift of treatments from hospital to pharmacies or vice versa (Fig. 2.1). As an example, the total hospital consumption of cytostatics from L01 and L02 ATC groups increased in France between 2004 and 2008 (Besse et al. 2012) but decreased in Spain (Franquet-Griell et al. 2017b) and Portugal (Santos et al. 2017) in the period 2010–2015 and 2007–2015, respectively. Therefore, obtaining data from several years is a way to determine time trends and also permits to identify the compounds that are being administered in a given period and then maybe discontinued because of the shift of prescriptions or elimination from the market. In addition to that, current consumption patterns used for PEC estimations permit compound prioritization for more targeted monitoring studies.

2.3.5 How Data Have to Be Organized?

Consumption data is generally provided as the number of pills, capsules, injections, or other presentations of a specific drug. Knowing the concentration of each active ingredient in a certain formulation, the total consumption of each cytostatic is calculated in kg per year. This data needs to be further normalized to μg/inhab/day to compare consumption patterns with other countries or regions. Data can be compiled and organized in an Excel format as indicated in Table 2.1. It is advisable to organize the different parameters in such a way that introducing the raw consumption data, the PEC values are automatically calculated and figured.

2.4 Uncertainties in the Analysis of PECs

In the calculation of PECs, there are several uncertainties which are related to the input parameters of the PEC formula. These uncertainties can produce a bias in the PEC calculation and can alter the results obtained and thus, their applicability. The uncertainty of each PEC parameter and how data has to be treated or adapted according to each case are discussed in the following sections. Among the different equations to calculate the PECs in surface waters, Eq. 2.4 is the most accurate and will be considered as it reflects the real conditions of discharge and water cycling.

Table 2.1 Datasheet to calculate PECs in WWTP effluents and rivers

					2010			2011			2012		
Total consumption (g/day)	PEC WWTPeff (ng/L)	PEC river (ng/L)	Total consumption (g/day)	PEC WWTPeff (ng/L)	PEC river (ng/L)	Total consumption (g/day)	PEC WWTPeff (ng/L)	PEC river (ng/L)					
G03: Sex hormones and modulators of the genital system	G03HA01	Cyproterone	0.33	0.15	72	20	2.04	70	19	1.9	59	17	1.7
H02: Corticosteroids for systemic use	H02AB07	Prednisone	0.5	0.02	608	300	30.1	626	310	31	635	314	31
L01: Antineoplastic agents	L01BC06	Capecitabine	0.11	0.15	2,258	213	21	2,272	214	21	1,863	175	17
	L01XX05	Hydroxycarbamide	0.5	0.02	1,598	791	79	1,741	862	86	1,697	840	84
	L01XE01	Imatinib	0.25	0.06	244	58	5.8	259	61	6.1	262	62	6
	L01BC05	Gemcitabine	0.1	0.40	97	5.9	0.6	89	5.4	0.5	87	5.3	0.5
	L01XE05	Sorafenib	0.51	0.85	73	5.6	0.5	73	5.6	0.5	62	4.8	0.5
L02: Endocrine therapy	L02AB01	Megestrol	0.78	0.11	619	433	43	614	429	42	538	376	37
	L02BA01	Tamoxifen	0.13	0.93	133	1.2	0.1	136	1.2	0.1	135	1.2	0.1
	L02BB01	Flutamide	0.1	0.10	259	23.5	2.3	208	18	1.9	152	14	1.4
	L02BB03	Bicalutamide	0.55	0.03	302	163.5	16	302	163	16	263	142	14
L03: Immunostimulants	L03AX03	BCG vaccine	0.5	0.00	2.6	1.3	0.1	2.7	1.4	0.1	2.3	1.2	0.1
	L03AX13	Glatiramer, acetate	0.5	0.00	8.6	4.4	0.4	9.9	5.0	0.5	10	5.5	0.5
L04: Immunosuppressants	L04AA06	Mycophenolic acid	0.63	0.41	5,092	1,910.4	191	5,408	2,028	203	5,553	2,083	208
	L04AD01	Cyclosporine	0.001	0.00	330	0.3	0.03	321.46	0.3	0.03	305	0.3	0.03
	L04AX01	Azathioprine	0.02	0.02	553	11.0	1.10	606.00	12.0	1.2	632	12	1.2
	L04AA13	Leflunomide	0.5	0.03	44	21.6	2.16	46.24	22.7	2.27	46	23	2.3

Examples are extracted from the study of Franquet-Griell et al. (2015)

2.4.1 Consumption Data

The consumption of anticancer drugs on a countrywide basis is in the order of tones/yr in European countries (Besse et al. 2012; Booker et al. 2014). However, the consumption patterns of individual cytostatics (within and between ATC groups) can vary among countries for the treatment of the same cancer pathology. The first uncertainty refers to the compounds selected for the calculation of PECs. Most of the published papers refer to a reduced and preselected number of compounds, which correspond basically to the L01 group of the ATC classification. Besides, data on global consumption of cytostatics at a national level is scarce, and often only one prescription route is considered (hospital or pharmacies). Differences are thus expected as the consumption of cytostatics in pharmacies is generally higher than in hospitals, being the former more representative of the consumption patterns. If not properly defined, this produces another uncertainty in the calculation and comparison of PECs among countries. For comparison purposes, the most widely consumed cytostatics are indicated in Table 2.2, where their consumption per capita (μg/inhab/day) according to data from pharmacies sales, hospital administration, or both along the European Union is compiled. For example, the dispensing of cytostatics from L01 ATC group in pharmacies varied between 253 and 494 μg/inhab/day in Spain during the period 2010–2015 (Franquet-Griell et al. 2017b). However, much higher consumption values were reported for this ATC group in Catalonia (633–596 μg/inhab/day) for a similar period of time (2010–2012) (Franquet-Griell et al. 2015). In Germany, the consumption of L01 cytostatics in pharmacies and hospitals was of 704 μg/inhab/day in 2012 (Kümmerer et al. 2016). In Portugal, the minimum and maximum global consumption records of L01 cytostatics were 496 and 784 μg/inhab/day, respectively, for the period 2007–2015 (Santos et al. 2017). From 90 pharmaceuticals listed in Table 2.2, the ones with the highest consumption were capecitabine (L01BC06), hydroxycarbamide (L01XX05), and fluorouracil (L01BC02). At European level, there is the consumption data for only four cytostatics (cyclophosphamide, fluorouracil, capecitabine, and carboplatin), being capecitabine the most consumed cytostatic accounting with 258.5 μg/inhab/day (Johnson et al. 2013). The consumptions of the other cytostatics are much lower, by at least a factor of 10 (Table 2.2).

2.4.2 Excretion Factor (F_{exc})

The fraction of a cytostatic excreted is a value that has great implications in the PEC estimations. Cytostatics are characterized by being excreted through urine or feces at high rates, either as a parental compound, metabolite, or conjugate. The high excretion rates will mean that a high fraction will be discharged to the sewerage system. Several publications are available on the metabolism of pharmaceuticals, and different excretion factors are reported for each drug (Besse et al. 2012;

Table 2.2 Consumption of cytostatics (μg/inhab/day) in different European Union countries in pharmacies (p), in hospitals (h), or both (p+h). Compounds are ordered by ATC code

		France		Germany	UK			Switzerland	Spain			Catalonia	Portugal	
		Besse[a]	Coetsier[b]	Kummerer[c]	Rowney[d]	Booker[e]	Johnson[f]	Buerge[g]	Ortiz de Garcia[h]	Martín[i]	Franquet-Griell[j]	Franquet-Griell[k]	Santos[l]	
		2012	2009	2016	2009	2014	2008	2006	2013	2014	2015	2015	2007–2015	
ATC code	Name	p+h	p+h	p+h	p+h	h	p+h?	nd	p+h	1 hospital	p	p+h	p	h
L01AA01	Cyclophosphamide	12.7	–	11.78	20.64	40.00	–	20.64	74.6	4.7	1.5	1.54	(min) 9.91	(max) 10.88
L01AA02	Chlorambucil	0.34	–	0.04	–	–	–	–	–	–	0.0003	0.07	0.05	0.07
L01AA03	Melphalan	0.20	–	0.08	–	–	–	–	–	–	0.0001	0.04	0.06	0.10
L01AA06	Ifosfamide	4.28	5.3	4.73	–	0.99	–	4.50	20.5	3.6	–	2.28	3.79	4.93
L01AA09	Bendamustine	–	–	0.65	–	–	–	–	–	–	–	0.02	0.00	0.02
L01AB01	Busulfan	0.001	–	0.02	–	–	–	–	–	–	0.0004	0.003	0.02	0.03
L01AC01	Thiotepa	–	–	0.01	–	–	–	–	–	–	–		0.00	0.00
L01AD02	Lomustine	0.13	–	0.03	–	–	–	–	–	–	–	–	0.01	0.03
L01AD04	Streptozocin	0.35	–	–	–	–	–	–	–	–	–	0.13	0.05	0.09
L01AD05	Fotemustine	0.05	–	0.0010	–	–	–	–	–	–	–	0.01	0.00	0.01
L01AX03	Temozolomide	2.22	–	2.82	0.84	0.99	–	–	–	–	–	1.37	1.07	1.45
L01AX04	Dacarbazine	1.22	–	0.82	–	–	–	–	–	–	0.001	0.01	0.40	0.99
L01BA01	Methotrexate	3.10	–	9.90	–	1.00	–	–	–	1.5	4.2	3.19	1.79	2.31
L01BA03	Raltitrexed	0.001	–	–	–	–	–	–	–	–	–	0.0009	0.00	0.00
L01BA04	Pemetrexed	1.55	–	1.63	–	0.99	–	–	–	–	–	1.11	0.51	1.41
L01BB02	Mercaptopurine	3.94	–	2.48	–	–	–	–	–	–	0.008	2.82	1.30	2.21
L01BB03	Thioguanine	0.09	–	0.13	–	–	–	–	–	–	0.0005	0.01	0.02	0.06
L01BB04	Cladribine	0.001	–	0.0010	–	–	–	–	–	–	–	0.0006	0.00	0.00
L01BB05	Fludarabine	0.23	–	0.06	0.2	–	–	–	–	–	0.01	0.04	0.04	0.08
L01BB06	Clofarabine	0.01	–	–	–	–	–	–	–	–	–	0.0002	0.00	0.00
L01BB07	Nelarabine	–	–	0.02	–	–	–	–	–	–	–	0.0005	0.00	0.01

L01BC01	Cytarabine	5.55	–	4.56	–	0.99	–	–	–	2.2	0.00004	0.42	4.35	5.07
L01BC02	Fluorouracil	71.95	–	66.47	–	12.00	46	–	–	45.4	0.04	0.70	51.58	72.64
L01BC03	Tegafur	1.55	–	0.11	–	–	–	–	–	–	2.8	6.74	–	–
L01BC05	Gemcitabine	15.74	–	16.27	14.2	6.99	–	–	51.2	6.7	–	12.0	7.62	9.64
L01BC06	Capecitabine	213.16	–	204.07	311.8[n]	182	79	–	–	–	236 (2010)	280	164	248
L01BC07	Azacitidine	–	–	0.37	–	–	–	–	–	–	–	0.29	0.05	0.44
L01CA01	Vinblastine	0.03	–	0.003	–	–	–	–	–	–	0.00002	0.0004	0.01	0.01
L01CA02	Vincristine	0.01	–	0.003	–	–	–	–	–	–	0.000002	0.00004	0.01	0.01
L01CA03	Vindesine	–	–	0.0003	–	–	–	–	–	–	–	–	0.00	0.00
L01CA04	Vinorelbine	0.54	–	0.31	–	–	–	–	–	0.12	–	0.21	0.24	0.30
L01CA05	Vinflunine	–	–	0.06	–	–	–	–	–	–	–	0.02	0.00	0.00
L01CB01	Etoposide	1.71	–	1.19	–	0.99	–	–	–	0.40	0.1	0.08	0.90	1.08
L01CD01	Paclitaxel	1.61	–	1.50	–	–	–	–	–	0.80	–	1.67	0.70	1.25
L01CD02	Docetaxel	1.14	–	0.65	–	–	–	–	–	0.29	–	0.94	0.54	0.67
L01CD04	Cabazitaxel	–	–	0.01	–	–	–	–	–	–	–	0.0005	0.00	0.01
L01CX01	Trabectedin	0.001	–	0.0003	–	–	–	–	–	–	–	0.00009	0.00	0.00
L01DB01	Doxorubicin	0.70	–	0.34	1.52	–	–	–	–	0.16	0.0005	0.06	0.38	0.42
L01DB03	Epirubicin	0.73	–	0.54	0.9	–	–	–	–	0.25	0.0004	0.01	0.35	0.46
L01DB06	Idarubicin	0.01	–	0.0034	–	–	–	–	–	–	0.000001	0.00009	0.00	0.02
L01DB07	Mitoxantrone	0.01	–	0.01	–	–	–	–	–	–	–	0.0002	0.01	0.01
L01DC01	Bleomycin	0.04	–	0.03	–	–	–	–	–	–	–	8×10^{-7}	–	–
L01DC03	Mitomycin	0.12	–	0.13	–	–	–	–	0.776	0.094	0.002	0.01	0.09	0.11
L01XA01	Cisplatin	0.94	–	0.65	1.090	–	–	–	–	–	–	0.50	0.45	0.60
L01XA02	Carboplatin	3.47	–	3.23	8.008	3.00	–	–	–	–	0.002	0.04	1.97	2.41
L01XA03	Oxaliplatin	1.39	–	0.85	0.430	0.99	–	–	–	–	–	1.37	0.68	1.05
L01XB01	Procarbazina	–	–	0.44	–	–	–	–	–	–	–	–	0.27	0.42
L01XC02	Rituximab	3.02	–	3.85	–	–	–	–	–	–	–	2.29	1.09	3.01

(continued)

Table 2.2 (continued)

		France		Germany	UK			Switzerland	Spain			Catalonia	Portugal	
		Besse[a]	Coetsier[b]	Kummerer[c]	Rowney[d]	Booker[e]	Johnson[f]	Buerge[g]	Ortiz de Garcia[h]	Martín[i]	Franquet-Griell[j]	Franquet-Griell[k]	Santos[l]	
		2012	2009	2016	2009	2014	2008	2006	2013	2014	2015	2015	2007–2015	
ATC code	Name	p+h	p+h	p+h	p+h	h	p+h?	nd	p+h	1 hospital	p	p+h	p	h
L01XC03	Trastuzumab	2.33	–	3.30	–	–	–	–	–	–	–	1.86	1.15	3.22
L01XC04	Alemtuzumab	0.03	–	0.01	–	–	–	–	–	–	–	0.004	0.00	0.01
L01XC06	Cetuximab	2.28	–	1.77	–	–	–	–	–	–	–	2.02	0.69	1.75
L01XC07	Bevacizumab	3.62	–	3.91	–	–	–	–	–	–	–	1.47	0.34	1.48
L01XC08	Panitumumab	0.05	–	0.29	–	–	–	–	–	–	–	0.20	0.00	0.11
L01XD03	Methyl aminolevulinate	0.10	–	0.11	–	–	–	–	–	–	0.2	0.11	0.01	0.02
L01XD04	Aminolevulinic acid	–	–	0.03	–	–	–	–	–	–	0.2	0	0.00	0.03
L01XE01	Imatinib	36.28	–	48.19	–	10.00	–	–	–	–	11	33.6	24.08	41.79
L01XE02	Gefitinib	–	–	1.94	–	–	–	–	–	–	–	1.40	0.00	2.78
L01XE03	Erlotinib	6.18	–	4.49	–	2.00	–	–	–	–	–	3.74	1.50	5.35
L01XE04	Sunitinib	0.83	–	1.16	–	–	–	–	–	–	–	0.50	0.21	0.41
L01XE05	Sorafenib	–	–	13.75	–	5.00	–	–	–	–	–	9.16	0.09	6.51
L01XE06	Dasatinib	–	–	0.85	–	–	–	–	–	–	–	0.90	0.19	1.45
L01XE07	Lapatinib	4.82	–	10.48	–	1.92	–	–	–	–	–	8.590	0.00	8.59
L01XE08	Nilotinib	2.44	–	11.10	–	–	–	–	–	–	–	7.00	0.02	4.89
L01XE09	Temsirolimus	0.05	–	0.02	–	–	–	–	–	–	–	0.01	1.07	1.45
L01XE10	Everolimus	–	–	0.13	–	–	–	–	–	–	–	0.01	0.01	0.20
L01XE11	Pazopanib	–	–	7.08	–	–	–	–	–	–	–	0.43	0.00	5.41
L01XX05	Hydroxycarbamide	283.88	–	230.82	–	32.99	–	–	–	–	237	221	206	300
L01XX08	Pentostatin	0.001	–	0.0002	–	–	–	–	–	–	–	9×10^{-5}	0.00	0.00
L01XX09	Miltefosine	0.01	–	0.02	–	–	–	–	–	–	–	4×10^{-4}	0.01	0.04

L01XX11	Estramustine	11.94	–	5.24	–	–	–	–	–	–	1.0	2.96	2.55	12.31
L01XX14	Tretinoin	0.14	–	0.10	–	–	–	–	–	–	0.08	0.08	0.13	018
L01XX17	Topotecan	0.01	–	0.01	–	–	–	–	–	–	–	0.003	0.00	0.00
L01XX19	Irinotecan	1.93	–	1.29	–	–	–	–	–	0.34	–	1.28	1.35	2.03
L01XX23	Mitotane	9.70	–	5.21	–	1.92	–	–	–	–	3.4	3.38	2.23	4.53
L01XX25	Bexarotene	0.98	–	0.99	–	–	–	–	–	–	–	0.83	0.71	0.94
L01XX27	Arsenic trioxide	0	–	–	–	–	–	–	–	–	–	7×10^{-4}	0.00	0.00
L01XX32	Bortezomib	0.01	–	0.01	–	–	–	–	–	–	–	0.01	0.00	0.01
L01XX35	Anagrelide	0.05	–	–	–	–	–	–	–	–	0.04	0.06	0.01	0.03
L02AA01	Dietilestilbestrol	–	–	–	–	–	–	–	–	–	–	2×10^{-4}	–	–
L02AA04	Fosfestrol	–	–	–	–	–	–	–	–	–	–	3×10^{-4}	–	–
L02AB01	Megestrol	–	–	–	–	–	–	–	79.32	–	72	77.7	20.95	29.48
L02AE01	Buserelin	0.002	–	–	–	–	–	–	–	–	0.001	0.001	0.00	0.00
L02AE02	Leuprorelin	0.13	–	–	–	–	–	–	–	–	0.2	0.24	0.04	0.13
L02AE03	Goserelin	0.05	–	–	–	–	–	–	–	–	0.02	0.02	0.06	0.07
L02AE04	Triptorelin	0.08	–	–	–	–	–	–	–	–	0.08	0.10	0.05	0.06
L02BA01	Tamoxifen	15.65	15.11	–	–	–	–	–	14.98	–	19	17.7	19.17	22.10
L02BA02	Toremifene	0.04	–	–	–	–	–	–	–	–	–	0.03	0.00	0.52
L02BA03	Fulvestrant	0.28	–	–	–	–	–	–	–	–	0.87	0.09	0.08	0.55
L02BB01	Flutamide	21.63	–	–	–	–	–	–	115.85	–	14	27.7	3.03	46.15
L02BB03	Bicalutamide	35.82	–	–	–	–	–	–	–	–	36	38.1	37.66	43.20
L02BG03	Anastrozole	1.32	–	–	–	–	–	–	–	–	0.4	0.46	0.37	0.80
L02BG04	Letrozole	1.42	–	–	–	–	–	–	–	–	2.4	2.30	0.71	1.62
L02BG06	Exemestane	7.56	–	–	–	–	–	–	–	–	6.8	7.44	2.18	3.49
L02BX03	Abiraterone	–	–	–	–	–	–	–	–	–	–	2.27	–	–
L03AA02	Filgrastim	–	–	–	–	–	–	–	–	–	–	9×10^{-3}	–	–
L03AA10	Lenograstim	–	–	–	–	–	–	–	–	–	–	3×10^{-4}	–	–
L03AA13	Pegfilgrastim	–	–	–	–	–	–	–	–	–	–	0.01	–	–
L03AB03	Interferon gamma	–	–	–	–	–	–	–	–	–	0.00002	8×10^{-3}	–	–

(continued)

Table 2.2 (continued)

		France		Germany	UK			Switzerland	Spain			Catalonia	Portugal	
		Besse[a]	Coetsier[b]	Kummerer[c]	Rowney[d]	Booker[e]	Johnson[f]	Buerge[g]	Ortiz de Garcia[h]	Martín[i]	Franquet-Griell[j]	Franquet-Griell[k]	Santos[l]	
		2012	2009	2016	2009	2014	2008	2006	2013	2014	2015	2015	2007–2015	
ATC code	Name	p+h	p+h	p+h	p+h	h	p+h?	nd	p+h	1 hospital	p	p+h	p	h
L03AB04	Interferon alfa–2a	–	–	–	–	–	–	–	–	–	0.00002	1×10^{-4}	–	–
L03AB05	Interferon alfa–2b	–	–	–	–	–	–	–	–	–	0.00003	9×10^{-5}	–	–
L03AB07	Interferon beta–1a	–	–	–	–	–	–	–	–	–	–	2×10^{-3}	–	–
L03AB08	Interferon beta–1b	–	–	–	–	–	–	–	–	–	–	0.01	–	–
L03AB10	Peginterferon alfa–2b	–	–	–	–	–	–	–	–	–	–	1×10^{-3}	–	–
L03AB11	Peginterferon alfa–2a	–	–	–	–	–	–	–	–	–	–	2×10^{-3}	–	–
L03AC01	Aldesleukin	–	–	–	–	–	–	–	–	–	–	4×10^{-5}	–	–
L03AX03	BCG vaccine	–	–	–	–	–	–	–	–	–	–	0.33	–	–
L03AX13	Glatiramer acetate	–	–	–	–	–	–	–	–	–	–	1.29	–	–
L03AX16	Plerixafor	–	–	–	–	–	–	–	–	–	–	2×10^{-4}	–	–
L04AA04	Antithymocyte immunoglobulin	–	–	–	–	–	–	–	–	–	–	2×10^{-4}	–	–
L04AA06	Mycophenolic acid	–	–	–	–	–	–	–	–	–	720	704	531	756
L04AA10	Sirolimus	–	–	–	–	–	–	–	–	–	0.07	0.11	0.08	0.17
L04AA13	Leflunomide	–	–	–	–	–	–	–	–	–	7.6	5.99	0.22	0.66
L04AA18	Everolimus	–	–	–	–	–	–	–	–	–	0.1	0.11	0.01	0.20
L04AA23	Natalizumab	–	–	–	–	–	–	–	–	–	–	0.44	–	–
L04AA24	Abatacept	–	–	–	–	–	–	–	–	–	–	0.37	–	–
L04AA25	Eculizumab	–	–	–	–	–	–	–	–	–	–	0.08	–	–
L04AA27	Fingolimod	–	–	–	–	–	–	–	–	–	–	1×10^{-3}	–	–
L04AB01	Etanercept	–	–	–	–	–	–	–	–	–	–	1.94	–	–
L04AB02	Infliximab	–	–	–	–	–	–	–	–	–	–	1.63	–	–

L04AB04	Adalimumab	–	–	–	–	–	–	–	–	–	–	1.07	–	–
L04AB05	Certolizumab pegol	–	–	–	–	–	–	–	–	–	–	0.09	–	–
L04AB06	Golimumab	–	–	–	–	–	–	–	–	–	–	0.01	–	–
L04AC03	Anakinra	–	–	–	–	–	–	–	–	–	–	0.40	–	–
L04AC05	Ustekinumab	–	–	–	–	–	–	–	–	–	–	0.03	–	–
L04AC07	Tocilizumab	–	–	–	–	–	–	–	–	–	–	0.50	–	–
L04AC08	Canakinumab	–	–	–	–	–	–	–	–	–	–	6×10^{-4}	–	–
L04AD01	Cyclosporine	–	–	–	–	–	–	–	–	–	35	42	1.82	10.07
L04AD02	Tacrolimus	–	–	–	–	–	–	–	–	–	2.6	2.25	0.85	2.20
L04AX01	Azathioprine	–	–	–	–	–	–	–	–	–	112	78.6	0.52	0.67
L04AX04	Lenalidomide	–	–	–	–	–	–	–	–	–	–	0.19	–	–

	Europe[m]	European mean value	Germany	UK	Italy	Netherlands	Austria	Denmark	Switzerland	Sweden	Norway	Finland	Czech Republic
L01AA01	Cyclophosphamide	10.4	8.4	–	3.4	2.7	13.4	5.7	12.7	35.7	9.2	2.3	10.2
L01BC02	Fluorouracil	29.7	51.9	–	3.2	18.5	41	2.6	30.3	30.2	27	–	19.8
L01BC06	Capecitabine	258	183	133	147	393	–	388	170	226	116	372	493
L01XA02	Carboplatin	3	3.4	2.1	–	–	1	1	–	1.5	–	–	8.4

nd not described, *na* not available, *5FU* fluorouracil

[a] Besse et al. (2012); [b] Coetsier et al. (2009); [c] Kümmerer et al. (2016); [d] Rowney et al. (2009); [e] Booker et al. (2014); [f] Johnson et al. (2008); [g] Buerge et al. (2006); [h] Ortiz de García et al. (2013); [i] Martín et al. (2014); [j] Franquet-Griell et al. (2017b); [k] Franquet-Griell et al. (2015); [l] Santos et al. (2017); [m] Johnson et al. (2013); [n] Value that corresponds to the consumption of capecitabine + 5FU

Whirl-Carrillo et al. 2012). The observed differences are probably explained by genomically distinct metabolizing capacities, as well as differences in the routes of administration, sex, age, and health status of the studied subjects. The metabolism and excretion data of cytostatics can be also obtained from databases: the Base Claude Bernard (https://www.bcbdexther.fr/), the Micromedex Drugdex® databank, the International Agency for Research on Cancer (IARC database), the BC Cancer Agency database, and the DrugBank database, among others.

2.4.3 Removal Efficiency at WWTP (F_{WWTP})

As occurs with the excretion factor, different values for the elimination of drugs in wastewater treatment plants are published in the literature (Besse et al. 2012; Chang et al. 2011; Fan et al. 2011). Removal rates can range from 11 to 90% (Fan et al. 2011; Ortiz de García et al. 2013), and thus, it is important to use the most accurate value in a given site. F_{WWTP} can be an empirical value or a theoretical one. Empirical F_{WWT} can be very different among studies as the elimination efficiency can vary in different WWTPs depending on locations of the served population, capacity, configuration, and type of treatment, in operating parameters, and in hydraulic and solid retention times. Other factors such as meteorological conditions, sampling procedure (grab, composite or flow proportional (Ort et al. 2010)), and sampling period (seasonality) can also affect the empirical F_{WWTP}. Whenever possible, specific experimental data of the WWTP operating in the study area should be used, and if several values are available, the mean or median values can be used. On the other hand, the F_{WWTP} can be also assessed by the SimpleTreat model described in the European Union System for the Evaluation of Substances (EUSES) (Strujis 2014), which considers the most important processes like volatilization, mixing, adsorption, and degradation. EPI Suite (United States Environmental Protection Agency (U.S. EPA) 2013) also enables estimating the removal efficiency based on the physical-chemical parameters of each compound.

2.4.4 Water Consumption and Inhabitants

Water consumption can range from 3.75 m^3/cap/yr to 287.1 m^3/cap/yr in countries such as Uganda and Canada, respectively (Food and Agriculture Organization of the United Nations (FAO) 2000). Therefore, water consumption has to be specified to obtain accurate PEC values, and it is recommended to use the mean water consumption in the study area or at a country-based level. Water consumption is easy to obtain at national scale from the ministries, the National Statistical Institutes, or the European Statistical System and the Eurostat. What is important in the calculation of PECs is that cytostatic consumption data (e.g., hospital, regional, or national levels) matches with the water consumption and inhabitants of the same area. This way,

PEC estimations are referred to a specific region/facility for which consumption data has been obtained, such as a hospital effluent, municipality, river basin, or province. Besides, national consumption data can be reallocated to river basins or provinces by proportionally recalculating the consumptions according to the inhabitants of a specific location. This has been done in the study of Franquet-Griell and co-workers, who calculated the PECs at river basin scale from global consumption data gathered for all around Spain (Franquet-Griell et al. 2017b). The same approach was used to calculate the PECs for different territorial zones in Portugal (NUTS level II regions) from data obtained at a national scale (Santos et al. 2017).

2.4.5 Dilution Factor (DF)

Probably *DF* is the parameter in the equation of utmost importance in the estimation of PECs in rivers. *DF* refers to the dilution from WWTP effluents to surface waters. Changes in this value can vary the results by more than 100-fold. As indicated before, the EMA proposes a *DF* of 10 (European Medicines Agency (EMA) 2006). However, this value can vary in orders of magnitude when considering the different river regimes around the world. This can be overcome by using the *DF* values obtained by Keller et al. in 2014, who calculate the median dilution factor for each country, based on a model that divides the terrestrial surface in fractions of $0.5° \times 0.5°$ (equivalent to 55×55 km at the equator) and take into consideration the river flows, the water consumption, and the population for a certain country (Keller et al. 2014). The results obtained vary between 0.0050 in Qatar and 94,463 in Suriname. Spain got a value of 25.92, which is greater than the default value of the EMA, and this value was used to calculate the PEC_{river} in Spain (Franquet-Griell et al. 2017b).

In the real world, however, the variations of the flow of the river can also be very broad, such as in the Mediterranean areas, where the flow and the frequency of precipitation are very irregular. Moreover, the flow of the river can vary from the source, along the middle course and to the mouth of the river. Since the PEC values at a country base level do not account for differences in river dynamics and seasonal or river basin-based variations, the refinement of the *DF* would allow the adjustment of PECs for a specific hydrographic basin or river portion to determine the most affected areas. These new *DF* can be calculated by adapting the formula from Keller et al. (2014):

$$DF = \frac{Qr \times 31536000}{Inhab_{\text{basin}} \times W_{basin}} \tag{2.5}$$

where

- Q_r (m^3/s) is the flow of a specific river. The flow data can be collected from the hydrographic confederations of each basin, administrative bodies, or water agencies depending on how river waters are managed in each country, and consider geographic and seasonal variability along the basin. Maximum, minimum and mean river flows can be used to better estimate PEC variability. The flow data reflect the withdrawal of water in each area.

- $Inhab_{basin}$ is the population in the basin area (inhabitants).
- W_{basin} is the consumption of water per capita in the basin, including domestic and industrial use (m^3/inhab/year). If this data is not available, water consumption at a national scale (100–200 m^3/inhab/year) can be used.
- 31,536,000 are seconds per year used to convert units.

To calculate PEC_{river} at basin-scale, the *DF* derived from high, mean, and low flows in each river basin should be applied to Eq. 2.5, and consumption of anticancer drugs needs to be proportional to the population in the studied area. The closest the data is to the region under analysis, the more accurate will be the PEC estimation.

2.5 Comparison Between PEC and MEC Values in European Countries

The validity of PECs is confirmed when the estimated values are compared with measured environmental concentrations (MECs) obtained in wastewaters and river waters monitoring studies from a given geographical area for which the PECs have been calculated. There are fluctuations in the concentration of pharmaceuticals in wastewater and surface waters, as highlighted by Ort et al. (2009) and Verlicchi et al. (2012, 2014). This adds an additional difficulty in the PEC/MEC validation. Several criteria have been used to appraise whether the PECs tend to overestimate or underestimate MECs. Among them, Coetsier et al. in 2009 proposed a ranking scheme in which values 0.2<PEC/MEC<1, PEC are acceptable and slightly underestimated; for those with 1<PEC/MEC<4, PEC results are acceptable but slightly overestimated, whereas for 4<PEC/MEC<8, overestimated results are given (Coetsier et al. 2009). However, other authors proposed 0.5<PEC/MEC<2 as acceptable (Ort et al. 2009; Verlicchi et al. 2014). In this chapter, the reliability of PECs according to MECs has been reported for several cytostatic drugs in wastewater effluents and surface waters (principally in a river). Table 2.3 displays the concentrations of cytostatic compounds (in ng/L) – MEC values in WWTP effluents from several European countries and the PECs values. Note, however, that the PEC calculation has been reported only in a few sites and often does not include all the cytostatic compounds but rather preselected drugs.

The studies performed in Catalonia (NE Spain) and Portugal are the most comprehensive in the prediction of the concentrations of cytostatic drugs in sewage effluents and surface waters as provide F_{exc} and F_{WWTP} data for a large number of drugs to be used to calculate the PECs in other studies (Franquet-Griell et al. 2015).

These studies provide data for 132 and 171 cytostatics consumed in hospitals and pharmacies during the period 2012–2015 and identify mycophenolic acid as the main cytostatic present in wastewaters and river waters. This finding was corroborated experimentally by monitoring river waters (Franquet-Griell et al. 2016, 2017a) and wastewaters (Franquet-Griell et al. 2017c) and demonstrated the ubiquity of mycophenolic acid in the regions where the PECs were calculated.

In two WWTPs from the Barcelona area, 20 compounds were detected in WWTP influents in the range 0.7–356 ng/L, and only cyclophosphamide (<4–5 ng/L) and megestrol (<3–20 ng/L) were detected in effluents with an acceptable PEC/MEC ratio according to Franquet-Griell et al. (2015) for cyclophosphamide and slightly overestimated for megestrol (Gómez-Canela et al. 2014). This study highlights the difficulty to calculate the PEC/MEC ratio when PEC_{eff} is very low and/or limits of detection (LOD) of the analytical methods are high, which precludes their determination.

Negreira et al. reported the occurrence of 13 cytostatic drugs and 4 metabolites in various WWTPs (cyclophosphamide, ifosfamide, methotrexate, capecitabine, doxorubicin, irinotecan, tamoxifen, temozolomide, gemcitabine, etoposide, paclitaxel, imatinib, and erlotinib) at median levels between 6.3 and 8.9 ng/L (Negreira et al. 2014). Comparing these results with those PECs reported (Franquet-Griell et al. 2015), the ratio PEC/MEC yields good results for cyclophosphamide, ifosfamide, and methotrexate, however, overestimated for capecitabine and doxorubicin (see Table 2.3).

In Catalan WWTPs, cyclophosphamide, ifosfamide, methotrexate, vincristine, etoposide, paclitaxel, docetaxel, tamoxifen, and azathioprine were detected at 22.9–175.1 ng/L in influents and at <1.3–<75.7 ng/L in effluents (Ferrando-Climent et al. 2013) and in general correspond to the PEC estimations of Franquet-Griell et al. (2015). Similar results were observed in the surface waters and wastewaters from Spain (Franquet-Griell et al. 2015; Negreira et al. 2013). This overestimation might be due to the fact that PEC calculations are made at a regional scale (Catalonia), while in a specific WWTP, the loads can vary according to the activities of the area. Some compounds such as vincristine, etoposide, and docetaxel had PEC_{eff} similar or higher than their limits of detection (LODs), and then the PEC/MEC ratio could not be calculated (Table 2.3). Finally, PEC/MEC ratios of cytostatics with low PEC_{eff} and/or high LODs of the reported analytical method could not be estimated (Ferrando-Climent et al. 2013; Franquet-Griell et al. 2015; Gómez-Canela et al. 2014; Negreira et al. 2014).

Cyclophosphamide, ifosfamide, and tamoxifen are among the main compounds for which the PEC/MEC ratios have been calculated. Coetsier et al. investigated the discharge of pharmaceutical products through a conventional biological sewage treatment plant in Alès (France) and reported a PEC/MEC ratio between 6.3 and >0.2 for ifosfamide and tamoxifen, respectively, with a clear overestimation for ifosfamide, whereas for tamoxifen, the PEC was acceptable and slightly underestimated (Coetsier et al. 2009). In another study, Kümmerer et al. evaluated

Table 2.3 Concentrations of cytostatic compounds (in ng/L) in Catalonia (Spain), France, the United Kingdom, and Germany, their respective PECs and the PECs vs. MECs. Compounds are ordered by ATC code

Catalonia (Spain)										
ATC code	Cytostatic	$PECs_{eff}$ (Franquet-Griell et al. 2015)	$WWTP_{eff}$ (Gómez-Canela et al. 2014)	PEC/MEC	$WWTP_{eff}$ (Negreira et al. 2014)	PEC/MEC	$WWTP_{eff}$ (Negreira et al. 2014)	PEC/MEC	$WWTP_{eff}$ (Ferrando-Climent et al. 2013)	PEC/MEC
L01AA01	Cyclophosphamide	2.94	<4–5	<0.73–0.58	8.8	0.33	6.3	0.47	15.7	0.2
L01AA02	Chlorambucil	0.01	N.D.	–	N.D.	–	N.D.	–	N.D.	–
L01AA03	Melphalan	0.05	N.D.	–	N.D.	–	N.D.	–	N.D.	–
L01AA06	Ifosfamide	8.76	N.D.	–	N.D.	–	8.9	0.98	<1.3	>6.7
L01AX03	Temozolomide	2.17	N.D.	–	N.D.	–	<4.2	>0.52	N.D.	–
L01BA01	Methrotexate	1.10	N.D.	–	N.D.	–	<1.8	>0.61	<4.1	>0.3
L01BB05	Fludarabine	0.18	N.D.	–	N.D.	–	N.D.	–	N.D.	–
L01BC05	Gemcitabine	5.51	N.D.	–	N.D.	–	<9.3	>0.59	N.D.	–
L01BC06	Capecitabine	201	N.D.	–	N.D.	–	7.7	26.1	N.D.	–
L01CA01	Vinblastine	1e–5	N.D.	–	N.D.	–	N.D.	–	N.D.	–
L01CA02	Vincristine	3e–5	N.D.	–	N.D.	–	N.D.	–	<23.5	<PEC
L01CB01	Etoposide	0.53	N.D.	–	N.D.	–	<40	>LOD	<75.7	<PEC
L01CD01	Paclitaxel	2.1	N.D.	–	N.D.	–	<4	>0.52	<8.7	>0.24
L01CD02	Docetaxel	0.97	N.D.	–	N.D.	–	N.D.	–	<12.7	<PEC
L01DB01	Doxorubicin	0.24	N.D.	–	N.D.	–	<2.4	>10	N.D.	–
L01DB03	Epirubicin	0.01	N.D.	–	N.D.	–	N.D.	–	N.D.	–
L01XE01	Imatinib	60.6	N.D.	–	N.D.	–	<120	>0.5	N.D.	–
L01XE03	Erlotinib	0.55	N.D.	–	N.D.	–	<3.4	>0.16	N.D.	–
L01XX19	Irinotecan	4.8	N.D.	–	< 1.2	>3.97	N.D.	–	N.D.	–

L02AE02	Leuprolide	<PEC	N.D.	–	N.D.	–	N.D.	–	N.D.	–
L02AE03	Goserelin	0.16	N.D.	–	N.D.	–	N.D.	–	N.D.	–
L02BA01	Tamoxifen	1.22	N.D.	–	113.5	4.4e–4	<3.0	>0.41	28.7	0.04
L02BG01	Aminoglutethimide	<PEC	N.D.	–	N.D.	–	N.D.	–	N.D.	–
L04AX01	Azathioprine	11.9	N.D.	–	N.D.	–	N.D.	–	<6.1	>1.9
G03AC05	Megestrol	18.6	<3–20	<6.2–0.93	N.D.	–	N.D.	–	N.D.	–
G03HA01	Cyproterone	19.0	N.D.	–	N.D.	–	N.D.	–	N.D.	–
H02AB07	Prednisone	40.9	N.D.	–	N.D.	–	N.D.	–	N.D.	-

France

ATC code	Cytostatic	PEC_{eff} (ng/L) (Coetsier et al. 2009)	$WWTP_{eff}$ (ng/L) (Coetsier et al. 2009)	PEC/MEC
L01AA06	Ifosfamide	24	<3.8	6.3
L02BA01	Tamoxifen	22	<5.8–102	>3.8–0.2

Germany

ATC code	Cytostatic	PEC_{eff} (ng/L) (Kümmerer & Al-Ahmad 2010)	$WWTP_{eff}$ (ng/L) (Kümmerer & Al-Ahmad 2010)	PEC/MEC
L01AA01	Cyclophosphamide	5.6	56	0.1
L01AA06	Ifosfamide	10.9	109	0.1

UK

ATC code	Cytostatic	PEC_{eff} (ng/L) (Johnson et al. 2013)	$WWTP_{eff}$ (ng/L) (Llewellyn et al. 2011)	PEC/MEC
L01AA01	Cyclophosphamide	70.2	0.19–3.5	369–20

NA not analyzed, *ND* not detected

*Median of the cytostatic concentrations in 12 Spanish WWTPs the presence of cyclophosphamide and ifosfamide in wastewater and surface water from Germany reporting levels of 56 and 109 ng/L, respectively (Table 2.3) (Kümmerer & Al-Ahmad 2010). Comparing these levels with those PEC_{eff} reported in the same study, PEC/MEC ratios were of 0.1, being PEC acceptable and slightly underestimated according to Coetsier et al. (2009). Similarly, Llewellyn et al. report levels between 0.19 and 3.5 ng/L of cyclophosphamide and ifosfamide in a WWTP effluent from the United Kingdom (UK) (Llewellyn et al. 2011). In this study, predicted environmental concentrations were not calculated, but when compared to the PEC_{eff} value of cyclophosphamide of 70.2 ng/L (Table 2.3) (Johnson et al. 2013), the PEC result of cyclophosphamide was overestimated (Coetsier et al. 2009).

Not much information has been published in the presence of cytostatic compounds in surface waters and the comparison with their predicted concentrations. In general, cytostatics are present at trace levels in river such as cyclophosphamide (0.05–10 ng/L) (Buerge et al. 2006; Metcalfe et al. 2003; Usawanuwat et al. 2014; Zuccato et al. 2000), ifosfamide (0.05–41 ng/L) (Buerge et al. 2006; Valcárcel et al. 2011), fluorouracil (578 ng/L) (Usawanuwat et al. 2014), hydroxycarbamide (788 ng/L) (Usawanuwat et al. 2014), and tamoxifen (5.8–147 ng/L) (Isidori et al. 2016; López-Serna et al. 2012; Negreira et al. 2013). In Spain, Valcárcel et al. determined the presence of ifosfamide in Guadarrama River (a tributary of the Tagus in Madrid region, Spain) at the level of 41 ng/L (Valcárcel et al. 2011). In another study, tamoxifen was detected at 18.9 ng/L in the Ebro River basin (NE Spain) (López-Serna et al. 2012). Taking into account a recent study about the estimated levels (PECs) of 78 anticancer drugs in Spanish river basins calculated from consumption data in pharmacies during the period 2010–2015 (Franquet-Griell et al. 2017b), the PEC/MEC ratio for ifosfamide and tamoxifen was calculated. According to Coetsier et al. (2009) study, it can be demonstrated that PECs for both compounds were acceptable and slightly underestimated. Another study compared the concentrations of cytostatic drugs in Besòs River (a small river in Catalonia, NE Spain) with a refined PECs based on Besòs river flow. In this work, authors concluded that PEC/MEC ratios showed reliable results for several drugs (mycophenolic acid, megestrol, ifosfamide, and cyclophosphamide), and most of the non-detected compounds (doxorubicin, fludarabine, goserelin, leuprolide, melphalan, epirubicin, and daunorubicin) corresponded to those with low PEC (Franquet-Griell et al. 2017a).

EMA proposes to perform a risk assessment when the PEC value is >0.01 μg/L. Whereas the PEC estimation provides a theoretical value based on the probability of a compound to be present in wastewaters of river waters, the MEC value provides evidence of the approach. MEC data proves the validity of PECs and highlights the importance to perform a risk assessment for some prioritized compounds, based on toxicological data using several species which is described in detail in previous studies (Franquet-Griell et al. 2015; Franquet-Griell et al. 2017b; Santos et al. 2017). Then, the risk quotient can be calculated to provide information on the environmental impact of this new type of water contaminants.

2.6 Conclusions

It is clear from the published studies that PEC calculation permits to better prioritize compounds which have a high probability to be detected in surface waters. Moreover, PEC calculations together with acute and chronic toxicological data enable environmental risk assessment of compounds more prone to be present in river waters. PEC values represent average concentrations likely to be found in water according to the consumption data of a specific area. In contrast, monitoring studies provide data of cytostatics in wastewaters and rivers at a given time or period. To have accurate PECs, and appropriate PEC/MEC validation, we highlight the need to calculate the PECs in the same area where the monitoring will be carried out. In addition, since the concentrations of cytostatic compounds in water can vary substantially according to the monitoring period, it is important to have a high sampling frequency. Accordingly, to compare PECs with MECs, minimum and maximum concentrations of each compound detected in the river should be considered. Most studies converge in indicating that PEC/MEC is underestimated, which could be due to the low removal in WWTP or the use of inaccurate DF for a particular river. In hotspot areas, such as downstream rivers flowing through populated areas, the concentrations detected can be much higher than predicted. In conclusion, it can be confirmed that PECs is a useful tool to identify and prioritize the cytostatic drugs that are likely to be found in the environment, but, as a model, it is necessary to validate the results obtained with empirical data. As a rule of thumb, it is recommended that monitoring is done following PEC calculations because then it is possible to detect compounds which were not suspected to be found in water, as the case for mycophenolic acid (Franquet-Griell et al. 2016).

Acknowledgments The authors thank the Spanish Ministerio de Sanidad, Servicios Sociales e Igualdad for kindly providing the consumption data of cytostatic drugs. Financial support from Ministerio de Economía y Competitividad is also acknowledged under the project CTM2014-60199-P and the FPI Grant BES-2012-053000. On the Portuguese side, this work was financially supported by: (i) project PTDC/EAM-AMB/31297/2017, funded by FEDER funds through COMPETE2020 – Programa Operacional Competitividade e Internacionalização (POCI) and by national funds (PIDDAC) through FCT/MCTES; (ii) project UID/EQU/00511/2019 – Laboratory for Process Engineering, Environment, Biotechnology and Energy – LEPABE funded by national funds through FCT/MCTES (PIDDAC); (iii) project "LEPABE-2-ECO-INNOVATION" – NORTE-01-0145-FEDER-000005, funded by Norte Portugal Regional Operational Programme (NORTE 2020), under PORTUGAL 2020 Partnership Agreement, through the European Regional Development Fund (ERDF).

References

Besse JP, Latour JF, Garric J (2012) Anticancer drugs in surface waters. What can we say about the occurrence and environmental significance of cytotoxic, cytostatic and endocrine therapy drugs? Environ Int 39:73–86

Booker V, Halsall C, Llewellyn N, Johnson A, Williams R (2014) Prioritising anticancer drugs for environmental monitoring and risk assessment purposes. Sci Total Environ 473–474:159–170

Brystol-Myers Squibb Company (2010) Hydroxyurea safety data sheet. 2017: http://www.msdsexplorer.com/PDFsFiles/7151.pdf

Buerge IJ, Buser HR, Poiger T, Müller MD (2006) Occurrence and fate of the cytostatic drugs cyclophosphamide and ifosfamide in wastewater and surface waters. Environ Sci Technol 40:7242–7250

Carter SB, Franklin TJ, Jones DF, Leonard BJ, Mills SD, Turner RW, Turner WB (1969) Mycophenolic acid: An anti-cancer compound with unusual properties. Nature 223:848–850

Chang H, Wan Y, Wu S, Fan Z, Hu J (2011) Occurrence of androgens and progestogens in wastewater treatment plants and receiving river waters: Comparison to estrogens. Water Res 45:732–740

Coetsier CM, Spinelli S, Lin L, Roig B, Touraud E (2009) Discharge of pharmaceutical products (PPs) through a conventional biological sewage treatment plant: MECs vs PECs? Environ Int 35:787–792

Dun B, Sharma A, Teng Y, Liu H, Purohit S, Xu H, Zeng L, She JX (2013) Mycophenolic acid inhibits migration and invasion of gastric cancer cells via multiple molecular pathways. PLoS ONE 8

European Chemicals Agency (ECHA) (2003) Technical guidance document on risk assessment. https://echa.europa.eu/documents/10162/16960216/tgdpart2_2ed_en.pdf

European Medicines Agency (EMA) (2006) Guideline on the environmental risk assessment of medicinal products for human use. EMEA/CHMP/SWP/4447/00 http://www.ema.europa.eu/docs/en_GB/document_library/Scientific_guideline/2009/10/WC500003978.pdf

Fan Z, Wu S, Chang H, Hu J (2011) Behaviors of glucocorticoids, androgens and progestogens in a municipal sewage treatment plant: Comparison to estrogens. Environ Sci Technol 45:2725–2733

Ferrando-Climent L, Rodriguez-Mozaz S, Barceló D (2013) Development of a UPLC-MS/MS method for the determination of ten anticancer drugs in hospital and urban wastewaters, and its application for the screening of human metabolites assisted by information-dependent acquisition tool (IDA) in sewage samples. Anal Bioanal Chem 405:5937–5952

Food and Agriculture Organization of the United Nations (FAO) (2000) AQUASTAT database. 2017: http://www.fao.org/home/en/

Franquet-Griell H, Gómez-Canela C, Ventura F, Lacorte S (2015) Predicting concentrations of cytostatic drugs in sewage effluents and surface waters of Catalonia (NE Spain). Environ Res 138:161–172

Franquet-Griell H, Ventura F, Boleda MR, Lacorte S (2016) Do cytostatic drugs reach drinking water? the case of mycophenolic acid. Environ Pollut 208:532–536

Franquet-Griell H, Cornadó D, Caixach J, Ventura F, Lacorte S (2017a) Determination of cytostatic drugs in Besòs River (NE Spain) and comparison with predicted environmental concentrations. Environ Sci Pollut Res 24:6492–6503

Franquet-Griell H, Gómez-Canela C, Ventura F, Lacorte S (2017b) Anticancer drugs: Consumption trends in Spain, prediction of environmental concentrations and potential risks. Environ Pollut 229:505–515

Franquet-Griell H, Pueyo V, Silva J, Orera VM, Lacorte S (2017c) Development of a macroporous ceramic passive sampler for the monitoring of cytostatic drugs in water. Chemosphere 182:681–690

Garcia-Ac A, Segura PA, Gagnon C, Sauvé S (2009) Determination of bezafibrate, methotrexate, cyclophosphamide, orlistat and enalapril in waste and surface waters using on-line solid-phase extraction liquid chromatography coupled to polarity-switching electrospray tandem mass spectrometry. J Environ Monit 11:830–838

Generalitat de Catalunya (GENCAT) (2016) Incidence of cancer in Catalonia and European countries. 2017: http://cancer.gencat.cat/web/es/professionals/estadistiques/Monografia-2016.pdf

Gómez-Canela C, Ventura F, Caixach J, Lacorte S (2014) Occurrence of cytostatic compounds in hospital effluents and wastewaters, determined by liquid chromatography coupled to high-resolution mass spectrometry. Anal Bioanal Chem 406:3801–3814

Isidori M, Lavorgna M, Russo C, Kundi M, Žegura B, Novak M, Filipič M, Mišík M, Knasmueller S, de Alda ML, Barceló D, Žonja B, Česen M, Ščančar J, Kosjek T, Heath E (2016) Chemical and toxicological characterisation of anticancer drugs in hospital and municipal wastewaters from Slovenia and Spain. Environ Pollut 219:275–287

Johnson AC, Jürgens MD, Williams RJ, Kümmerer K, Kortenkamp A, Sumpter JP (2008) Do cytotoxic chemotherapy drugs discharged into rivers pose a risk to the environment and human health? An overview and UK case study. J Hydrol 348:167–175

Johnson AC, Oldenkamp R, Dumont E, Sumpter JP (2013) Predicting concentrations of the cytostatic drugs cyclophosphamide, carboplatin, 5-fluorouracil, and capecitabine throughout the sewage effluents and surface waters of Europe. Environ Toxicol Chem 32:1954–1961

Keller VDJ, Williams RJ, Lofthouse C, Johnson AC (2014) Worldwide estimation of river concentrations of any chemical originating from sewage-treatment plants using dilution factors. Environ Toxicol Chem 33:447–452

Kosjek T, Heath E (2011) Occurrence, fate and determination of cytostatic pharmaceuticals in the environment. Trends Anal Chem 30:1065–1087

Kümmerer K, Al-Ahmad A (2010) Estimation of the cancer risk to humans resulting from the presence of cyclophosphamide and ifosfamide in surface water. Environ Sci Pollut Res 17:486–496

Kümmerer K, Haiß A, Schuster A, Hein A, Ebert I (2016) Antineoplastic compounds in the environment—substances of special concern. Environ Sci Pollut Res 23:14791–14804

Law BK (2005) Rapamycin: an anti-cancer immunosuppressant? Crit Rev Oncol Hematol 56:47–60

Lenz K, Koellensperger G, Hann S, Weissenbacher N, Mahnik SN, Fuerhacker M (2007) Fate of cancerostatic platinum compounds in biological wastewater treatment of hospital effluents. Chemosphere 69:1765–1774

Llewellyn N, Lloyd P, Jürgens MD, Johnson AC (2011) Determination of cyclophosphamide and ifosfamide in sewage effluent by stable isotope-dilution liquid chromatography-tandem mass spectrometry. J Chromatogr A 1218:8519–8528

López-Serna R, Petrović M, Barceló D (2012) Occurrence and distribution of multi-class pharmaceuticals and their active metabolites and transformation products in the Ebro River basin (NE Spain). Sci Total Environ 440:280–289

Majd N, Sumita K, Yoshino H, Chen D, Terakawa J, Daikoku T, Kofuji S, Curry R, Wise-Draper TM, Warnick RE, Guarnaschelli J, Sasaki AT (2014) A Review of the Potential Utility of Mycophenolate Mofetil as a Cancer Therapeutic. J Cancer Res 2014:12

Martín J, Camacho-Muñoz D, Santos JL, Aparicio I, Alonso E (2011) Simultaneous determination of a selected group of cytostatic drugs in water using high-performance liquid chromatography-triple-quadrupole mass spectrometry. J Sep Sci 34:3166–3177

Martín J, Camacho-Munõz D, Santos JL, Aparicio I, Alonso E (2014) Occurrence and ecotoxicological risk assessment of 14 cytostatic drugs in wastewater. Water Air Soil Pollut 225

Metcalfe CD, Koenig BG, Bennie DT, Servos M, Ternes TA, Hirsch R (2003) Occurrence of neutral and acidic drugs in the effluents of canadian sewage treatment plants. Environ Toxicol Chem 22:2872–2880

National Cancer Institute (NCI) (2017a) NCI Dictionary of Cancer Terms. https://www.cancer.gov/publications/dictionaries/cancer-terms?cdrid=270742

National Cancer Institute (NCI) (2017b) Immunosuppression. https://www.cancer.gov/about-cancer/causes-prevention/risk/immunosuppression

Negreira N, López de Alda M, Barceló D (2013) On-line solid phase extraction-liquid chromatography-tandem mass spectrometry for the determination of 17 cytostatics and metabolites in waste, surface and ground water samples. J Chromatogr A 1280:64–74

Negreira N, de Alda ML, Barceló D (2014) Cytostatic drugs and metabolites in municipal and hospital wastewaters in Spain: Filtration, occurrence, and environmental risk. Sci Total Environ 497-498:68–77

Olalla A, Negreira N, López de Alda M, Barceló D, Valcárcel Y (2018) A case study to identify priority cytostatic contaminants in hospital effluents. Chemosphere 190:417–430

Ort C, Hollender J, Schaerer M, Siegrist H (2009) Model-based evaluation of reduction strategies for micropollutants from wastewater treatment plants in complex river networks. Environ Sci Technol 43:3214–3220

Ort C, Michael GL, Reungoat J, Mueller JF (2010) Sampling for PPCPs in wastewater systems: Comparison of different sampling modes and optimization strategies. Environ Sci Technol 44:6289–6296

Ortiz de García S, Pinto Pinto G, García Encina P, Irusta Mata R (2013) Consumption and occurrence of pharmaceutical and personal care products in the aquatic environment in Spain. Sci Total Environ 444:451–465

Parrella A, Lavorgna M, Criscuolo E, Russo C, Fiumano V, Isidori M (2014) Acute and chronic toxicity of six anticancer drugs on rotifers and crustaceans. Chemosphere 115:59–66

Roche (2012) Safety Data Sheets. Access at 2017.: https://www.roche.com/sustainability/what_we_do/for_communities_and_environment/environment/safety_data_sheets-row.htm

Rowney NC, Johnson AC, Williams RJ (2009) Cytotoxic drugs in drinking water: a prediction and risk assessment exercise for the Thames catchment in the United Kingdom. Environ Toxicol Chem 28:2733–2743

Santos MSF, Franquet-Griell H, Lacorte S, Madeira LM, Alves A (2017) Anticancer drugs in Portuguese surface waters – estimation of concentrations and identification of potentially priority drugs. Chemosphere 184:1250–1260

Strujis J (2014) SimpleTreat 4.0: a model to predict fate and emission of chemicals in wastewater treatment plants: background report describing the equations

United States Environmental Protection Agency (U.S. EPA) (2013) Exposure assessment tools and models. EPI Suite v4.1. 2018: https://www.epa.gov/tsca-screening-tools/epi-suitetm-estimation-program-interface

Usawanuwat J, Boontanon N, Boontanon SK (2014) Analysis of three anticancer drugs (5-fluorouracil, cyclophosphamide and hydroxyurea) in water samples by HPLC-MS/Ms. Int J Adv Agric Environ Eng 1:72

Valcárcel Y, González Alonso S, Rodríguez-Gil JL, Gil A, Catalá M (2011) Detection of pharmaceutically active compounds in the rivers and tap water of the Madrid Region (Spain) and potential ecotoxicological risk. Chemosphere 84:1336–1348

Verlicchi P, Al Aukidy M, Galletti A, Petrovic M, Barceló D (2012) Hospital effluent: Investigation of the concentration and distribution of pharmaceuticals and environmental risk assessment. Sci Total Environ 430:109–118

Verlicchi P, Al Aukidy M, Jelic A, Petrović M, Barceló D (2014) Comparison of measured and predicted concentrations of selected pharmaceuticals in wastewater and surface water: A case study of a catchment area in the Po Valley (Italy). Sci Total Environ 470–471:844–854

Weissbrodt D, Kovalova L, Ort C, Pazhepurackel V, Moser R, Hollender J, Siegrist H, McArdell C (2009) Mass flows of X-ray contrast media and cytostatics on hospital wastewater. Environ Sci Technol 43:4810–4817

Whirl-Carrillo M, McDonagh EM, Hebert JM, Gong L, Sangkuhl K, Thorn CF, Altman RB, Klein TE (2012) Pharmacogenomics knowledge for personalized medicine. Clin Pharmacol Ther 92:414–417

WHO (2017) Incidences of cancer in the world. http://www.who.int/mediacentre/factsheets/fs297/en/

World Health Organization (WHO) (2017) ATC/DDD Index 2017. https://www.whocc.no/atc_ddd_index/

Young AR, Grew R, Holmes MGR (2003) Low flows 2000: A national water resources assessment and decision support tool. Water Sci Technol:119–126

Zhang J, Chang VWC, Giannis A, Wang JY (2013) Removal of cytostatic drugs from aquatic environment: a review. Sci Total Environ 445–446:281–298

Zuccato E, Calamari D, Natangelo M, Fanelli R (2000) Presence of therapeutic drugs in the environment. Lancet 355:1789–1790

Part II
Environment/Wastewater Treatment

Chapter 3
Hospitals and Pharmacies as Sources of Contamination by Cytostatic Pharmaceuticals: Long-Term Monitoring in the Czech Republic

Lucie Blahova, Lenka Dolezalova, Jan Kuta, Sarka Kozakova, and Ludek Blaha

Abstract Two of the important contamination sources by antineoplastic drugs (AD) are hospitals and pharmacies. Unwanted releases have been documented during all steps of the preparation and administration of these hazardous drugs to patients leading to contamination of both working places and outside environment (transfer by aerosols, contaminated materials, or water after cleaning). Here we present results of a long-term project from 21 hospitals in the Czech Republic (971 samples; 2008–2016) investigating the contamination by major AD-5-fluorouracil (FU), cyclophosphamide (CP) and platinum drugs (total Pt as a sum of broadly used Pt-based drugs cisplatin, oxaliplatin, and carboplatin). In general, lower median levels of contamination have been found in pharmacies, which could be attributed to personnel education and higher safety working standards. On the other hand, surface contamination in other hospital areas exceeded the suggested threshold guidance values (TGVs) in up to 40% of samples depending on the monitored drug (TGVs being 67, 12, and 38 pg per square cm for CP, Pt, and FU, respectively). The highest values, maxima exceeding 29,000 and 49,000 pg per square cm for CP and FU, respectively, have repeatedly been found in outpatient clinics. The monitoring and discussions with responsible managers promoted the implementation of proper procedures and technologies that resulted in an overall decrease of the contamination during the monitored period.

Keywords Occupational exposure · Surface contamination · 5-Fluorouracil · Cyclophosphamide · Platinum drugs · LC-MS/MS · ICP MS

L. Blahova · J. Kuta · L. Blaha (✉)
Faculty of Science, RECETOX, Masaryk University, Brno, Czech Republic
e-mail: blaha@recetox.muni.cz

L. Dolezalova · S. Kozakova
Masaryk Memorial Cancer Institute, Brno, Czech Republic

E. Heath et al. (eds.), *Fate and Effects of Anticancer Drugs in the Environment*,
https://doi.org/10.1007/978-3-030-21048-9_3

3.1 Introduction

Hospitals and pharmacies, where antineoplastic drugs are being prepared and administered to oncology patients, are one of the primary sources of contamination of these hazardous chemicals. Health and environmental risks related to handling antineoplastic drugs were discussed in many studies since the 1970s when serious side effects of chemotherapy (secondary malignancies) were first reported.

Several studies addressed the risks related to occupational exposures to antineoplastic drugs. Concentrations of ADs were reported from hospital samples, including air (Kiffmeyer et al. 2002; Wittgen et al. 2006; Odraska et al. 2011; Panahi et al. 2016), various working surfaces (walls, benches, shelves, doors, telephones or floors in pharmacies, clinics, sanitary rooms, toilets) (Hedmer et al. 2005; Castiglia et al. 2008; Fabrizi et al. 2012; Nussbaumer et al. 2012; Odraska et al. 2014), as well as other materials such as clothes, linens, external packages of the drugs, etc. (Sessink et al. 1992; Fleury-Souverain et al. 2014). All these types of contamination represent serious exposure, especially to hospital workers (pharmacists and nurses preparing and administering ADs, surgical teams in the operating rooms, physicians or other hospital staff such as waste handlers, laundry workers, custodial workers, etc.) (Kromhout et al. 2000; da Silva et al. 2016). Occupational exposure was repeatedly confirmed also by biological monitoring of ADs in the urine of health-care professionals (Sessink et al. 1994; Ndaw et al. 2010; Fabrizi et al. 2016). However, other exposure scenarios to ADs are of concern to both human and environmental health including, e.g., release of ADs from hospitals through wastewaters (Kümmerer 2009; Verlicchi et al. 2010; Kosjek and Heath 2011) or their release via air (Panahi et al. 2016). In addition, 75% of oncology patients are outpatients, receiving their treatment at oncology wards and returning home after administration, which also spreads the contamination (Kopp et al. 2013; Yuki et al. 2015). This may lead to exposures of family members (including highly susceptible groups like children and pregnant women) as well as to additional release into the environment but these new issues will require further research attention.

Major progress in the field of occupational safety was recorded after the implementation of protective measures like centralization of drug preparation in specialized hospital pharmacies equipped with biological safety cabinets or positive/negative pressure isolators. The most hazardous activities were localized to well-controlled areas. Nevertheless, the overall contribution of the measures outlined above to health and environmental risk protection is still not fully clarified (Kiffmeyer et al. 2002; Sessink 2011). In addition, the measures mostly focused on large hospital facilities and pharmacies, which often follow the best practices and risk management measures within QA/QC protocols and have well-trained staff (Acampora et al. 2005; Crickman and Finnell 2006; Yoshida et al. 2013). Available studies (Wick et al. 2003; Sottani et al. 2012; Kopp et al. 2013) seem to indicate that pharmacies have in general similar or lower (median) levels of surface contamination in comparison to other hospital areas, where ADs are administered such as outpatient clinics, sanitary rooms, etc.

Diverse ADs are being currently used to treat different types of malignancies, new types of therapies are being developed and the amount of AD consumption is increasing (Nussbaumer et al. 2011). It is thus impossible to investigate contamination levels of all individual ADs, and feasible prioritized approach focuses on monitoring of those ADs that are used in the highest quantities and/or may pose highest risks. Recent reviews by Lancharro et al. (2016) showed that cyclophosphamide (CP, broadly used IARC class 1 carcinogen) is the most commonly monitored AD. Other surrogate markers of AD contamination include 5-fluorouracil (FU, nucleotide analogue used frequently and in the high quantities (Schierl et al. 2009)) or total platinum (Pt) reflecting contamination by platinum-based drugs used in large quantities and some being classified as IARC Group 2A (Gorná et al. 2011; Nussbaumer et al. 2012). The most commonly used exposure assessment of ADs focuses on surfaces and collection of wipe samples (Hon et al. 2014; Jeronimo et al. 2015). This allows adequate standardization and comparability of the results.

The objective of the present chapter is to discuss the results of the long-term research project (results from 2008–2016) running in the Czech Republic, which focuses on the surface contamination of hospitals and hospital pharmacies by major antineoplastic drugs, i.e., cyclophosphamide and platinum-based drugs. Recently, 5-fluorouracil has been added among the monitored compounds, and the first results from 2015 to 2016 are presented. We describe the overall outcomes and long-term trends, compare levels of ADs in pharmacy and hospital areas, and discuss possible drivers beyond the contamination and possible risk assessment tools and management measures.

3.2 Study Design and Methods

The long-term research project (2008–2016) covered 28 pharmacies (i.e., 60% of all 47 pharmacies in the country holding accreditation for an AD preparation). During individual years, between 1 and 19 pharmacies were included in the study based on a voluntary basis. Since 2010, hospital areas (outpatient clinics, bedrooms, sanitary rooms, and nurse offices) were also monitored but with lower frequency, which allowed us to derive only preliminary conclusions.

3.2.1 Wipe Sampling and Sample Preparation

Sampling was performed by the staff in individual hospitals according to written instructions and video manual (https://muni.cz/go/a0d852). Samples were collected at the end of the working hours prior to regular cleaning of the facilities. Sample locations included floors, tables as well as door/fridge handles, phones, keyboards, etc., in pharmacies and hospital areas. The selection of places for wiping was not harmonized, and it reflected practices and actual needs of individual facilities.

Wipe-sampled area (most commonly 30 × 30 cm; 900 cm^2) was wiped using the non-woven swab moistened with acetate buffer (0.75 ml, 20 mM, pH = 4). Besides the wipe samples, field blank samples were collected to check for possible cross-contamination.

3.2.2 *Analyses of ADs*

Sample preparation involved the extraction in acetate buffer with follow-up LC-MS/MS and ICP-MS analysis as previously described elsewhere (Odraska et al. 2011; Odraska et al. 2013; Odraska et al. 2014). The concentrations of CP (and FU since 2015) along with appropriate internal standards have been measured by the optimized high-performance liquid chromatography method using UPLC Waters Acquity with tandem quadrupole mass spectrometer Waters XEVO TQ-S (Waters, Manchester, U.K.). The used column was ACQUITY UPLC BEH C18 (Waters), and the mobile phase consisted of 0.1% formic acid in water (A) and methanol (B) running in gradient (5% B 0–1 min; 5–20% B 1–2 min; 20–85% B 2–8.5 min; 85% B 8.5–10 min with 4 min of equilibration). The concentration of Pt (a marker of Pt-containing drugs—cisplatin, oxaliplatin, and carboplatin—are the most commonly used in the Czech Republic) was determined by inductively coupled plasma mass spectrometry (Agilent 7500ce and 7700x, Agilent Technologies, Japan). Internal standards were applied to correct for injection volume errors, matrix effects, and signal drift of the analytical determination (CP D4 for quantification of CP; rhenium (185Re) and bismuth (209Bi) for quantification of Pt, and FU 13C, 15 N2 for quantification of FU). All the concentrations have been expressed as pg/cm^2, and the detection limits were 1.0, 0.2, and 7 pg/cm^2 for CP, Pt, and FU, respectively.

3.3 Results and Discussion

The overall summary of detected AD concentrations in pharmacies (2008–2016) and hospitals (2010–2016) is presented in Table 3.1. In total, 971 samples have been collected and 757, 740, and 323 analyses of CP, Pt, and FU were performed, respectively (field blank samples not included). Results of FU analyses are only briefly discussed in continuation for comparison since the lower number of samples was analyzed so far.

Results in Table 3.1 are sorted according to different areas with variable procedures and operations. In pharmacies, two separated areas have been covered and include (i) preparation rooms (i.e., isolated areas, where concentrated ADs are being handled, diluted, and prepared for patients including laminar flow isolator areas) and (ii) storage area, where ADs are being received from suppliers and stored. Hospital areas include (i) outpatient clinics, where ADs are being administered to patients,

Table 3.1 Results of CP and Pt monitoring in pharmacies and hospitals of the Czech Republic during 2008–2016

	Cyclophosphamide				Platinum			
	N/Npos.	Median	Mean	Min/max	N/Npos.	Median	Mean	Min/max
Pharmacies								
Preparation room								
Tables	160/104	7.6	468	<1/33,853	157/105	1.3	118	<0.2/5333
Floors	96/64	7.1	62	<1/638	76/51	1.0	6.5	<0.2/84
Others[a]	80/41	2.3	123	<1/4656	65/45	1.2	20	<0.2/450
Storage area								
Tables	115/33	<1	28	<1/1466	98/43	<0.2	78	<0.2/7343
Floors	81/25	<1	17	<1/235	63/30	<0.2	3.7	<0.2/57
Others[a]	36/11	<1	55	<1/1184	38/22	0.8	3.0	<0.2/23
Hospital areas								
Outpatients clinic								
Tables	23/20	20	216	<1/2742	18/15	1.7	39	<0.2/510
Floors	33/32	271	751	<1/5628	32/32	71	346	0.97/5390
WC	18/14	4.2	55	<1/597	17/16	586	863	<0.2/4220
Facilities for nurses								
Tables	37/18	<1	32	<1/741	62/27	<0.2	7.7	<0.2/227
Administrative areas								
Pharmacies + hospital areas								
Others[a]	56/11	<1	4.5	<1/142	37/7	<0.2	0.3	<0.2/4.1
TOTAL	757/373	1.2	212	<1/33,853	740/466	1.0	82	<0.2/7343

Columns show a total number of analyzed samples (N); number of positives, i.e., above the detection limit (Npos.) and basic statistics (median, mean, minimum, and maximum). All concentrations are in pg/cm^2. Less than (<) values—below the limit of detection
[a]Others—door and fridge handles, phones, keyboards, displays

and (ii) facilities for nurses with no access to patients (such as sanitary and manipulation rooms). The final samples covered administrative areas (offices), where no ADs are being handled and thus low contamination is expected. Within individual areas (Table 3.1), floors as well as surfaces in contact with hands have been studied including "tables" (where standardized 900 cm^2 sampling was possible) and "others" (fridge handles, phones, and keyboards, where a variable area has been wiped).

Table 3.1 shows several trends and differences. Considering all the analyzed samples (row "TOTAL" in Table 3.1), 49% and 63% of the samples contained CP and Pt (above the LOD), respectively. The cleanest surfaces (low detection frequency and low concentrations of CP and Pt) were found in offices (administrative areas), where median values were below 1 pg/cm^2 for both CP and Pt. This is in agreement with other studies (Wick et al. 2003; Hedmer et al. 2005) demonstrating that the spread of contamination may be prevented by application of proper standards and procedures (such as working in zones with different levels of contamination). Within pharmacies, average detection frequency for both CP and Pt ranged between 29% and 69% for different types of areas and wipes. On the contrary, very high detection frequency (78–100%) for both CP and Pt was detected in outpatient clinics of hospitals, where less controlled procedures and protective measures are being implemented in comparison to pharmacies.

Maxima of CP (up to more than 33,000 pg/cm^2) were recorded in the preparation room of the hospital pharmacy, where ADs are being openly handled and prepared for patients. However, accidental extremes of elevated concentrations of carcinogenic CP have also been detected on tables and floors in an outpatient clinic (Table 3.1). Contrarily, the highest levels of Pt (ng/cm^2) were detected at various places in both pharmacies and hospital areas. The differences in profiles of CP and Pt suggest that selecting one or few surrogate markers may not reflect the overall status of contamination. Anticancer drugs are a chemically heterogeneous class with a wide range of masses and hydro/lipophilicity; therefore some studies suggest the need to focus on monitoring of a larger group of ADs (Maeda and Miwa 2013; Fabrizi et al. 2016). The suggested problem of systematically elevated contamination at outpatient clinics is especially of concern when comparing median and average values of contamination (Table 3.1). The highest median (271 pg/cm^2) and arithmetic mean (751 pg/cm^2) values for CP were found on the floors in outpatient clinics, that were about an order of the magnitude higher than concentrations detected in preparatory rooms and storage areas of pharmacies. Similarly, for Pt, highly elevated median and average levels were found in outpatient clinics with concentrations one to two orders of magnitude higher in comparison with pharmacy areas. The highest mean value 863 pg/cm^2 has been found on surfaces at the toilet, where ADs may be released by patients through urine as well as from contaminated excreta like sweat and vomit. High levels of contamination on the toilet floors have previously been reported also in other studies (Kromhout et al. 2000; Kopp et al. 2013).

Our results may be compared with other larger scale surveys. For example, contamination by CP, Pt, FU, and other ADs among the individual oncology settings in hospital in Germany has been described by Kopp et al. (2013). From 375 samples, the highest detection frequency was observed for FU (94% above the LOD, $N = 153$) and Pt (88%, $N = 172$), followed by CP (55%, $N = 73$). The median contamination of FU was 8.1 pg/cm^2 with the highest concentration (14,556 pg/cm^2) from the floor of the therapy room. For Pt and CP, the median concentrations were 1.6 and 0.4 pg/cm^2, respectively, with the maxima 714 pg/cm^2 and 2230 pg/cm^2, respectively. Another study from Italy (Castiglia et al. 2008) showed a wide range of contamination levels for CP and FU with very high medians of 18.83 μg/dm^2 (i.e., 188,300 pg/cm^2) and 0.086 μg/dm^2 (8600 pg/cm^2). In a study by Schierl et al. (2009), over 1000 wipe samples taken from 102 German hospital and retail pharmacies were analyzed for contamination of Pt and FU with medians 0.4 and 4.96 pg/cm^2, respectively. The maximum value was found for Pt on storage shelves and boxes (23,068 pg/cm^2) and for FU on the surface of transfer chambers (253,333 pg/cm^2) (Schierl et al. 2009).

The trends detected in the overall results from long-term 2008–2016 monitoring (Table 3.1, discussed above) remain apparent also when looking in detail on a subset of recent 2015–2016 data presented in Fig. 3.1. Here, an example of contamination of tables (i.e., places in direct contact with hands) is shown for CP, Pt, and FU. The

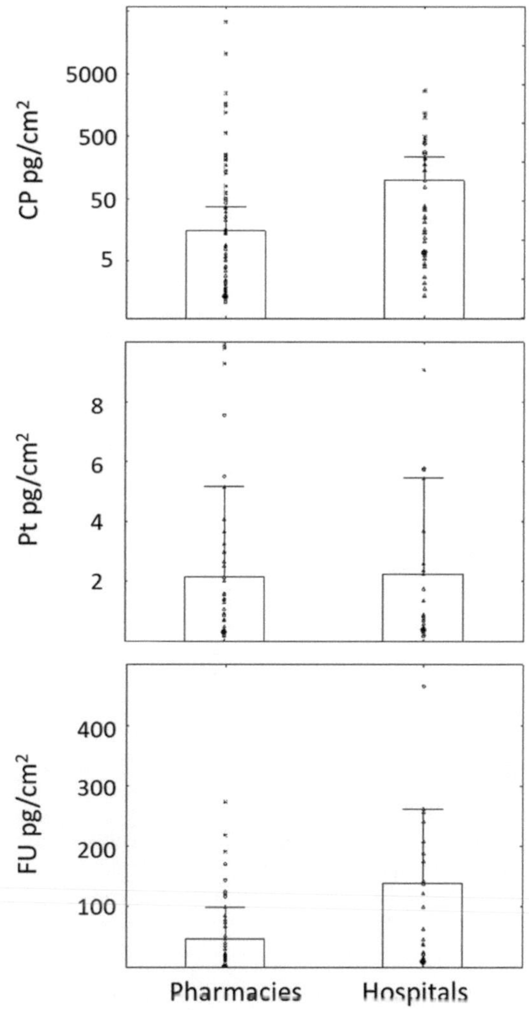

Fig. 3.1 Contamination of table surfaces in pharmacies and hospitals of the Czech Republic during 2015 and 2016 by cyclophosphamide (CP, upper panel; log-scale Y-axis), platinum (Pt, central panel), and 5-fluorouracil (FU, bottom panel). Bars show median values, error bars 25–75% range (extremes and outlier values are not shown for clarity)

contamination with all three ADs was either comparable or systematically elevated in hospital areas if compared to the well-controlled areas within pharmacies.

Nevertheless, during the long-term monitoring within the Czech Republic, systematic improvement of the situation and lowering of the surface contamination levels has been recorded. Figure 3.2 shows temporal trends in CP (panel A) and Pt (panel B) concentrations on tables in all monitored pharmacies. Hospitals are not discussed because of the shorter period covered and the lower number of analyzed samples. The decreasing trend was observed especially in the case of CP, and it was also apparent in localized data (Fig. 3.2C), where surface contamination from one of the most broadly monitored pharmacies is presented. For Pt, no such trend was observed, which could be attributed to generally lower contamination levels in

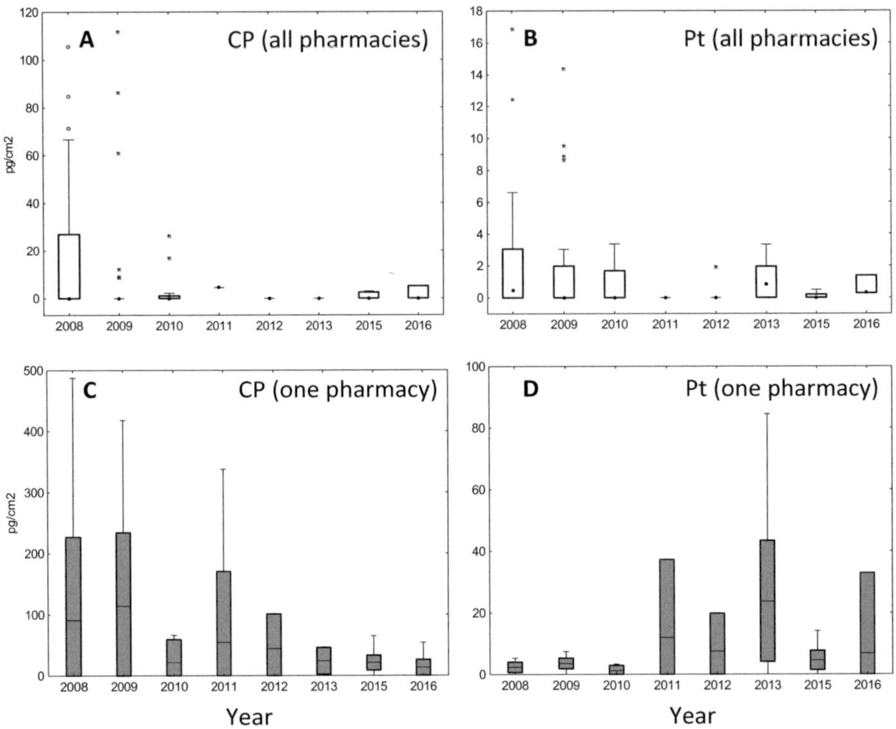

Fig. 3.2 Temporal trends in surface contamination in Czech Republic pharmacies. Panels A (CP) and B (Pt) show pooled data for all Czech Republic pharmacies focusing on table surfaces out of the preparatory rooms (receiving, storage, and sending areas). Panels C and D show CP and Pt in a single pharmacy that has been thoroughly monitored over the years (all surface samples were pooled). For panels A–B: central point shows median value, boxes are 25–75% range, and error bars show non-outlier range. For panels C–D: central line shows arithmetic mean, boxes are 95% confidence interval of mean, error bars are non-outlier ranges

absolute values (compared to CP) as well as by potential other sources of Pt than ADs. Similarly, Yoshida et al. (2013) reported a decrease in contamination by about 80% at CP, Pt, and FU following the application of control measures like careful use of personal protective equipment, improved training, or maintenance. Effective reduction of workplace contamination by CP and Pt in the outpatient clinic was also reported in our previous study after improvements of working and cleaning procedures (Odraska et al. 2013). Further, lower contamination by CP in pharmacies and patient care areas was also reported during the long-term monitoring (2008–2014) in multiple Canadian centers (Poupeau et al. 2016), where the 75th percentile was reduced from 50 to 3 pg/cm^2.

The long-term data also allowed for the investigation of possible drivers beyond the surface contamination. Amount of ADs being handled in a given pharmacy could be considered among the major possible causes assuming that larger quantities of ADs used might be accompanied by elevated contamination. However, the

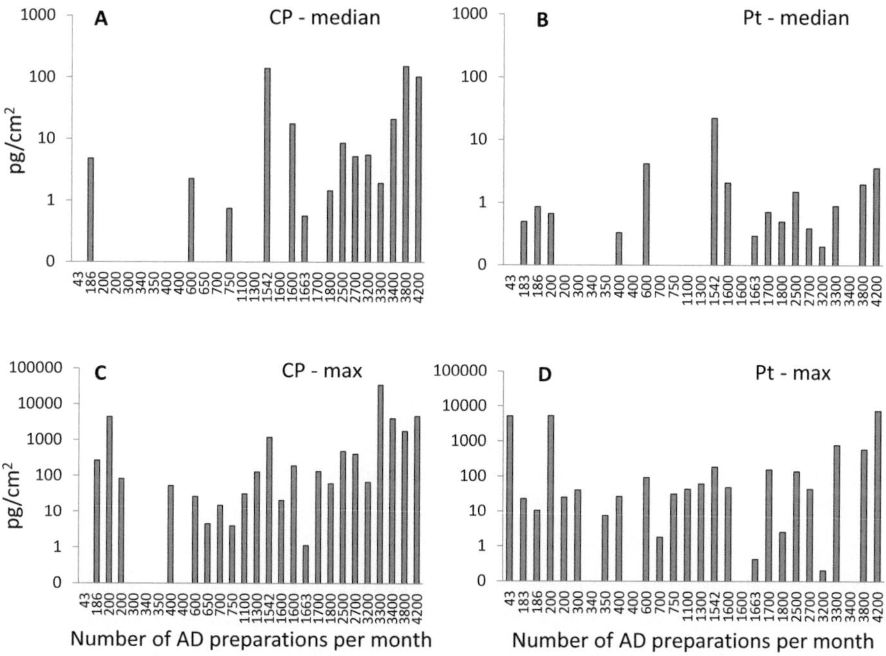

Fig. 3.3 Indicators of surface contamination (median values and maxima of concentrations) for CP (panels A and C) and Pt (panels B, D) in pharmacies relative to the size of the pharmacy—values on X-axes show numbers of AD preparations prepared monthly; CP/Pt concentrations (in pg/cm^2) are given on Y-axes. Each column corresponds to one pharmacy

distribution of maxima or median values observed in individual pharmacies in the present study (Fig. 3.3A–D) did not show any apparent trend for CP or Pt. Indicators of the surface contamination by ADs were rather randomly distributed, irrespectively of the numbers of ADs prepared in individual pharmacies (ranging across an order of magnitude). This finding is in agreement with available literature. For example, no significant correlation was found between the contamination levels in 102 German pharmacies and the amounts of handled Pt and FU (Schierl et al. 2009). Similarly, another German study of surface contamination in 28 outpatient oncology healthcare centers showed no correlation between contamination by FU, Pt, CP, and other ADs (gemcitabine, ifosfamide, methotrexate, docetaxel, and paclitaxel) and the amount of drug handled in these centers (Kopp et al. 2013). Interestingly, in a study by Sessink et al. (1994), a correlation between contamination of protective gloves and amount of drug prepared was observed for FU but was found to be insignificant for CP and methotrexate.

An important issue that remains a matter of discussion is the actual characterization of risks associated with AD surface contamination. Based on current paradigm, it is impossible to set a level of exposure to carcinogenic chemicals (like many ADs) that could be considered safe for human health. Consequently, the only management option is keeping the exposure at the lowest possible level. Nevertheless,

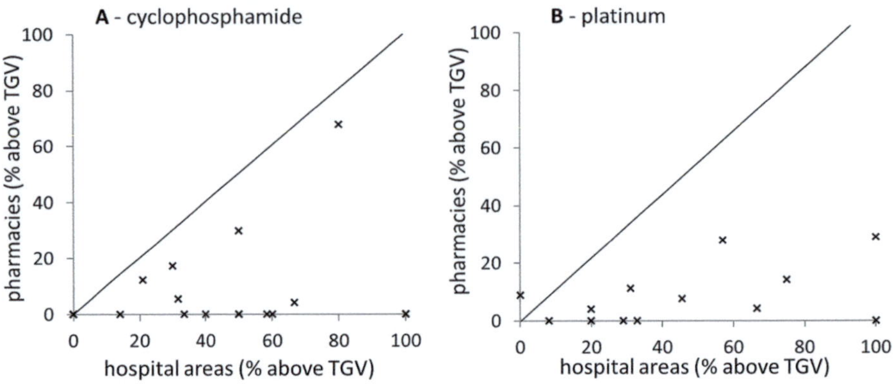

Fig. 3.4 Exceedance of threshold guidance values (TGVs) (% from all the samples analyzed) for CP (panel A) and Pt (panel B) in hospitals from the Czech Republic (2008–2016) aggregated data. The graphs show paired comparisons from individual facilities with % exceedance in hospital areas and pharmacies on X- and Y-axes, respectively

recommendations and threshold guidance values are discussed among scientists and regulators including the US National Institute for Occupational Safety and Health, American Society of Health-System Pharmacists or International Society of Oncology Pharmacy Practitioners (NIOSH; ASHP 2006; ISOPP 2007). For example, the US Pharmacopeia (USP) has recently proposed a new guideline for CP and recommended as acceptable levels lower than 1 ng per square cm.

Along with other proposals (Bouwman-Boer et al. 2015), our monitoring uses so-called threshold guidance values (TGVs) derived by a statistical approach. TGV derived within our monitoring represents a concentration (pg/cm^2) of the 75th percentile of all measured contamination levels from different surfaces analyzed within the project CYTO during 2006–2010, which included mainly pharmacies. Correspondingly, those samples where concentrations are below the TGV (i.e., about 75% of all values) are considered acceptable. Higher concentrations (i.e., above the TGV) indicate the need for management actions or improvement of procedures. We use TGVs, of 67 pg/cm^2 for CP, which is well comparable to 100 pg/cm^2 suggested as 90th percentile TGV by other authors (Sessink 2011; Kiffmeyer et al. 2013). TGV used for Pt is 12 pg/cm^2, which is slightly higher than a 75th percentile TGV (4 pg/cm^2) proposed by Schierl et al. (2009).

An example of the TGV interpretation for CP and Pt is shown in Fig. 3.4, which compares situations (frequency in % of TGV exceedances) in pharmacies vs. other hospital areas from the same facility. Clearly, TGV exceedance frequency for CP and Pt was rarely above 30% in pharmacies (Y-axis). On the other hand, in hospital areas (X-axis) TGVs were commonly exceeded in more than 50% of the samples, reaching in extremes up to 100% (i.e., all samples analyzed from a given hospital had elevated surface concentrations of ADs above the TGV). A similar trend was observed also for the aggregated data including all samples: for CP in pharmacies 104 out of 568 analyzed samples exceeded the TGV (18%), while for samples from hospital areas, the TGV exceedance rate was doubled (36–60% out of 167 samples). This again confirms higher health risks from exposures to ADs for nurses, physicians, and other hospital workers and potentially also to family members accompanying the patients during the visits in hospitals (Lancharro et al. 2016).

Various studies investigated the efficiency of different prevention and risk mitigation measures such as closed-system drug transfer device for preparation and administration (CSTD) (Wick et al. 2003; Sessink et al. 2011), that were shown to significantly reduce the spread of contamination (Lancharro et al. 2016). Decreased contamination by ADs was reported also after implementation of multiple measures, e.g., working in zones with separated use of protective gloves, proper housekeeping, use of multichannel administration sets or automated self-cleaning toilet seats, and others (Odraska et al. 2011 2013, 2014). Recent guidelines from different countries and organizations for safe handling of cytostatic drugs are listed in a recent review (Lancharro et al. 2016). Recommendations for routine cleaning and decontamination or waste disposal were also recently provided by NIOSH (https://www.cdc.gov/niosh/topics/antineoplastic/). However, available studies demonstrate that contamination of AD cannot be completely eliminated (Lancharro et al. 2016).

3.4 Conclusions and Perspectives

Contamination of the indoor environment (pharmacies and hospital areas) by ADs represents a major source of contamination and a health threat to hospital workers, where, for example, risks of miscarriage have been proven by meta-analyses of multiple occupational studies (Dranitsaris et al. 2005; Connor et al. 2014). Traditional research mostly focused on pharmacies, where ADs are openly handled (Sessink et al. 2011). However, published studies (Lancharro et al. 2016) as well as the results presented in this chapter point to the importance of other open hospital areas, where implementation of safety procedures is more complicated than in pharmacy facilities. Our long-term study demonstrates that periodic monitoring, where results are being provided directly to managers of pharmacies and hospital health safety officers, leads to rising of general awareness and improvements of practices, which is then reflected in decreasing trends of surface contamination by ADs. In addition to the discussed CP, Pt, and FU, further research efforts should focus on levels and fate of other classes of ADs (e.g., on compounds like sunitinib, imatinib, everolimus, or fluoropyrimidine), which are nowadays being broadly used. They are also often prescribed to outpatients for oral administration at home. Thus, patient households represent another source of AD release to the environment, which has only recently attracted attention of researchers (Yuki et al. 2015; Böhlandt et al. 2017). In summary, systematic assessment of contamination (i.e., exposure levels) to widely administered ADs provides invaluable data for continuous evaluation of risks and improvements of occupational and public health as well as the environment.

Acknowledgments The research has been supported by voluntary contributions from all monitoring participants, Czech Health Research Council project (NV18-09-00188) and infrastructure grants from the Ministry of Education C.R. LM2015051 and CZ.02.1.01/0.0/0.0/16_013/0001761.

References

Acampora A, Castiglia L, Miraglia N et al (2005) A case study: surface contamination of cyclophosphamide due to working practices and cleaning procedures in two Italian hospitals. Ann Occup Hyg 49:611–618. https://doi.org/10.1093/annhyg/mei029

ASHP (2006) ASHP, guidelines on handling hazardous drugs. Am J Health Syst Pharm 63:1172–1193

Böhlandt A, Sverdel Y, Schierl R (2017) Antineoplastic drug residues inside homes of chemotherapy patients. Int J Hyg Environ Health 220:757–765. https://doi.org/10.1016/j.ijheh.2017.03.005

Bouwman-Boer Y et al (2015) Occupational safety and health in practical pharmaceutics. In: Occupational safety and health in practical pharmaceutics. Springer, Cham

Castiglia L, Miraglia N, Pieri M et al (2008) Evaluation of occupational exposure to antiblastic drugs in an Italian Hospital Oncological Department. J Occup Health 50:48–56. https://doi.org/10.1539/joh.50.48

Connor TH, Lawson CCPM, McDiarmid MA (2014) Reproductive health risks associated with occupational exposures to antineoplastic drugs in health care settings: a review of the evidence. J Occup Environ Med 56:901–910

Crickman R, Finnell DS (2006) Chemotherapy Safe Handling. Limiting nursing exposure with a hazardous drug control program. Clin J Oncol Nurs 21:73–78

da Silva CBP, Julio IP, Donadel GE, Martins I (2016) UPLC-MS/MS method for simultaneous determination of cyclophosphamide, docetaxel, doxorubicin and 5-fluorouracil in surface samples. J Pharmacol Toxicol Methods 82:68–73. https://doi.org/10.1016/j.vascn.2016.08.004

Dranitsaris G, Johnston M, Poirier S et al (2005) Are health care providers who work with cancer drugs at an increased risk for toxic-events? A systematic review and meta analysis of the literature. J Clin Oncol Off J Am Soc Clin Oncol 23:548S–548S

Fabrizi G, Fioretti M, Rocca LM, Curini R (2012) DESI-MS 2: a rapid and innovative method for trace analysis of six cytostatic drugs in health care setting. Anal Bioanal Chem 403:973–983. https://doi.org/10.1007/s00216-011-5626-7

Fabrizi G, Fioretti M, Mainero Rocca L (2016) Dispersive solid-phase extraction procedure coupled to UPLC-ESI-MS/MS analysis for the simultaneous determination of thirteen cytotoxic drugs in human urine. Biomed Chromatogr 30:1297–1308. https://doi.org/10.1002/bmc.3684

Fleury-Souverain S, Nussbaumer S, Mattiuzzo M, Bonnabry P (2014) Determination of the external contamination and cross-contamination by cytotoxic drugs on the surfaces of vials available on the Swiss market. J Oncol Pharm Pract 20:100–111. https://doi.org/10.1177/1078155213482683

Gorná L, Doležalová L, Oravec M, et al (2011) Monitoring povrchové kontaminace cytotoxickými lé č ivy v nemocni č ních lékárnách České republiky (Monitoring of cytotoxic drugs in environment of hospital pharmacies in the Czech Republic)

Hedmer M, Georgiadi A, Rämme Bremberg E et al (2005) Surface contamination of cyclophosphamide packaging and surface contamination with antineoplastic drugs in a hospital pharmacy in Sweden. Ann Occup Hyg 49:629–637. https://doi.org/10.1093/annhyg/mei042

Hon C-Y, Barzan C, Astrakianakis G (2014) Identification of knowledge gaps regarding healthcare workers exposure to antineoplastic drug: review of literature, North America versus Europe. Saf Health Work 5:169–174

ISOPP (2007) International Society of Oncology Pharmacy Practitioners. Standards of practice: safe handling of cytotoxics. J Oncol Pharm Pract 13:1–81

Jeronimo M, Colombo M, Astrakianakis G, Hon CY (2015) A surface wipe sampling and LC-MS/MS method for the simultaneous detection of six antineoplastic drugs commonly handled by healthcare workers. Anal Bioanal Chem 407:7083–7092. https://doi.org/10.1007/s00216-015-8868-y

Kiffmeyer T, Kube C, Opiolka S et al (2002) Vapour pressures, evaporation behaviour and airborne concentrations of hazardous drugs: implications for occupational safety. Pharm J 268:331–337

Kiffmeyer TK, Tuerk J, Hahn M et al (2013) Application and assessment of a regular environmental monitoring of the antineoplastic drug contamination level in pharmacies-the MEWIP project. Ann Occup Hyg 57:444–455. https://doi.org/10.1093/annhyg/mes081

Kopp B, Schierl R, Nowak D (2013) Evaluation of working practices and surface contamination with antineoplastic drugs in outpatient oncology health care settings. Int Arch Occup Environ Health 86:47–55. https://doi.org/10.1007/s00420-012-0742-z

Kosjek T, Heath E (2011) Occurrence, fate and determination of cytostatic pharmaceuticals in the environment. TrAC Trends Anal Chem 30:1065–1087. https://doi.org/10.1016/j.trac.2011.04.007

Kromhout H, Hoek F, Uitterhoeve R et al (2000) Erratum: postulating a dermal pathway for exposure to anti-neoplastic drugs among hospital workers. Applying a conceptual model to the results of three workplace surveys. Ann Occup Hyg 44:551–560. PII: S003487800000508

Kümmerer K (2009) The presence of pharmaceuticals in the environment due to human use – present knowledge and future challenges. J Environ Manag 90:2354–2366. https://doi.org/10.1016/j.jenvman.2009.01.023

Lancharro PM, De Castro-Acuna Iglesias N, Gonzalez-Barcala FJ, Gonzalez JDM (2016) Evidence of esposure to cytostatic drugs in healthcare staff: a review of recent literature. Farm Hosp 40:604–621. https://doi.org/10.7399/fh.2016.40.6.9103

Maeda S, Miwa Y (2013) Multicomponent high-performance liquid chromatography/tandem mass spectrometry analysis of ten chemotherapeutic drugs in wipe samples. J Chromatogr B Anal Technol Biomed Life Sci 921-922:43–48. https://doi.org/10.1016/j.jchromb.2013.01.014

Ndaw S, Denis F, Marsan P et al (2010) Biological monitoring of occupational exposure to 5-fluorouracil: urinary α-fluoro-β-alanine assay by high performance liquid chromatography tandem mass spectrometry in health care personnel. J Chromatogr B Anal Technol Biomed Life Sci 878:2630–2634. https://doi.org/10.1016/j.jchromb.2010.02.011

NIOSH Preventing occupational exposure to antineoplastic and other hazardous drugs in health care settings [citado 15-04-2015]. Disponible en: http://www.cdc.gov/niosh/docs/2004-165/

Nussbaumer S, Bonnabry P, Veuthey JL, Fleury-Souverain S (2011) Analysis of anticancer drugs: a review. Talanta 85:2265–2289. https://doi.org/10.1016/j.talanta.2011.08.034

Nussbaumer S, Geiser L, Sadeghipour F et al (2012) Wipe sampling procedure coupled to LC-MS/MS analysis for the simultaneous determination of 10 cytotoxic drugs on different surfaces. Anal Bioanal Chem 402:2499–2509. https://doi.org/10.1007/s00216-011-5157-2

Odraska P, Dolezalova L, Piler P et al (2011) Utilization of the solid sorbent media in monitoring of airborne cyclophosphamide concentrations and the implications for occupational hygiene. J Environ Monit 13:1480–1487. https://doi.org/10.1039/c0em00660b

Odraska P, Dolezalova L, Kuta J et al (2013) Evaluation of the efficacy of additional measures introduced for the protection of healthcare personnel handling antineoplastic drugs. Ann Occup Hyg 57:240–250. https://doi.org/10.1093/annhyg/mes057

Odraska P, Dolezalova L, Kuta J et al (2014) Association of surface contamination by Antineoplastic drugs with different working conditions in hospital pharmacies. Arch Environ Occup Health 693:148–158. https://doi.org/10.1080/19338244.2013.763757

Panahi D, Azari M, Akbari ME et al (2016) Development of a new method for sampling and monitoring oncology staff exposed to cyclophosphamide drug. Environ Monit Assess 188. https://doi.org/10.1007/s10661-016-5255-x

Poupeau C, Tanguay C, Plante C et al (2016) Pilot study of biological monitoring of four antineoplastic drugs among Canadian healthcare workers. J Oncol Pharm Pract 23:323. https://doi.org/10.1177/1078155216643860

Schierl R, Böhlandt A, Nowak D (2009) Guidance values for surface monitoring of antineoplastic drugs in german pharmacies. Ann Occup Hyg 53:703–711. https://doi.org/10.1093/annhyg/mep050

Sessink PJ (2011) Environmental contamination with cytostatic drugs: past, present and future. Saf Consid Oncol Pharm Special Ed:3–5

Sessink PJM, Boer KA, Scheefhals APH (1992) Occupational exposure to antineoplastic agents at several departments in a hospital. Int Arch Occup Environ Health 64:105–112

Sessink PJ, Van de Kerkhof MC, Anzion RB et al (1994) Environmental contamination and assessment of exposure to antineoplastic agents by determination of cyclophosphamide in urine of exposed pharmacy technicians: is skin absorption an important exposure route? Arch Environ Health 49:165–169. https://doi.org/10.1080/00039896.1994.9940377

Sessink PJM, Connor TH, Jorgenson JA, Tyler TG (2011) Reduction in surface contamination with antineoplastic drugs in 22 hospital pharmacies in the US following implementation of a closed-system drug transfer device. J Oncol Pharm Pract 17:39–48. https://doi.org/10.1177/1078155210361431

Sottani C, Porro B, Imbriani M, Minoia C (2012) Occupational exposure to antineoplastic drugs in four Italian health care settings. Toxicol Lett 213:107–115. https://doi.org/10.1016/j.toxlet.2011.03.028

Verlicchi P, Galletti A, Petrovic M, Barceló D (2010) Hospital effluents as a source of emerging pollutants: an overview of micropollutants and sustainable treatment options. J Hydrol 389:416–428. https://doi.org/10.1016/j.jhydrol.2010.06.005

Wick C, Slawson MH, Jorgenson J, Tyler LS (2003) Using a closed-system protective device to reduce personnel exposure to antineoplastic agents. Am J Health Pharm 60:2314–2320

Wittgen BPH, Kunst PWA, Perkons WR et al (2006) Assessing a system to capture stray aerosol during inhalation of nebulized liposomal cisplatin. J Aerosol Med 19:385–391

Yoshida J, Koda S, Nishida S et al (2013) Association between occupational exposure and control measures for antineoplastic drugs in a pharmacy of a hospital. Ann Occup Hyg 57:251–260. https://doi.org/10.1093/annhyg/mes061

Yuki M, Ishida T, Sekine S (2015) Secondary exposure of family members to cyclophosphamide after chemotherapy of outpatients with Cancer: a pilot study. Oncol Nurs Forum 42:665–671. https://doi.org/10.1188/15.ONF.42-06AP

Chapter 4
Tamoxifen: Occurrence, Fate, Transformation Products, and Non-Conventional Treatment Technologies

Carlos Escudero-Oñate, Sara Rodríguez-Mozaz, and Laura Ferrando-Climent

Abstract Tamoxifen is a non-selective estrogen receptor modulator (SERM) used to treat breast cancer as well as a prophylactic agent in women with significant risk of developing the disease. After intake, it is partially metabolized in the liver and both tamoxifen and a series of metabolites are excreted. Hydroxylation plays a key role in the metabolism, and therefore hydroxylated metabolites such as 4-hydroxy-tamoxifen and endoxifen are excreted. After excretion, tamoxifen and metabolites enter the sewer system *via* hospital and domestic sewage, and since they are not totally removed in the conventional wastewater treatment plants (WWTPs), they are further discharged into natural water bodies. In fact, several studies have reported the presence of this compound in hospital and urban wastewater effluents as well as in surface waters at concentration levels of ng/L. The presence of this drug in the aquatic environment may be considered as a threat to organisms due to its known toxicity, endocrine disruption effects, and bioaccumulation potential. In this chapter, the occurrence of tamoxifen, and its metabolites, in water bodies is reviewed and its transformation in non-conventional wastewater treatments as well as its environmental risk is evaluated.

Keywords Tamoxifen · Wastewater · Metabolites · Transformation products · White rot fungi · Advanced oxidation processes · Environmental risk assessment

C. Escudero-Oñate
Norwegian Institute for Water Research (NIVA), Oslo, Norway

S. Rodríguez-Mozaz
Catalan Institute for Water Research (ICRA), H2O Building, Scientific and Technological Park of the University of Girona, Girona, Spain

L. Ferrando-Climent (✉)
Tracer Technology Department, Fluid Flow and Environmental Technology Section, Institute for Energy Technology (IFE), Kjeller, Norway
e-mail: laura.ferrando-climent@ife.no

E. Heath et al. (eds.), *Fate and Effects of Anticancer Drugs in the Environment*,
https://doi.org/10.1007/978-3-030-21048-9_4

4.1 Introduction

In the last decades, pharmaceutical compounds have become a concerning group of emerging pollutants according to a large number of studies (Ebele et al. 2017; Ferrando-Climent et al. 2014; Johnson et al. 2008; Pereira et al. 2017; Petrie et al. 2015; Sangion and Gramatica 2016; Sui et al. 2015; Vasquez et al. 2014). Pharmaceutical compounds are substances specifically designed to exert target effects in the human body, but still a lot of research is being performed to know how these substances can affect other non-targeted organisms in the natural environment as well as their indirect impact on human health. Among the different groups of pharmaceutical compounds, special attention deserves the group of chemotherapeutic drugs, compounds employed in the treatment of cancer diseases. They are known as anticancer drugs and have been shown to exert remarkable cytotoxic, genotoxic, mutagenic, carcinogenic, endocrine disrupting, and/or teratogenic effects in several organisms. These effects are intended for cancer treatment since they are mainly designed to disrupt or prevent cellular proliferation, usually interfering in processes as relevant as DNA synthesis. However, such hazardous substances entering the environment can pose a serious threat to non-target organisms.

Even with the chemotherapy treatments being administrated mainly at hospital facilities, it is usual that patients leave the hospital just few hours after receiving the treatment. Additionally, there are types of cancer such as prostate cancer whose treatments are delivered in ambulatory facilities or even at home. This means that anticancer drugs, their metabolites, and related compounds (biomarkers, etc.) are excreted by humans via hospital and also by domestic wastewaters, being the immediate recipient of the urban sewage system. It is worth noting the sometimes underrated role of veterinary facilities when assessing the point sources of these kinds of substances to the environment. These premises are also a relevant source of anticancer drugs since most of these drugs are frequently used in veterinary applications, mainly in the treatment of cancer in dogs and cats (Withrow et al. 2012).

Pharmacokinetics is also important when it comes to assessing the products that might eventually enter the environment. When patients follow a medicinal treatment, they are administered the active pharmaceutical ingredient (API) and excipients. However, patients rarely excrete only the unmodified drug. Instead, they excrete a blend of APIs and metabolites. Thus, a comprehensive and thorough screening on the occurrence of pharmaceutical compounds in the environment should not only address the presence of the target compound but shall also include a set of plausible metabolites and transformation products (TPs). Currently, there is still a gap in knowledge about the occurrence and fate of anticancer drugs in the environment. This becomes even more relevant if it is taken into consideration that the consumption of this kind of substances has been increasing, and in the near future a further increase on the consumption trend is foreseen due to the growth in the number of cancer patients. To properly assess the increasing environmental and human risk of anticancer drugs in the environment, more information needs to be gathered about their presence, toxicity, bioaccumulation properties, and persistence. The fact that most of these drugs are transported through the sewage and the evidence of the presence of these compounds in water bodies indicate that the current wastewater treatment plants (WWTPs) are not efficient in their removal, and therefore there is a clear need for the development of improved water treatment schemes.

Among the different anticancer drugs employed nowadays, special attention deserves tamoxifen (TAM). TAM was initially referred to as an anti-estrogen since it blocks estrogen receptors in breast tissue, thus reducing the effects caused by estrogen. However, TAM also acts as an agonist for estrogen receptors in certain regions of the body such as the endometrium, liver, and bone; thus, it was classified as a selective estrogen receptor modulator (SERM) (Mikelman et al. 2017). This substance has been very frequently employed to treat breast cancer, and as a prophylactic measure in women with a significant risk of developing that disease (DellaGreca et al. 2007). TAM is released in the sewer system through hospitals and/or domestic wastewater. The substance has been reported to exhibit a poor biodegradability in regular wastewater treatment plants and, therefore, is just poorly removed from sewage water (Ferrando-Climent et al. 2015; Ferrando-Climent et al. 2014). Due to its poor removal, TAM was already detected in the effluents from the wastewater treatment plant, a vector to enter the environment (Ferrando-Climent et al. 2013; Ferrando-Climent et al. 2014; Kosjek and Heath 2011; Negreira et al. 2013).

To date, several studies have pointed out also the occurrence of this compound in hospital effluents and in surface water in concentrations ranging from 21 to 200 ng/L (see Table 4.1). The presence of this drug in the aquatic environment is of special concern due to its known toxicity, endocrine disruption effects, and bioaccumulation potential (Jean et al. 2012). Besides, recent studies have revealed that the occurrence of TAM with other anticancer drugs in real water samples provokes a toxic effect of the cocktail higher than the contribution of the individual toxicity in a synergistic manner (Mater et al. 2014).

The occurrence of TAM in environmental water bodies indicated the inefficient performance of the existing WWTPs. Therefore, there is a need to develop alternative water treatment techniques and/or upgrade the existing ones. Tertiary treatments, such as chlorination of the final effluent, advanced oxidation processes (AOP), and/or the use of non-conventional bioprocesses may contribute to the enhancement of the overall removal of TAM from wastewater.

The objective of this chapter is to review the occurrence of TAM and its TPs in the environment, assess the performance of non-conventional water treatment technologies available nowadays, and provide a risk assessment of the discharge of this substance to water bodies.

4.2 Occurrence of Tamoxifen and Its Human Metabolites

A total of 16 studies published between 2004 and 2017 are listed in Table 4.1, which represents information about the levels of TAM and TPs found in different types of water (wastewater, surface water, and groundwater), as well as information about the country where the study was conducted and the analytical method applied. Studies

Table 4.1 Occurrence of TAM and TPs in different water bodies and streams

Compound	Location	Hospital Effluent ng L^{-1}	WWTP Influent ng L^{-1}	WWTP Effluent ng L^{-1}	Surface water ng L^{-1}	Groundwater ng L^{-1}	Analytical Methodology	References
TAM	Sweeden	–	13–24	–	–	–	SPE(Oasis) + UPLC(BEH C18)-MS/MS	Grabic et al. (2012)
TAM	China	0.2– 8.2	0.28	–	–	–	SPE(Oasis) + UPLC(BEH C18)-MS/MS	Liu et al. (2010)
TAM	France	–	–	< 102	< 25	–	SPE(STRATA™ X) + HPLC (Ascentis C18)-MS/MS	Coetsier et al. (2009)
TAM	Germany	–	–	–	–	6–16.5	SPE(Oasis) + HPLC(Polaris C18-Ether)-MS/MS	Reh et al. (2013)
TAM	UK	–	–	–	27–212	–	SPE(STRATA X) + HPLC(Luna2 C18)-MS/MS	Roberts and Thomas (2006)
TAM	UK	–	–	< 42	–	–	SPE(STRATA X) + HPLC(Luna2 C18)-MS/MS	Ashton et al. (2004)
TAM	Spain	–	–	–	0.5–25.1	–	SPE(Oasis) + HPLC(Luna C18)-Orbitrap	Franquet-Griell et al. (2017)
TAM	Spain	–	–	–	3.5–11.7	–	SPE(STRATA X) + HPLC(Synergi Polar-RP)-MS/MS	Iglesias et al. (2014)
TAM	Spain	–	–	–	–	12.2–223	on-line SPE-HPLC-QTRAP	López-Serna et al. (2013)
TAM	Spain	–	nd–17.2	–	–	–	on-line SPE-HPLC-QTRAP	Negreira et al. (2013)
TAM	Spain	15–170	15–58	11–42	12–38	–	SPE(Oasis) + UPLC(HSS T3)-Qtrap	Ferrando-Climent et al. (2014)
TAM	Spain	45–970	–	–	–	–	SPE(Oasis) + UPLC(HSS T3)-Qtrap	Ferrando-Climent et al. (2015)
TAM	Spain	26–133	30–58	–	–	–	SPE(Oasis) + UPLC(HSS T3)-QTRAP	Ferrando-Climent et al. (2013)

OH-TAM	Spain	D	–	–	–	–		Ferrando-Climent et al. (2013)
Endoxifen	Spain	D	–	–	–	–		Ferrando-Climent et al. (2013)
TAM	Spain	–	110–147	nd180.6	–	–	on-line SPE-HPLC-QTRAP	Negreira et al. (2014)
OH-TAM	Spain	–	–	nd–5.8	–	–		Negreira et al. (2014)
OH-D-TAM	Spain	–	–	91.6	–	–		Negreira et al. (2014)
TAM	Spain	nd–7.4	6.7–15	nd	–	–	on-line SPE-HPLC-QTRAP	Isidori et al. (2016)
OH-Tam	Spain	blq	blq–7.7	blq	–	–		Isidori et al. (2016)
OH-D-Tam	Spain	nd–11	bld–75	14	–	–		Isidori et al. (2016)
TAM	Slovenia	blq–10	11–61	7.1	–	–	on-line SPE-HPLC-QTRAP	Isidori et al. (2016)
OH-Tam	Slovenia	blq–10	blq–35	nd	–	–		Isidori et al. (2016)
OH-D-Tam	Slovenia	nd	nd–66	nd	–	–		Isidori et al. (2016)
TAM	Japan	4–44	nd–38	nd–62	nd–533	–	SPE(Oasis) + UPLC (BEH C18)-MS/MS	Azuma et al. (2016)
OH-Tam	Japan	nd–7	nd–4	nd–25	nd	–		Azuma et al. (2016)

OH-Tam 4 Hydroxy tamoxifen, *OH-D-Tam* 4-hydroxy-N-desmethyltamoxifen, *Endoxifen* 4,4-dihydroxy desmethyltamoxifen, *D* Detected but not quantified, *nd* Non-detected have been performed all over the world, where Spain with nine published papers is the country providing most of the data available on TAM occurrence in the environment, followed by the UK with two studies. In other countries, such as Sweden, France, Germany, the Netherlands, Norway, Slovenia, China, and Japan, only one study was published. TAM has been measured both in hospital and urban raw wastewaters in seven and nine studies, respectively. Values found in both types of wastewater effluents were similar, ranging from low ng/L up to a maximum of 970 ng/L in hospital wastewater (Ferrando-Climent et al. 2015) and 147 ng/L in a WWTP influent (Negreira et al. 2014), both in Spain. In some of the studies (Table 4.1), special attention was paid to the efficiency of a conventional WWTP in terms of removal of this compound (Azuma et al. 2016; Ferrando-Climent et al. 2015; Isidori et al. 2016; Negreira et al. 2014). Based on the data collected, TAM can be classified as a recalcitrant compound because it is poorly removed in WWTPs where its concentrations remain almost unaltered. Moreover, eight studies report the occurrence of TAM in natural environment; six consider surface water, and only two provided information for groundwater. Values of TAM in these natural waters were quite similar and even higher than those found in treated wastewater: up to 533 ng/L of TAM was detected in surface waters in Japan (Azuma et al. 2016) and up to 223 ng/L in groundwaters in Spain (López-Serna et al. 2013). This can be attributed to other sources of pollution besides WWTP, such as farms and veterinary hospitals discharge since TAM is used for hormonal treatment in such facilities (Papich 2016).

The occurrence and fate of TAM in the whole water cycle (i.e., in hospital wastewater, influent and effluent wastewater of an urban WWTP, and surface waters receiving its effluents) have been evaluated in two occasions—in Spain (Ferrando-Climent et al. 2014) and Japan (Azuma et al. 2016). Their findings suggest the importance of upgrading or complementing the current treatment methodologies in WWTPs in the future in order to reduce the environmental risk that contaminants such as TAM can pose.

Not only prescribed pharmaceuticals but also their metabolites can access the sewage system and, after incomplete removal in WWTP, the natural environment too. In addition, both parent compounds and metabolites can be further transformed by abiotic and biotic processes in the natural environment. TAM metabolites have thus attracted the interest of researchers lately although only few studies have reported its presence in the environment. The study of Ferrando-Climent et al. was the first one reporting the detection of the metabolite 4-hydroxy-tamoxifen (OH-TAM) and 4,4-dihydroxy desmethyltamoxifen (endoxifen) in wastewater (Ferrando-Climent et al. 2013).

Negreira et al. (2014) measured OH-TAM and also 4-hydroxy-N-desmethyltamoxifen (OH-D-TAM) in WWTP effluents at a concentration slightly lower than TAM itself. Isidori et al. (2016), also detected both metabolites in hospital and urban wastewaters in Spain and Slovenia, though OH-D-Tam was detected at the highest concentration (up to 75 ng/L in WWPT influents in Spain) (Isidori et al. 2016). OH-Tam was also detected in wastewaters in Japan although in this case levels in treated wastewater were as high as in the raw wastewater (Azuma et al. 2016).

In all the analytical methods (Table 4.1), TAM and TPs were analyzed together with other contaminants of emerging concern such as other cancer drugs. All of the multi-residue methodologies performed the pre-concentration of the target compounds by means of solid phase extraction (SPE), where OASIS cartridges were mostly used, followed by STRATA X and PLRP-s. The latest SPE cartridges mentioned here were used in four methodologies, where SPE was coupled on-line to LC-MS instruments (Table 4.1). TAM and metabolites were analyzed in all cases by liquid chromatography coupled to a mass spectrometry system.

4.3 Conventional and Non-Conventional Technologies: Transformation Products

Most of the urban wastewater treatment plants operating nowadays were designed to remove organic matter (that may cause oxygen depletion in the receiving water bodies), as well as to reduce the content on nutrients such as nitrogen and phosphorous (that can lead to eutrophication processes). As previously discussed, after the intake and metabolizing of the pharmaceuticals, regularly they are flushed by the toilet and get into the urban sewage, where they are transported to the WWTP. In some cases, they are successfully degraded in the conditions of the treatment plant or undergo sorption processes into different fractions, such as the sludge, that contribute to their overall removal from water (Blair et al. 2015; Gottschall et al. 2012; Musson and Townsend 2009). The scarce literature available to date points out that anticancer drugs have in most cases very low or even no biodegradability in the conventional activated sludge technology (Kosjek and Heath 2011; Kümmerer and Al-Ahmad 1997; Kümmerer et al. 1997; Lenz et al. 2007). Since they are not efficiently removed from the wastewater, these substances can reach the environment, which acts as a final sink of urban sewage effluents.

Based on the limitations of conventional wastewater treatments to remove anticancer drugs and other recalcitrant micropollutants, different technologies and strategies for their removal are currently being studied. Tertiary treatments (placed after the regular biological processes employed in WWTPs) can contribute substantially to decontamination of the waste effluent when the biological treatment alone is unable to remove efficiently micropollutants load (Metcalf and Eddy Inc 2003). In fact, the objective of tertiary treatments is usually improving the quality of the effluent before being discharged into the aquatic environment, particularly when it is released into a highly sensitive or fragile ecosystem (estuaries, low-flow rivers, coral reefs, etc.). There are many types and combinations of tertiary treatment processes which can be used in a WWTP. Among them, disinfection is one of the most widely used and is usually placed at the end of the process, with the goal of substantially decrease the microbiological load in the wastewater *prior* to its discharge to the environment.

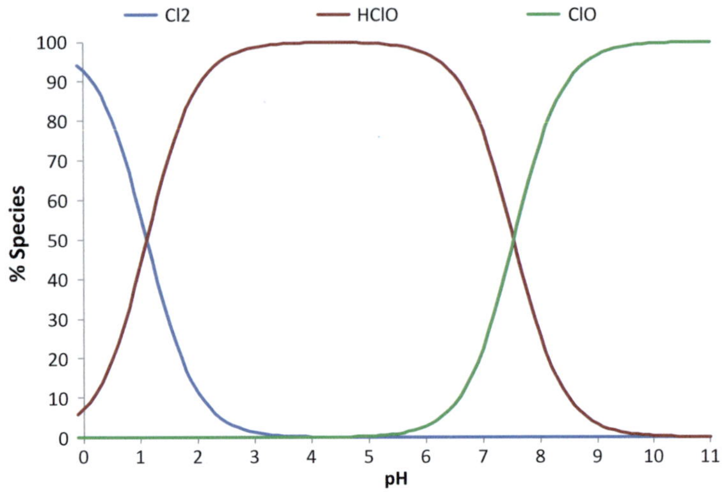

Fig. 4.1 Relative distribution of the main aqueous chlorine species as a function of pH

Among the most extensive disinfection methods existing nowadays, those employing chlorination are the most frequent. Chlorination of the final effluent has been largely used for inactivating or destroying pathogens right before the effluent is disposed of in the recipient water body. In water, the speciation of chlorine is pH dependent (Fig. 4.1). The most relevant species at circumneutral pH is the hypochlorous acid (HClO), which is responsible for most of the reactions that take place with organic matter. The hypochlorite anion in water establishes equilibrium between different active chlorine species, mainly chlorine (Cl_2), hypochlorous acid (HClO), and hypochlorite ClO^-. The amount of each species is strongly dependent on physico-chemical parameters of the solution such as temperature and ionic strength.

Hypochlorous acid exhibits high selectivity towards organic micropollutants, and its reactivity is usually restricted to limited sites (reducing, nucleophilic, and unsaturated sites) (Deborde and von Gunten 2008). The reactivity of TAM and their relevant metabolites, 4-hydroxy-tamoxifen and 4-hydroxy-N-desmethyl-tamoxifen, in free chlorine-containing water was explored by Negreira et al. (2015). The authors found that TAM was just poorly transformed under the applied chlorination conditions. However, the two main metabolites, 4-hydroxy-tamoxifen and 4-hydroxy-N-desmethyl-tamoxifen, were rapidly degraded, each one yielding seven chlorinated byproducts. The authors concluded that the byproducts found can be also formed under the chlorination conditions regularly employed for disinfection purposes as tertiary treatment. An additional QSAR assessment of the potential aquatic toxicity indicated that the toxicity of the by-products increased in respect to the parent compounds (TAM and its major active metabolites, 4-hydroxy-tamoxifen and 4-hydroxy-N-desmethyl-tamoxifen).

Non-conventional biological treatments such as those based on the use of white rot fungi (WRF) have also been evaluated for the degradation of pharmaceutical compounds with promising results. WRF such as *Trametes versicolor* have shown an excellent biodegradation capacity towards recalcitrant or relatively inert compounds, making them good candidates to be assessed in future bioremediation schemes. The success of these type of microorganisms is connected to their unspecific oxidative enzymatic system, which includes lignin-modifying enzymes, especially laccases and peroxidases, but also to their intracellular enzymatic complexes (e.g., cytochrome P450) (Asgher et al. 2008). Among the different set of chemical reactions that might be involved in the biotransformation of pharmaceutical compounds by these microorganisms, especially relevant are hydroxylation, formylation, deamination, and dehalogenation (Harms et al. 2011). The removal of TAM from hospital wastewater using the WRF *T. versicolor* has been previously studied and discussed by Ferrando-Climent et al. (2015). The authors found out that TAM was completely removed from wastewater through a combined sorption-biodegradation process. The authors also identified the formation of two compounds, belonging to two hydroxylated positional isomers. TAM and their TPs showed no toxicity towards bacteria *V. fischeri* at the concentrations employed in the authors' experimental set.

In addition to the aforementioned techniques, other water treatment approaches based on advanced oxidation processes have been extensively evaluated for the removal of organic pollutants in order to achieve better water quality. These processes refer to a set of chemical treatment procedures designed to remove organic (and occasionally inorganic) substances from water and wastewater by oxidation through reactions with hydroxyl radicals ($^{\bullet}OH$). AOPs have demonstrated their effectiveness not only for disinfection purposes but also for the removal of a large number of compounds, becoming useful tertiary treatment in wastewater treatment plants (Klavarioti et al. 2009). Hydroxyl radicals are very reactive and short-life oxidants species (10 μs at a 10^{-4} M concentration), which act in a non-selective way (von Gunten 2003a, b). The formation of $^{\bullet}OH$ radicals is a very complex process, which can take place according to a variety of different reaction mechanisms. However, these radicals need to be generated on site in order to drive oxidation/reduction reactions with the organic molecules present in wastewater. Hydroxyl radicals might be generated by using different treatments such as ultraviolet radiation/hydrogen peroxide (UV/H_2O_2), ozone/hydrogen peroxide (O_3/H_2O_2), ultraviolet radiation/ozone (UV/O_3), Fenton's process ($Fe(II)/H_2O_2$), titanium dioxide/ultraviolet radiation (TiO_2/UV), etc. In the case of ozone-based processes, the disintegration of ozone in water produces $^{\bullet}OH$ radicals, with higher oxidation potential than ozone itself.

The mixture of the different radical species presented above and their intrinsic reactivity towards organic matter leads normally to a set of transformation routes that end up in a cocktail of oxidized TPs. So far, scarce are the studies that have evaluated the potential removal of anticancer drugs by AOPs (Chen et al. 2008; DellaGreca et al. 2007; Zhang et al. 2013).

Ferrando-Climent et al. (2017) have reported the degradation of TAM from an environmentally relevant concentration (100 μg L^{-1}) under different AOP schemes including (i) O_3, (ii) O_3/UV, (iii) O_3/H_2O_2, (iv) UV, and (v) UV/H_2O_2. The authors identified three reaction pathways that lead to eight TPs from the exposure of TAM to different AOPs. Despite the fast degradation of TAM noticed, a general increase in the toxicity of the effluent in the time-course experiments was observed. As in the case of Negreira et al. (2015), when exploring the transformation of TAM under chlorination conditions, Ferrando-Climent et al. (2017) pointed out the same paramount conclusion in their TAM-AOP study: despite the pollutant is successfully removed, some other compounds with higher toxicity are generated along the treatment. Such a statement should be extrapolated to future studies of micropollutants degradation in terms of a thorough and comprehensive appraisal of the environmental risk.

Although the processes aforementioned—AOP and the non-conventional biological treatment—also have other purposes besides the removal of micropollutants, both of them may nowadays be implemented in the existing WWTPs (Fig. 4.2). The most frequent configuration, which is already in use in several WWTPs, involves placing the AOPs after the biological process as a tertiary treatment (Fig. 4.2). A second potential configuration would involve substituting the conventional biological treatment, based in activated sludge, by a non-conventional biological process (MBR, fungi, algae, and/or consortium of microorganisms). In the third option of configuration, the activated sludge process of the regular scheme of the WWTP would be upgraded using sequential changes of microbial community composition installing sequential bioreactors. In this approach, another bioreactor (based on non-conventional biological process) would be installed right after the secondary clarifier (Fig. 4.2; Sarkar et al. 2016). Other possibilities might involve placing the AOP units right after the primary clarifier and before the biological treatment (Fig. 4.2). All the configurations described could improve the biodegradability of micropollutants (Mark M. Benjamin and Lawler 2013).

4.4 Environmental Risk Assessment

As it has been described in this chapter, TAM may be released into the environment through various waste streams, usually through sewage system. Different types of sources should be considered: (i) indirect sources such as the down-the-drain releases from patients using the pharmaceutical for cancer therapy (domestic, ambulatory, and hospital facilities), (ii) the incorrect disposal of unused drugs into household wastewater, (iii) veterinary facilities, and (iv) direct sources such as the released during manufacture or formulation. However, there is no information available regarding concrete discharges of TAM from manufacturing or formulation as well as there is no sufficient data to estimate the potential output of TAM from research facilities (basic research applications). It is expected, based on

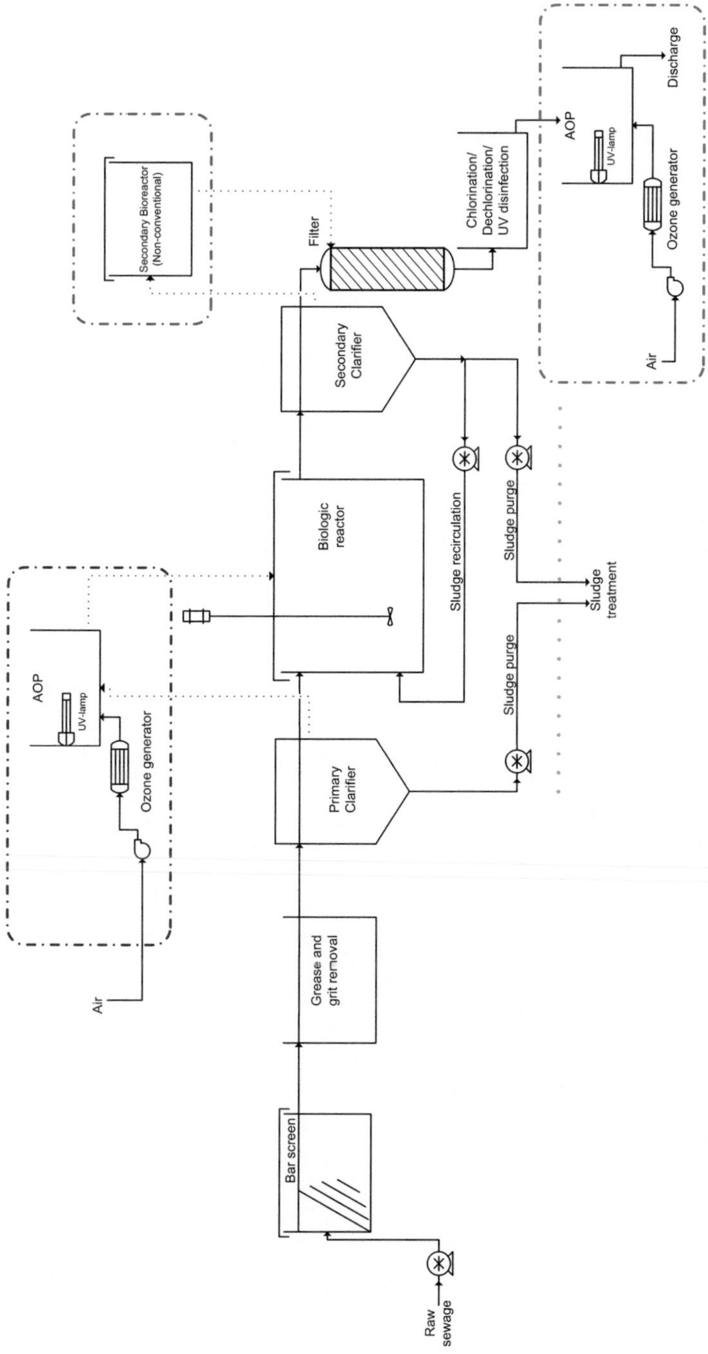

Fig. 4.2 Layout of a conventional WWTP with the potential upgrades: (i) AOP placed in conventional configuration—green line, (ii) AOP placed before biological treatment—blue line, and (iii) secondary bioreactor with non-conventional biological treatment—red line

concentrations typically used for research purposes, that this is not a considerable source of TAM in water (Environment and Climate Change Agency of Canada, 2015).

Generally, to estimate the environmental risk of a pharmaceutical drug in it is only the aquatic compartment which is an object of the attention, since human medicines may be excreted partly or wholly unchanged by patients, and subsequently enter into the aquatic system (sewage and later surface waters). In the case of TAM, the particularly high value for the partition coefficient (logP between 6 and 6.5) leads to the thinking that the pharmaceutical compound can be mostly found in the solid and/or particulate phase. However, the partition of TAM regardless or not of the presence of solids in the water (wastewater or natural waters) has never been studied.

Eventually, the environmental risk of pharmaceuticals is assessed by the parameter risk quotient (RQ). The RQ is calculated as the ratio between the predicted environmental concentration (PEC) of the substance in the aquatic environment and the predicted no effect concentration (PNEC), i.e., the concentration, based on the available test results, predicted to cause no effect on the organisms living there. Usually, a "worst-case" risk quotient is presented and as such our PEC information is based on the European country with the highest per capita use (Astra Zeneca datasheet). Three categories according to the RQ are draft out:

- $PEC/PNEC \leq 0.1$: Use of the substance has been considered to result in insignificant environmental risk.
- $0.1 < PEC/PNEC \leq 1$: Use of the substance has been considered to result in low environmental risk.
- $1 < PEC/PNEC \leq 10$: Use of the substance has been considered to result in moderate environmental risk.
- $PEC/PNEC > 10$: Use of the substance has been considered to result in high environmental risk.

The RQ for TAM based on PEC values resulted in an insignificant environmental risk for several species (Astra Zeneca; Environmental risk assessment data; Tamoxifen 2012). However, when the RQ was assessed by other authors using the measured environmental concentration (MEC) instead of the PEC values, low or moderate environmental risk was found for some trophic levels in different countries (Ferrando-Climent et al. 2014; Orias et al. 2015).

Ferrando-Climent et al. (2014) found that TAM poses a potential hazard to the aquatic environment since it cumulates several risk factors for aquatic ecosystems: (i) RQ > 1 for some trophic levels, (ii) proved toxicity, (iii) suspected endocrine disruption effects, and (iv) high bioaccumulation potential. TAM showed RQ values: (i) below 1 for species belonging to different trophic levels such *Acartia tonsa* or the Rainbow trout gonad, (ii) from 1 to 100 for the *Pimephales promelas* fish, and (iii) from 10 to 100 for *S. capricornutum* microalgae (Ferrando-Climent et al. 2014). On the other hand, Orias et al. studied the RQ for TAM in surface waters of several countries (Spain, China, United Kingdom, and France) using the measured environmental concentration already reported by several authors. RQ values higher than 1 were found in river samples from the UK and China (Orias et al. 2015). The data available concerning the RQ of TAM remains, however, limited due to the scarcity of data regarding environmental concentrations compared to other pharmaceutical residues, such as diclofenac, ibuprofen, etc.

Additionally, it should be highlighted that the calculation of the PEC does not take into account indirect ecotoxicity linked to bioaccumulation (Orias et al. 2015), and TAM has been found to be highly bioaccumulative and poorly biodegradable (Jean et al. 2012). When it comes to the environmental risk of the TPs or metabolites of TAM, there is no information available since very few studies have reported their environmental concentrations as well as no ecotoxicological assessment has been performed yet. However, some studies point out that the TPs from TAM seems to be more toxic than the TAM itself (Besse et al. 2012).

4.5 Conclusions

TAM can pose a potential risk to the aquatic environment. The transformation of this substance after intake and excretion as well as its transformation into different products should be taken into account to have accurate information about the occurrence of this substance in the environment.

Improved water treatment schemes are required to achieve better quality standards in the effluents of wastewater treatment plants. A holistic approach, far beyond the assessment of the removal of the target compound, is needed. Special care should be taken when choosing a tertiary treatment since the formation of TPs could lead to an overall increase of the toxicity of the effluent. Extensive chemical characterization of the effluent along with ecotoxicity evaluation is desirable and recommended.

References

Asgher M, Bhatti HN, Ashraf M, Legge RL (2008) Recent developments in biodegradation of industrial pollutants by white rot fungi and their enzyme system. Biodegradation 19:771–783. https://doi.org/10.1007/s10532-008-9185-3

Ashton D, Hilton M, Thomas KV (2004) Investigating the environmental transport of human pharmaceuticals to streams in the United Kingdom. Sci Total Environ 333:167–184. https://doi.org/10.1016/j.scitotenv.2004.04.062

Astra Zeneca (2012) Environmental risk assessment data; Tamoxifen

Azuma T et al (2016) Detection of pharmaceuticals and phytochemicals together with their metabolites in hospital effluents in Japan, and their contribution to sewage treatment plant influents. Sci Total Environ 548–549:189–197. https://doi.org/10.1016/j.scitotenv.2015.12.157

Besse JP, Latour JF, Garric J (2012) Anticancer drugs in surface waters. What can we say about the occurrence and environmental significance of cytotoxic, cytostatic and endocrine therapy drugs? Environ Int 39:73–86. https://doi.org/10.1016/j.envint.2011.10.002

Blair B, Nikolaus A, Hedman C, Klaper R, Grundl T (2015) Evaluating the degradation, sorption, and negative mass balances of pharmaceuticals and personal care products during wastewater treatment. Chemosphere 134:395–401. https://doi.org/10.1016/j.chemosphere.2015.04.078

Chen Z, Park G, Herckes P, Westerhoff P (2008) Physicochemical treatment of three chemotherapy drugs: irinotecan, tamoxifen, and cyclophosphamide. J Adv Oxid Technol 11:254–260

Coetsier CM, Spinelli S, Lin L, Roig B, Touraud E (2009) Discharge of pharmaceutical products (PPs) through a conventional biological sewage treatment plant: MECs vs PECs? Environ Int 35:787–792. https://doi.org/10.1016/j.envint.2009.01.008

Deborde M, von Gunten U (2008) Reactions of chlorine with inorganic and organic compounds during water treatment—kinetics and mechanisms: a critical review. Water Res 42:13–51. https://doi.org/10.1016/j.watres.2007.07.025

DellaGreca M, Iesce MR, Isidori M, Nardelli A, Previtera L, Rubino M (2007) Phototransformation products of tamoxifen by sunlight in water. Toxicity of the drug and its derivatives on aquatic organisms. Chemosphere 67:1933–1939. https://doi.org/10.1016/j.chemosphere.2006.12.001

Ebele AJ, Abou-Elwafa Abdallah M, Harrad S (2017) Pharmaceuticals and personal care products (PPCPs) in the freshwater aquatic environment. Emer Contamin 3:1–16. https://doi.org/10.1016/j.emcon.2016.12.004

Ferrando-Climent L, Cruz-Morató C, Marco-Urrea E, Vicent T, Sarrà M, Rodriguez-Mozaz S, Barceló D (2015) Non conventional biological treatment based on Trametes versicolor for the elimination of recalcitrant anticancer drugs in hospital wastewater. Chemosphere 136:9–19. https://doi.org/10.1016/j.chemosphere.2015.03.051

Ferrando-Climent L, Gonzalez-Olmos R, Anfruns A, Aymerich I, Corominas L, Barceló D, Rodriguez-Mozaz S (2017) Elimination study of the chemotherapy drug tamoxifen by different advanced oxidation processes: transformation products and toxicity assessment. Chemosphere 168:284–292. https://doi.org/10.1016/j.chemosphere.2016.10.057

Ferrando-Climent L, Rodriguez-Mozaz S, Barceló D (2013) Development of a UPLC-MS/MS method for the determination of ten anticancer drugs in hospital and urban wastewaters, and its application for the screening of human metabolites assisted by information-dependent acquisition tool (IDA) in sewage samples. Anal Bioanal Chem 405:5937–5952. https://doi.org/10.1007/s00216-013-6794-4

Ferrando-Climent L, Rodriguez-Mozaz S, Barceló D (2014) Incidence of anticancer drugs in an aquatic urban system: from hospital effluents through urban wastewater to natural environment. Environ Pollut 193:216–223. https://doi.org/10.1016/j.envpol.2014.07.002

Franquet-Griell H, Cornadó D, Caixach J, Ventura F, Lacorte S (2017) Determination of cytostatic drugs in Besòs River (NE Spain) and comparison with predicted environmental concentrations. Environ Sci Pollut Res 24:6492–6503. https://doi.org/10.1007/s11356-016-8337-y

Gottschall N et al (2012) Pharmaceutical and personal care products in groundwater, subsurface drainage, soil, and wheat grain, following a high single application of municipal biosolids to a field. Chemosphere 87:194–203. https://doi.org/10.1016/j.chemosphere.2011.12.018

Grabic R, Fick J, Lindberg RH, Fedorova G, Tysklind M (2012) Multi-residue method for trace level determination of pharmaceuticals in environmental samples using liquid chromatography coupled to triple quadrupole mass spectrometry. Talanta 100:183–195. https://doi.org/10.1016/j.talanta.2012.08.032

Harms H, Schlosser D, Wick LY (2011) Untapped potential: exploiting fungi in bioremediation of hazardous chemicals. Nat Rev Microbiol 9:177–192. https://doi.org/10.1038/nrmicro2519

Iglesias A, Nebot C, Vázquez BI, Coronel-Olivares C, Abuín CMF, Cepeda A (2014) Monitoring the presence of 13 active compounds in surface water collected from rural areas in Northwestern Spain. Int J Environ Res Public Health 11:5251–5272. https://doi.org/10.3390/ijerph110505251

Isidori M et al (2016) Chemical and toxicological characterisation of anticancer drugs in hospital and municipal wastewaters from Slovenia and Spain. Environ Pollut 219:275–287. https://doi.org/10.1016/j.envpol.2016.10.039

Jean J et al (2012) Identification and prioritization of bioaccumulable pharmaceutical substances discharged in hospital effluents. J Environ Manag 103:113–121. https://doi.org/10.1016/j.jenvman.2012.03.005

Johnson AC, Jürgens MD, Williams RJ, Kümmerer K, Kortenkamp A, Sumpter JP (2008) Do cytotoxic chemotherapy drugs discharged into rivers pose a risk to the environment and human health? An overview and UK case study. J Hydrol 348:167–175. https://doi.org/10.1016/j.jhydrol.2007.09.054

Klavarioti M, Mantzavinos D, Kassinos D (2009) Removal of residual pharmaceuticals from aqueous systems by advanced oxidation processes. Environ Int 35:402–417. https://doi.org/10.1016/j.envint.2008.07.009

Kosjek T, Heath E (2011) Occurrence, fate and determination of cytostatic pharmaceuticals in the environment. TrAC Trends Anal Chem 30:1065–1087

Kümmerer K, Al-Ahmad A (1997) Biodegradability of the anti-tumour agents 5-fluorouracil, cytarabine, and gemcitabine: impact of the chemical structure and synergistic toxicity with hospital effluent. Acta Hydrochim Hydrobiol 25:166–172. https://doi.org/10.1002/aheh.19970250402

Kümmerer K, Steger-Hartmann T, Meyer M (1997) Biodegradability of the anti-tumour agent ifosfamide and its occurrence in hospital effluents and communal sewage. Water Res 31:2705–2710

Lenz K et al (2007) Monitoring, removal and risk assessment of cytostatic drugs in hospital wastewater. Water Sci Technol 56:141–149

Liu X, Zhang J, Yin J, Duan H, Wu Y, Shao B (2010) Analysis of hormone antagonists in clinical and municipal wastewater by isotopic dilution liquid chromatography tandem mass spectrometry. Anal Bioanal Chem 396:2977–2985. https://doi.org/10.1007/s00216-010-3531-0

López-Serna R, Jurado A, Vázquez-Suñé E, Carrera J, Petrović M, Barceló D (2013) Occurrence of 95 pharmaceuticals and transformation products in urban groundwaters underlying the metropolis of Barcelona, Spain. Environ Pollut 174:305–315. https://doi.org/10.1016/j.envpol.2012.11.022

Mark M. Benjamin, Lawler DF (2013) Water quality engineering: physical/chemical treatment processes.

Mater N, Geret F, Castillo L, Faucet-Marquis V, Albasi C, Pfohl-Leszkowicz A (2014) In vitro tests aiding ecological risk assessment of ciprofloxacin, tamoxifen and cyclophosphamide in range of concentrations released in hospital wastewater and surface water. Environ Int 63:191–200. https://doi.org/10.1016/j.envint.2013.11.011

Metcalf & Eddy Inc (2003) Wastewater engineering: treatment and reuse. Fourth edition/revised by George Tchobanoglous, Franklin L. Burton, H. David Stensel. Boston: McGraw-Hill, [2003] ©2003

Mikelman S, Mardirossian N, Gnegy ME (2017) Tamoxifen and amphetamine abuse: Are there therapeutic possibilities? J Chem Neuroanat 83:50–58. https://doi.org/10.1016/j.jchemneu.2016.08.004

Musson SE, Townsend TG (2009) Pharmaceutical compound content of municipal solid waste. J Hazard Mater 162:730–735. https://doi.org/10.1016/j.jhazmat.2008.05.089

Negreira N, de Alda ML, Barceló D (2014) Cytostatic drugs and metabolites in municipal and hospital wastewaters in Spain: filtration, occurrence, and environmental risk. Sci Total Environ 497-498:68–77. https://doi.org/10.1016/j.scitotenv.2014.07.101

Negreira N, López de Alda M, Barceló D (2013) On-line solid phase extraction-liquid chromatography-tandem mass spectrometry for the determination of 17 cytostatics and metabolites in waste, surface and ground water samples. J Chromatogr A 1280:64–74. https://doi.org/10.1016/j.chroma.2013.01.031

Negreira N, Regueiro J, López de Alda M, Barceló D (2015) Transformation of tamoxifen and its major metabolites during water chlorination: identification and in silico toxicity assessment of their disinfection byproducts. Water Res 85:199–207. https://doi.org/10.1016/j.watres.2015.08.036

Orias F et al (2015) Tamoxifen ecotoxicity and resulting risks for aquatic ecosystems. Chemosphere 128:79–84. https://doi.org/10.1016/j.chemosphere.2015.01.002

Papich MG (2016) Tamoxifen citrate. In: Saunders handbook of veterinary drugs, 4th edn. W.B. Saunders, St. Louis, pp 758–759. https://doi.org/10.1016/B978-0-323-24485-5.00535-0

Pereira AMPT, Silva LJG, Laranjeiro CSM, Meisel LM, Lino CM, Pena A (2017) Human pharmaceuticals in Portuguese rivers: the impact of water scarcity in the environmental risk. Sci Total Environ 609:1182–1191. https://doi.org/10.1016/j.scitotenv.2017.07.200

Petrie B, Barden R, Kasprzyk-Hordern B (2015) A review on emerging contaminants in wastewaters and the environment: current knowledge, understudied areas and recommendations for future monitoring. Water Res 72:3–27. https://doi.org/10.1016/j.watres.2014.08.053

Reh R, Licha T, Geyer T, Nödler K, Sauter M (2013) Occurrence and spatial distribution of organic micro-pollutants in a complex hydrogeological karst system during low flow and high flow periods, results of a two-year study. Sci Total Environ 443:438–445. https://doi.org/10.1016/j.scitotenv.2012.11.005

Roberts PH, Thomas KV (2006) The occurrence of selected pharmaceuticals in wastewater effluent and surface waters of the lower Tyne catchment. Sci Total Environ 356:143–153. https://doi.org/10.1016/j.scitotenv.2005.04.031

Sangion A, Gramatica P (2016) Hazard of pharmaceuticals for aquatic environment: Prioritization by structural approaches and prediction of ecotoxicity. Environ Int 95:131–143. https://doi.org/10.1016/j.envint.2016.08.008

Sarkar S, Tribedi P, Ghoah P, Saha T, Sil AK (2016) Sequential changes of microbial community composition during biological wastewater treatment in single unit waste stabilization system. Waste Biomass Valoriz 2016:1–11. https://doi.org/10.1007/s12649-015-9471-3

Sui Q, Cao X, Lu S, Zhao W, Qiu Z, Yu G (2015) Occurrence, sources and fate of pharmaceuticals and personal care products in the groundwater: a review. Emer Contamin 1:14–24. https://doi.org/10.1016/j.emcon.2015.07.001

Vasquez MI, Lambrianides A, Schneider M, Kümmerer K, Fatta-Kassinos D (2014) Environmental side effects of pharmaceutical cocktails: what we know and what we should know. J Hazard Mater 279:169–189. https://doi.org/10.1016/j.jhazmat.2014.06.069

von Gunten U (2003a) Ozonation of drinking water: part I. Oxidation kinetics and product formation. Water Res 37:1443–1467

von Gunten U (2003b) Ozonation of drinking water: part II. Disinfection and by-product formation in presence of bromide, iodide or chlorine. Water Res 37:1469–1487

Withrow S, Vail D, Page R (2012) Withrow and MacEwen's small animal clinical oncology, 5th edn. 768

Zhang J, Chang VWC, Giannis A, Wang J-Y (2013) Removal of cytostatic drugs from aquatic environment: a review. Sci Total Environ 445-446:281–298. https://doi.org/10.1016/j.scitotenv.2012.12.061

Web References

http://www.ec.gc.ca/ese-ees/default.asp?Lang=En&n=2D71C125-1, consulted 15/06/2018

Chapter 5
Chlorination By-Products of Anticancer Drugs

Juan Carlos Guillen, Božo Žonja, and Miren López de Alda

Abstract Cytostatic drugs and their metabolites enter the aquatic environment mainly through the excretion of urine and faeces from chemotherapy patients into the public sewer system and can eventually reach tap water if they are not properly eliminated during waste and drinking water treatment processes.

Chlorine is globally the most used chemical disinfectant in wastewater treatment plants as well as in the pretreatment of hospital effluents prior to their discharge into the public sewage system, mostly because of its low cost. Since aqueous chlorine is not capable of complete mineralization of many anthropogenic compounds, numerous disinfection by-products may be formed due to oxidation/substitution reactions. Such reactions can happen during wastewater treatment processes and due to the discharge of chlorinated waters (e.g., tap water) or bleach into the sewage system. Very little is also known about their potential transformation into other chemical species, which might be even more toxic than the parent drugs.

In recent years, several studies have demonstrated that anticancer drugs can yield a series of by-products when they come in contact with aqueous chlorine. Understanding the chemical fate of these by-products is an important first step to understand their environmental significance. Therefore, the main aim of this chapter is to comprehensively review the existing literature on the reactivity of anticancer drugs in the presence of free chlorine and the formation of their by-products.

Keywords Cytostatic residues · disinfection by products · transformation products · chemotherapy agents · water treatment

J. C. Guillen · B. Žonja · M. López de Alda (✉)
Water, Environmental and Food Chemistry Unit, Department of Environmental Chemistry, Institute of Environmental Assessment and Water Research (IDAEA-CSIC), Barcelona, Spain
e-mail: miren.lopezdealda@idaea.csic.es

E. Heath et al. (eds.), *Fate and Effects of Anticancer Drugs in the Environment*,
https://doi.org/10.1007/978-3-030-21048-9_5

5.1 Introduction

Cytostatic drugs are pharmaceuticals which have been widely used for chemotherapy. Due to their mutagenic, cytotoxic, and teratogenic properties, more and more concerns have been expressed regarding their occurrence in the environment (Ferk et al. 2009, Zounkova et al. 2010). They have been often detected in the aquatic environment at sub-ng L^{-1} levels, which is low in terms of an immediate threat, but chronic exposure can have delayed toxic effects due to their interference (direct or indirect) with the DNA (Elersek et al. 2016, Isidori et al. 2016, Kosjek and Heath 2011). As all pharmaceuticals, cytostatic drugs and their metabolites are excreted from the human body *via* urine and faeces. Hospital effluents are rarely pretreated prior to their discharge into the public sewer system and are thereby considered potential hotspots (Zhang et al. 2013). From there, and through the sewage system, they end up in wastewater treatment plants (WWTPs) and if not properly removed ultimately in the environment.

Cytostatic pharmaceuticals can be transformed by abiotic processes in engineered systems like (waste) water treatment processes. These processes include, but are not limited to, ozonation, degradation with UV/H_2O_2, or chlorination/chloramination. On the other hand, in the natural environment, they can be transformed either by hydrolysis or they can be degraded as the result of sunlight-induced photodegradation. However, the degradation and transformation of pharmaceuticals do not necessarily mean that they will lose their pharmacological activity (Zhu et al. 2015).

In the WWTPs and in drinking water facilities and hospital effluents, it is common to use **chlorination** as a disinfection process which uses either chlorine, chloramines, or chlorine dioxide as oxidant reagents. This is generally considered a water disinfection engineered process, but the high usage of chlorine-based chemicals such as sodium hypochlorite (liquid bleach; NaOCl) in hospitals can also lead to the formation of chlorination by-products in hospital effluents as they do not have a specifically designed treatment. Once in wastewater, chlorine disinfectants can react with the present organic matter and, thus, form organochlorine compounds. One of the compounds group of concern are halogenated organic compounds absorbable on activated carbon (AOX), which are persistent in the environment and were shown to exhibit toxic effects towards aquatic organisms (Emmanuel et al. 2004). Chlorine is the most widely used chemical oxidant for the disinfection of drinking water mainly due to its relatively low cost. In drinking water treatment, chlorination is performed in order to limit the growth of heterotrophic organisms in the distribution systems. It is used for both initial pre-treatment, to start initial disinfection, and post-treatment (Deborde and von Gunten 2008).

For water treatment, both gaseous chlorine and hypochlorite are typically used. When dissolved in water, both gaseous chlorine and hypochlorite form a weak hypochlorous acid. This acid can partially dissociate to form hypochlorite ions. However, the presence of one or the other species (hypochlorous acid *vs.*

hypochlorite ion) is pH dependent (Lopez et al. 2001). Both species can react with organic micropollutants, but hypochlorous acid is the dominant reactive species during chlorination (Deborde and von Gunten 2008). As reported by Deborde and von Gunten (2008), reactions of hypochlorous acid with organic compounds occur *via* three transformation pathways: (i) oxidation reactions; (ii) addition reactions to the unsaturated bond; and (iii) electrophilic substitution reactions. Having these pathways in mind, they concluded that the hypochlorous acid has high selectivity towards organic compounds. However, aqueous chlorine is a mild oxidant and it cannot completely mineralize organic contaminants. Consequently, various by-products can be formed as a result of oxidation, addition, or substitution reactions. The by-products formed in the disinfection process can sometimes be even more toxic than their parent compounds (Bedner and MacCrehan 2006), or can form highly toxic *so-called* disinfection by-products (DBPs) (Richardson 2005). Such example is the formation of the most genotoxic and cytotoxic DBPs known (iodo-DBPs) during treatment with chlorine (or monochloramine) from the, generally non-toxic, iodinated X-ray contrast media (ICM) (Duirk et al. 2011, Plewa et al. 2004, Postigo and Richardson 2014, Richardson et al. 2008). Many DBPs are formed as a result of chlorine interaction with natural/dissolved organic matter (NOM). They can also be easily formed from micropollutants like pharmaceuticals, since many of them have activated aromatic rings that can react with oxidants like chlorine (Postigo and Richardson 2014). Some of the recent examples include antibiotics like sulfamethoxazole (Dodd and Huang 2004, Huang et al. 2008), the anti-inflammatory drug diclofenac (Quintana et al. 2010, Soufan et al. 2012), the lipid regulator gemfibrozil (Bulloch et al. 2012, Krkošek et al. 2011), antacids like cimetidine (Buth et al. 2007) or antineoplastics like etoposide (Negreira et al. 2015a), erlotinib (Negreira et al. 2015b), tamoxifen (Negreira et al. 2015c) or vinca alkaloids (Negreira et al. 2016).

As alternatives to chlorination, chloramine and chlorine dioxide are also used for wastewater disinfection. **Chloramination** is an alternative to chlorination because it rather inhibits the formation of trihalometanes (THMs) and haloacetic acids (HAAs) (Krasner et al. 2013). However, it has been reported that it is directly linked with the formation of carcinogenic nitrosamines like N-nitrosodimethylamine (NDMA). Much like the case of chlorination, these species are formed from natural organic matter precursors or micropullutants. Several studies have shown that pharmaceuticals that contain dimethylamine groups can produce NDMA under chloramination (Postigo and Richardson 2014, Richardson et al. 2007, Richardson and Ternes 2011). Some comprehensive studies aiming at identifying precursors which contain the dimethylamine group and can form NDMA have reported that the antacid ranitidine (out of many) has a strong potential to form NDMA *via* nucleophilic substitution (Le Roux et al. 2011, Shen and Andrews 2011). Finally, **chlorine dioxide** tends to form fewer halogenated DBPs. As a result, when pharmaceuticals are exposed to chlorine dioxide, the majority of by-products which are formed are the result of oxidation and not halogenation (Postigo and Richardson 2014).

5.2 Chlorination of Cytostatic Drugs

5.2.1 *Etoposide*

Etoposide is a semisynthetic derivative of podophyllotoxin that exhibits antitumor activity. It is used in combination with other chemotherapeutic agents in the treatment of refractory testicular tumours, as the first-line treatment in patients with small cell lung cancer, and to treat other malignancies such as lymphoma, non-lymphocytic leukaemia, and glioblastoma multiforme (DrugBank-Etoposide).

Chlorination of etoposide was reported and described in detail by Negreira et al. (2015a). They found that etoposide reacted very quickly in water containing free chlorine. This led to the formation of two oxidation by-products in a few seconds. They were identified by ultra-high performance liquid chromatography (UHPLC) coupled with high-resolution hybrid quadrupole-Orbitrap tandem mass spectrometry (Fig. 5.1). Additionally, it was seen that reaction rates were pH-dependent and depended on the concentration of free chlorine in the water sample and the type of matrix. At slightly acidic pH (6), the reaction was much more favourable than at neutral (7) to basic (8) pH. Etoposide (empirical formula $C_{29}H_{32}O_{13}$) transforms primarily to DBP1 ($C_{28}H_{30}O_{13}$) as the main by-product in reactions conditions when free chlorine is not in excess. However, in the presence of excess chlorine, DBP1 is quickly degraded to DBP2 ($C_{26}H_{26}O_{13}$), whose concentrations remain constant throughout.

The main by-products of etoposide are shown in Fig. 5.1, together with their corresponding fragmentation patterns. MS/MS fragmentation of the first by-product (called DBP1) yielded fragments similar to that observed for the parent compound—etoposide. In more details, both etoposide and its DBP1 lose the glycoside ring together with the formation of a double bond in the central molecule. This is followed by the cleavage of the lactone ring with the corresponding loss of the carboxyl group. Finally, the methyl group that remained attached to the molecule after the lactone ring break-up is lost. By studying the fragmentation pattern, the authors saw that neither the DBP1 nor the DBP2 contained chlorine in their structure. The first by-product was formed as a result of demethylation in one of the anisol groups of the etoposide leading to the corresponding phenol. The second by-product (DBP2) is suggested to originate from double demethylation plus the lactone ring opening, generating the corresponding carboxylic acid (Negreira et al. 2015a). In this case, the authors identified the first by-product at the highest level of identification confidence (Schymanski et al. 2014) since it was possible to purchase the original standard of this by-product. This by-product had been identified previously as a human metabolite. This is not uncommon for environmental by-products (Osorio et al. 2014, Zonja et al. 2014, Zonja et al. 2016). The DBP1, 3′-O-demethyletoposide, confirmed with a commercial standard, was identified as a metabolite of etoposide in human plasma and urine by Cai et al. (1999).

The analysis of etoposide and its by-product DBP1 in various real water samples by on-line SPE-LC-MS/MS confirmed the presence of the by-product in one raw

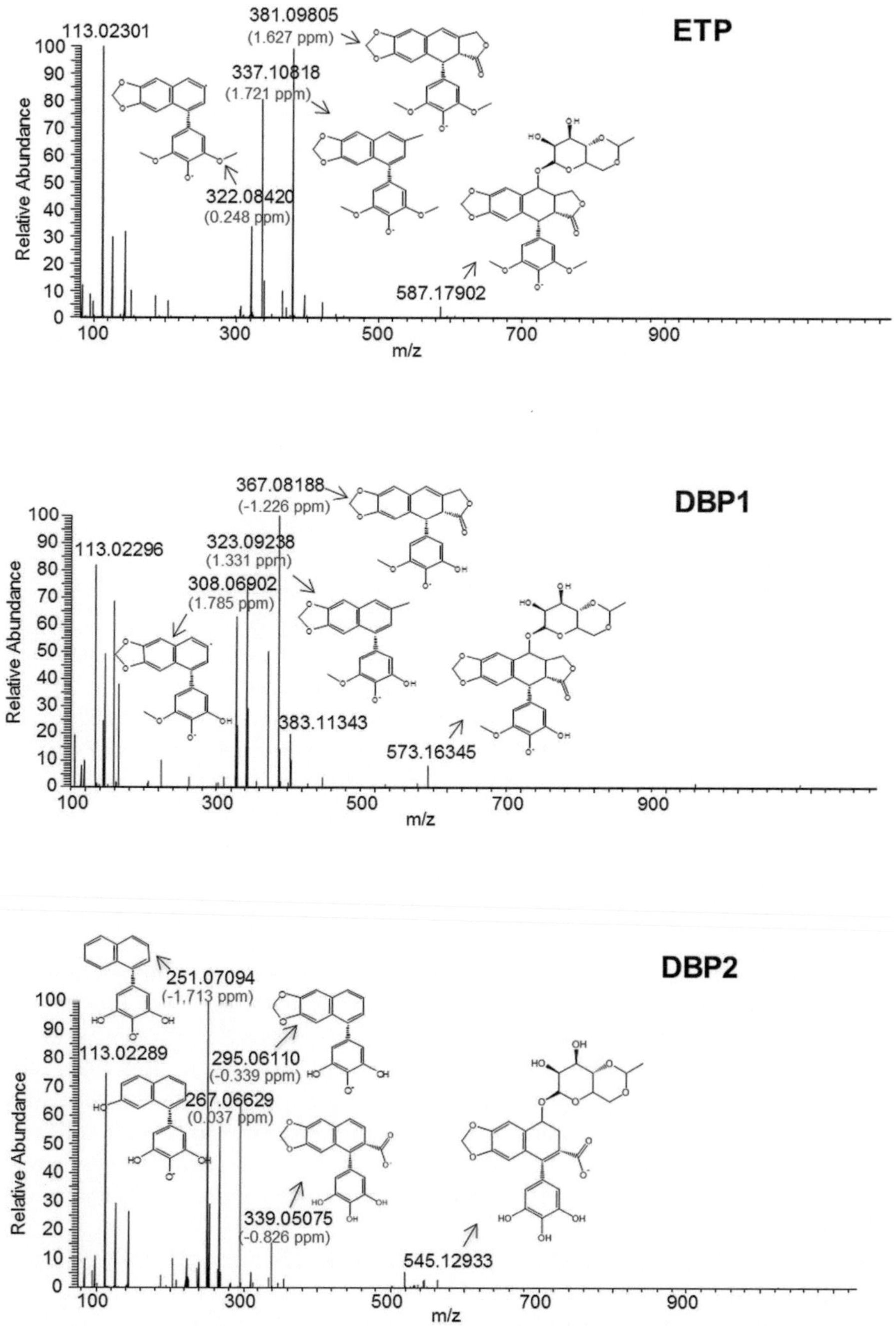

Fig. 5.1 Accurate MS/MS spectra and fragmentation pattern of etoposide (ETP) and its by-products DBP1 and DBP2. (Reproduced from Negreira et al. (2015a) with permission of Elsevier. The name abbreviations of the by-products were changed from original BP to DBP in order to have the same nomenclature in the chapter text)

wastewater sample at 33 ng L^{-1} and in several river waters at 14–31 ng L^{-1}, whereas etoposide was not detected in any sample. Although this by-product of chlorination is a known minor human metabolite of etoposide, it was not possible to undoubtedly demonstrate the process involved in its formation (human metabolism *vs.* chlorination) (Negreira et al. 2015a).

5.2.2 *Erlotinib*

Erlotinib hydrochloride is a drug used to treat non-small cell lung cancer, pancreatic cancer, and several other types of cancer (DrugBank-Erlotinib). Chlorination of erlotinib was also investigated and described in detail by Negreira et al. (2015b). In initial experiments, when the reaction was performed in ultrapure water, erlotinib was almost completely degraded after 1 h in the presence of free chlorine. But in experiments using real wastewater, it showed a much lower degradation yield, most likely, due to the competition between erlotinib and the organic matter of the sample for chlorine (Negreira et al. 2015b).

The authors used a commonly used screening approach which allowed the detection of 19 by-products. MS/MS fragmentation was the starting point used in the identification of the by-products of erlotinib by comparing the fragmentation pattern of the parent compound and the by-products detected.

Among the detected by-products, six compounds, which are shown in Figure 5.2 (DBP-428A, DBP-428B, DBP-428C, DBP-444, DBP-462A, and DBP-462B), corresponded to chlorinated derivatives of erlotinib. In all cases, the presence of chlorine atoms was confirmed by the existence of the characteristic chlorine isotopic pattern (Negreira et al. 2015b).

As can be seen from Fig. 5.2, DBP-428A (m/z 428.1371), DBP-428B (m/z 428.1370), and DBP-428C (m/z 428.1368) corresponded to three isomers of mono-chlorinated erlotinib. Their MS/MS spectra showed identical fragmentation patterns for DBP-428A and DBP-428B, suggesting a closer structural relationship between these two isomers, which was also supported by the closer elution of both compounds. The authors likewise reported that the exact position of chlorine in the ring could not be determined on the basis of their MS/MS spectra. But when they considered the presence of the ortho-/para- director groups amino and ethynyl, they concluded that the electrophilic aromatic substitution was likely to occur in ortho- to the amino and to the ethynyl moieties (position 2), in ortho- to the ethynyl and para- to the amino groups (position 4), and/or in ortho- to the amino and para- to the ethylnyl groups (position 6).

Other two by-products, DBP-462A (m/z 462.0979) and DBP-462B (m/z 462.0977), matched the molecular formula of di-chlorinated derivatives of erlotinib, but their intensities in the sample were low and the authors reported they were not able to perform their MS/MS fragmentation. However, they did confirm their presence by studying the full MS scan data, which showed the presence of two chlorine atoms.

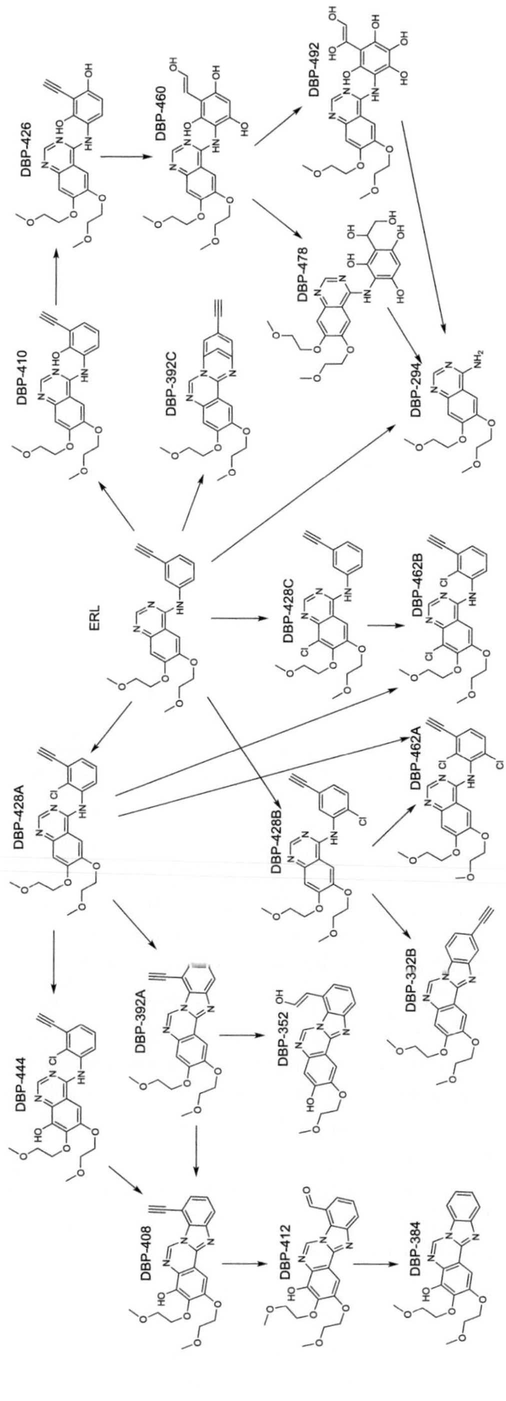

Fig. 5.2 Proposed transformation pathway for erlotinib in chlorine-containing water samples. (Reproduced from Negreira et al. (2015b) with permission of Elsevier. The name abbreviations of the by-products were changed from original TP to DBP in order to have the same nomenclature in the chapter text)

As a general rule, it is important to note that erlotinib was transformed into new products mainly *via* chlorination and hydroxylation reactions (Fig. 5.2).

In order to determine the kinetics of these reactions, time-course profiles of formation of these by-products were monitored for over 8 h in chlorinated wastewater samples. These experiments showed that these by-products can be formed under typical wastewater disinfection conditions, and in this context the authors stressed out the importance of the results generated since information on the transformation pathway of erlotinib provides an overview of its fate in the aquatic environment, and can therefore be used for further monitoring studies.

5.2.3 *Tamoxifen*

Tamoxifen is a selective estrogen receptor modulator (SERM) with tissue-specific activity for the treatment and prevention of estrogen receptor positive breast cancer. It is primarily indicated for the treatment of metastatic breast cancer in women and men and ductal carcinoma in situ (DrugBank-Tamoxifen).

Chlorination of tamoxifen and its major active human metabolites, 4-hydroxy-tamoxifen and 4-hydroxy-N-desmethyl-tamoxifen, was also studied and described in detail by Negreira et al. (2015c). Under chlorination conditions, the authors found that tamoxifen was stable, whereas both studied human metabolites degraded rapidly and were transformed into several chlorinated by-products.

In concordance with the lack of degradation observed for TAM, no DBPs could be detected for the parent drug after the chlorination experiments, but several potential DBPs were detected for its major metabolites, namely OH-TAM and OH-D-TAM.

In total, seven OH-TAM by-products were identified (Figure 5.3).

The majority of the identified by-products had a characteristic chlorine isotopic pattern. Out of them, three were mono-chlorinated derivatives, whereas the rest four DBPs were di-chlorinated by-products. The identification of by-products was rather simple due to the MS/MS similarities between the parent compound and the by-products. However, it was not possible to determine the exact position of the chlorine(s) on the molecule beyond any doubt. All by-products resulted as derivatives of the parent compound either *via* hydroxylation or chlorination or both. However, the authors presented plausible evidence that the substitution reactions occurred in the phenolic ring. In addition to the substitution reactions which occurred on the phenolic ring, N-demethylation was only observed for the DBP-440. This suggested that the phenanthrene ring formed in DBP-454 might be involved in this reaction.

As expected, OH-D-TAM had similar chlorination by-products as OH-TAM. In total, seven DBPs were tentatively identified (Fig. 5.4). The majority of the by-products identified had a characteristic chlorine isotopic pattern, where two were mono-chlorinated derivatives and the remaining five DBPs were di-chlorinated by-products. The identification of the by-products was rather simple,

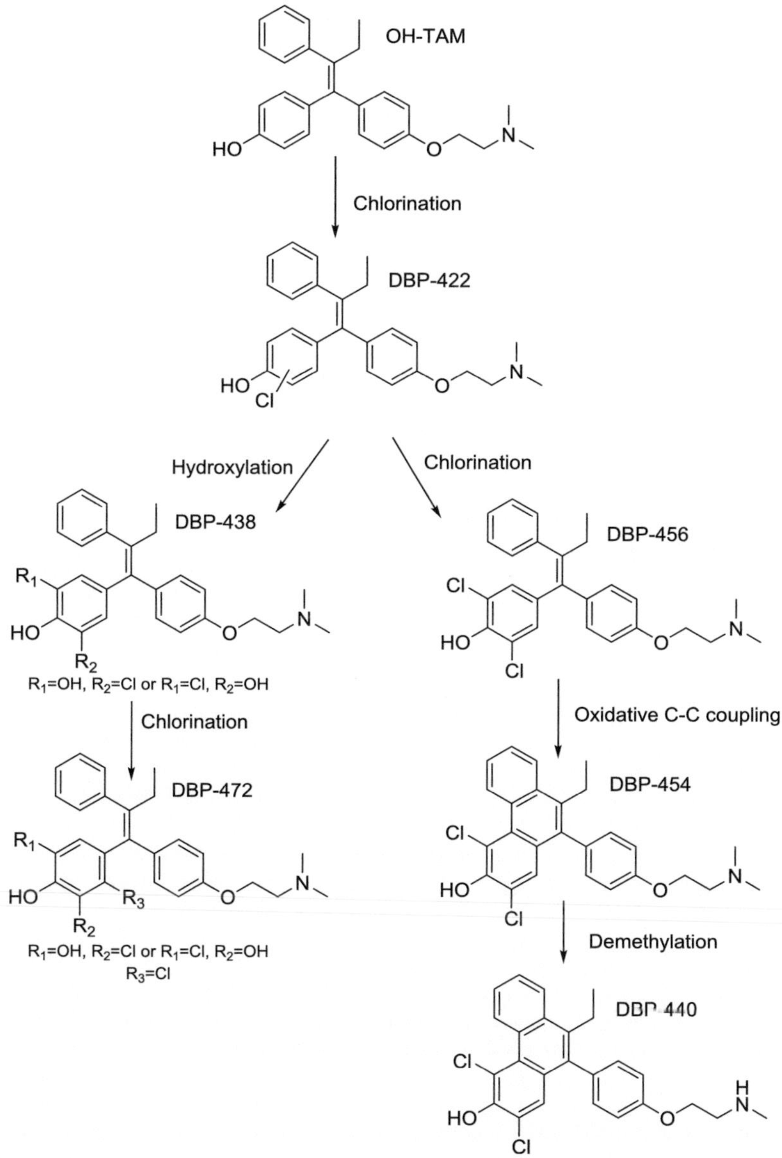

Fig. 5.3 Proposed transformation pathway for OH-TAM in chlorine-containing water samples. (Reproduced from Negreira et al. (2015c) with permission from Elsevier)

but, again, it was not possible to determine the exact position of the chlorine(s) on the molecule beyond any doubt. All by-products resulted as derivatives of the parent compound *via* either hydroxylation or chlorination or both. However, it was not possible to always confirm the structure of the compounds since in some cases it was

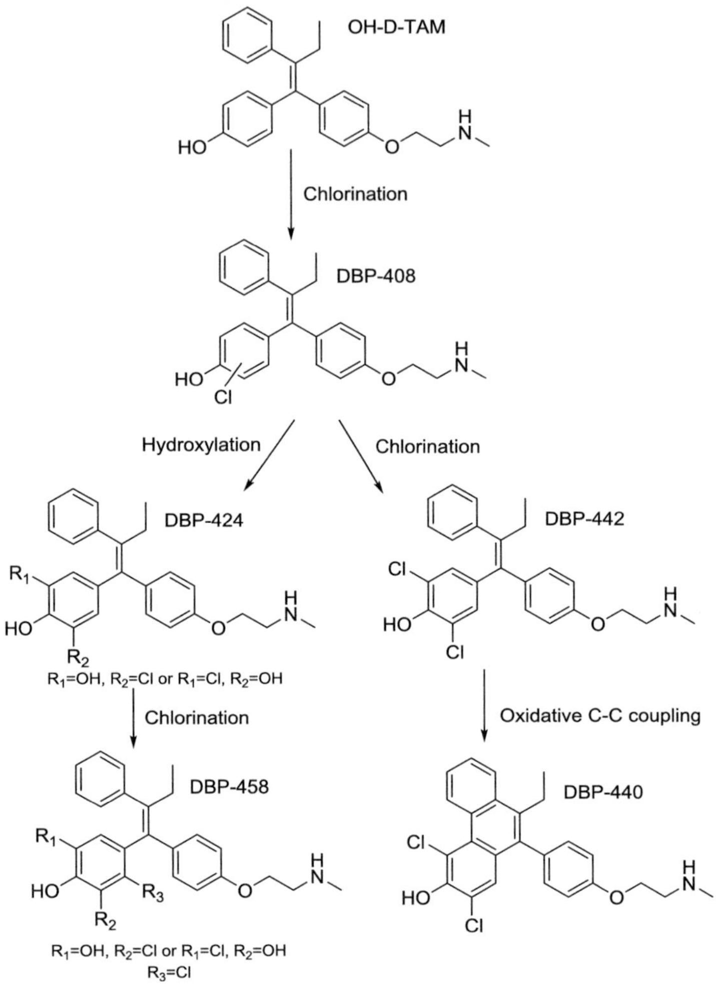

Fig. 5.4 Proposed transformation pathway for OH-D-TAM in chlorine-containing water samples. (Reproduced from Negreira et al. (2015c) with permission of Elsevier)

not possible to acquire MS/MS spectra due to the low intensity of the protonated ion $[M+H]^+$ of the parent by-product.

The lack of reactivity of the parent drug (TAM) is structurally related when compared to the reactivity of its human metabolites due to the presence of the hydroxyl groups in the latter compounds. The presence of the hydroxyl group in the position 4 of the phenyl ring of the metabolites acts as a strongly activating group towards electrophilic aromatic substitution by donating electrons to the ring (Negreira et al. 2015c).

In addition to the identification of the structure of the by-products, Negreira et al. performed QSAR assessment of their potential aquatic toxicity (2015c). They

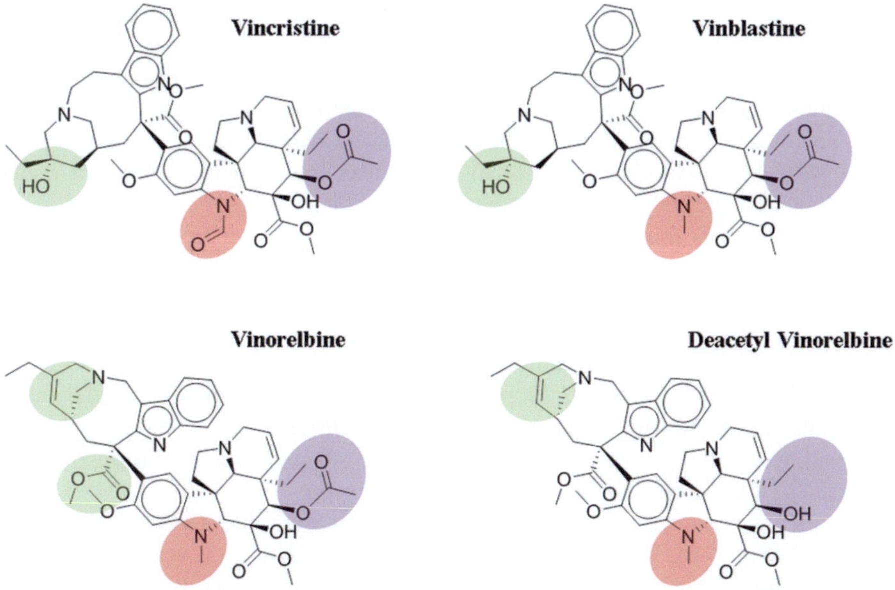

Fig. 5.5 Structures of the four vinca alkaloids that were used for chlorination experiments in Negreira et al. (2016)

discovered that by-products toxicity significantly increased in comparison with the parent compound. Therefore, laboratory-based toxicity studies followed by the assessment of the risk of these by-products in aquatic environments are a prudent way to go.

5.2.4 Vinca Alkaloids

Vinca alkaloids are a group of indole-indoline dimers which are alkaloids obtained from plants of the VINCA genus. They are the second-most-used class of anticancer drugs (Moudi et al. 2013). These compounds are administered alone or in combination with other drugs to treat non-small cell lung cancer, breast cancer, bladder cancer, lymphomas, and leukemia, among others (DrugBank-Vincas).

The chlorination of the vinca alkaloids vincristine, vinblastine, vinorelbine, and its major active human metabolite, deacetyl vinorelbine, was investigated and described in detail by Negreira et al. (2016).

Vinca alkaloids are comprised of two distinct parts, namely, the catharanthine moiety and the vindoline moiety (Fig. 5.5). The only difference between vinblastine (VBL) and vincristine (VCN) is that the former has a methyl group on the indole nitrogen of the vindoline skeleton, while VCN has a formyl group instead. On the other hand, vinorelbine (VRB) differs from VBL in the catharanthine subunit by the dehydration of the piperidine ring and the shortening of the bridge between C-9′ and N-6′ by one carbon. The human metabolite deacetyl vinorelbine (dea-VRB) is the result of the 4-O-deacetylation of VRB in the vindoline part of the molecule.

In total, 65 DBPs were tentatively identified in chlorinated ultrapure water for the four parent compounds. Many of these DBPs were chlorinated derivatives (20 of them were mono-chlorinated, eight di-chlorinated, and two tri-chlorinated compounds) and, therefore, present a potential cause of concern.

Vincristine was found to be degraded in water containing free chlorine much slower than the other vinca alkaloids studied. In addition, several by-products were identified, but none of them was formed by incorporation of chlorine atoms in their structures. This suggested that the degradation of VCN in the presence of free chlorine proceeds only *via* oxidation reactions (Negreira et al. 2016).

In the case of *vinblastine*, 22 potential DBPs were identified with the most abundant ones being products of hydrolysis. In addition, some of the by-products detected were chlorinated compounds. These compounds had the lowest signal intensities (if compared to all 22 detected by-products).

As for *vinorelbine*, a total of 17 potential DBPs were detected and tentatively identified (Fig. 5.6). The most abundant by-product, DBP-383, was the result of demethylation of the parent compound. Other formation reactions included hydroxylation (DBP-392 (two isomers)). Out of the 17 DBPs, 11 were identified as chlorinated compounds, but without revealing the position of the chlorine atom. Out of these chlorinated derivatives, DBP-400 was the most abundant, formed as a result of the replacement of one hydrogen by one chlorine atom followed by demethylation. The three isomers of DBP-400 corresponded to the loss of the methyl group in different positions of the molecule or different positions of the chlorine atoms.

Finally, in the case of *deacetyl vinorelbine*, the authors were able to detect and tentatively identify 20 by-products. Among them, up to 13 were found to contain chlorine in their structure. The most abundant by-product, DBP-379, was formed as a result of chlorination in ortho- position to the methoxy group of the parent dea-VRB, followed by additional demethylation (Negreira et al. 2016). In a similar way to VRB, oxidation in ortho- to the nitrogen atom on the piperidine ring on the catharanthine moiety and chlorination in a double bond led to its corresponding by-product. All in all, similar by-products were detected when compared the chlorination of VRB and dea-VRB, with the difference of one acetyl group which VRB losses as part of its human metabolism.

Due to the high number of mono-chlorinated, di-chlorinated, and tri-chlorinated compounds detected in the laboratory experiments, additional studies are necessary. It would be necessary to assess the occurrence of these novel DBPs in the aquatic environment and to evaluate their potential (cyto)toxic effects (Negreira et al. 2016).

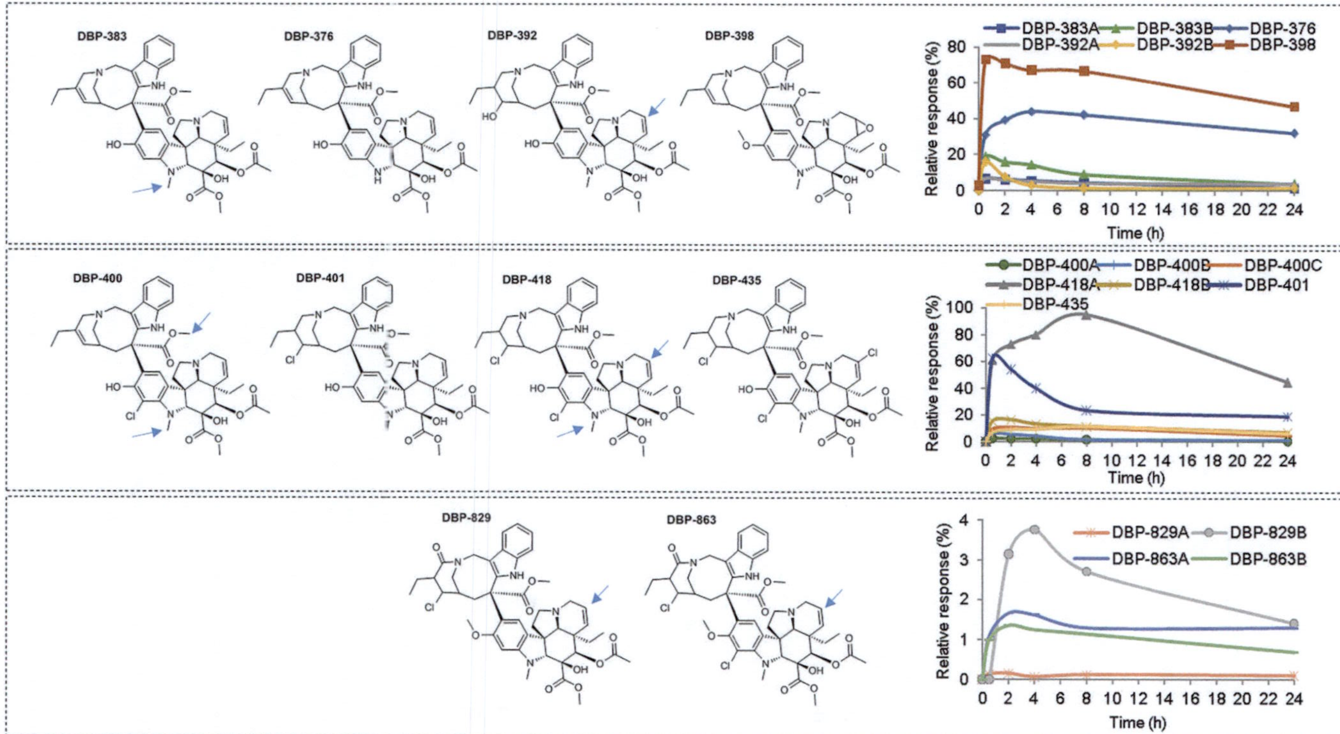

Fig. 5.6 Proposed structures and time-course profiles of VRB by-products. Initial concentrations of the parent compound and free chlorine were 100 μg L^{-1} and 1 mg L^{-1}, respectively. Arrows indicate potential alternative positions for isomers. (Reproduced from Negreira et al. (2016) with permission of Elsevier)

5.3 Conclusions

Disinfection processes like chlorination applied to waste and drinking waters can lead to the formation of numerous halogenated and non-halogenated by-products of the contaminants present in the aquatic environment. Both kinds of by-products, but especially the former, can represent a risk for the environment and human health. This issue has been studied in the case of several contaminants of emerging concern including pharmaceuticals. However, the particular group of pharmaceuticals used as chemotherapy agents in the treatment of cancer has been very seldom investigated, despite their recognised toxicity. The studies performed so far with etoposide, erlotinib, tamoxifen, and vinca alkaloids have shown the formation of numerous DBPs, some of which have been found to be present in real waste and surface waters (e.g., the etoposide DBP 3′-O-desmethyl etoposide). In addition, the application of QSAR models for the risk assessment of the by-products, in some cases, e.g., in the case of etoposide and tamoxifen, has shown greater toxicity for the by-products, if compared to the parent compounds. Therefore, the performance of additional studies with other anticancer drugs, first at the laboratory level, and then investigating the presence of the identified by-products in the real world together with their potential toxicity, appears necessary for appropriate risk assessment and protection of the human and environmental health against this class of contaminants. To this end, the information gathered in this type of studies as regards molecular formula and accurate masses, if added to databases, should be very useful in future screening methods.

References

Bedner M, MacCrehan WA (2006) Transformation of acetaminophen by chlorination produces the toxicants 1,4-benzoquinone and N-acetyl-p-benzoquinone imine. Environ Sci Technol 40 (2):516–522

Bulloch DN, Lavado R, Forsgren KL, Beni S, Schlenk D, Larive CK (2012) Analytical and biological characterization of halogenated gemfibrozil produced through chlorination of wastewater. Environ Sci Technol 46(10):5583–5589

Buth JM, Arnold WA, McNeill K (2007) Unexpected products and reaction mechanisms of the aqueous chlorination of cimetidine. Environ Sci Technol 41(17):6228–6233

Cai X, Woo MH, Edick MJ, Relling MV (1999) Simultaneous quantitation of etoposide and its catechol metabolite in human plasma using high-performance liquid chromatography with electrochemical detection. J Chromatogr B Biomed Sci Appl 728(2):241–250

Deborde M, von Gunten U (2008) Reactions of chlorine with inorganic and organic compounds during water treatment-Kinetics and mechanisms: a critical review. Water Res 42(1-2):13–51

Dodd MC, Huang CH (2004) Transformation of the antibacterial agent sulfamethoxazole in reactions with chlorine: kinetics, mechanisms, and pathways. Environ Sci Technol 38 (21):5607–5615

DrugBank-Erlotinib DrugBank, https://www.drugbank.ca/drugs/DB00530. Accessed 4 July 2018)

DrugBank-Etoposide DrugBank, https://www.drugbank.ca/drugs/DB00773. Accessed 4 July 2018

DrugBank-Tamoxifen DrugBank, https://www.drugbank.ca/drugs/DB00675. Accessed 4 July 2018

DrugBank-Vincas DrugBank, https://www.drugbank.ca/categories/DBCAT000714. Accessed 4 July 2018

Duirk SE, Lindell C, Cornelison CC, Kormos J, Ternes TA, Attene-Ramos M, Osiol J, Wagner ED, Plewa MJ, Richardson SD (2011) Formation of toxic iodinated disinfection by-products from compounds used in medical imaging. Environ Sci Technol 45(16):6845–6854

Elersek T, Milavec S, Korošec M, Brezovsek P, Negreira N, Zonja B, de Alda ML, Barceló D, Heath E, Ščančar J, Filipič M (2016) Toxicity of the mixture of selected antineoplastic drugs against aquatic primary producers. Environ Sci Pollut Res 23(15):14780–14790

Emmanuel E, Keck G, Blanchard J-M, Vermande P, Perrodin Y (2004) Toxicological effects of disinfections using sodium hypochlorite on aquatic organisms and its contribution to AOX formation in hospital wastewater. Environ Int 30(7):891–900

Ferk F, Mišík M, Grummt T, Majer B, Fuerhacker M, Buchmann C, Vital M, Uhl M, Lenz K, Grillitsch B, Parzefall W, Nersesyan A, Knasmüller S (2009) Genotoxic effects of wastewater from an oncological ward. Mutat Res Genet Toxicol Environ Mutagen 672(2):69–75

Huang CH Dodd MC, Shah AD (2008) Reaction and transformation of antibacterial agents with aqueous chlorine under relevant water treatment conditions. Fate of pharmaceuticals in the environment and in water treatment systems, 261–289.

Isidori M, Lavorgna M, Russo C, Kundi M, Žegura B, Novak M, Filipič M, Mišík M, Knasmueller S, de Alda ML, Barceló D, Žonja B, Česen M, Ščančar J, Kosjek T, Heath E (2016) Chemical and toxicological characterisation of anticancer drugs in hospital and municipal wastewaters from Slovenia and Spain. Environ Pollut 219:275–287

Kosjek T, Heath E (2011) Occurrence, fate and determination of cytostatic pharmaceuticals in the environment. TrAC Trends Anal Chem 30(7):1065–1087

Krasner SW, Mitch WA, McCurry DL, Hanigan D, Westerhoff P (2013) Formation, precursors, control, and occurrence of nitrosamines in drinking water: A review. Water Res 47 (13):4433–4450

Krkošek WH, Koziar SA, White RL, Gagnon GA (2011) Identification of reaction products from reactions of free chlorine with the lipid-regulator gemfibrozil. Water Res 45(3):1414–1422

Le Roux J, Gallard H, Croué JP (2011) Chloramination of nitrogenous contaminants (pharmaceuticals and pesticides): NDMA and halogenated DBPs formation. Water Res 45(10):3164–3174

Lopez A, Mascolo G, Ciannarella R, Tiravanti G (2001) Formation of volatile halogenated by-products during chlorination of isoproturon aqueous solutions. Chemosphere 45(3):269–274

Moudi M, Go R, Yien CYS, Nazre M (2013) Vinca alkaloids. Int J Prev Med 4(11):1231

Negreira N, López de Alda M, Barceló D (2015a) Degradation of the cytostatic etoposide in chlorinated water by liquid chromatography coupled to quadrupole-Orbitrap mass spectrometry: identification and quantification of by-products in real water samples. Sci Total Environ 506–507:36–45

Negreira N, Regueiro J, López de Alda M, Barceló D (2015b) Degradation of the anticancer drug erlotinib during water chlorination: non-targeted approach for the identification of transformation products. Water Res 85:103–113

Negreira N, Regueiro J, López de Alda M, Barceló D (2015c) Transformation of tamoxifen and its major metabolites during water chlorination: identification and in silico toxicity assessment of their disinfection byproducts. Water Res 85:199–207

Negreira N, Regueiro J, López de Alda M, Barceló D (2016) Reactivity of vinca alkaloids during water chlorination processes: identification of their disinfection by-products by high-resolution quadrupole-Orbitrap mass spectrometry. Sci Total Environ 544:635–644

Osorio V, Imbert-Bouchard M, Zonja B, Abad JL, Perez S, Barcelo D (2014) Simultaneous determination of diclofenac, its human metabolites and microbial nitration/nitrosation transformation products in wastewaters by liquid chromatography/quadrupole-linear ion trap mass spectrometry. J Chromatogr A 1347:63–71

Plewa MJ, Wagner ED, Richardson SD, Thruston AD Jr, Woo YT, McKague AB (2004) Chemical and biological characterization of newly discovered iodoacid drinking water disinfection byproducts. Environ Sci Technol 38(18):4713–4722

Postigo C, Richardson SD (2014) Transformation of pharmaceuticals during oxidation/disinfection processes in drinking water treatment. J Hazard Mater 279:461–475

Quintana JB, Rodil R, López-Mahía P, Muniategui-Lorenzo S, Prada-Rodríguez D (2010) Investigating the chlorination of acidic pharmaceuticals and by-product formation aided by an experimental design methodology. Water Res 44(1):243–255

Richardson SD (2005) New disinfection by-product issues: emerging DBPs and alternative routes of exposure. Global Nest 7(1):43–60

Richardson SD, Fasano F, Ellington JJ, Crumley FG, Buettner KM, Evans JJ, Blount BC, Silva LK, Waite TJ, Luther GW, McKague AB, Miltner RJ, Wagner ED, Plewa MJ (2008) Occurrence and mammalian cell toxicity of iodinated disinfection byproducts in drinking water. Environ Sci Technol 42(22):8330–8338

Richardson SD, Plewa MJ, Wagner ED, Schoeny R, DeMarini DM (2007) Occurrence, genotoxicity, and carcinogenicity of regulated and emerging disinfection by-products in drinking water: a review and roadmap for research. Mutat Res Rev Mutat Res 636(1-3):178–242

Richardson SD, Ternes TA (2011) Water analysis: emerging contaminants and current issues. Anal Chem 83(12):4616–4648

Schymanski EL, Jeon J, Gulde R, Fenner K, Ruff M, Singer HP, Hollender J (2014) Identifying small molecules via high resolution mass spectrometry: communicating confidence. Environ Sci Technol 48(4):2097–2098

Shen R, Andrews SA (2011) Demonstration of 20 pharmaceuticals and personal care products (PPCPs) as nitrosamine precursors during chloramine disinfection. Water Res 45(2):944–952

Soufan M, Deborde M, Legube B (2012) Aqueous chlorination of diclofenac: kinetic study and transformation products identification. Water Res 46(10):3377–3386

Zhang J, Chang VWC, Giannis A, Wang J-Y (2013) Removal of cytostatic drugs from aquatic environment: a review. Sci Total Environ 445-446:281–298

Zhu B, Zonja B, Gonzalez O, Sans C, Pérez S, Barceló D, Esplugas S, Xu K, Qiang Z (2015) Degradation kinetics and pathways of three calcium channel blockers under UV irradiation. Water Res 86:9–16

Zonja B, Aceña J, Jelic A, Petrovic M, Pérez Solsona S, Barceló D (2014) Transformation products of emerging contaminants in the environment: analysis, processes, occurrence, effects and risks. Nollet, D.A.L.a.L.M.L. (ed), pp. 303–324, John Wiley and Sons Ltd.

Zonja B, Delgado A, Abad JL, Pérez S, Barceló D (2016) Abiotic amidine and guanidine hydrolysis of lamotrigine-N2-glucuronide and related compounds in wastewater: the role of pH and N2-substitution on reaction kinetics. Water Res 100:466–475

Zounkova R, Kovalova L, Blaha L, Dott W (2010) Ecotoxicity and genotoxicity assessment of cytotoxic antineoplastic drugs and their metabolites. Chemosphere 81(2):253–260

Chapter 6
Cytostatic Drug Residues in Wastewater Treatment Plants: Sources, Removal Efficiencies and Current Challenges

Lida Ioannou-Ttofa and Despo Fatta-Kassinos

Abstract The discharge of pharmaceutical compounds present in wastewater into the aquatic environment has been a source of discussion and concern in scientific and regulatory communities for the last decades. However, cytostatic drugs used in chemotherapy have received less attention than other pharmaceuticals, despite the fact that they may possess cytotoxic, genotoxic, mutagenic, carcinogenic, endocrine-disrupting, and/or teratogenic effects toward several organisms. Hospital wastewater effluents are the major source of cytostatic drugs, while also urban wastewater effluents receive a substantial contribution of excreted drugs as the result of out-patient's treatment. Their removal efficiency during wastewater treatment processes has been found to vary significantly, depending both on the compounds' physico-chemical properties and on the treatment process applied. In general, biological processes do not achieve high removal efficiencies for these compounds, since many of them are poorly biodegradable. Consequently, they are continuously released into the aquatic and terrestrial environment at trace levels. However, there is still a lack of knowledge regarding both their presence in the natural environment – which depends on consumption patterns, the excretion fraction, and the effectiveness of the wastewater treatment – and the possible risks to humans and to the environment, requiring further investigation toward this direction. The aim of this chapter is to thoroughly review the removal efficiencies and mechanisms of cytostatic drugs in wastewater treatment plants, as well as what is required for future research, in view of the current concerns related to the induction of toxic effects in aquatic and terrestrial organisms by these compounds.

Keywords Biological treatment · Conventional activated sludge · Cytostatic drug · Wastewater treatment plant

---

L. Ioannou-Ttofa
Nireas-International Water Research Center, University of Cyprus, Nicosia, Cyprus

D. Fatta-Kassinos (✉)
Nireas-International Water Research Center, University of Cyprus, Nicosia, Cyprus

Department of Civil and Environmental Engineering, University of Cyprus, Nicosia, Cyprus
e-mail: dfatta@ucy.ac.cy

E. Heath et al. (eds.), *Fate and Effects of Anticancer Drugs in the Environment*,
https://doi.org/10.1007/978-3-030-21048-9_6

6.1 Introduction

According to the International Agency for Research on Cancer (IARC) of the World Health Organization (WHO), there are ca. 14.1 million new cancer cases and 8.2 million cancer-related deaths per year worldwide (data from 2012), while 32.6 million people are living with cancer (within 5 years of diagnosis); making cancer the second leading cause of death (21%) after cardiovascular illness (48%) (IARC 2017; WHO 2017). Since the incidence of cancer is increasing all over the world, the production and consumption of cytostatic drugs have consequently augmented in the last years (Franquet-Griell et al. 2016).

Like other pharmaceuticals, many cytostatic drugs can be excreted as unchanged or partially metabolized species via urine and/or feces and further directly discharged into the sewage system not only from hospitals but also from household discharges, since the outpatient treatment of cancer has considerably increased over the past decades (Kosjek and Heath 2011; Fraquet-Griell et al. 2015; Heath et al. 2016). Some of these drugs have been found to be poorly biodegradable and therefore can resist conventional biological processes (Johnson et al. 2008; Booker et al. 2014). However, there is still a lack of information about the fate and occurrence of most of them during conventional treatment (Martín et al. 2014). Nevertheless, according to the findings of the studies until now, it seems that there is a high probability that cytostatic drugs – both parent compounds and their metabolites – can reach the environment (i.e., surface water, groundwater, and even drinking water) (Liu et al. 2010; Zhang et al. 2013; Ferrando-Climent et al. 2014; Kümmerer et al. 2016).

Despite their low environmental concentrations (at sub-ng/L levels) compared to other groups of pharmaceuticals, such as antibiotics, analgesic and psychiatric drugs, among others (Kosjek and Heath 2011; Isidori et al. 2016), the cytostatic drugs require special attention because they are mostly non-selective in their modes of action, affecting both cancerous and non-cancerous fast dividing cells and often cause severe systemic side effects (Allwood et al. 2002; Lutterbeck et al. 2015). These drugs have high pharmacological potency and possess potent cytotoxic, genotoxic, mutagenic, carcinogenic, endocrine disruptor and/or teratogenic effects in several organisms, since they have been designed to disrupt or prevent cellular proliferation, by interfering directly or indirectly with DNA (Besse et al. 2012). Moreover, under chronic exposure they can induce subtle genetic and cell cycle changes in aquatic fauna and flora (Booker et al. 2014). The main concern over these toxic compounds is their potential occurrence in freshwater systems, even at trace levels, which are then abstracted as a potable water supply, presenting therefore a major risk of human exposure, as well as posing a wider risk to aquatic fauna and flora (Rowney et al. 2009; Booker et al. 2014). Recent studies have revealed that mixtures of cytostatic drugs occurring in real wastewater possess a higher toxicological effect compared with the individual drugs (Mater et al. 2014), highlighting the importance of developing ways to better manage these compounds.

Studies on the presence of cytostatic drugs in the environment are still quite scarce, compared to the studies of other groups of pharmaceuticals and for most of these compounds the environmental burden is unknown, which is partially explained by the lack of analytical methods capable of determining them at the low concentration levels ($\leq$ ng/L) in the environmental matrices (Martín et al. 2011; Ferrando-Climent et al. 2013; Martín et al. 2014). Another reason is the fact that these drugs are highly toxic to humans, making thus the various relevant laboratories and their research staff reluctant to analyze these compounds, since additional safety protocols and measures are needed (Heath et al. 2016). Moreover, various difficulties exist in the chromatographic analysis of environmental samples, which are related to the lack of knowledge, in many cases, of the form(s) of the compound which is present in the sample, producing hence analytical data that are only relevant to the amount of the free compound ignoring the amount in which it may be present in other forms (e.g., chelated and conjugated form) (Ternes et al. 1999; Michael et al. 2017).

The studies available in the scientific literature mainly focus on hospital effluents (Kümmerer and Al-Ahmad 1997; Mahnik et al. 2007; Lenz et al. 2007a; Yin et al. 2010; Kovalova et al. 2012; Zhang et al. 2013; Ferrando-Climent et al. 2013), where cytostatic residues are detected at quite high concentration levels (ng/L–μg/L), while also urban wastewater effluents and surface water have been recently investigated (Roberts and Thomas 2006; Coetsier et al. 2009; Martín et al. 2011, 2014; Gómez-Canela et al. 2012, 2014; Rabii et al. 2014; Negreira et al. 2014; Ferrando-Climent et al. 2014; Azuma et al. 2015; Isidori et al. 2016). Urban wastewater treatment plants (WWTPs) typically include different treatment processes (i.e., mechanical, physicochemical/chemical, and biological), which may affect the fate of the various pharmaceuticals, including cytostatic drugs, in different ways and, consequently, the spread of the drugs itself and/or their metabolites in the environment. The most consumed cytostatic drugs, i.e., *cyclophosphamide*, *ifosfamide*, *tamoxifen*, and *methotrexate*, have been found in the biologically treated wastewater effluents at the level of few nanograms per liter (i.e., n.d.-102 ng/L) (Steger-Hartmann et al. 1997; Buerge et al. 2006; Roberts and Thomas 2006; Coetsier et al. 2009; Martín et al. 2011, 2014; Gómez-Canela et al. 2012; Negreira et al. 2014; Ferrando-Climent et al. 2014). The cytostatic drugs, whose removal through WWTPs has been investigated by the scientific community, are summarized in Fig. 6.1.

Accordingly, the objectives of this study were to thoroughly review and evaluate the removal efficiencies and removal mechanisms of cytostatic drugs in WWTPs, and define the current challenges faced in relation to their presence in the environment.

6.2 Fate of Cytostatic Drugs in WWTPs

Most of the WWTPs in the developed countries consist of a preliminary, a primary, and a secondary stage, while sometimes a tertiary and/or disinfection stage is applied as a post-treatment step. During the last years the efficiency of conventional WWTPs in removing various cytostatic drugs is being investigated and the results of these

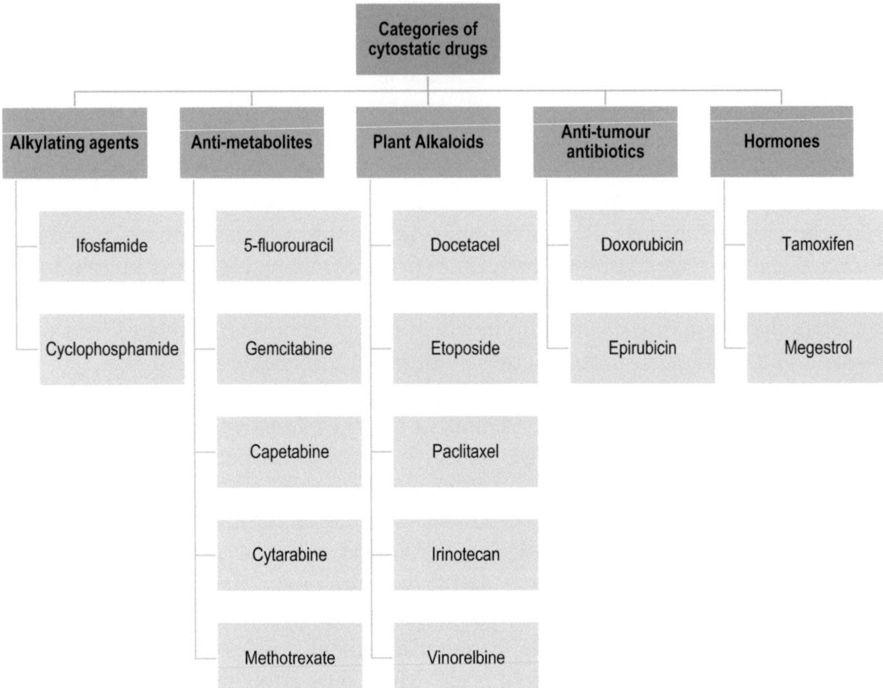

Fig. 6.1 Cytostatic drugs reported to be partially/totally degraded through conventional biological processes according to the available literature

studies are presented in the following paragraphs. However, the influence of the various treatment processes of WWTPs (i.e., mechanical, physicochemical/chemical, and biological) and the operational conditions (e.g., hydraulic retention time (HRT), sludge retention time (SRT), etc.) on the fate and spread of cytostatic drugs (both parent compounds and their metabolites) in the environment is a key element which is still missing from the scientific literature.

6.2.1 *Removal of Cytostatic Drugs by Conventional Biological Treatment*

Pharmaceutical residues can be removed from the aqueous phase in a conventional WWTP due to several processes, which can be either biotic (i.e., biotransformation/biodegradation) or non-biotic/abiotic (i.e., sorption, photolysis, hydrolysis) (Radjenović et al. 2009).

The performance of WWTPs applying biological treatment for removing cytostatic drugs, according to the available scientific literature, is summarized in Table 6.1, in which it is highlighted that some of them were found to be quite

Table 6.1 Removal of cytostatic drugs through conventional biological processes

Cytostatic drugs	Biological treatment details	Country	Description	Influent concentration	Effluent concentration	% Removal	References
Alkylating agents							
Ifosfamide	**WWTP (CAS)**	Switzerland	*Treatment stages*: mechanical, biological (activated sludge), chemical treatment (phosphate precipitation by iron salts, no chlorination), and subsequent sand filtration	15–1.4 ng/L	6–1.7 ng/L	up to 87%	Buerge et al. (2006)
	WWTP (CAS)	China	–	16.4 ng/L	9.0 ng/L	45%	Yin et al. (2010)
	WWTP (CAS)	Spain	–	3.5 ng/L	1.2 ng/L	65.7%	Martín et al. (2011)
	4 WWTPs (CAS)	Spain	*Treatment stages*: pretreatment, primary treatment (settling), and secondary treatment (activated sludge)	6.5–19 ng/L Mean: 12 ng/L	4–16 ng/L Mean: 10.5 ng/L	up to 67%	Martín et al. (2014)
	12 WWTPs (CAS) PE of each WWTP: 30,000–1,500,000	Spain	*Treatment stages*: primary and secondary biological treatment	2.2–27.9 ng/L Median: 4.6 ng/L	2.5–15.9 ng/L Median: 8.9 ng/L	up to 69.5%	Negreira et al. (2014)
Cyclophosphamide	**WWTP (CAS)**	Germany	–	6–143 ng/L	6–17 ng/L	up to 88%	Steger-Hartmann et al. (1997)
	WWTP (CAS)	Switzerland	*Treatment stages*: mechanical, biological (activated sludge),	11–4 ng/L	10–2 ng/L	up to 50%	

(continued)

Table 6.1 (continued)

Cytostatic drugs	Biological treatment details	Country	Description	Influent concentration	Effluent concentration	% Removal	References
			chemical treatment (phosphate precipitation by iron salts, no chlorination), and subsequent sand filtration				Buerge et al. (2006)
	WWTP (CAS)	China	–	14.5 ng/L	8.5 ng/L	41%	Yin et al. (2010)
	WWTP (CAS)	Spain	–	13.1 μg/L	< 0.35 ng/L (MDL)	100%	Gómez-Canela et al. (2012)
	WWTP (CAS) Flow rate: 40,000–50,000 m^3/d	Spain	*Note*: not only domestic sewage water but also wastewater from different sources (e.g., health centers, industrial zone, etc.) are discharging in this WWTP	8–26 ng/L	7–25 ng/L	up to 12.5%	Ferrando-Climent et al. (2014)
	2 WWTPs (CAS) *WWTPA* PE: 2,843,750 Flow rate: 525,000 m^3/d	Spain	*Note*: the treated urban effluents are finally discharged to the Mediterranean sea	4–10 ng/L	4–5 ng/L	up to 50%	Gómez-Canela et al. (2014)
	WWTPB PE: 2,275,000 Flow rate: 420,000 m^3/d			4 ng/L	n.a.	n.a.	

	12 WWTPs (CAS)	Spain	*Treatment stages*: primary and secondary biological treatment	2.4–43.8 ng/L	2.5–25 ng/L	up to 62%	Negreira et al. (2014)
	PE of each WWTP: 30,000–1,500,000			Median: 5.8 ng/L	Median: 6.3 ng/L		
	1 large WWTP (CAS/tertiary treatment)	Spain	*Note*: this WWTP receives mainly domestic water, including also hospital and industrial effluents and rainwater	7.1–17.3 ng/L	4.1–11.7 ng/L		
	PE: 1.7 million Flow rate: 420,000 m^3/d		*Treatment stages*: primary and secondary biological treatment (with nutrients elimination), and part of the water (about 50 hm^3/year) is intermittently subjected to further tertiary treatment to produce reclaimed water for different uses (e.g., watering and cleaning of the city, maintenance of the nearby river, ecological flow, etc.)		Median: 8.8 ng/L	Median: 43%	
	2 WWTPs *WWTPA (CAS)*	Canada	*Treatment stages of WWTPA*: screening (remove particles of more than 25 mm), removing abrasive materials (e.g., sand, heavy particles, etc.), physico-chemical treatment (eliminate suspended particles and reduce phosphorus), coagulation, and sedimentation	17–22 ng/L	18–21 ng/L	up to 18%	Rabii et al. (2014)
	WWTPB (CAS + UV)		*Treatment stages of WWTPB*: treatment stages of WWTPA following by an additional UV disinfection step *Note*: the treated effluents discharged in a local river	< LOD	< LOD	–	

(continued)

Table 6.1 (continued)

Cytostatic drugs	Biological treatment details	Country	Description	Influent concentration	Effluent concentration	% Removal	References
	2 WWTPs (CAS)	Slovenia and Spain	–	19–27 ng/L	17 ng/L	10%	Isidori et al. (2016)
	Slovenian WWTP						
	PE: 360,000						
	Flow rate: 80,000 m^3/d						
	SRT: 15–20 d						
	HRT: 21 h						
	Spanish WWTP			<2.3 ng/L (LOD)	<2.3 ng/L (LOD)	–	
	PE: 1,700,000						
	Flow rate: 234,000 m^3/d						
	SRT: 15–20 d						
	HRT: 8–12 h						
Antimetabolites							
5-Fluorouracil	**2 WWTPs**	Slovenia and Spain	–	31 ng/L	<0.5 ng/L (LOD)	100%	Isidori et al. (2016)
	Slovenian WWTP						
	PE: 360,000						
	Flow rate: 80,000 m^3/d						
	SRT: 15–20 d						
	HRT: 21 h						
	Spanish WWTP			3.5 ng/L (LOD)	<0.5 ng/L (LOD)	–	
	PE: 1,700,000						
	Flow rate: 234,000 m^3/d						
	SRT: 15–20 d						
	HRT: 8–12 h						

Gemcitabine	**WWTP (CAS)**	Spain		9.3 ng/L	7 ng/L	25%	Martín et al. (2011)
	4 WWTPs (CAS)		*Treatment stages*: pretreatment, primary treatment (settling), and secondary (activated sludge) treatment	39–52 ng/L	65–88 ng/L	–	Martín et al. (2014)
				Mean: 44 ng/L	Mean: 76 ng/L		
	2 WWTPs	Slovenia and Spain	–	n.d. –61 ng/L	< 0.7 ng/L (LOD)	100%	Isidori et al. (2016)
	Slovenian WWTP						
	PE: 360,000						
	Flow rate: 80,000 m^3/d						
	SRT: 15–20 d						
	HRT: 21 h						
	Spanish WWTP			<0.7 ng/L (LOD)	<0.7 ng/L (LOD)	–	
	PE: 1,700,000						
	Flow rate: 234,000 m^3/d						
	SRT: 15–20 d						
	HRT: 8–12 h						
Capecitabine	**12 WWTPs (CAS)**	Spain	*Treatment stages*: primary and secondary biological treatment	5.6–72.6 ng/L	5.1–36 ng/L	19%–89%	Negreira et al. (2014)
	PE of each WWTP: 30,000–1,500,000			Median: 25.6 ng/L	Median: 7.7 ng/L		
	2 WWTPs	Slovenia and Spain	–	158 ng/L	<0.5 ng/L (LOD)	100%	Isidori et al. (2016)
	Slovenian WWTP						
	PE: 360,000						
	Flow rate: 80,000 m^3/d						

(continued)

Table 6.1 (continued)

Cytostatic drugs	Biological treatment details	Country	Description	Influent concentration	Effluent concentration	% Removal	References
	SRT: 15–20 d						
	HRT: 21 h						
	Spanish WWTP			<0.7 ng/L (LOD)	<0.5 ng/L (LOD)	–	
	PE: 1,700,000						
	Flow rate: 234,000 m^3/d						
	SRT: 15–20 d						
	HRT: 8–12 h						
Cytarabine	**4 WWTPs (CAS)**	Spain	*Treatment stages*: pretreatment, primary treatment (settling), and secondary (activated sludge) treatment	44–464 ng/L	10–190 ng/L	47%–64%	Martín et al. (2014)
				Mean: 151 ng/L	Mean: 65 ng/L		
	WWTP (CAS)	–	–	n.a.	n.a.	24%	Kümmerer et al. (2016)
Methotrexate	**WWTP (CAS)**	Spain	*Note*: not only domestic sewage water but also wastewater from different sources (e.g., health centers, industrial zone, etc.) are discharging in this WWTP	11–23 ng/L	n.d. –6 ng/L	up to 100%	Ferrando-Climent et al. (2014)
	Flow rate: 40,000–50,000 m^3/d					(45%–100%)	
	4 WWTPs (CAS)	Spain	*Treatment stages*: pretreatment, primary treatment (settling), and secondary (activated sludge) treatment	7–56 ng/L	< 0.08 ng/L (LOD)	100%	Martín et al. (2014)
				Mean: 22 ng/L			
				Median: 13 ng/L			

	12 WWTPs (CAS) PE of each WWTP: 30,000–1,500,000	Spain	*Treatment stages*: primary and secondary biological treatment	2.6–18.1 ng/L Median: 9.6 ng/L	n.d.	100%	Negreira et al. (2014)
	2 WWTPs *WWTPA (CAS)*	Canada	*Treatment stages of WWTPA*: screening (remove particles of more than 25 mm), removing abrasive materials (e.g., sand, heavy particles, etc.), physico-chemical treatment (eliminate suspended particles and reduce phosphorus), coagulation, and sedimentation	20–60 ng/L	13–53 ng/L	12%–37%	Rabii et al. (2014)
	WWTPB (CAS + UV)		*Treatment stages of WWTPB*: treatment stages of WWTPA following by an additional UV disinfection step *Note*: the treated effluents discharged in a local river	17–20 ng/L	20–22 ng/L	–	
	2 WWTPs *Slovenian WWTP*: PE: 360,000 Flow rate: 80,000 m^3/d SRT: 15–20 d HRT: 21 h	Slovenia and Spain	–	29–303 ng/L	< 0.5 ng/L (LOD)	100%	Isidori et al. (2016)
	Spanish WWTP: PE: 1,700,000 Flow rate: 234,000 m^3/d SRT: 15–20 d HRT: 8–12 h			8.3–29 ng/L	<0.5 ng/L (LOD)	100%	

(continued)

Table 6.1 (continued)

Cytostatic drugs	Biological treatment details	Country	Description	Influent concentration	Effluent concentration	% Removal	References
Plant alkaloids							
Docetaxel	**WWTP (CAS)** Flow rate: 40,000–50,000 m^3/d	Spain	*Note*: not only domestic sewage water but also wastewater from different sources (e.g., health centers, industrial zone, etc.) are discharging in this WWTP	65–219 ng/L	<3.8 ng/L (LOD)	~100%	Ferrando-Climent et al. (2014)
Etoposide	**WWTP (CAS)**	Spain	-	15 ng/L	3.4 ng/L	77%	Martín et al. (2011)
	4 WWTPs (CAS)	Spain	*Treatment stages*: pretreatment, primary treatment (settling), and secondary (activated sludge) treatment	15–47 ng/L Mean: 30.5 ng/L	<2.95 ng/L (LOD)	100%	Martín et al. (2014)
Paclitaxel	**WWTP (CAS)** Flow rate: 40,000–50,000 m^3/d	Spain	*Note*: not only domestic sewage water but also wastewater from different sources (e.g., health centers, industrial zone, etc.) are discharging in this WWTP	n.d. –18 ng/L	<2.7 ng/L (LOD)	100%	Ferrando-Climent et al. (2014)
Irinotecan	**1 large WWTP (CAS/tertiary treatment)** PE: 1.7 million Flow rate: 420,000 m^3/d	Spain	*Note*: this WWTP receives mainly domestic water, including also hospital and industrial effluents and rainwater *Treatment stages*: primary and secondary biological treatment (with nutrients elimination), and part of the water (about 50 hm^3/year) is intermittently subjected to further tertiary treatment to produce reclaimed water for different uses (e.g., watering and cleaning of the city, maintenance of the nearby river, ecological flow, etc.)	8.8–21.3 ng/L Median: 12.4 ng/L	n.d.	100%	Negreira et al. (2014)

	2 WWTPs *Slovenian WWTP* PE: 360,000 Flow rate: 80,000 m^3/d SRT: 15–20 d HRT: 21 h	Slovenia and Spain	–	n.d. –49 ng/L	< 0.4 ng/L (LOD)	100%	Isidori et al. (2016)
	Spanish WWTP PE: 1,700,000 Flow rate: 234,000 m^3/d SRT: 15–20 d HRT: 8–12 h			<1.4 ng/L (LOD)	<0.4 ng/L (LOD)	100%	
Antitumor antibiotics							
Doxorubicin	**WWTP (CAS)**	Spain	–	4.5 ng/L	< 4.3 ng/L (MDL)	100%	Martín et al. (2011)
	12 WWTPs (CAS) PE of each WWTP: 30,000–1,500,000	Spain	*Treatment stages*: primary and secondary biological treatment	2.5–2.7 ng/L Median: 2.6 ng/L	n.d.	100%	Negreira et al. (2014)
	4 WWTPs (CAS)	Spain	*Treatment stages*: pretreatment, primary treatment (settling), and secondary (activated sludge) treatment	<4.15 ng/L (LOD)	20.3–42.4 ng/L Mean: 31.4 ng/L	–	Martín et al. (2014)
Epirubicin	**WWTP (CAS)**	Spain	–	<3.8 ng/L (LOD)	<0.7 ng/L (LOD)	–	Martín et al. (2011)
	3 WWTPs (CAS)	Spain	–	< 2.77 ng/L (LOD)	< 2.77 ng/L (LOD)	–	Gómez-Canela et al. (2012)

(continued)

Table 6.1 (continued)

Cytostatic drugs	Biological treatment details	Country	Description	Influent concentration	Effluent concentration	% Removal	References
	2 WWTPs	Canada					Rabii et al. (2014)
	WWTPA (CAS)		*Treatment stages of WWTPA*: screening (remove particles of more than 25 mm), removing abrasive materials (e.g., sand, heavy particles, etc.), physico-chemical treatment (eliminate suspended particles and reduce phosphorus), coagulation, and sedimentation	< 18 ng/L (LOD)	< 18 ng/L (LOD)	–	
	WWTPB (CAS + UV)		*Treatment stages of WWTPB*: treatment stages of WWTPA following by an additional UV disinfection step	< 18 ng/L (LOD)	< 18 ng/L (LOD)	–	
			Note: the treated effluents discharged in a local river				
Hormones							
Tamoxifen	**WWTP (CAS + UV)**	United Kingdom	*Treatment stages*: screening, preliminary clarification, activated sludge treatment (sludge age: 2.4 days), trickling filter system, and UV disinfection prior to discharge	143–215 ng/L	n.a.	30%	Roberts and Thomas (2006)
	PE: 2.6 million						
	WWTP (CAS)	Spain	*Note*: not only domestic sewage water but also wastewater from different sources (e.g., health centers, industrial zone, etc.) are discharging in this WWTP	15–58 ng/L	11–42 ng/L	up to 43%	Ferrando-Climent et al. (2014)
	Flow rate: 40,000–50,000m^3/d						

	1 large WWTP (CAS/tertiary treatment)	Spain	*Note*: this WWTP receives mainly domestic water, including also hospital and industrial effluents and rainwater	Median: 179.2 ng/L	Median: 113.5 ng/L	37%	Negreira et al. (2014)
	PE: 1.7 million		*Treatment stages*: primary and secondary biological treatment (with nutrients elimination), and part of the water (about 50 hm^3/year) is intermittently subjected to further tertiary treatment to produce reclaimed water for different uses (e.g., watering and cleaning of the city, maintenance of the nearby river, ecological flow, etc.)				
	Flow rate: 420,000 m^3/d						
	2 WWTPs	Slovenia and Spain	–	11–61 ng/L	7.1 ng/L	35.5%	Isidori et al. (2016)
	Slovenian WWTP						
	PE: 360,000						
	Flow rate: 80,000 m^3/d						
	SRT: 15–20 d						
	HRT: 21 h						
	Spanish WWTP			15 ng/L	7.4 ng/L	50.6%	
	PE: 1,700,000						
	Flow rate: 234,000 m^3/d						
	SRT: 15–20 d						
	HRT: 8–12 h						

(continued)

Table 6.1 (continued)

Cytostatic drugs	Biological treatment details	Country	Description	Influent concentration	Effluent concentration	% Removal	References
Megestrol	**2 WWTPs (CAS)**	Spain	–	3–150 ng/L	3–20 ng/L	87%	Gómez-Canela et al. (2014)
	WWTPA						
	PE: 2,843,750						
	Flow rate: 525,000 m^3/d						
	WWTPB			3–220 ng/L	<3 ng/L (LOD)	100%	
	PE: 2,275,000						
	Flow rate: 420,000 m^3/d						

CAS conventional activated sludge, *n.a.* not available, *n.d.* non detected, *LOD* limit of detection, *MLQ* method limit of quantification, *P.E.* population equivalent, *MDL* methodological detection limits. "–" no data available resistant to biodegradation. The removal rates of the detected cytostatic drugs in WWTPs ranged from 10%–88% (*cyclophosphamide*, *ifosfamide, capecitabine, tamoxifen,* and *cytarabine*) to 77%–100% (*doxorubicin*, *doxetaxel*, *etoposide*, *gemcitabine, irinotecan*, *cyclophosphamide*, *paclitaxel*, *megestrol, 5-fluorouracil,* and *methotrexate*) (Buerge et al. 2006; Yin et al. 2010; Martín et al. 2011, 2014; Gómez-Canela et al. 2012; Ferrando-Climent et al. 2014; Rabii et al. 2014; Negreira et al. 2014; Isidori et al. 2016; Kümmerer et al. 2016). The overall removal efficiency of biological treatment clearly varies among different cytostatic compounds, and the efficiency seems to be associated with their physicochemical properties (e.g., hydrophobicity, water solubility, etc.) and as expected with their biodegradability (Zhang et al. 2013); however, it should be highlighted that operational details are not provided in most of the available studies on the fate and behavior of cytostatic residues during treatment in WWTPs, and hence no concrete conclusions can be drawn regarding this issue. In addition, no data exists on the influence of the various cytostatic drugs on the composition and function of the microbial community of the activated sludge of the various WWTPs, which is something that should be further examined in real WWTPs. Cytostatic drugs have been found to be generally polar and persistent, with high aquatic mobility and dissipation behavior in surface waters, while there are some exceptions, such as plant alkaloids and antitumor antibiotics, which favor adsorption to sewage sludge (Kosjek and Heath 2011).

The main groups of cytostatic drugs (i.e., alkylating agents, antimetabolites, plant alkaloids, antitumor antibiotics, and hormones) and their removal during conventional biological treatment are discussed in detail in the following sections.

6.2.1.1 Alkylating Agents (Cyclophosphamide and Ifosfamide)

Alkylating agents are the oldest class of cytostatic drugs still frequently used, and they play an important role in the treatment of several types of cancer (Kondo et al. 2010). Their mechanism of action involves attaching an alkyl group onto the DNA helix, inhibiting thus the replication of the DNA and thereby the cell division (Kümmerer et al. 2016). Due to the fact that they interact directly with the DNA, they account for various genotoxic and even teratogenic effects, causing high environmental concern (Buerge et al. 2006; Kümmerer et al. 2016). *Ifosfamide* and *cyclophosphamide* are the most widely consumed alkylating agents (Martín et al. 2011) and various studies dealing both with their removal through biological processes and their subsequent presence in the aquatic environment have been already published (Steger-Hartmann et al. 1997; Buerge et al. 2006; Martín et al. 2011, 2014; Ferrando-Climent et al. 2013; Gómez-Canela et al. 2014; Negreira et al. 2014; Rabii et al. 2014).

The maximum removal of *cyclophosphamide* has been reported for a WWTP in Spain (100% removal) (Gómez-Canela et al. 2012), while the lowest was observed in a WWTP in Slovenia and was equal to 10% (Isidori et al. 2016), as shown in Fig. 6.2, in which the minimum and maximum removals of the various cytostatic drugs through biological treatment are presented, including also the number of the

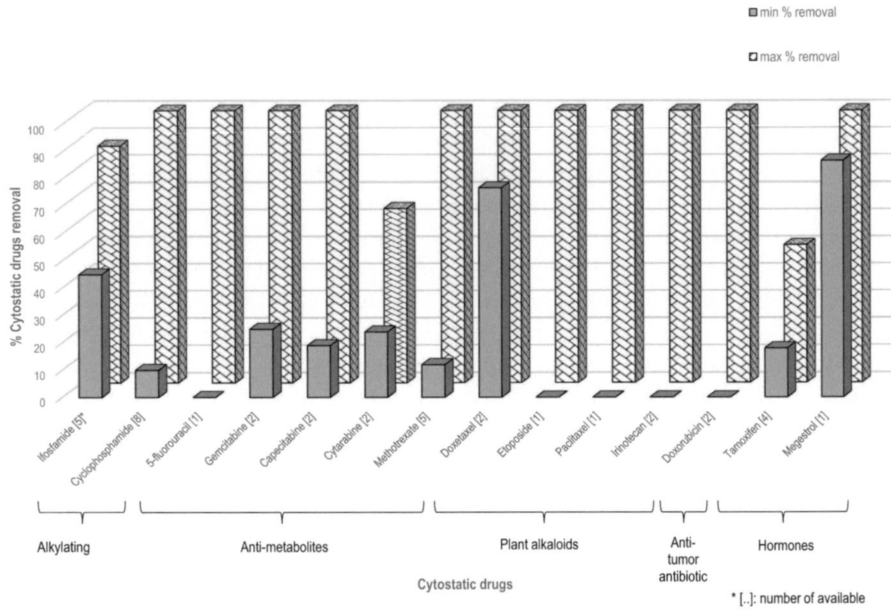

Fig. 6.2 % removal (minimum and maximum values) of cytostatic drugs through conventional biological processes

available studies. For *ifosfamide*, the lowest removal reported was in a WWTP in China (45% removal) (Yin et al. 2010), while the highest, up to 87%, in a WWTP in Switzerland (Buerge et al. 2006) (Fig. 6.2). *Ifosfamide* and *cyclophosphamide* were found in the influent streams of various WWTPs at concentration levels, lower than 2 ng/L (Castiglioni et al. 2005; Thomas et al. 2007; Martín et al. 2011; Isidori et al. 2016) and up to 130.1 ng/L (Ferrando-Climent et al. 2013) and 13.1 μg/L (Gómez-Canela et al. 2012), respectively. On the other hand, in the treated effluents their concentrations ranged from non-detected (n.d.) 0.09 ng/L (Llewellyn et al. 2011) up to 71 ng/L (Coetsier et al. 2009) for *ifosfamide* and from n.d. - 0.19 ng/L to 25 ng/L for *cyclophosphamide* (Ferrando-Climent et al. 2014; Negreira et al. 2014), as shown in Fig. 6.3, in which the minimum and maximum concentrations of cytostatic drugs identified in biologically treated effluents are shown. The different behavior of *ifosfamide* and *cyclophosphamide* through biological treatment can be attributed to their different physicochemical properties (i.e., $\log K_{ow}$, 0.86 and 0.63; water solubility, 3780 mg/L and 40,000 mg/L; pKa, 1.45 and 2.84, respectively). Moreover, it should be noted that in many of the available studies (Ternes 1998; Castiglioni et al. 2005; Coetsier et al. 2009; Busetti et al. 2009; Llewellyn et al. 2011; Ferrando-Climent et al. 2013; Negreira et al. 2013; Gómez-Canela et al. 2014; Azuma et al. 2015), these alkylating agents were detected only in the influent or only in the effluent flows of a WWTP, and thus no conclusions can be obtained regarding the efficiencies of these treatment plants in removing these compounds, minimizing

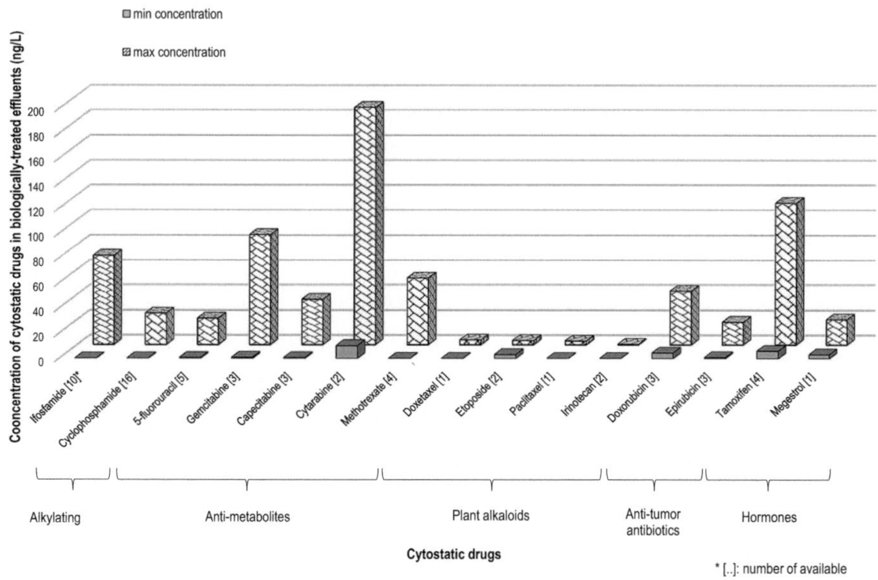

Fig. 6.3 Minimum and maximum residual concentrations of cytostatic drugs in the biologically treated effluents

significantly the number of the available studies investigating their removal through biological treatment.

It should be mentioned that only in the study Buerge et al. (2006), data regarding the treatment processes applied in the specific WWTP are available, in which a chemical treatment using phosphate for the precipitation of iron salts is applied after conventional activated sludge (CAS) treatment, followed by sand filtration, potentially explaining the highest efficiency observed compared to the other CAS WWTPs (Yin et al. 2010; Isidori et al. 2016). However, since this information is only available for this study (Buerge et al. 2006), this is only an assumption that requires further investigation. On the other hand, it is noted that *cyclophosphamide* was found in the effluent samples of a large Spanish WWTP (PE: 1.7 million and flow rate: 420,000 m^3/d), where tertiary treatment is applied, in the range of 4.1–11.7 ng/L, achieving only a mean removal of 43%, demonstrating its high resistance to biological, and even to tertiary treatment (Negreira et al. 2014).

It should be highlighted that in some studies, both alkylating agents were detected in the effluent streams, even at concentrations up to 71 ng/L in the case of ifosfamide (Thomas et al. 2007), while they were not detected in the influent samples or were detected in lower concentrations (Kümmerer and Al-Ahmad 1997; Thomas et al. 2007). This paradox may be probably explained by the fact that in the influent stream, these drugs are present also in the form(s) of conjugates, which obviously were not included in the determination of the free drugs, and thus the analytical methods available in each study were not able to identify these conjugates. During the biological treatment, these conjugates are potentially broken down, and consequently, the free form of these drugs is released, giving rise to higher concentrations in the treated effluents compared to that found in the inflow (Ternes et al. 1999; Michael et al. 2017).

The concentrations and the removal rates of these two alkylating agents vary significantly through the WWTPs worldwide, making it difficult to conclude if these compounds are highly resistant or not to biological treatment. However, the majority of the available studies so far shown that their removal rates through conventional processes are lower than 65%, showing that there is a need for further post-treatment. Moreover, the fact that both alkylating agents may possess teratogenic and mutagenic effects to various species (e.g., bacteria, yeast, rat/mouse, monkey, even human, in vivo or in vitro) according to IARC (2017) underlies the already mentioned need for further research and better management of these compounds.

6.2.1.2 Antimetabolites (5-Fluorouracil, Gemcitabine, Capecitabine, Cytarabine, and Methotrexate)

Induction of apoptotic death in cancer cells *via* genotoxic stress by chemotherapy remains the core of anticancer treatment all over the world. An important class of cytostatic drugs based on this principle, namely, antimetabolites, has been widely used for a variety of cancer therapies, including leukemia, breast, ovarian, and gastrointestinal cancers (Krynetskaia et al. 2008). More specifically, their mechanism of action is based on hindering cellular metabolism and thereby hindering the production of DNA (Kümmerer et al. 2016). It is noted that the fate, behavior, and removal rates of this group of cytostatic drugs in WWTPs has been scarcely investigated, according to the available scientific literature (Martín et al. 2011, 2014; Negreira et al. 2013, 2014; Gómez-Canela et al. 2014). Specifically, only five antimetabolites, namely, *5-fluorouracil*, *gemcitabine*, *capecitabine*, *cytarabine*, and *methotrexate*, have been investigated so far, as shown in Fig. 6.1.

5-Flouorouracil was not detected neither in the influent nor in the effluent wastewater of four Spanish WWTPs, one Swiss and one in Baltimore, according to the studies of Tauxe-Wuersch et al. (2006), Martín et al. (2011), (2014), and Jim et al. (2012). This can be probably explained by the fact that *5-flouorouracil* is rapidly metabolized by the human liver and produces biologically inactive metabolites, which are excreted with urine (Rosano 1998). However, in a Slovenian WWTP, *5-flouorouracil* was detected in influent wastewater at very low concentrations (i.e., lower than 3.1 ng/L), and the removal rate achieved was up to 100% after the biological treatment. The same was also observed in a Spanish WWTP (influent concentration: 3.5 ng/L, 100% removal) (Isidori et al. 2016) (Fig. 6.2), indicating that this highly toxic antimetabolite, which may possess teratogenic effects to various living organisms (IARC 2017), is well biodegradable.

Gemcitabine removal rates ranged from 25% (Martín et al., 2011) to 100% (Isidori et al. 2016) during biological treatment (Fig. 6.2), achieving effluent concentrations lower than 7 ng/L; however, only three studies investigated the biological removal of this drug from urban wastewater. In the study of Martín et al. (2014), *gemcitabine* was detected in higher concentrations in the treated effluents (65–88 ng/

L) of four Spanish WWTPs compared to the raw wastewater (39–52 ng/L). This may be attributed to the fact that in raw wastewater samples, this antimetabolite can be present in conjugate form(s) which cannot be detected by the analytical methods available, while these conjugates were probably broken down to *gemcitabine's* free form after biological treatment, and thus it was detected at higher concentrations in the treated effluents. Gemcitabine is more soluble in water (water solubility, 51,400 mg/L; logK_{ow}, -2.01) if compared to other cytostatic drugs and can be assumed that this drug is removed from the aqueous phase through biotransformation and/or biodegradation (and not through sorption onto sludge).

Capecitabine was detected in the influent samples of 14 WWTPs in concentrations ranging from 0.7 ng/L up to 158 ng/L, and its removal rates from 19% to 100% during CAS treatment (Negreira et al. 2014; Isidori et al. 2016) (Fig. 6.2); achieving residual concentrations of 0.5–36 ng/L (Fig. 6.2). It should be noted that according to the study of Azuma et al. (2015), *capecitabine* was not completely degraded neither after the application of chlorination as disinfection process of CAS treatment (effluent concentration: 6 ng/L), nor after the application of ozonation (effluent concentration: 2 ng/L), a fact that indicates its high resistance to such tertiary treatment.

For *cytarabine,* quite low elimination rates were reported, i.e., from 24% (Kümmerer et al. 2016) up to 64% (Martín et al. 2014), as shown in Fig. 6.2. However, only three studies have investigated this compound, so far. It is remarkable that *cytarabine* was detected in significantly high concentrations both in influent and effluent samples of four Spanish WWTPs, even up to 464 and 190 ng/L, respectively, according to Martín et al. (2014). These concentration levels are significantly higher compared to other antimetabolites investigated, which ranged from < 0.7 ng/L to 198 ng/L for the raw wastewater (with an exception in the case of *methotrexate*, 303 ng/L (Isidori et al. 2016)) and from < 0.5 ng/L to 53 ng/L for the treated effluents (Table 6.1). It should be also highlighted that in the study of Martín et al. (2011), *cytarabine* was detected in the effluent flow of a WWTP in Spain at higher concentration (14 ng/L) than in the influent flow (9.2 ng/L), probably due to its presence in the influent sample in conjugate form(s), which were further broken down at its free form after the treatment, as explained in detail before. Nevertheless, it should be noted that neither *cytarabine* nor *capecitabine* or *gemcitabine* may possess mutagenicity or teratogenicity, according to IARC (2017), classifying them as less harmful compounds, compared to other cytostatic drugs (i.e., alkylating agents and hormones).

Methotrexate is a cytostatic drug that was found to be sufficiently removed through conventional biological treatment (water solubility: 2600 mg/L, logK_{ow}: -1.85 and pKa: 4.7), achieving complete degradation in most of the available studies (Ferrando-Climent et al. (2014); Martín et al. (2014); Negreira et al. (2014); Isidori et al. (2016)), as shown in Fig. 6.2. Only in the study of Rabii et al. (2014) its removal was found to be low (12%). Its influent concentrations ranged from 2.6 to 303 ng/L, while its effluent concentrations from <0.08 to 53 ng/L (Fig. 6.3), according to the findings of the above studies. It should be highlighted that in the study of Ferrando-Climent et al. (2014), *methotrexate* was found at similar concentration levels in hospital effluents (n.d. - 19 ng/L) and WWTPs influents (n.d. - 23 ng/L), mainly due to the fact that this drug is also used for the treatment of various autoimmune diseases, apart from cancer, and it is frequently consumed and excreted by patients at home (Sweetman and Martindale 2011).

Based on the limited available literature, no concrete conclusions can be drawn regarding the fate and removal of these five antimetabolites through biological treatment, and hence further in-depth biodegradation studies are needed to confirm their degree of biodegradability.

6.2.1.3 Plant Alkaloids (Docetaxel, Etoposide, Paclitaxel, and Irinotecan)

Plant alkaloids are nitrogen-containing organic compounds derived from certain types of plants and used as cytostatic drugs for the treatment of various types of cancer (Pietsch et al. 2008). They are cell cycle-specific, meaning that they attack the cells during various phases of division. These cytostatic drugs seem to favor adsorption to sewage sludge during conventional biological treatment, revealing an enhanced bioaccumulation potential (Kosjek and Heath 2011). Four plant alkaloids, namely, *docetaxel*, *etoposide*, *paclitaxel*, and *irinotecan,* have been examined regarding their removal through biological processes.

The removal of *docetaxel* through biological treatment has been investigated only by the study of Ferrando-Climent et al. (2014). In this study, its complete removal ($logK_{ow}$: 2.83, insoluble to water, pKa: 12.02) was obtained in a CAS WWTP in Spain, with its influent concentration being in the range of 65–219 ng/L, while its effluent concentration was found to be lower than 3.8 ng/L, which was the limit of detection (LOD) of the analytical method applied (Ferrando-Climent et al. 2014). It should be noted that *docetaxel* was the only cytostatic drug investigated in this study, which was found at higher levels in influent sewage (i.e., 65–219 ng/L) than in hospital effluents (i.e., 61–79 ng/L). This might be attributed to its slow metabolism in human organism, since approx. 80% of this drug is excreted after the first 48 h and the patients have commonly left the hospitals.

High removals of *etoposide* through CAS treatment has been reported, i.e., from 77% up to 100%, by Martín et al. (2011) and (2014), as shown in Fig. 6.2. These high removals are probably attributed to its sorption onto sewage sludge, as reported in the study of Kosjek and Heath (2011) regarding the general behavior of the plant alkaloids in the WWTPs. The influent concentration of *etoposide* in the various WWTPs was in the level of 15–83 ng/L, while the effluent concentration was lower than 2.95–3.4 ng/L (Fig. 6.3) (Martín et al. 2011, 2014; Ferrando-Climent et al. 2013). It should be also noted that while *etoposide* was detected at high concentration levels in hospital effluents (up to 714 ng/L), it was not detected neither in the influent nor in the effluent flows of three Spanish and one Slovenian WWTPs (Ferrando-Climent et al. 2014; Negreira et al. 2014; Isidori et al. 2016). This can be explained by the fact that this drug is mainly used for cancer treatments which require patient's hospitalization.

Only one study investigated the occurrence of *paclitaxel* (logK_{ow}: 3.95, insoluble to water, pKa: 11.99) in WWTPs, which was detected in only one out of the three influent samples of a Spanish WWTP at 18 ng/L, while it was not detected in any of the effluent samples (< 2.7 ng/L (LOD)), suggesting a complete removal during the conventional biological treatment (Ferrando-Climent et al. 2014).

Elimination (100% removal) of another plant alkaloid, namely *irinotecan*, was achieved in a Spanish WWTP, in which tertiary treatment is applied, according to Negreira et al. (2014), with its highest influent concentration to be up to 21.3 ng/L. According to the authors' knowledge, this was the first time that *irinotecan* was quantified in urban wastewater samples, while it was also detected in hospital effluents at levels up to 730 ng/L (Gómez-Canela et al. 2014). *Irinotecan* was also found in the inflow of a Slovenian WWTP, at a concentration level up to 49 ng/L, according to Isidori et al. (2016), and it was also completely degraded through biological treatment (< 0.4 ng/L residual concentration).

Although high degradation through biological treatment was observed for all four plant alkaloids, according to the available studies described, considering the fact that for most of them only one or two studies exist in the literature, no solid conclusions regarding their occurrence, fate, removal mechanisms, and biodegradability can be drawn at the moment, and further investigation is deemed necessary to be conducted.

6.2.1.4 Antitumor Antibiotics (Doxorubicin and Epirubicin)

Antitumor antibiotics interfere with the DNA inside cells and stop or slow cancer cells from growing and division (Kümmerer et al. 2016). The removal of only two antitumor antibiotics (i.e., *doxorubicin* and *epirubicin*) in WWTPs has been investigated until now (Fig. 6.1).

Doxorubicin, which is a drug that may possess mutagenicity to various species (IARC 2017), was found to be completely eliminated during CAS treatment, according to the studies of Martín et al. (2011) and Negreira et al. (2014). The concentration of *doxorubicin* found in the influent flows of the WWTPs was at the same levels in both studies, i.e., 2.5–4.5 ng/L. In contrast, *doxorubicin* was detected in the effluent stream of four Spanish WWTPs at higher levels (20.3–42.4 ng/L), according to Martín et al. (2014), while it was not detected in the corresponding influent flow (< 4.15 ng/L (LOD)). This can be probably attributed to the phenomenon of the formation of conjugates, which was described in detail in the above sections.

Epirubicin was not detected in the influent wastewater of three Spanish WWTPs, according to the studies of Martín et al. (2011), Gómez-Canela et al. (2012), and (2014), while it was detected in the hospital effluents examined in the same studies (in the low concentration level of 4.5–6 ng/L), indicating that this cytostatic compound may probably be degraded or diluted during passage through the sewage grid (Gómez-Canela et al. 2014). This is also in agreement with the study of Rabii et al. (2014), as *epirubicin* was also not found neither in the influent nor in the effluent flows of the two WWTPs in Canada. However, it should be highlighted that the LOD of the analytical method used in this study was quite high compared to other studies (i.e., 18 ng/L).

The limited available literature shows that contradictory data exist regarding *doxorubicin* removal through CAS treatment, requiring further investigation to get a clear picture regarding its biodegradability. On the other hand, for *epirubicin* the first results show that it is degraded or diluted during passage through the grid, but more exhaustive research is needed, in order to understand and explain its behavior through WWTPs.

6.2.1.5 Hormones Used for Cancer Treatment (Tamoxifen and Megestrol)

Hormone therapy is a form of systemic therapy that works by interfering with the hormone system, in order to slow or stop the growth of cancer cells, and is used for the treatment of various hormone-dependent cancers, such as breast and prostate cancer. Two often consumed hormones, *tamoxifen* and *megestrol*, have been investigated in terms of their removal through biological treatment.

Tamoxifen (logK_{ow}, 7.88; water solubility, 16.7 mg/L; pKa, 8.87) is one of the most frequently used hormones for the treatment of breast cancer, and its removal through secondary and tertiary biological treatment was found to be insufficient, ranging from 18% to 50.6% (Roberts and Thomas 2006; Ferrando-Climent et al. 2014; Negreira et al. 2014; Isidori et al. 2016), as shown in Fig. 6.2. The concentration of *tamoxifen* in raw wastewater ranged from 3.5 ng/L (Negreira et al., 2013) to 15 ng/L (Ferrando-Climent et al. 2014; Isidori et al. 2016) and even up to 215 ng/L (Roberts and Thomas 2006), while its detected residual concentrations in the treated effluents were higher, between 5.8 ng/L and 113.5 ng/L (Fig. 6.3). Negreira et al. (2014) and Roberts and Thomas (2006) reported poor removal of *tamoxifen*, equal to 37% and 30%, respectively, by two WWTPs in Spain and United Kingdom, where tertiary treatment was applied, highlighting its high resistance to biodegradation and tertiary treatment, as well.

Megestrol (logK_{ow}: 3.2, water solubility: 2 mg/L) is the most common progestogen used in medicine and it was detected in the inflow of a Spanish WWTP at concentration levels of 3–150 ng/L, achieving a significant removal of 87% through CAS treatment (Gómez-Canela et al. 2014). In addition, *megestrol* was the only compound detected in a second WWTP examined in this study, at a maximum concentration of 220 ng/L in the influent, achieving a complete removal (100%) through biological treatment. The residual concentrations of *megestrol* in the secondary-treated effluents were between 3 and 20 ng/L (Fig. 6.3).

Tamoxifen appears to be poorly degradable through secondary and tertiary processes and would therefore require further investigation in terms of its occurrence and toxicity in the aquatic and terrestrial environment, since it may be considered also as endocrine disruptor. In addition, even though early indications of the one study available in the literature show that *megestrol* is highly biodegradable, further investigation on the fate/behavior of this hormone during biological treatment is needed, in order to confirm this outcome.

6.2.2 Removal of Cytostatic Drugs by Membrane Bioreactors

Advanced biological treatment processes, such as membrane bioreactors (MBRs), have been proved to be a remarkable alternative technique, operated at highly intensified biomass and high SRTs, favoring thus an enhanced biotransformation and mineralization of resistant pharmaceutical compounds (Ruel et al. 2011; Zhang et al. 2013). Although the investment and operational costs of MBRs are higher than CAS, the fact that they provide a more hygienic effluent, due to the membrane filtration (i.e., microfiltration (MF) or ultrafiltration (UF)), makes them an attractive alternative treatment technology. However, it should be noted that there is no available study in the scientific literature applying MBRs at real scale investigating the removal of cytostatic compounds, while only bench- and pilot-scale systems have been examined up to now.

The antimetabolite, *5-fluorouracil*, was found that can be almost completely eliminated within 24 h by a pilot-scale MBR system from hospital effluents, and the antitumor antibiotics, *doxorubicin*, *epirubicin*, and *daunorubicin*, were removed up to 90% from the liquid phase, mainly due to the adsorption to sewage sludge according to the study of Mahnik et al. (2007). On the other hand, a moderate elimination was displayed in an MBR pilot plant for hydrophilic *cisplatin* and *carboplatin* (i.e., 51% and 63% removal, respectively), mainly due to their irreversible adsorption to the activated sludge (Lenz et al. 2007b). Delgado et al. (2009) demonstrated that a lab-scale MBR system operated at aerobic/anoxic conditions (operational conditions: HRT: 32 h and SRT: 70 days) achieve 80% removal of both *cyclophosphamide* and its metabolite (i.e., 4-ketocyclophosphamide), due to both adsorption and degradation. According to Delgado et al. (2011), *cyclophosphamide* was also removed up to 80% in a pilot-scale MBR (MF membrane unit: pore size: 0.1 μm, surface area: 0.0055 m^2; operational conditions: HRT: 48 h and SRT: 50 days) from urban effluents, while on the other hand, residual toxicity was measured in the permeate stream, indicating that further post-treatment is required to eliminate effluents' toxicity before their discharge in the environment. In contrast, insignificant removal of *cyclophosphamide* (up to 20%) was achieved in the pilot-scale MBRs treating hospital wastewater, according to Kovalova et al. (2012) and Köhler et al. (2012).

In the study of Avella et al. (2010), it was reported that the fouling of the membranes in a cross-flow lab-scale MBR system (MF membrane unit: ceramic tubular membranes, surface area: 0.0055 m^2, pore size: 0.1 μm) increased significantly by the addition of *cyclophosphamide* and its mean metabolites in urban wastewater, compared with the fouling observed during the operation of a similar MBR system which was fed only with urban wastewater (i.e., without the addition of cytostatic drugs). The MBR's transmembrane pressure in the former setup increased for 3-folds with the addition of these drugs. Finally, in a recent study of Wu et al. (2017), the influence of the presence of eight cytostatic drugs (i.e., *azathioprine*, *cyclophosphamide*, *doxorubicin*, *epirubicin*, *flutamide*, *methotrexate*, *mitotane*, and *tamoxifen*) on the composition and function of the microbial community in a forward osmosis anaerobic MBR (FO-AnMBR) (operational conditions: SRT, 60 days, and HRT, 15–40 h) treating urban wastewater was investigated. The authors report that the presence of cytostatic drugs in the FO-AnMBR system caused the inhibition of microbial metabolism, decreasing the evenness of the microbial community and marginally changing community's composition, while the extracellular polymeric substances (EPS) concentration was increased.

A higher treatment efficiency in removing cytostatic drugs seems to be achieved through MBR systems at bench and pilot scale, in most of the cases mentioned above, if compared with CAS systems, while further investigation in real scale applications is required in order to have more concrete and reliable data. Further research is needed with regard to the MBR effluents' toxicity, which according to the authors' opinion is expected to be lower than the CAS-treated effluents, due to the filtration (MF or UF) applied in the MBR systems, but definitive data are needed to confirm this assumption too. However, in the study of Zhang et al. (2013), it was highlighted that tertiary treatment might be needed to facilitate the safe discharge of treated effluents from MBR systems, indicating not only the complete degradation of the parent compound(s) and/or their active metabolites, but also the elimination of their toxicity (short and long term). Moreover, it is necessary to conduct research on the influence of the various cytostatic drugs on the composition and function of the microbial community of the sludge of the advanced biological systems.

6.2.3 Removal of cytostatic drugs by disinfection processes

6.2.3.1 Chlorination

Chlorination is one of the most commonly used disinfection processes, mainly due to its low cost, and is commonly used as the final treatment step in the WWTPs, in order to remove the waterborne pathogens before discharging the effluents into receiving streams (Michael et al. 2013; Ferro et al. 2015). It is well known that chlorination may result in the formation of mutagenic/carcinogenic disinfection by-products deriving from the reaction of free chlorine with the various organic compounds present in wastewater, and some of these substances have been also proved to be carcinogenic in humans and animals (Amin et al. 2013). In contact with aqueous chlorine, pharmaceutical compounds may undergo oxidation reactions yielding disinfection by-products with different and even higher persistence and toxicity than their parent compounds (Negreira et al. 2015a). However, limited studies on the removal of cytostatic drugs through chlorination have been reported to date, making it difficult to reach conclusions regarding (i) the efficiency of this disinfection process in removing these compounds, (ii) the potential formation of disinfection by-products, and (iii) the harmful effects of the parent compounds and their by-products on various species (e.g., ecotoxicity, mutagenicity, teratogenicity, etc.).

Specifically, in the study of Azuma et al. (2015), the efficiency of two Japanese WWTPs using conventional activated sludge followed by chlorination in removing various pharmaceuticals, including two cytostatic drugs (i.e., *cyclophosphamide* and *capecitabine*), was investigated. The authors report the effluent concentrations of *cyclophosphamide* and *capecitabine*, after chlorination at 11 and 6 ng/L, respectively. However, no data regarding their concentrations in the influent stream are given in this study; hence, no conclusions regarding the removal efficiency of this process were discussed.

Considering the available bench-scale chlorination studies of cytostatic drugs, it was shown that a removal up to 60% of *erlotinib* in urban wastewater was achieved through chlorination (100 μg/L *erlotinib* and 1 mg/L chlorine) (Negreira et al. 2015a). In this study, 19 disinfection by-products of erlotinib, produced *via* chlorination and hydroxylation reactions, were identified, providing a first overview of the fate of this cytostatic drug in the aquatic environment, which can be further used for monitoring studies. In the study of Negreira et al. (2015b), it was found that *tamoxifen* was quite stable under chlorination of ultrapure water, while its major active metabolites, i.e., 4-hydroxy-tamoxifen and 4-hydroxy-N-desmethyl-tamoxifen, were rapidly degraded, each one yielding seven chlorinated by-products. The toxicity of these by-products was found to be significantly higher than that of the parent compounds, especially for the dechlorinated species, highlighting the risk these by-products might have on aquatic organisms. On the other hand, *etoposide* was found to react very quick (half-life time ($t_{1/2}$) around 4 min) in water containing free chlorine (1 mg/L of *etoposide* and 100 mg/L of chlorine) leading to the formation of two chlorinated by-products in only a few seconds, i.e., 3′-O-desmethyl etoposide (BP1) and BP2 ($C_{26}H_{25}O_{13}^{-}$), where its reaction rates depended on the pH (optimum pH: 6) and the concentration of free chlorine in water samples (Negreira et al. 2015c). The highly cytotoxic chlorinated by-product of *etoposide*, BP1, was also detected in environmental matrices, i.e., urban wastewater and river samples, at concentration levels of 14 and 33 ng/L, respectively, while *etoposide* was not detected in these samples (Negreira et al. 2015c), highlighting that the environmental risks of this disinfection by-product should be included in the future monitoring studies. The chlorination of *vincristine*, *vinblastine*, *vinorelbine*, and its major metabolite *4-O-deacetyl vinorelbine* in water and wastewater samples was investigated by Negreira et al. (2016). Under the examined chlorination conditions (i.e., 100 μg/L of each drug and 1 mg/L of free chlorine), *vincristine* was found to be quite stable, while *vinblastine*, *vinorelbine*, and *4-O-deacetyl vinorelbine* were quickly degraded. Among 65 disinfection by-products that were identified, 20 corresponded to mono-chlorinated compounds, 8 to dechlorinated, and 2 to tri-chlorinated compounds, with the latest being of major environmental concern. The rest by-products formed involved hydroxylation and oxidation reactions. According to a recent study of Yin et al. (2017), it was revealed that *methotrexate* could rapidly react ($t_{1/2}$ of 1.65 min) in chlorinated water (chlorine dose: 2 mg/L) at room temperature and under neutral conditions, leading to the formation of two by-products, namely 4-amino-3-chlorinated-N10-methylpteroylglutamic (monochloro-methotrexate) and 4-amino-3,5-dichloro-N10-methylpteroylglutamic (dichloro-methotrexate). These two chlorinated by-products were both detected in real hospital wastewater and it was found that both of them can inhibit the proliferation of zebrafish liver cells through S-phase arrest (Yin et al., 2017). This is also in agreement with the study of Roig et al. (2014), who report a rapid elimination of *methotrexate* in water samples, that was achieved through chlorination (chlorination conditions: 1 mg/L of *methotrexate* and 100 mg/L of free chlorine; methotrexate $t_{1/2}$ around 21 min).

In conclusion, the efficacy of chlorination, as disinfection step after CAS treatment, in removing the various cytostatic drugs, is still scarcely reported and should be extensively investigated in the near future, especially due to the formation of the various toxic chlorinated by-products, which may have higher toxic/cytotoxic/mutagenic properties than the parent compounds.

6.2.3.2 UV Irradiation

Ultraviolet (UV) irradiation has been increasingly used as disinfection process in various WWTPs worldwide as an alternative to chlorine. In direct photolysis, a molecule absorbing radiation may become unstable and subsequently decompose, while the indirect photolysis involves naturally occurring molecules, which generate strong reactive species (e.g., singlet oxygen, hydroxyl radicals (HO•), alkyl peroxyl radicals, etc.), that further react with various organic and inorganic compounds present in a certain water body (Rizzo et al. 2013; Michael et al. 2013).

Only two studies investigating the efficiency of UV irradiation at real scale as post-treatment of CAS treatment in degrading cytostatic drugs are available in the scientific literature, indicating insufficient removals for most of the cytostatic compounds examined, except *cyclophosphamide*, which was found to be completely removed (Roberts and Thomas 2006; Rabii et al. 2014). Specifically, according to the study of Roberts and Thomas (2006), *tamoxifen* was poorly removed by CAS treatment followed by UV disinfection, achieving a removal rate up to 30%. In the study of Rabii et al. (2014), the effectiveness of two Canadian WWTPs, i.e., one CAS WWTP and one tertiary WWTP (CAS + UV irradiation) in the degradation of various cytostatic drugs (i.e., *cyclophosphamide*, *gemcitabine*, *ifosfamide*, *methotrexate*, *irinotecan* and *epirubicin*), was investigated. An insignificant removal of *cyclophosphamide* during conventional biological treatment was achieved, up to 18%, while by applying the UV irradiation as disinfection process, complete removal of this drug was reached. The latter is unexpected from a chemical point of view, since *cyclophosphamide* molecule does not contain any aromatic ring or double C = C bonds, which can absorb photons under UV irradiation; and this high removal is not in accordance with the results of the studies where UV photolysis of *cyclophosphamide* was applied at bench scale (Kim et al. 2009; Česen et al. 2015; Zhang et al. 2017). In contrast, effluents' concentrations of *methotrexate* from a tertiary WWTP (CAS + UV irradiation) were found to be slightly higher (20–22 ng/L) than the influent concentrations (17–20 ng/L), indicating that probably the conjugates formed are broken down after tertiary treatment, giving a rise to higher concentrations of the drug's free form compared to that found in the inflow. The other cytostatic drugs (i.e., *gemcitabine*, *ifosfamide*, *irinotecan*, and *epirubicin*) were not detected neither in the raw wastewater nor in the treated effluents (Rabii et al. 2014).

Insignificant removal was achieved through the application of UV photolysis for the investigated alkylating agents in water or artificial wastewater, while high efficiency was achieved for the majority of antimetabolites and plant alkaloids. More specifically, in the study of Kim et al. (2009), a low degradation of *cyclophosphamide* was achieved (i.e., 12% degradation) with the application of UV irradiation (UV dose: 5201 mJ/cm^2). This is also in agreement with the results of a recent study of Zhang et al. (2017), where *cyclophosphamide* shown negligible photodegradation effect (<1% degradation) by direct UV photolysis (UV dose: 400 mJ/cm^2) in water, mainly due to the fact that the amides bonds in the molecule of *cyclophosphamide* cannot be cleaved efficiently through R-CO or CO-N bond breakage, since the resonance stability between N-C and C-O bonds is extremely high. The above is also in agreement with the findings of the study of Česen et al. (2015), where it was found that direct UV photolysis at bench scale was not efficient in degrading neither *cyclophosphamide* nor *ifosfamide* from artificial wastewater. In a more recent study of Česen et al. (2016), it was highlighted that during UV treatment of ultrapure water spiked with *cyclophosphamide* and *ifosfamide* (26.1 mg/L of each drug), four TPs of each drug were formed (i.e., Ndecl-CP, keto-CP, CP-TP138a, CP-TP138, IF-TP199, IF-TP275, IF-TP138, and IF-TP259), while none of these TPs were detected neither in the influent nor in the effluent samples taken from a Slovenian WWTP in the framework of this study. No degradation was observed also for *5-fluorouracil* (initial dose: 200 μg/L) in ultrapure water under bench-scale UV photolysis, according to Lin and Lin (2014); while a slight degradation, up to 24% (UV dose: 400 mJ/cm^2) was observed in the study of Zhang et al. (2017). Moreover, *melphalan*, *etoposide*, and *prednisone* were completely degraded under UV-C irradiation (after 1, 30, and 5 min, respectively), *gemcitabine* and *capecitabine* were almost completely eliminated (90% degradation) after 90 min of photolysis, while *cytarabine*, *ifosfamide*, and *cyclophosphamide* were found to be recalcitrant to UV-C irradiation, according to Franquet-Griell et al. (2016). Nevertheless, it should be highlighted that the addition of H_2O_2 during UV treatment was proved to be highly effective for improving the degradation of the above-mentioned cytostatic drugs, even up to 100% (Kim et al. 2009; Lin and Lin 2014; Česen et al. 2015; Česen et al. 2016; Zhang et al. 2017).

No solid conclusions can be obtained regarding the efficacy of UV irradiation as post-treatment of CAS process in the degradation of cytostatic drugs, since limited and contradictory data exist, until now, requiring further and more thorough investigation. In addition, it is of high importance to investigate the various TPs formed during this disinfection process, especially with regard to their toxicity compared to the parent compounds.

6.2.3.3 Ozonation

Ozone is a powerful oxidant and has been increasingly used for the treatment and disinfection of urban wastewater effluents, also as an alternative to chlorine. Ozone reacts with organic contaminants either by the direct reaction of molecular ozone or by the reaction of free radicals (mainly HO•) produced by the decomposition of ozone. Molecular ozone reacts selectively with unsaturated bonds, aromatic rings, and amino groups, while the hydroxyl radicals react unselectively with organic and inorganic compounds (Garcia-Ac et al. 2010).

Only one study applying ozonation at real scale as tertiary stage of biological treatment investigating the degradation of two cytostatic drugs (i.e., *cyclophosphamide* and *capecitabine*) has been published up to date. Azuma et al. (2015) found that in the effluent samples from a WWTP using ozonation as tertiary treatment, the mean concentrations of the cytostatic drugs examined were significantly lower than the concentrations detected in the effluent samples of a WWTP using chlorination as disinfection process. However, no data are available regarding their influent concentrations, in order to estimate their overall removal efficiency. For example, the effluent concentration of *cyclophosphamide* after chlorination was 11 ng/L, while after ozonation it was reduced to 7 ng/L. The same was observed for *capecitabine*, which was 6 ng/L after chlorination and 2 ng/L after ozonation, indicating the higher efficiency of ozonation compared to chlorination on the degradation of these compounds.

Moreover, considering the results obtained from bench-scale studies, *cyclophosphamide* (initial concentration: 10^{-3} mol/L) was found to significantly degrade up to 80% through ozonation (ozone dose, 30 mg/L, and pH, 9) in water during the first 10 min of treatment (Fernandez et al. 2010), while a medium degradation of *cyclophosphamide* and *ifosfamide* (i.e., 42 and 36%, respectively) was reached after 120 min of ozonation of water samples by applying a lower ozone dose of 10 mg/L, according to the study of Česen et al. (2015). Complete degradation (higher than 97%) of *cyclophosphamide*, *irinotecan*, *ifosfamide*, and *capecitabine* via ozonation (ozone gas concentration: 43.9 g/m^3) of real hospital wastewater was reported by Ferre-Aracil et al. (2016).

There is extremely limited information available in the scientific literature regarding the ozonation treatment of cytostatic drugs in real applications as disinfection process after CAS treatment, making it difficult to reach conclusions regarding its efficiency in degrading these compounds and the potential of the oxidation products formation along with their toxicity, which might be higher than those of the biologically treated effluents. However, from the first findings of the bench-scale studies, ozonation seems to be efficient in removing these drugs, while no data exist on the oxidation products formed and the toxicity of the treated effluents.

6.3 Discussion, Concluding Remarks and Current Challenges

Conventional biological wastewater treatment facilities are designed in order to eliminate the organic/inorganic content (e.g., carbon, nitrogen, phosphorus, etc.) of urban wastewater to a high percentage, but not pharmaceutical compounds, such as cytostatic drugs. Cytostatic drugs are compounds that are of high environmental relevance, due to their lack of specific mode of action – since they are non-selective – and due to the fact that they can be extremely harmful (i.e., fetotoxic, genotoxic, mutagenic, and teratogenic side effects) to all living eukaryotic organisms even at very low concentrations (e.g., sub-ng/L).

Several cytostatic drugs have been identified in secondary- and tertiary-treated effluents of various WWTPs all over the world, demonstrating their high resistance to biological and tertiary treatment. The overall removal efficiency of WWTPs varied among different cytostatic compounds, highly associated with their physico-chemical properties and biodegradability, which also differs significantly. However, it was noted that more comprehensive studies are required to thoroughly examine the behavior of cytostatic drugs under conventional biological treatment, focusing on the elimination processes taking place within the WWTPs, including sorption onto sewage sludge.

It is well known that in order to further improve the efficiency of WWTPs in sufficiently degrading the various microcontaminants, including cytostatic compounds from urban wastewater, two possible options exist: (i) the upgrade of current WWTPs and/or (ii) the application of advanced treatment technologies, such as advanced biological, advanced chemical oxidation and/or membrane filtration and separation processes. In general, it may be more environmentally friendly and cost-effective to directly improve the treatment efficiencies of the existing WWTPs by prolonging, for example, their HRTs and SRTs and/or combining nutrient removal stages, than adding an extra advanced treatment process, as post-treatment. However, in many cases the application of an advanced process might be the best and maybe also the only choice, in order to achieve a sufficient removal of those persistent and toxic compounds from wastewater, although it might be accompanied by higher environmental impacts and higher capital and operational cost. Moreover, more in-depth investigation should be obtained regarding the negative effects (i.e., chemical stress) of cytostatic compounds on the microbial community of the sludge of the biological systems.

The utilization of advanced biological treatment processes, such as MBR systems, exhibited a hopeful choice for effective treatment of cytostatic drugs. However, only bench- and pilot-scale systems have been examined until now, revealing a higher treatment efficiency compared with CAS systems. Further investigation in real-scale applications is required, in order to have more concrete and reliable results, with regard to the toxicity of the MBR-treated effluents.

On the other hand, advanced chemical oxidation processes, mainly applied as post-treatment of conventional treatment, can significantly improve the treatment efficiency, since the use of a strong oxidizing agent (i.e., HO•) can result in a high degree of wastewater treatment, including the breakdown of recalcitrant and toxic compounds, such as cytostatic drugs. These new advanced processes seem likely to be very effective in degrading such compounds; however, further verification is needed, since their efficacy has just started to be investigated in the last years, mainly at bench-scale applications. More extensive research is required in this field with regard to the pharmacologically active and/or potentially more toxic TPs formed during these oxidation processes.

Finally, environmental and economic analyses are considered necessary for the large-scale applications of the advanced biological and advanced chemical oxidation processes used for the removal of cytostatic residues from urban effluents in order to gain a clearer view of it all.

References

Allwood M, Stanley A, Wright P (2002) The cytotoxics handbook. Radcliffe Medical Press, England, p 512

Amin MM, Hashemi H, Bovini AM, Hung YT (2013) A review on wastewater disinfection. Int J Environ Heal Eng 2(1):22

Avella AC, Delgado LF, Görner T, Albasi C, Galmiche M, De Donato P (2010) Effect of cytostatic drug presence on extracellular polymeric substances formation in municipal wastewater treated by membrane bioreactor. Bioresource Technol 101(2):518–526

Azuma T, Ishiuchi H, Inoyama T, Teranishi Y, Yamaoka M, Sato T, Mino Y (2015) Occurrence and fate of selected anticancer, antimicrobial, and psychotropic pharmaceuticals in an urban river in a subcatchment of the Yodo River basin, Japan. Environ Sci Pollut R 22(23):18676–18686

Besse JP, Latour JF, Garric J (2012) Anticancer drugs in surface waters: what can we say about the occurrence and environmental significance of cytotoxic, cytostatic and endocrine therapy drugs? Environ Int 39(1):73–86

Booker V, Halsall C, Llewellyn N, Johnson A, Williams R (2014) Prioritising anticancer drugs for environmental monitoring and risk assessment purposes. Sci Total Environ 473:159–170

Buerge IJ, Buser HR, Poiger T, Müller MD (2006) Occurrence and fate of the cytostatic drugs cyclophosphamide and ifosfamide in wastewater and surface waters. Environ Sci Technol 40(23):7242–7250

Busetti F, Linge KL, Heitz A (2009) Analysis of pharmaceuticals in indirect potable reuse systems using solid-phase extraction and liquid chromatography–tandem mass spectrometry. J Chromatogr A 1216(31):5807–5818

Castiglioni S, Bagnati R, Calamari D, Fanelli R, Zuccato E (2005) A multiresidue analytical method using solid-phase extraction and high-pressure liquid chromatography tandem mass spectrometry to measure pharmaceuticals of different therapeutic classes in urban wastewaters. J Chromatogr A 1092(2):206–215

Česen M, Kosjek T, Laimou-Geraniou M, Kompare B, Širok B, Lambropolou D, Heath E (2015) Occurrence of cyclophosphamide and ifosfamide in aqueous environment and their removal by biological and abiotic wastewater treatment processes. Sci Total Environ 527:465–473

Česen M, Kosjek T, Busetti F, Kompare B, Heath E (2016) Human metabolites and transformation products of cyclophosphamide and ifosfamide: analysis, occurrence and formation during abiotic treatments. Environ Sci Poll Res Int 23(11):11209–11223

Coetsier CM, Spinelli S, Lin L, Roig B, Touraud E (2009) Discharge of pharmaceutical products (PPs) through a conventional biological sewage treatment plant: MECs vs PECs? Environ Intern 35(5):787–792

Delgado LF, Dorandeu C, Marion B, Gonzalez C, Faucet-Marquis V, Schetrite S, Albasi C (2009) Removal of a cytostatic drug by a membrane bioreactor. Desal Water Treat 9(1-3):112–118

Delgado LF, Faucet-Marquis V, Pfohl-Leszkowicz A, Dorandeu C, Marion B, Schetrite S, Albasi C (2011) Cytotoxicity micropollutant removal in a crossflow membrane bioreactor. Bioresource Technol 102(6):4395–4401

Fernández LA, Hernández C, Bataller M, Véliz E, López A, Ledea O, Padrón S (2010) Cyclophosphamide degradation by advanced oxidation processes. Water Environ J 24(3):174–180

Franquet-Griell H, Gómez-Canela C, Ventura F, Lacorte S (2015) Predicting concentrations of cytostatic drugs in sewage effluents and surface waters of Catalonia (NE Spain). Environ Res 138:161–172

Ferrando-Climent L, Rodriguez-Mozaz S, Barceló D (2013) Development of a UPLC-MS/MS method for the determination of ten anticancer drugs in hospital and urban wastewaters, and its application for the screening of human metabolites assisted by information-dependent acquisition tool (IDA) in sewage samples. Anal Bioanal Chem 405(18):5937–5952

Ferrando-Climent L, Rodriguez-Mozaz S, Barceló D (2014) Incidence of anticancer drugs in an aquatic urban system: from hospital effluents through urban wastewater to natural environment. Environ Pollut 193:216–223

Ferre-Aracil J, Valcárcel Y, Negreira N, de Alda ML, Barceló D, Cardona SC, Navarro-Laboulais J (2016) Ozonation of hospital raw wastewaters for cytostatic compounds removal. Kinetic modelling and economic assessment of the process. Sci Total Environ 556:70–79

Ferro G, Polo-López MI, Fernández-Ibáñez P (2015) Conventional and new processes for urban wastewater disinfection: effect on emerging and resistant microorganisms. In: Advanced treatment technologies for urban wastewater reuse. Springer, New York, pp 107–128

Franquet-Griell H, Ventura F, Boleda MR, Lacorte S (2016) Do cytostatic drugs reach drinking water? The case of mycophenolic acid. Environ Pollut 208:532–536

Garcia-Ac A, Broséus R, Vincent S, Barbeau B, Prévost M, Sauvé S (2010) Oxidation kinetics of cyclophosphamide and methotrexate by ozone in drinking water. Chemosphere 79(11):1056–1063

Gómez-Canela C, Cortés-Francisco N, Oliva X, Pujol C, Ventura F, Lacorte S, Caixach J (2012) Occurrence of cyclophosphamide and epirubicin in wastewaters by direct injection analysis–liquid chromatography–high-resolution mass spectrometry. Environ Sci Pollut Res 19(8):3210–3218

Gómez-Canela C, Ventura F, Caixach J, Lacorte S (2014) Occurrence of cytostatic compounds in hospital effluents and wastewaters, determined by liquid chromatography coupled to high-resolution mass spectrometry. Analyt Bioanalyt Chem 406(16):3801–3814

Heath E, Česen M, Negreira N, de Alda ML, Ferrando-Climent L, Blahova L, Nguyen TV, Adahchour M, Ruebel A, Llewellyn N, Ščančar J (2016) First inter-laboratory comparison exercise for the determination of anticancer drugs in aqueous samples. Environ Sci Pollut Res 23(15):14692–14704

International Agency for Research on Cancer (2017) The IARC register. http://globocan.iarc.fr/Pages/fact_sheets_cancer.aspx. Accessed 15 May 2017

Isidori M, Lavorgna M, Russo C, Kundi M, Žegura B, Novak M, Filipič M, Mišík M, Knasmueller S, de Alda ML, Barceló D (2016) Chemical and toxicological characterisation of anticancer drugs in hospital and municipal wastewaters from Slovenia and Spain. Environ Pollut 219:275–287

Jim TY, Bisceglia KJ, Bouwer EJ, Roberts AL, Coelhan M (2012) Determination of pharmaceuticals and antiseptics in water by solid-phase extraction and gas chromatography/mass spectrometry: analysis via pentafluorobenzylation and stable isotope dilution. Analyt Bioanalyt Chem 403(2):583–591

Johnson AC, Jürgens MD, Williams RJ, Kümmerer K, Kortenkamp A, Sumpter JP (2008) Do cytotoxic chemotherapy drugs discharged into rivers pose a risk to the environment and human health? an overview and UK case study. J Hydrol 348(1):167–175

Kim I, Yamashita N, Tanaka H (2009) Photodegradation of pharmaceuticals and personal care products during UV and UV/H_2O_2 treatments. Chemosphere 77(4):518–525

Köhler C, Venditti S, Igos E, Klepiszewski K, Benetto E, Cornelissen A (2012) Elimination of pharmaceutical residues in biologically pre-treated hospital wastewater using advanced UV irradiation technology: a comparative assessment. J Hazard Mater 239:70–77

Kondo N, Takahashi A, Ono K, Ohnishi T (2010) DNA damage induced by alkylating agents and repair pathways. J Nucleic Acids 2010:7

Kosjek T, Heath E (2011) Occurrence, fate and determination of cytostatic pharmaceuticals in the environment. TrAC Trend Anal Chem 30(7):1065–1087

Kovalova L, Siegrist H, Singer H, Wittmer A, McArdell CS (2012) Hospital wastewater treatment by membrane bioreactor: performance and efficiency for organic micropollutant elimination. Environ Sci Technol 46(3):1536–1545

Krynetskaia N, Xie H, Vucetic S, Obradovic Z, Krynetskiy E (2008) High mobility group protein B1 is an activator of apoptotic response to antimetabolite drugs. Mol Pharmacol 73(1):260–269

Kümmerer K, Al-Ahmad A (1997) Biodegradability of the anti-tumour agents 5-Fluorouracil, cytarabine, and gemcitabine: impact of the chemical structure and synergistic toxicity with hospital effluent. CLEAN-Soil Air Water 25(4):166–172

Kümmerer K, Haiß A, Schuster A, Hein A, Ebert I (2016) Antineoplastic compounds in the environment - substances of special concern. Environ Sci Pollut R 23(15):14791–14804

Lenz K, Koellensperger G, Hann S, Weissenbacher N, Mahnik SN, Fuerhacker M (2007a) Fate of cancerostatic platinum compounds in biological wastewater treatment of hospital effluents. Chemosphere 69(11):1765–1774

Lenz K, Mahnik SN, Weissenbacher N, Mader RM, Krenn P, Hann S, Koellensperger G, Uhl M, Knasmüller S, Ferk F, Bursch W (2007b) Monitoring, removal and risk assessment of cytostatic drugs in hospital wastewater. Water Sci Technol 56(12):141–149

Lin HHH, Lin AYC (2014) Photocatalytic oxidation of 5-fluorouracil and cyclophosphamide via UV/TiO_2 in an aqueous environment. Water Res 48:559–568

Liu X, Zhang J, Yin J, Duan H, Wu Y, Shao B (2010) Analysis of hormone antagonists in clinical and municipal wastewater by isotopic dilution liquid chromatography tandem mass spectrometry. Anal Bioanal Chem 396(8):2977–2985

Llewellyn N, Lloyd P, Jürgens MD, Johnson AC (2011) Determination of cyclophosphamide and ifosfamide in sewage effluent by stable isotope-dilution liquid chromatography–tandem mass spectrometry. J Chromatogr A 1218(47):8519–8528

Lutterbeck CA, Machado ÊL, Kümmerer K (2015) Photodegradation of the antineoplastic cyclophosphamide: a comparative study of the efficiencies of UV/H_2O_2, UV/Fe^{2+}/H_2O_2 and UV/TiO_2 processes. Chemosphere 120:538–546

Mahnik SN, Lenz K, Weissenbacher N, Mader RM, Fuerhacker M (2007) Fate of 5-fluorouracil, doxorubicin, epirubicin, and daunorubicin in hospital wastewater and their elimination by activated sludge and treatment in a membrane-bio-reactor system. Chemosphere 66(1):30–37

Martín J, Camacho-Muñoz D, Santos JL, Aparicio I, Alonso E (2011) Simultaneous determination of a selected group of cytostatic drugs in water using high-performance liquid chromatography-triple-quadrupole mass spectrometry. J Sep Sci 34(22):3166–3177

Martín J, Camacho-Muñoz D, Santos JL, Aparicio I, Alonso E (2014) Occurrence and ecotoxicological risk assessment of 14 cytostatic drugs in wastewater. Water Air Soil Pollut 225(3):1896

Mater N, Geret F, Castillo L, Faucet-Marquis V, Albasi C, Pfohl-Leszkowicz A (2014) In vitro tests aiding ecological risk assessment of ciprofloxacin, tamoxifen and cyclophosphamide in range of concentrations released in hospital wastewater and surface water. Environ Int 63:191–200

Michael I, Rizzo L, McArdell CS, Manaia CM, Merlin C, Schwartz T, Dagot C, Fatta-Kassinos D (2013) Urban wastewater treatment plants as hotspots for the release of antibiotics in the environment: a review. Water Res 47(3):957–995

Michael C, Bayona JM, Lambropoulou D, Agüera A, Fatta-Kassinos D (2017) Two important limitations relating to the spiking of environmental samples with contaminants of emerging concern: How close to the real analyte concentrations are the reported recovered values? Environ Sci Pollut R 24(17):15202–15205

Negreira N, de Alda ML, Barceló D (2013) On-line solid phase extraction–liquid chromatography–tandem mass spectrometry for the determination of 17 cytostatics and metabolites in waste, surface and ground water samples. J Chromatogr A 1280:64–74

Negreira N, de Alda ML, Barceló D (2014) Cytostatic drugs and metabolites in municipal and hospital wastewaters in Spain: filtration, occurrence, and environmental risk. Sci Total Environ 497:68–77

Negreira N, Regueiro J, de Alda ML, Barceló D (2015a) Degradation of the anticancer drug erlotinib during water chlorination: Non-targeted approach for the identification of transformation products. Water Res 85:103–113

Negreira N, Regueiro J, de Alda ML, Barceló D (2015b) Transformation of tamoxifen and its major metabolites during water chlorination: identification and in silico toxicity assessment of their disinfection byproducts. Water Res 85:199–207

Negreira N, de Alda ML, Barceló D (2015c) Degradation of the cytostatic etoposide in chlorinated water by liquid chromatography coupled to quadrupole-Orbitrap mass spectrometry: Identification and quantification of by-products in real water samples. Sci Total Environ 506: 36-45

Negreira N, Regueiro J, de Alda ML, Barceló D (2016) Reactivity of vinca alkaloids during water chlorination processes: Identification of their disinfection by-products by high-resolution quadrupole-Orbitrap mass spectrometry. Sci Total Environ 544:635–644

Pietsch J, Günther J, Henle T, Dreßler J (2008) Simultaneous determination of thirteen plant alkaloids in a human specimen by SPE and HPLC. J Sep Sci 31(13):2410–2416

Rabii FW, Segura PA, Fayad PB, Sauvé S (2014) Determination of six chemotherapeutic agents in municipal wastewater using online solid-phase extraction coupled to liquid chromatography-tandem mass spectrometry. Sci Total Environ 487:792–800

Radjenović J, Petrović M, Barceló D (2009) Fate and distribution of pharmaceuticals in wastewater and sewage sludge of the conventional activated sludge (CAS) and advanced membrane bioreactor (MBR) treatment. Water Res 43(3):831–841

Rizzo L, Manaia C, Merlin C, Schwartz T, Dagot C, Ploy MC, Michael I, Fatta-Kassinos D (2013) Urban wastewater treatment plants as hotspots for antibiotic resistant bacteria and genes spread into the environment: a review. Sci Total Environ 447:345–360

Roberts PH, Thomas KV (2006) The occurrence of selected pharmaceuticals in wastewater effluent and surface waters of the lower Tyne catchment. Sci Total Environ 356(1):143–153

Roig B, Marquenet B, Delpla I, Bessonneau V, Sellier A, Leder C, Thomas O, Bolek R, Kummerer K (2014) Monitoring of methotrexate chlorination in water. Water Res 57:67–75

Rosano TG (1998) In: Ellenhorn MJ, Schonwald S, Ordog G, Wasserberger J (eds) Ellenhorn's medical toxicology: diagnosis and treatment of human poisoning. Williams and Wilkins, Baltimore, MD

Rowney NC, Johnson AC, Williams RJ (2009) Cytotoxic drugs in drinking water: a prediction and risk assessment exercise for the Thames catchment in the United Kingdom. Environ Ttoxicol Chem 28(12):2733–2743

Ruel SM, Choubert JM, Esperanza M, Miège C, Madrigal PN, Budzinski H, Le Ménach K, Lazarova V, Coquery M (2011) On-site evaluation of the removal of 100 micro-pollutants through advanced wastewater treatment processes for reuse applications. Water Sci Technol 63(11):2486–2497

Steger-Hartmann TKKHA, Kümmerer K, Hartmann A (1997) Biological degradation of cyclophosphamide and its occurrence in sewage water. Ecotox Environ Safe 36(2):174–179

Sweetman SC, Martindale W (2011) Martindale: the complete drug reference. Pharmaceutical Press, London

Tauxe-Wuersch A, De Alencastro LF, Grandjean D, Tarradellas J (2006) Trace determination of tamoxifen and 5-fluorouracil in hospital and urban wastewaters. Int J Environ Anal Chem 86(7):473–485

Ternes TA (1998) Occurrence of drugs in German sewage treatment plants and rivers. Water Res 32(11):3245–3260

Ternes TA, Kreckel P, Mueller J (1999) Behaviour and occurrence of estrogens in municipal sewage treatment plants-II. Aerobic batch experiments with activated sludge. Sci Total Environ 225(1):91–99

Thomas KV, Dye C, Schlabach M, Langford KH (2007) Source to sink tracking of selected human pharmaceuticals from two Oslo city hospitals and a wastewater treatment works. J Environ Monitor 9(12):1410–1418

World Health Organization (2017) The WHO register. http://www.who.int/en/. Accessed 20 June 2017.

Wu Y, Wang X, Tay MQX, Oh S, Yang L, Tang C, Cao B (2017) Metagenomic insights into the influence of salinity and cytostatic drugs on the composition and functional genes of microbial community in forward osmosis anaerobic membrane bioreactors. Chem Eng J 326:462–469

Yin J, Shao B, Zhang J, Li K (2010) A preliminary study on the occurrence of cytostatic drugs in hospital effluents in Beijing, China. Bull Environ Contam Toxicol 84(1):39

Yin J, Niu Y, Shao B (2017) Products of methotrexate during chlorination. J Environ Sci 55:100–108

Zhang J, Chang VW, Giannis A, Wang JY (2013) Removal of cytostatic drugs from aquatic environment: a review. Sci Total Environ 445:281–298

Zhang Y, Zhang J, Xiao Y, Chang VW, Lim TT (2017) Direct and indirect photodegradation pathways of cytostatic drugs under UV germicidal irradiation: process kinetics and influences of water matrix species and oxidant dosing. J Hazard Mater 324:481–488

Chapter 7
Degradation and Elimination of Anticancer Drugs by Water and Wastewater Treatment – Toxicity and Biodegradability Before and After the Treatment

Carlos Alexandre Lutterbeck, Ênio Leandro Machado, and Klaus Kümmerer

Abstract Anticancer drugs are chemically spoken a broad group of pharmaceuticals especially designed to treat cancer. Many have been widely used for chemotherapy for decades. They are intrinsically toxic. After administration the active compound is excreted together with its metabolites because of incomplete mineralization in the human or animal body. Thereby they end up in hydrosphere and the pedosphere. In the hydrosphere they are present at the microgram per litre range or below. Therefore, they are part of the so-called micropollutants. During the last 20 years, researchers have focused their attention on the environmental fate of anticancer drugs as well as on the risks that these compounds may pose to humans and the environment. In general, these compounds are characterized by a poor environmental biodegradability and often they are not completely removed by conventional wastewater treatments. So called advanced oxidation processes (AOPs), i.e. additional oxidative treatment of wastewater treatment plant effluents has therefore been taken into consideration to solve the problem. Such processes are also used as a final treatment for the treatment of potable water. Among them is the treatment with UV light. Some advanced oxidation processes (AOPs) have been shown to eliminate pharmaceuticals in general at a high degree. However, often incomplete mineralization results in the formation of unwanted transformation products (TPs) of unknown chemical structure, toxicity and fate. In some cases, it was shown that TPs were easier to biodegrade compared to the parent compounds. Nevertheless, it was also found that many TPs are not biodegradable and are more toxic or exhibiting a different toxicity profile than the parent compounds.

C. A. Lutterbeck (✉) · Ê. L. Machado
Graduate Program in Environmental Technology, Universidade de Santa Cruz do Sul, Santa Cruz do Sul, Rio Grande do Sul, Brazil
e-mail: carlosalexandre@unisc.br

K. Kümmerer
Sustainable Chemistry and Material Resources, Institute of Sustainable and Environmental Chemistry, Faculty of Sustainability, Leuphana University of Lüneburg, Lüneburg, Germany

E. Heath et al. (eds.), *Fate and Effects of Anticancer Drugs in the Environment*,
https://doi.org/10.1007/978-3-030-21048-9_7

The present chapter provides an overview of different treatments used (chlorination, ozonation, photo treatment and other nonconventional treatments) to remove anticancer drugs from water (surface, distilled and ultrapure) and wastewater. It evaluates their efficiency based on the degrees of elimination, mineralization, biodegradation and toxicity of the parent compounds as well as possibly formed TPs.

Keywords Anticancer Drug · Degradation · Elimination · Wastewater Treatment · Toxicity · Biodegradability · Transformation Product

7.1 Introduction

All over the world, the number of cancer patients is dramatically increasing. According to a report from the World Health Organization (WHO), cancer accounted for 8.8 million deaths worldwide in 2015 (WHO 2015). Furthermore, it is also believed that the number of new cases will rise by about 70% over the next 2 decades (WHO 2015). While the number of studies on pharmaceuticals in the environment has increased considerably in recent years, one important class has, somewhat surprisingly, been overlooked in this discourse: anticancer drugs. Nowadays, approximately 4000 active pharmaceutical compounds, used in human and veterinary drugs, are available on the European market (Mompelat et al. 2009), and among them are approximately 100 anticancer drugs (Kümmerer et al. 2016).

Anticancer drugs are a group of molecules with highly diverse chemical structural formula which is related to different modes of action. Many have been widely used in chemotherapy for decades. They are especially designed to treat cancer and certain autoimmune diseases. Because of their none-selective mode of action, that is, because they affect both cancerous and all fast-dividing cells, anticancer drugs have been regarded as potentially carcinogenic, genotoxic, mutagenic and teratogenic compounds (Allwood et al. 2002; Eitel et al. 1999). After administration, anticancer drugs are metabolized in the human body. Parent compounds and metabolites are excreted by humans into the sewage system. Elimination of anticancer drugs by conventional wastewater treatments is often incomplete and inefficient like for other pharmaceuticals and chemicals (Wang and Lin 2014; Zhang et al. 2013). It has been shown that many anticancer drugs are not biodegradable (Kümmerer et al. 1996, 1997; Yu et al. 2006; Lutterbeck et al. 2015a). The first studies confirming the presence of anticancer drugs in aquatic systems were already published in the 1980s (Aherne et al. 1985; Richardson and Bowron 1985). Further studies were released in the early 1990s (Aherne et al. 1990), and these kinds of investigations continue till today (Besse et al. 2012; Santana-Viera et al. 2016, 2017; Olalla et al. 2018). After reaching surface waters, parent compounds and their metabolites can undergo several physicochemical/biological processes (e.g. dilution, hydrolysis, biodegradation, photolysis and sorption to river bed sediments). Some can lead to the formation of transformation products (TPs), which might be even more harmful and persistent than parent compounds (Mahmoud et al. 2013; Pérez-Estrada et al. 2008).

Some typical examples of anticancer drugs that are among the most consumed and therefore frequently detected in different aquatic compartments are: Methotrexate (MTX), Tamoxifen (TAM), Cyclophosphamide (CYC), Ifosfamide (IF), 5-Fluorouracil (5-FU), Doxorubicin (DOX) and Cytarabine (CYT).

Methotrexate was introduced onto the pharmaceutical market in the 1940s, and today is still widely used at high doses for the chemotherapy of various forms of cancer (bronchial, breast and ovarian cancer, lymphomas, leukaemia) (Rubino 2001). Up to 90% of the unchanged drug can be excreted in urine and faeces (Lutterbeck et al. 2015a). Some studies reveal the presence of MTX in hospital and wastewater treatment plant (WWTP) effluents, and even in surface waters at a concentration range of ng/L (Halling-Sørensen et al. 1998; Castiglioni et al. 2005; Yin et al. 2010; Besse et al. 2012).

Tamoxifen is a non-steroidal antiestrogen that has been used to treat breast cancer for many years and was detected in urban and hospital wastewater samples as well as in surface waters in the range of ng/L (Ashton et al. 2004; Ferrando-Climent et al. 2013; Thomas and Hilton 2004). TAM is considered as a pro-drug since it is known to exert its pharmacological effect through its major active metabolites, 4-hydroxy-tamoxifen and 4-hydroxy-N-desmethyl-tamoxifen, which are mainly excreted in the urine in the days following administration (Negreira et al. 2015a).

Cyclophosphamide is an alkylating agent that has been in clinical use since the late 1950s to treat various forms of cancer (bronchial, breast, and ovarian cancer, lymphomas, leukaemia) and is also used as an immunosuppressant for autoimmune diseases and after organ transplantations. It acts by inhibiting and altering DNA replication. After application, about 20% of the parent compound is released unchanged through renal excretion (Rowney et al. 2009). Several studies have already detected the presence of CYC in different water compartments in concentrations ranging from ng/L to μg/L (Buerge et al. 2006; Yin et al. 2010; Gómez-Canela et al. 2013; Booker et al. 2014; Santos et al. 2017; Olalla et al. 2018).

Ifosfamide is an analogue of CYC with a similar spectrum of antitumour activity but higher toxicity potential and financial costs than CYC (Mioduszewska et al. 2016). Typical excretion values of un-metabolized IF vary between 27% and 50% (Boos et al. 1991). Many authors have already reported the presence of IF in different water matrices, such as hospital effluents, municipal WWTPs and surface waters in concentrations from low ng/L up to μg/L (Kümmerer et al. 1997; Steger-Hartmann et al. 1996; Yin et al. 2010; Buerge et al. 2006; Ternes 1998).

5-Fluorouracil is an antimetabolite that works by inhibiting DNA synthesis and consequently altering the tumour growth. It has been used for more than 50 years to treat solid tumours like colorectal, breast, skin, bladder and lung cancers (Mahnik et al. 2007). Approximately 15% of the parent compound of 5-FU is excreted unchanged (Zhang et al. 2013). 5-FU has already been detected in the aquatic environment at concentrations ranging from ng/L up to μg/L (Kovalova et al. 2009; Mahnik et al. 2004, 2007; Weissbrodt et al. 2009).

Doxorubicin is an anticancer drug that belongs to the group of anthracyclines and has been prescribed since the 1960s and is being frequently used in the treatment of haematological and solid neoplasms, including acute leukaemia, high-grade lymphoma, breast cancer, and bladder cancer (Mahnik et al. 2006). Approximately 3.5–5.7% of administered DOX is excreted in unmetabolized form via urine within 24 h (Dorr and VonHoff 1994). DOX has been detected in hospital effluents in concentrations ranging from low ng/L up to μg/L (Lenz et al. 2007; Mahnik et al. 2006; Yin et al. 2010).

Cytarabine, a pyrimidine analogue, is used to treat certain types of leukaemia (acute non-lymphocytic leukaemia, acute lymphocytic leukaemia, chronic myelocytic leukaemia). Approximately 10% of CYT is excreted by urine (Zhang et al. 2013), and it has been detected in WWTPs influents, effluents and surface water in concentrations up to 9.2, 14 and 13 ng/L, respectively (Martín et al. 2011; Kosjek and Heath 2011).

Because of the presence of these compounds in the environment, researchers focused their attention on the environmental fate of anticancer drugs as well on the risks that these compounds may pose to the environment, and humans once they reach drinking water resources. Despite the fairly low consumption rates and low environmental concentrations, especially in comparison to other groups of pharmaceuticals (such as antibiotics, antipsychotics, pain killers and beta-blockers), it is impossible to establish threshold values for the lowest concentrations of these compounds that would have a certain effect on aqueous biota (Kosjek and Heath 2011). Regardless, some authors believe that, due to their toxicological properties and poor biodegradability (Toolaram et al. 2014), anticancer drugs might have adverse effects on all eukaryotic organisms even at very low concentrations (Besse et al. 2012; Johnson et al. 2008).

The elimination of anticancer drugs and other pharmaceuticals by conventional wastewater treatment, such as conventional activated sludge process and biological filtration, is often incomplete and inefficient since these drugs are recalcitrant to biodegradation and have low adsorption capacity (Wang and Lin 2014; Zhang et al. 2013). Furthermore, oxidative methods are often applied for the production of drinking water for hygienically reasons. However, if chemicals and pharmaceuticals, among them highly toxic anticancer drugs, are present in the potable water, their fate is also of high interest. In this sense, advanced treatments, namely Advanced Oxidation Processes (AOPs), have been considered as a promising alternative treatment technique since they are capable of certain parent compounds' degradation. However, often incomplete mineralization results in the formation of unwanted TPs with unknown chemical structure, toxicity, and fate (Haddad et al. 2015). It was hypothesized that in case of incomplete mineralization by only partial oxidation of the parent compound, resulting TPs would be better biodegradable compared to the parent compounds (De la Cruz et al. 2012). However, that seems not to be true in general.

Despite two decades of extensive research, there is still a big lack of knowledge concerning the environmental fate of the anticancer drugs, their metabolites, and TPs stemming from (advanced) effluent treatment and formed in environmental processes concerning possible risks resulting from their presence in aquatic environments (Kümmerer et al. 2016). Moreover, the few studies on the (eco)toxicity of anticancer drugs performed before and after the treatments are based on the acute effects in living organisms. Unfortunately, this kind of tests has almost no environmental relevance and can underestimate the toxicity of chemicals that show mainly long-term toxicity, for example antibiotics or anticancer drugs (Backhaus and Grimme 1999; Froehner et al. 2000; Henschel et al. 1997). For this reason, there is a need for a more comprehensive toxicological assessment of anticancer drugs and their TPs in general and for studies involving mainly long-term assays (chronic) and tests in particular, which may provide further insights concerning the cytotoxic, genotoxic and mutagenic potentialities of these compounds. It is crucial to understand the behaviour of anticancer drugs in the environment (sorption, hydrolysis, biodegradation, natural photolysis) and to look for alternative treatments that could remove these compounds, improve their biodegradability and/or reduce their toxicity.

Therefore, this chapter evaluates the efficiency of different water and wastewater treatment alternatives based on the degrees of elimination and mineralization, and on the biodegradability and toxicity of the parent compounds as well as possibly formed TPs.

7.2 Chlorination

Chlorine has been widely used in WWTPs and in the pre-treatment of hospital effluents prior to their discharge into the sewage system for reasons of disinfection (Zhang et al. 2013; Prasse et al. 2015). In some countries, it is also used for additional treatment of WWTPs effluent, e.g. for wastewater treatment used for irrigation, depending on the raw wastewater quality and application demands (Pedrero et al. 2010; Norton-Brandão et al. 2013). During the disinfection process, depending on the concentration of organic compounds and on the applied chlorine doses, the pharmaceuticals present in the water can react with free chlorine and form chlorinated TPs (Fig. 7.1) with different persistence and toxicity if compared to their parent compounds (Negreira et al. 2015b; Yin et al. 2017). Therefore, the identification of possible chlorination products of pharmaceuticals as well as the understanding of the chemical fate of these TPs is of particular importance since the products formed during the disinfection might be more recalcitrant and harmful than the parent compounds (Bedner and MacCrehan 2006; Prasse et al. 2015). Up until now only few studies investigated the degradation of anticancer drugs by chlorination as well as the toxicity of by-products formed during the process.

One of the first studies that investigated the removal of anticancer drugs by chlorination including the toxicity of by-products formed during the degradation process was conducted by Roig et al. (2014). The authors investigated the degradation of MTX, a mutagenic and teratogenic anticancer drug from the subgroup of the antimetabolites. The results obtained by Roig et al. (2014) showed that chlorination eliminated 99 % of the parent compound (starting concentration 1 mg/L) after 120 min. However, no significant dissolved organic carbon (DOC) removal was observed at the end of the treatment, indicating the formation of TPs. Analysis of

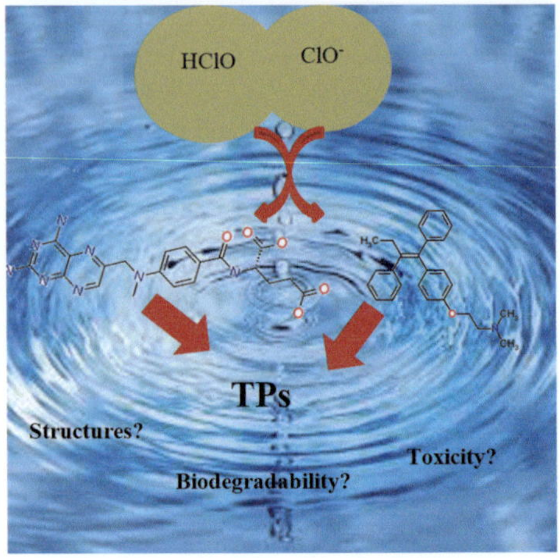

Fig. 7.1 Chlorination has been used in studies involving the degradation of anticancer drugs

LC-MS revealed the formation of a monochlorinated TP of MTX. However, that could by far not fully account for the measured DOC. According to the authors, in silico analysis presented no clear evidence that the possible by-products formed after chlorination might be more toxic than MTX itself. Nevertheless, since a number of alerts are altered after chlorination, it cannot be excluded that the toxicity of these TPs might be modulated compared with the parent compound.

In the same manner, Yin et al. (2017) also investigated the fate of MTX (2 mg/L) during aqueous chlorination in laboratory batch experiments. The parent compound was almost completely eliminated (98.2 %) within 10 min. The authors identified the presence of two MTX chlorinated derivatives, monochloro-MTX and dichloro-MTX. Both TPs were also detected in real hospital wastewaters. Yin et al. (2017) also evaluated the effects of MTX and monochloro-MTX on the cell cycle progression *in vitro* using a zebrafish liver cell line. The results indicated that both compounds (MTX and monochloro-MTX) could inhibit the proliferation of zebrafish liver cells by S phase arrest and their effects on the cell cycle profile did not differ significantly. MTX treatment caused a significant increase in the number of S phase cells accompanied by a disappearance of G2/M population, when compared with the control treatment. Cells treated with monochloro-MTX also exhibited a significant increase in S phase with a corresponding decrease in G2/M phase, and the effect was as prominent as those treated with MTX. For both compounds, the cell cycle profile determined at a concentration of 10 μmol/L was very similar to that of 1 μM. No dose–effect relationship was observed between these two concentrations. Nevertheless, considering the limitations of cell cycle analysis and uncertainty of QSAR analysis, the authors recommend further *in vitro* studies with different test modes should investigate the toxic effects of MTX chlorination products for a better understanding and risk assessment.

Negreira et al. (2015a) investigated the degradation of TAM and its major active metabolites 4-hydroxy-tamoxifen (OH-TAM) and 4-hydroxy-N-desmethyl-tamoxifen (OH-D-TAM) during chlorination experiments in ultrapure water and real wastewater. Under the studied chlorination conditions, TAM remained fairly unchanged after 90 min of reaction (94 ± 4%), whereas OH-TAM and OH-D-TAM concentrations rapidly decreased to around 6% and 3%, respectively, after only 2 min. A total of 13 chlorinated TPs were tentatively identified. The authors also found that these by-products can be formed under typical wastewater disinfection conditions. In silico analysis of the acute aquatic toxicity showed that the identified by-products of OH-TAM were up to 9- and 18-fold more toxic than TAM itself against the microcrustacean *Daphnia magna* and the fish *Pimephales promelas*, respectively, whereas the by-products of OH-D-TAM were up to 32 and 110 times more toxic against *D. magna* and *P. promelas*, respectively, than OH-D-TAM.

Both examples, MTX and TAM, demonstrate that chlorination of waters and wastewaters containing drugs does neither necessarily result in complete mineralization nor detoxification.

7.3 Advanced Oxidation Processes

Advanced (photo) oxidation processes are based on the generation and use of powerful and highly reactive transitory species, mainly the hydroxyl radicals (HO•). It can be generated photochemically (by UV or solar light) or by other means such as ozone, electrochemical oxidation and ultrasound, which can effectively oxidize organic matter (Glaze et al. 1987). In recent years, many studies have examined the application of photo-assisted systems to remove anticancer drugs from different water matrices (Zhang et al. 2013). Table 7.1 summarizes the results reported by the studies discussed in this chapter.

7.3.1 Ozonation

Ozonation is one of the most often investigated AOPs in the treatment of waters and wastewaters. Ozone is a strong oxidant and is therefore expected to mineralize micropollutants upon treatment. In some countries such as Germany, it is used as final treatment for potable water in order to stabilize the drinking water microbiologically during its flow through the pipes to the consumer. Within the last decade, ozonation as the only treatment of effluents was investigated but increasingly also as part of integrated systems for the treatment of different wastewaters (Margot et al. 2013; Lee et al. 2014). The removal of organic contaminants present in wastewaters by ozonation process occurs either by direct reaction of the contaminant with molecular ozone which would resemble the Criegee mechanism known from organic chemistry and which needs unsaturated double bonds to be present in the molecule, or by indirect reaction with free radicals, mainly hydroxyl radicals (OH•), generated by the decomposition of ozone (Garcia-Ac et al. 2010).

Table 7.1 Toxicity and biodegradability results of the investigated anticancer drugs before and after the treatments by different methods

Compound	Treatment method	Primary elimination	TOC/ DOC removal	Identification of TPs	Biodegradability	Toxicity (change/endpoint)	References
	Chlorination						
MTX		99% after 120 min	Not significant	1TP	–	No increased carcinogenicity, genotoxicity and mutagenicity for the TPs in comparison with MTX	Roig et al. (2014)
MTX		98.2% within 10 min	–	2TPs	–	No clear evidence that the chlorinated products of MTX might be more toxic than the parent compound	Yin et al. (2017)
TAM		Not significant	–	–	–	–	Negreira et al. (2015b)
OH-TAM		94% in 2 min	–	7TPs	–	High increase of the toxicity (acute toxicity up to 9- and 18-fold higher against *Daphnia magna* fish *Pimephales promelas*, respectively.	Negreira et al. (2015b)
OH-D-TAM		97% in 2 min	–	7TPs	–	High increase of the toxicity (acute toxicity up to 32 and 110 times higher against *D. magna* and *P. promelas*, respectively)	Negreira et al. (2015b)
	Ozonation						
CYC	O_3	100% within 10 min	< 50% in 30 min	–	–	Increase of the acute toxicity (from non-toxic to 3.4 ±3.0 TU)	Lin et al. (2015)
IF	O_3	100% within 10 min	< 50% in 30 min	–	–	Increase of the acute toxicity (from non-toxic to 2.7±2.2 TU)	Lin et al. (2015)
5-FU	O_3	100% within 10 min	< 50% in 30 min	–	–	Non-toxic prior and after the treatment	Lin et al. (2015)

Mixture (CYC + IF + 5-FU + PEN)	O_3	100% within 20 min	40% in 120 min			Increase of the acute toxicity (from non-toxic to TU=16.3 at pH 11 and TU=7.5 at pH 7)	Lin et al. (2015)
TAM	O_3	99% within 20–30 min	–	4TPs	–	High increase of the acute toxicity (from almost not toxic to 2% EC_{50})	Ferrando-Climent et al. (2017)
TAM	O_3/H_2O_2	99% within 20–30 min	–	4TPs	–	High increase of the acute toxicity (from almost not toxic to 15% EC_{50})	Ferrando-Climent et al. (2017)
	Photolysis						
TAM	UV	90% after 4 h	–	3TPs	–	Increase of the acute toxicity (from almost not toxic to 33% EC_{50})	Ferrando-Climent et al. (2017)
TAM	SSL	50% after 1 month	–	–	–	Decrease of the chronic toxicity (from EC_{50} of 8.1 x 10^{-4} to EC_{50} of 9.6 x 10^{-3})	DellaGreca et al. (2007)
5-FU	UV	100% after 32 min	18% after 256 min	3TPs	Increase of biodegradability	Significant reduction of the toxicity (chronic luminescence inhibition from 50% to 5%)	Lutterbeck et al. (2016)
5- FU	SSL	32% after 256 min	Not significant	3TPs	No differences prior and after treatment	No differences prior and after treatment	Lutterbeck et al. (2016)
5- FU	UV	100% between 15 and 30 min	50% 480 min	–	–	Decrease of the toxicity (from an acute inhibition of 65% to around 25%)	Governo et al. (2017)
CYC	UV	70% after 2880 min	Not significant	–	–	Parent compound and TPs were not toxic	Česen et al. (2016)
IF	UV	70% after 2880 min	Not significant	–	–	Parent compound and TPs were not toxic	Česen et al. (2016)

(continued)

Table 7.1 (continued)

Compound	Treatment method	Primary elimination	TOC/ DOC removal	Identification of TPs	Biodegradability	Toxicity (change/endpoint)	References
	Advanced (photo)oxidation processes						
DOX	UV/TiO_2	100% in 30 min	–	14 TPs	–	Slight decrease after 120 min (acute toxicity reduction of 3%)	Calza et al. (2014)
MTX	UV/TiO_2	100% in 30 min	–	8 TPs	–	Significant decrease after 180 min (acute toxicity reduction of 90%)	Calza et al. (2014)
MTX	UV/H_2O_2	100% in 16 min	65% in 256 min	6 TPs	Reduction of biodegradability	Significant decrease of the toxicity (chronic luminescence inhibition (from 61% to 21%) and in the growth inhibition (from 34% to 2%)	Lutterbeck et al. (2015a)
TAM	UV/H_2O_2	90% after 4 h	–	3TPs	–	Increase of the acute toxicity (from almost not toxic to 35% EC_{50})	Ferrando-Climent et al. (2017)
CYT	UV/H_2O_2	100% in 120 min	60% after 120 min	–	–	Increase of the acute toxicity (from almost non-toxic up to inhibitions of around 90%)	Ocampo-Pérez et al. (2010)
CYT	UV/$K_2S_2O_8$	98% after 60 min	10% after 120 min	–	–	Increase of the acute toxicity (from between 5-12% up to inhibitions of around 18%)	Ocampo-Pérez et al. (2010)
CYT	SSL/TiO_2	100% within 45 min	80% after 360 min	8TPs	–	Decrease of the acute toxicity (from 16% inhibition to 5%)	Koltsakidou et al. (2017a)
CYT	SSL/TiO_2/H_2O_2	100% within 45 min	80% after 360 min	4TPs	–	–	Koltsakidou et al. (2017a)
CYT	SSL/TiO_2/$S_2O_8^{2-}$	100% within 45 min	80% after 360 min	6TPs	–	–	Koltsakidou et al. (2017a)

5- FU	UV/H_2O_2	100% after 15 min	50% in 120 min	–	–	Decrease of the toxicity (from acute an inhibition of 65% to around 20%)	Governo et al. (2017)
5- FU	$UV/Fe^{2+}/H_2O_2$	100% after 8 min	67% in 120 min	–	–	Decrease of the toxicity (from acute an inhibition of 65% to around 9%)	Governo et al. (2017)
5- FU	SSL /Fe^{3+}/H_2O_2	100% after 180 min	25% in 360 min	2TPs	–	Slight decrease of the acute toxicity (from 3% inhibition to 1%)	Koltsakidou et al. (2017b)
5- FU	SSL/$Fe^{3+}/S_2O_8^{2-}$	100% after 120 min	40 in 360 min	3TPs	–	Increase of the acute toxicity (from 6% inhibition to 30%)	Koltsakidou et al. (2017b)
5- FU	SL/$[Fe(C_2O_4)_3]_3^-/H_2O_2$	100% after 60min	65% in 360 min	1TP	–	Slight decrease of the acute toxicity (from 2% inhibition to 1%)	Koltsakidou et al. (2017b)
5- FU	UV/H_2O_2	99% in 8 min	41% in 256 min	4TPs	Increase of biodegradability	Significant reduction of the toxicity (chronic luminescence inhibition (from 51% to 3%) and growth inhibition (from 20% to 0%)	Lutterbeck et al. (2015c)
5- FU	UV/TiO_2	100% in 240 min	100% after 1440 min	–	–	No differences prior and after treatment	Lin and Lin (2014)
CYC	UV/TiO_2	100% in 240 min	45% after 960 min	–	–	Increase of the toxicity (no values available)	Lin and Lin (2014)
CYC	UV/TiO_2	100% after 180 min	36% after 180 min	3TPs	–	Parent compound and TPs were toxic	Lai et al. (2015)
IF	UV/TiO_2	100% after 180 min	47% after 180 min	8TPs	–	Parent compound and TPs were toxic	Lai et al. (2015)
CYC	UV/H_2O_2	100% in 8 min	72.5% after 256 min	5TPs	–	Non-toxic prior and after the treatment	Lutterbeck et al. (2015b)
CYC	UV/TiO_2	100% in 32 min	89.6% after 256 min	4TPs	–	Non-toxic prior and after the treatment	Lutterbeck et al. (2015b)

(continued)

Table 7.1 (continued)

Compound	Treatment method	Primary elimination	TOC/DOC removal	Identification of TPs	Biodegradability	Toxicity (change/endpoint)	References
	Nonconventional treatments and integrated systems						
CYC	*T. versicolor*	Not significant	–	–	–	Parent compound and TPs were toxic	Ferrando-Climent et al. (2015)
IF	*T. versicolor*	Not significant	–	–	–	Parent compound and TPs were toxic	Ferrando-Climent et al. 2015
TAM	*T. versicolor*	92% in 60 min	–	–	–	Parent compound and TPs were not toxic	Ferrando-Climent et al. 2015
TAM	O_3/UV	100% within 30 min	–	6TPs	–	High increase of the toxicity (from almost not toxic to 1% EC_{50})	Ferrando-Climent et al. 2017
CYC	bioreactor + UV/H_2O_2/O_3	99% in 120 min	97% after 120 min	–	–	–	Česen et al. (2015)
IF	bioreactor + UV/H_2O_2/O_3	94% in 120 min	97% after 120 min	–	–	–	Česen et al. (2015)

Over the past years, an increasing number of researches have focused their activities to investigate the efficacy of ozone-based systems to remove anticancer drugs from wastewaters (Garcia-Ac et al. 2010; Ferre-Aracil et al. 2016; Li et al. 2016; Somensi et al. 2012). However, in most cases, these studies investigated only the primary elimination of the compounds as well as the reaction kinetics and operational parameters and did not consider the formation of TPs in the treatments. Furthermore, if TPs were investigated, rarely the toxicity of the TPs generated during ozonation was investigated.

Lin et al. (2015) investigated the removal of CYC, IF, and 5-FU from wastewaters by ozonation at an initial concentration of 5 mg/L. Besides, the authors also evaluated the toxicity of the compounds before and after the degradation experiments. The EC_{50} value was expressed as a percentage (%, v/v) of the initial sample, where the EC_{50} was the concentration that caused a 50% reduction in the bioluminescence of *Vibrio fischeri*. Then, a toxic unit (TU), which was calculated by dividing 100 by the EC_{50}, was used to present the results. The study revealed that CYC, IF and 5-FU were eliminated within 10 min, whereas total organic carbon (TOC) reduction of less than 50% was obtained within 30 min, indicating the formation of stable TPs. The highest primary elimination and TOC reduction were obtained at pH 11. Furthermore, Microtox assays showed significant toxicity increases of the two photolytic solutions of CYC and IF after 30 min ozonation, indicating that the TPs formed during the process were more toxic than the parent compound. 5-FU presented absence of toxicity *prior* and after ozonation against *V. fischeri*. The authors also investigated the ozonation of a solution that contained a mixture of compounds to simulate wastewater. This solution contained besides CYC, IF and 5-FU, the vasodilator drug pentoxifylline (PEN). Also at pH 11, all the parent compounds were completely removed after 20 min, while a TOC reduction of around 40% was obtained after 2 h only. Despite the removal of the compounds after 20 min, a significant increase of the acute toxicity was observed after 2 h, suggesting that toxic TPs were formed.

The elimination of TAM (100 μg/L) by O_3 and O_3/H_2O_2 was investigated by Ferrando-Climent et al. (2017). A removal percentage of TAM higher than 99% was achieved after 20–30 min treatment. Moreover, the authors also observed an increased acute toxicity trend within TAM treatment towards *V. fischeri*, from being almost not toxic to reach a value of 2% EC_{50} at the end of the experiment (after 240 min). This toxic effect of the ozonized solution was mainly attributed to 4 TPs that were formed along the experiments.

The examples presented here include one drug (TAM) that was studied in both treatments (chlorination and ozonation). Different endpoints of toxicity were used. The results demonstrate that not only chlorination of waters containing drugs failed to completely mineralize the parent compounds, but also that ozonation results in the formation of TPs that are in some cases even more toxic than the parent compounds.

7.3.2 *Photo-Assisted Treatments*

7.3.2.1 UV and Simulated Sunlight (SSL) Photolysis

Photolysis, e.g. by UV light in technical systems (e.g. potable water treatment, wastewater treatment) and natural waters by irradiation with sunlight, has been studied in order to understand the significance of this degradation pathway for anticancer drugs (Wang and Lin 2014; Mahmoud and Kümmerer 2012).

Ferrando-Climent et al. (2017) studied the elimination of TAM (100 μg/L) by UV photolysis. The obtained results showed a primary elimination of around 88% after 120 min. The authors identified 3 TPs that were formed during the photolysis experiments. It should be highlighted that not one of the TPs formed by the UV degradation of TAM is also formed in the chlorination and ozonation studies discussed above, what indicates different degradations pathways of TAM associated with the treatments. Moreover, the toxicity against *V. fischeri* of the photolytic mixture was higher compared to TAM itself. According to the authors, the increase in the toxicity of the solution at the end of the UV photolysis demonstrates that despite the almost complete removal of TAM, a total mineralization was not achieved and thus the toxicity might be considered as a result of the presence of a cocktail of TPs that are being generated from TAM.

DellaGreca et al. (2007) investigated the photodegradation of TAM (80 mg/L) by SSL. Primary elimination was about 50% after over a month of irradiation. Four photo TPs were identified, one of them (TP 370) was also found by Ferrando-Climent et al. (2017) in UV photolysis experiments. Only one TP presented a statistically significant higher acute toxicity against the rotifer *Brachionus calyciflorus* and the crustacean *Thamnocephalus platyurus* in comparison to TAM. The photolytic mixture (formed after 1 month of sunlight irradiation) showed the lowest EC_{50} values for *B. calyciflorus* and *T. platyurus*, while no effect was found at the maximum concentration tested (10 mg/L) for *D. magna*. When considering the chronic assays, again the irradiated mixture was clearly less toxic towards *B. calyciflorus* and *C. dubia* than the parent compound, suggesting that the partial degradation of TAM (50%) led to the formation of less toxic TPs.

Lutterbeck et al. (2016) conducted a study focusing on the degradation of 5-FU (starting concentration of 20 mg/L) by UV and Xe lamps (the latter "simulating" sunlight irradiation). The authors verified a complete primary elimination of 5-FU after 32 min using the UV lamp and a partial degradation (32%) after 256 min using the Xe lamp. No significant DOC removal was observed in the experiments performed with the Xe lamp, whereas a mineralization degree of 18% was achieved after a treatment time of 256 min by the UV lamp. Three TPs were formed and identified during the photodegradation experiments. The authors also found that the partial degradation of 5-FU did not improve the biodegradability of the reaction mixture, whereas the UV treatment, which led to the complete elimination of the parent compound after 256 min, significantly improved the biodegradability (BOD after 28 days increased from 1.73 to 5.03 mg O_2/L). Samples resulting from treatments with Xe lamp showed no significant differences of the parent compound compared to the mixture resulting from photolysis in terms of toxicity while samples submitted to photolysis with the UV lamp showed statistically significant reductions in terms of chronic toxicity towards *V. fischeri*. Lastly, in silico analysis showed that only one TP presented a positive alert in a rule-based expert system for genotoxicity.

Governo et al. (2017) investigated the degradation of 5-FU (50 mg/L) in water by UV/VIS photolysis. According to the authors, 5-FU was completely eliminated after 15–30 min of irradiation by direct photolysis and about 50% of the 5-FU was mineralized after 8 h of reaction. Bioassays using *V. fischeri* showed a reduced toxicity of the photolytic mixture in comparison to the parent compound which is in accordance with the findings reported by Lutterbeck et al. (2016).

A research performed by Česen et al. (2016) investigated the removal of CYC and IF by UV irradiation as well as the ecotoxicity and genotoxicity of both parent compounds and their human metabolites/transformation products (TPs) carboxy-cyclophosphamide (CPCOOH), ketocyclophosphamide (ketoCP) and N-dechloroethyl-cyclophosphamide (NdCP) as individual compounds and as mixture. After 48 h of treatment, the primary elimination of CYC and IF were around 70% and 63%, respectively, whereas no significant DOC removal was obtained under this extended irradiation time. This very slow degradation of the compounds as well as the absence of DOC removal are in line with the results of previous studies and might be attributed to the very low absorbance of CYC and IF at wavelengths above 200 nm (Lutterbeck et al. 2016; Ofiarska et al. 2016). Česen et al. (2016) reported similar EC_{50} values of 17.1 mg/L and 11.5 mg/L for CPCOOH and the mixture, respectively; in experiments performed with the cyanobacteria *S. leopoliensis* suggesting that unknown TPs formed during UV irradiation are not toxic to the test organism. Only CPCOOH exhibited significant genotoxic activity in the absence (2.75 ± 0.12) and the presence (2.04 ± 0.47) of S9 metabolic activation.

Ocampo-Pérez et al. (2010) studied the photodegradation of CYT. The authors observed that after 120 min UV irradiation only 10% of the initial CYT (10 mg/L) was degraded.

7.3.2.2 Combined Advanced (Photo)oxidation Processes

This section presents the results obtained in studies involving UV irradiation combined with different oxidizing/catalyst agents.

UV and Oxidants

Ferrando-Climent et al. (2017) investigated the degradation of TAM by the combined use of O_3 and UV (O_3/UV). The results showed a complete removal of the parent compound within 30 min and the formation and identification of six TPs. Results of the Microtox assays presented a dramatical increase of the acute toxicity towards *V. fischeri* within the first minutes of the experiments reaching a final value of 1% EC_{50}, whereas TAM is not present. So, the authors attributed the higher toxicity to the presence of 3 TPs that were formed during the experiments and remained in the final solution. In the same research, Ferrando-Climent et al. (2017) also studied the elimination of TAM (100 μg/L) by UV/H_2O_2. Results showed that TAM was almost completely eliminated (around 90%) after 10 min and 3 TPs were formed within photoperoxidation experiments (the same ones formed in the previously described UV photolysis experiments). Ferrando-Climent et al. (2017) observed a higher acute toxicity of the photolytic mixture when compared to the parent compound in experiments performed with *V. fischeri*. According to the authors, despite the fast elimination of TAM, low mineralization degrees lead to the formation of intermediates which were more toxic to *V. fischeri* than the parent compound.

Lutterbeck et al. (2015a) investigated the removal of MTX by UV/H_2O_2. A full primary elimination was achieved after 16 min, whereas a DOC reduction of around 65% was observed after 256 min. The authors identified 6 TPs already described in previous studies (Kosjek et al. 2015; Calza et al. 2014; Kiffmeyer et al. 1997). Results of the further Closed Bottle Test (CBT) indicated that the photolytic mixture was less biodegradable than the parent compound itself. Statistically significant toxicity reductions were observed in the chronic luminescence inhibition (from 61% to 21%) and in the growth inhibition of *V. fischeri* (from 34% to 2%). The authors concluded that the intermediates formed during the treatment, although not biodegradable, are less toxic against *V. fischeri* than the parent compound.

Ocampo-Pérez et al. (2010) observed that after 120 min of UV irradiation only 10% of the initial CYT (10 mg/L) was degraded, while by the UV/H_2O_2 treatment the parent compound was almost completely eliminated after 120 min and a TOC removal of 60% was achieved, which demonstrates that the OH radicals formed by the degradation of H_2O_2 significantly improved the degradation process. The acute inhibition of *V. fischeri* was higher after treatment, indicating the formation of TPs that were more toxic than CYT. However, the surplus H_2O_2 is toxic and could have contributed to the increased inhibition of luminescence of *V. fischeri*. In experiments performed with UV using $K_2S_2O_8$ as an oxidant, Ocampo-Pérez et al. (2010) report a primary elimination of 98% after 60 min and a TOC reduction of 10% after 120 min. However, differently from the UV/H_2O_2 treatment, the inhibition of *V. fischeri* during UV/$K_2S_2O_8$ treatment remained virtually constant even at longer treatment times. These results indicate that the TPs generated by the UV/$K_2S_2O_8$ treatment are not excessively toxic and sulphate ion generated as a final product did not inhibit the bacteria *V. fischeri*.

Governo et al. (2017) investigated the degradation of 5-FU (50 mg/L) in water by UV/H_2O_2 and UV/Fe^{2+}/H_2O_2 treatments. According to the authors, the UV/H_2O_2 and UV/Fe^{2+}/H_2O_2 processes achieved a fully primary elimination after 15 min and 8 min, respectively, whereas for the mineralization these values were of around 50% and 67% only (after 120 min), respectively. Acute toxicity assays using *V. fischeri* showed that the TPs generated from both processes were less toxic than the parent compound, although the authors had not monitored and identified their structures.

The removal of 5-FU (20 mg/L) from water by UV/H_2O_2 was also investigated by Lutterbeck et al. (2015c). A primary elimination of 99% was observed after 8-min treatment, while only 41% mineralization was attained after 256 min, indicating the formation of TPs. The authors identified 4 TPs. Results of the following CBT experiments showed a noticeable increase of the biodegradability of the photolytic mixture formed after the treatment when compared to the parent compound (BOD after 28 days increased from 0.15 to 5.69 mg O_2/L). Furthermore, both the chronic toxicity (luminescence) and growth inhibition against *V. fischeri* were significantly reduced, while only one of the four TPs formed during the UV/H_2O_2 process presented mutagenicity alerts by QSAR models. However, this TP was a transient species that was formed and eliminated after 16 min.

The degradation of 5-FU (30 mg/L) by photo-assisted treatment was also the focus of the research conducted by Koltsakidou et al., (2017b). The SSL/Fe^{2+}/H_2O_2, SSL/Fe^{3+}/$S_2O_8^{2-}$ and SSL/$[Fe(C_2O_4)_3]_3^{-}$/H_2O_2 processes completely removed the parent compounds after 180 min, 120 min and 60 min, respectively. The ferrioxalate's complex process achieved the highest mineralization rate after 360 min (65 %), followed by SSL/Fe^{3+}/$S_2O_8^{2-}$ (40 %) and SSL/Fe^{2+}/H_2O_2 (25 %). The authors identified the formation of 3 TPs during treatments: 3 in the SSL/Fe^{3+}/$S_2O_8^{2-}$, 2 TPs formed during the SSL/Fe^{2+}/H_2O_2 and 1 formed during the SSL/$[Fe(C_2O_4)_3]_3^{-}$/H_2O_2 treatment. Results of the acute toxicity assays indicated an increased toxicity during the first stages of the reaction for both the SSL/Fe^{3+}/H_2O_2 and SSL/Fe^{3+}/$S_2O_8^{2-}$ treatments towards *V. fischeri*. According to Koltsakidou et al. (2017b) at this time, the TPs formed were still present in the solution. Consequently, the toxicity profile indicates that the TPs' content during the procedure exhibits more toxic effects than the parent compound for those two treatments. However, the toxicity levels at the end of its treatment are lower than the toxicity levels at the first stages of the reaction. On the other hand, SSL/$[Fe(C_2O_4)_3]^{3-}$/H_2O_2 treatment presented negligible toxicity towards *V. fischeri* at all stages of the photocatalytic reaction. This fact might be associated with the speed of the reaction, since the SSL/$[Fe(C_2O_4)_3]^{3-}$/H_2O_2 treatment degraded 5-FU faster and resulted in the presence of only one TP, which was quickly further degraded. It is well known that the efficiency of the photo-Fenton process can be further enhanced by using organic carboxylic acids to complex Fe (III) (Pignatello et al. 2006). According to the authors, the TPs formed at the first stages of SSL/Fe^{3+}/H_2O_2 and SSL/Fe^{3+}/$S_2O_8^{2-}$ processes are probably responsible for the acute toxicity of the solutions. Therefore, Koltsakidou et al. (2017b) concluded that the ferrioxalate photo-Fenton (SSL/[Fe $(C_2O_4)_3]^{3-}$/H_2O_2) treatment was the most efficient, since faster degradation of 5-FU and higher mineralization percentages were achieved, resulting in the formation of only one TP and lower acute toxicity in comparison to the other two treatments.

Photocatalysis

The photocatalytic (UV/TiO_2) degradation of MTX and DOX was investigated by Calza et al. (2014). MTX (15 mg/L) and DOX (15 mg/L) were fully eliminated after 30 min irradiation. The authors identified 8 TPs of MTX and 14 of DOX and addressed the acute toxicity towards *V. fischeri*. It was found that in case of DOX, inhibition of luminescence increased and reached the maximum after 15 min of treatment time (43%), then slowly decreased and a slight decrease could be observed after 120 min. According to the authors, this maximum toxicity coincides with the time point at which most of the TPs reached highest concentrations. In the case of MTX, the toxicity remained at the maximum level until 30 min of irradiation and then decreased until 120 min of irradiation, where it was very low. Thus, since MTX was already completely eliminated after 10 min, the observed high toxicity after 30 min of treatment was associated with the TPs that were generated. At longer irradiation times (45 min), there was a reduction of the toxicity to 40%, further reduced by longer irradiation time. Interestingly, the authors found a high initial acute toxicity of MTX towards *V. fischeri*, which is in contradiction with results obtained in previous studies. Lutterbeck et al. (2014) did not observe any acute toxicity in experiments performed with concentrations up to 100 mg/L of MTX, whereas Henschel et al. (1997) obtained a high acute EC_{50} of 1220 mg/L only in assays also carried out with *V. fischeri*.

The degradation of CYT by simulated solar assisted photocatalysis using TiO_2 was investigated by Koltsakidou et al. (2017a). On the one hand, results of the toxicity experiments showed a low toxic effect of the parent compound towards *V. fischeri*. On the other hand, the authors verified that the TPs generated during the first minute of photocatalytic treatment (SSL/TiO_2) presented higher toxicity levels, reaching its highest toxic effect at 30 min of irradiation. According to the authors, this rise in toxicity can be attributed to the formation of more toxic reaction intermediates than the parent compound or due to synergistic effects among the formed TPs. After 45 min the toxicity decreased continuously leading to a very low toxicity level after 360 min of SSL/TiO_2 treatment. Therefore, according to the authors, the photocatalytic SSL/TiO_2 degradation of CYT can lead to a complete elimination of the acute toxicity against *V. fischeri*.

Koltsakidou et al. (2017a) studied also photocatalysis with SSL in presence of TiO_2 and additional oxidants. The results of the experiments carried out with SSL/TiO_2/H_2O_2 and SSL/TiO_2/$S_2O_8^{2-}$ showed a full elimination of CYT (30 mg/L) within 45 min. However, even treatment times of 360 min were not sufficient to achieve a complete mineralization (80%), demonstrating that the photocatalytic treatment also needs prolonged irradiation times for complete mineralization in the case of SSL as a light source. A total of 11 different TPs were identified. It is important to mention that the TPs formed during the different reactions differed in their structures.

The photocatalytic degradation of 5-FU (27.6 mg/L) and CYC (27.6 mg/L) was investigated by Lin and Lin (2014). On one hand, the obtained results showed a full elimination of both parent compounds after 240 min. On the other hand, a full mineralization of 5-FU was obtained only after 1440 min, while for CYC the mineralization degrees were of only 45% after a treatment time of 960 min. Microtox acute toxicity results showed that the toxicity towards *V. fischeri* gradually increased during the 960 min of irradiation of CYC. According to Lin and Lin (2014), this phenomenon appears to be compound-dependent, since the same trend was not observed in the case of 5-FU.

Lai et al. (2015) studied the photocatalytic degradation and transformation of IF and CYC solutions prepared with distilled water and an initial concentration of 20 mg/L. The results showed that the parent compounds were completely eliminated after 180 min, while TOC removals of only 47% and 36% for IF and CYC, respectively, were achieved after longer treatment times (360 min). Eight IF and three CYC TPs were formed and identified. The acute toxicity against *V. fischeri* increased during the initial steps of the experiments and then gradually decreased. After 360 min of reaction time, the toxicity decreased to zero. Lai et al. (2015) monitored also the release of chloride (Cl^-) anion during the degradation of the parent compounds and its distribution during the change of toxicity throughout the reaction period. Based on the comparison of chlorine distribution in the target compounds, degradation TPs and dechlorination with the toxicity trends, the authors concluded that the acute toxicity most likely results from the formation of chlorinated TPs.

The photo (catalytic) degradation of CYC by UV/H_2O_2 and UV/TiO_2 was the focus of the study carried out by Lutterbeck et al. (2015b). The obtained results showed that more than 99% of the parent compound was removed by the UV/H_2O_2 and UV/TiO_2 reactions in 8 min and 32 min, respectively. Mineralization degrees of 72.5% for the UV/H_2O_2 reactions and of 89.6% for the UV/TiO_2 reactions were achieved after 256 min. Five TPs were detected during the UV/H_2O_2 reactions whereas four were found in the UV/TiO_2 treatments. All the TPs generated during the experiments were identified and are also formed during CYC metabolism in the human body. None of them showed significant acute or chronic toxicity as well as growth inhibition against *V. fischeri*.

In conclusion, the results of the studies presented in this section (photo-assisted treatments) confirmed the general findings that have been received for chlorination and ozonation (incomplete mineralization, different kinetics depending on the type of compound and treatment, formation of toxic TPs).

7.3.3 Other Nonconventional Treatments and Integrated Systems

Ferrando-Climent et al. (2015) investigated the degradation of IF, TAM and CYC using a biological treatment based on the fungus *Trametes versicolor*. Besides the experiments, two types of control experiments were run. One of them was used to assess the potential photodegradation of the micropollutants as well as the matrix effect onto the contaminants from the experimental conditions. Another one consisted in sterilized cultures (autoclave, 121 °C for 30 min) under identical conditions to those of the experimental cultures that were used to evaluate potential fungal sorption processes that could be taking place in time courses degradation experiments. The amount of adsorbed pollutant was determined from the difference in the CYC, IF and TAM concentration between the non-inoculated and sterilized control neglecting the probable change in the type of organic matter resulting from sterilization. Results showed that CYC and IF were neither degraded nor sorbed by *T. versicolor* at 10 mg/L. However, TAM was eliminated by 92% and 99% after 1 h and 9 days, respectively, in the experiments performed with synthetic wastewater spiked at 0.3 mg/L and treated with *T. versicolor*. According to the authors, the removal observed can be mainly attributed to sorption processes since the experiment with sterilized biomass showed 83% and 94% of elimination from the water after 24 h and 9 days, respectively. Two compounds (TAM hydroxylated positional isomers) were identified as derived from biotransformation of TAM through the *T. versicolor* activity. TAM and the by-products formed during the experiments with *T. versicolor* showed no toxic effects towards *V. fischeri* at the tested concentrations. Surprisingly, results of acute toxicity assays performed with abiotic controls of CYC and IF as well as the samples taken at the initial time and at the end of the synthetic water experiments with *T. versicolor*, showed a high toxicity against *V. fischeri*. These results are in contradiction with some previously published results, in which no acute toxic effect was observed even at high concentrations (120 mg/L) and an EC_{10} 74.1 mg/L was observed in the chronic assays (Lutterbeck et al. 2014).

Although classified as a cytotoxic compound (alkylating agent), CYC is a pro-drug that is not active itself unless it undergoes metabolic activation in the liver, what can explain the absence of mutagenicity (Lutterbeck et al. 2015b). Whether this holds also for ecotoxicological endpoints is not known yet. The research performed by Česen et al. (2015) investigated the degradation of IF (10 μg/L) and CYC (10 μg/L) by the combined use of biological (attached growth biomass in a flow-through bioreactors) and abiotic treatments ($UV/H_2O_2/O_3$). Results of the biological treatment showed removal values of 59% and 35% for CP and IF, respectively. Furthermore, the authors found out that despite similar removal rates of the parent compounds by the AOP alone (after 120 min) (99% and 94% for CYC and IF, respectively) and by the combined use of biological and abiotic treatments, the DOC removal of the integrated system was very high (97%) when compared to DOC removal by AOP (9%) alone. So, Česen et al. (2015) suggest that the majority of compounds present in the samples were mineralized by combined treatment and only a low amount of TPs were formed. However, they did not investigate the formation of potentially formed TPs.

7.4 Discussion

7.4.1 Efficiency of the Treatments

Because of the use of different degradation experiments, generalizations about the treatments efficiency are difficult. However, based on the results presented in this chapter, it seems evident that the ozonation presents the worst results regarding reductions of the toxicity. In the studies addressed here, although ozone quickly eliminated the parent compounds, it was not efficient to attain high mineralization degrees, so stable TPs were formed, which were responsible for the increase in the toxicity. On the other hand, the photo-assisted methods presented the best results when considering the toxicity reductions. This fact might be related to high mineralization degrees achieved by these treatments, lowering the generation of toxic TPs. Nevertheless, one should be careful with generalizations, since all the treatment methods presented here have advantages and disadvantages. Even in cases where relative high mineralization degrees (> 60%) were achieved by photo-assisted methods (UV/H_2O_2), an increased toxicity or reduced biodegradability was observed after the experiments (Ocampo-Pérez et al. 2010; Lutterbeck et al. 2015a). Therefore, the operational conditions of the treatments should be optimized in order to set the best concentration, treatment time, catalyst, pH, etc., since an excess of hydrogen peroxide, for example, can add toxicity, retard the reaction, reduce the mineralization degree and increase the costs of the process. In the same way, although higher catalyst loadings may produce an increased availability of active sites, an excess of TiO_2 reduces the light penetration leading to a plateau or even a decrease in the mineralization rate. Another disadvantage is that TiO_2 has to be removed after treatment. Therefore, the optimization of operational conditions through aspects such as exposure time, concentration of oxidant reagents as well as the concentrations of the compounds in the wastewaters is of crucial importance to achieve higher mineralization degrees and consequently reduce the formation of TPs so that both cost and environmental impacts are minimized.

Long treatment times needed for the high degree of mineralization of the parent compounds and TPs suggest that a practical application is rather limited if higher volumes of effluent or raw water have to be treated. On the other hand, the results clearly demonstrate that even at very long treatment times, further improvement is not always achieved, since in some cases only a little increase in the degree of mineralization is reached after a certain time point.

Finally, the treatment methods addressed in this chapter that reached high mineralization rates and in this way minimized the generation of TPs were carried out over longer treatment times (hours) and in bench scale, i.e. far from realistic conditions found in WWTPs.

7.4.2 *Biodegradability of TPs*

Regarding biodegradability, many parent compounds are not biodegradable or only biodegraded at low degrees. Mineralization of parent compounds resulting from advanced treatment is most often incomplete. However, experimental results addressing the impact of advanced treatment on biodegradability of the formed reaction products are still rare and available data show that it is not common to achieve higher biodegradability of the TPs formed within advanced treatment, i.e. the TPs generated by the AOPs have similar or even lower biodegradation rates than the parent compounds. This is at the first glance surprising as most often hydroxylated compounds result from oxidative advanced treatment, which are often thought to have an improved biodegradability because of the increased bioavailability. Nevertheless, one has to be aware that biodegradability is an enzymatic process which includes other factors too. One could also be the presence of other organic material that is easier to degrade than the TPs, often present in low concentrations. However, results of tests working in the concentration level at a few milligrams per litre and excluding other carbon sources such as the Closed Bottle test (OECD 301D), which in turn should favour adaption of microorganisms and thereby favouring positive results, do not confirm this. For example, it was observed that in case of MTX (Lutterbeck et al. 2015a) the photolytic mixture was less biodegradable than the parent compound, whereas a significant higher biodegradation degree of 5-FU was observed after the UV and UV/H_2O_2 (Lutterbeck et al. 2015c, 2016). On the other hand, no differences were observed for 5-FU prior and after treatment with SSL (Lutterbeck et al. 2016). In this way, it is not possible to assume that AOPs, even after longer treatment times, will increase the biodegradability of the compounds in general. According to the authors' opinion, biodegradation assays should complement treatment studies and thereby allow for a better evaluation of the efficiency of different treatment methods.

7.4.3 Toxicity of TPs

As one can see in Table 7.1, in 14 cases of the studies reported in the present chapter (Negreira et al. 2015b; Lin et al. 2015; Ferrando-Climent et al. 2017; Ocampo-Pérez et al. 2010; Koltsakidou et al. 2017b; Lin and Lin 2014) an increase of the toxicity was observed after the advanced treatment, indicating that the formed TPs were more toxic than the parent compound. Contrarily, in 12 cases (DellaGreca et al. 2007; Lutterbeck et al. 2015a, c, 2016; Governo et al. 2017; Calza et al. 2014; Koltsakidou et al. 2017a, b), a reduction of the toxicity after the advanced treatment was reported. Finally, in 14 cases (Roig et al. 2014; Yin et al. 2017; Lin et al. 2015; Lutterbeck et al. 2015b, 2016; Česen et al. 2016; Lin and Lin 2014; Ferrando-Climent et al. 2015) no toxicity differences were observed before and after the treatments related to the toxicity endpoints used, either because the parent compounds were not toxic or because the treatments did not increase the initial toxicity already existing of the compounds. It is noteworthy to highlight that mostly the toxicity evaluation involved acute assays with *V. fischeri*.

Having a closer look onto the treatment methods, chlorination and ozonation were not able to reduce the toxicity of the investigated anticancer drugs. In 2 out of 4 studies, chlorination increased the toxicity (Negreira et al. 2015b) while in 4 out of 5 cases, the toxicities observed after the O_3 experiments were higher (Lin et al. 2015; Ferrando-Climent et al. 2017). When considering the photo-assisted methods, only in six studies out of 21, the formed TPs were more toxic than the parent compounds (Ferrando-Climent et al. 2017; Ocampo-Pérez et al. 2010; Koltsakidou et al. 2017b; Lin and Lin 2014; Koltsakidou et al. 2017a, b; Lutterbeck et al. 2015c) and in 12 cases toxicity decreased (DellaGreca et al. 2007; Lutterbeck et al. 2015a, 2016; Governo et al. 2017; Calza et al. 2014). Regarding the nonconventional and integrated treatments, in one case an increased toxicity was observed after the reactions (Ferrando-Climent et al. 2017).

The results discussed in the present chapter show that toxicity is also compound dependent. In studies involving TAM and its metabolites, from 9 studies, 7 indicated a higher toxicity of the TPs after the treatment. In the research carried out by Ferrando-Climent et al. (2015), both TAM and the TPs were not toxic, while DellaGreca et al. (2007) observed a toxicity reduction after the treatment with SSL. When considering MTX, in two studies a significant decrease of the toxicity was verified at the end of the experiments (Lutterbeck et al. 2015a and Calza et al. 2014), while in another two studies QSAR analysis revealed no increased carcinogenicity, genotoxicity and mutagenicity for the TPs in comparison with MTX (Roig et al. 2014 and Yin et al. 2017). Studies involving CYT showed that in 1 experiment (Koltsakidou et al. 2017a) a reduction of the toxicity was achieved whereas in another 2 an increased toxicity was verified (Ocampo-Pérez et al. 2010). DOX showed a slight toxicity decrease in one study (Calza et al. 2014). Nevertheless, among all the investigated anticancer drugs, 5-FU was the compound, which presented the highest toxicity reductions involving different experiments and endpoints. In 7 cases toxicity reductions were achieved at the end of the treatments. Finally, CYC presented an increased acute toxicity at the end of the experiments in two investigations while IF in only one (Lin and Lin 2014 and Lin et al. 2015). In all other cases, the toxicities of both compounds remained similar before and after the treatments.

It was also found that treatment time can be crucial for avoiding an increase in toxicity ("window of toxicity"). Due to the short exposure time of the acute tests, long-term effects present in substances with well-known bactericidal properties, such as antibiotics and anticancer drugs, may not be detected in short-term assays, which, in turn, may then lead to an underestimation of the toxic potential of these compounds (Backhaus and Grimme 1999; Kümmerer et al. 2004). Furthermore, the ecotoxicological significance of that test is not clear. There is a severe lack of studies that include endpoints of mutagenicity, genotoxicity and endocrine disruption. In general, it is difficult to compare data, as many different more or less common endpoints and tests were used for toxicity assessments. So, in cases of the studies reported here where "no toxicity" was found this holds only for the endpoint investigated but does not allow for a general risk assessment. Therefore, stating the absence of toxicity has to be done very carefully. In this sense, a broader approach, involving additional endpoints relevant for human health such as mutagenicity, genotoxicity and endocrine disruption, different organisms (from different trophic levels) and focusing on long exposure times (chronic assays) and in vivo assays, is necessary for a better understanding of the toxicological relevance of the treatment of effluents containing anticancer drugs and their metabolites as well as their transformation in the environment, e.g. by light or bacteria. Moreover, as the residues of these compounds in the environment often occur as complex mixtures, testing the mixtures to verify possibly synergistic or antagonistic effects of molecules with similar mode of action is also an important issue for a more detailed risk assessment.

7.5 Knowledge Gaps and Limitations

The treatment methods discussed led to contradictory results, not allowing for a general conclusion about its efficiency and showing that a compound and treatment specific assessment is necessary. The results presented here do not allow for a general prediction about the efficiency of the treatments related to individual compounds. One should be aware that the degradation experiments were carried out using seven different compounds at different initial concentrations that have different chemical and biological properties. Even studies using the same treatment method (e.g. photodegradation or ozonation) were performed under different operational conditions, with treatment times varying from a few minutes to hours, using different ozone doses, different lamps (with different intensities), and varying the type and catalyst doses. All these variables hinder the comparison among the treatments and make the assessment of its efficiency a challenging task.

The chosen endpoint to assess toxicity seems to be crucial too. So the absence of toxicity does not necessarily mean that there is "no" toxicity any more after the treatment. That demonstrates that technical treatment faces several challenges not just the formation of possible toxic TPs but also that this may depend on treatment time, the completeness of analytical identification of TPs and their toxicity. Also the needed long treatment times (at least several hours in most cases for complete mineralization if reached at all) present a significant hurdle for the practical implementation of advanced treatments. Furthermore, most of the studies are conducted under laboratory conditions (i.e. clear water) and not with native wastewater or WWTP effluents. These contain additional organic and inorganic compounds, which have an impact on kinetics, efficacy and type and number of TPs formed. These issued are not well studied up until now and therefore are still not well understood. These challenges are therefore often neglected in discussions on the practical implementation of advanced treatment.

The integration of different treatment methods, combining biotic and abiotic treatment systems should be more investigated, as they could present a promising approach such as the one described by Česen et al. (2015). The authors verified that despite similar primary elimination degrees by AOP alone and by the combined use of a biological treatment followed by AOP, mineralization increases significantly, where attained by the integrated treatment (DOC removal from 9% to 97%). However, the authors did not assess the fate of TPs. Therefore, it is not clear whether the high removal rates did correspond with high reduction of toxicity and high degradation rates of TPs or just their sorption to sludge. The combined application of different treatment alternatives, such as the advanced oxidation processes with a biological treatment, should be further investigated, in order to verify the feasibility of the processes to be used as pre- or post-treatment and ensure the complete mineralization of non-biodegradable and toxic pollutants and to achieve high mineralization degrees that might reduce as much as possible the formation of TPs.

7.6 Conclusions

Several anticancer drugs have been found to be recalcitrant to biodegradation and detected in the environment. Implementation of advanced treatment was a focus of research in order to better understand the potential of such approaches to resolve the issue of micropollutants including anticancer drugs in waters and wastewaters. Implementing such advanced treatment into practice, i.e. for the treatment of real wastewater (WWTP effluent) has already been undertaken. However, based on the findings described in this chapter, it seems to be a very challenging task, especially considering the fact that the experiments discussed here were performed in laboratories and under controlled conditions, i.e. disregarding the synergistic or antagonist effects of the mixtures of compounds, the presence and absence of microbes, reactive oxygen species, photosensitizers and quenchers, chemical oxidative agents, non-specific organic matter and excretion of the compounds.

Many studies applied different conditions within the same or similar treatments and also many different treatments are described in the literature. TPs resulting from oxidative advanced treatments are not generally better biodegradable than a given parent compound and can be even more toxic. Therefore, there is a need for a treatment and compound-specific assessment. A standardization of treatment conditions would be desirable for a better assessment on the one hand. However, the amount and type of TPs formed as well their properties such as toxicity and biodegradability, polarity and further environmental fate are strongly governed by experimental conditions set. These can change at different places in the environment and thereby critical conditions and outcomes may be missed meaning risks could be underestimated or overestimated.

In summary, the results presented in this chapter demonstrate that at least for practical application knowledge on primary elimination and conditions favouring highly desirable mineralization is by far insufficient to select the appropriate treatment method and treatment time.

Considering the increased demand for existing and new anticancer drugs worldwide and the lack of safe threshold limits for the toxic effects that these compounds can pose to the human health and to other non-target organisms and given the fact that only 20% of countries do have effluent treatment, the presence of anticancer drugs and its metabolites in the environment should be a topic of suitable measures to reduce or event prevent input of these compounds at the source. This should go along with such measure for other pharmaceuticals and chemicals in general. That includes creating concern not only for the scientific community but also for the general public.

Acknowledgements Carlos Alexandre Lutterbeck thanks the Brazilian "Conselho Nacional de Desenvolvimento Científico e Tecnológico" (CNPq) for their financial support (Grant Nr. 152009/2016-7 and Grant N. 157933/2018-0). Ênio Leandro Machado acknowledges the financial support provided by CNPq-Project 307257/2015-0 and CNPq-Project 307599/2018-3.

References

Aherne GW, English J, Marks V (1985) The role of immunoassay in the analysis of microcontaminants in water samples. Ecotoxicol Environ Saf 9(1):79–83

Aherne GW, Hardcastle A, Nield AH (1990) Cytotoxic drugs and the aquatic environment: estimation of bleomycin in river and water samples. J Pharm Pharmacol 42(10):741–742

Allwood M, Stanley A, Wright P (2002) The cytotoxics handbook, 4th edn. Radcliffe Medical Press, England

Ashton D, Hilton M, Thomas KV (2004) Investigating the environmental transport of human pharmaceuticals to streams in the United Kingdom. Sci Total Environ 333(1):167–184

Backhaus T, Grimme LH (1999) The toxicity of antibiotic agents to the luminescent bacterium *Vibrio fischeri*. Chemosphere 38(14):3291–3301

Bedner M, MacCrehan WA (2006) Transformation of Acetaminophen by Chlorination Produces the Toxicants 1,4-Benzoquinone and N-Acetyl-p-benzoquinone Imine. Environ Sci Technol 40(2):516–522

Besse J-P, Latour J-F, Garric J (2012) Anticancer drugs in surface waters: what can we say about the occurrence and environmental significance of cytotoxic, cytostatic and endocrine therapy drugs? Environ Int 39(1):73–86

Booker V, Halsall C, Llewellyn N, Johnson A, Williams R (2014) Prioritising anticancer drugs for environmental monitoring and risk assessment purposes. Sci Total Environ 473–474 (0):159–170

Boos J, Welslau U, Ritter J, Blaschke G, Schellong G (1991) Urinary excretion of the enantiomers of ifosfamide and its inactive metabolites in children. Cancer Chemother Pharmacol 28(6):455–460

Buerge IJ, Buser HR, Poiger T, Muller MD (2006) Occurrence and fate of the cytostatic drugs cyclophosphamide and ifosfamide in wastewater and surface waters. Environ Sci Technol 40(23):7242–7250

Calza P, Medana C, Sarro M, Rosato V, Aigotti R, Baiocchi C, Minero C (2014) Photocatalytic degradation of selected anticancer drugs and identification of their transformation products in water by liquid chromatography–high resolution mass spectrometry. J Chromatogr A 1362(0):135–144

Castiglioni S, Bagnati R, Calamari D, Fanelli R, Zuccato E (2005) A multiresidue analytical method using solid-phase extraction and high-pressure liquid chromatography tandem mass spectrometry to measure pharmaceuticals of different therapeutic classes in urban wastewaters. J Chromatogr A 1092(2):206–215

Česen M, Kosjek T, Laimou-Geraniou M, Kompare B, Širok B, Lambropolou D, Heath E (2015) Occurrence of cyclophosphamide and ifosfamide in aqueous environment and their removal by biological and abiotic wastewater treatment processes. Sci Total Environ 527:465–473

Česen M, Eleršek T, Novak M, Žegura B, Kosjek T, Filipič M, Heath E (2016) Ecotoxicity and genotoxicity of cyclophosphamide, ifosfamide, their metabolites/transformation products and their mixtures. Environ Pollut 210:192–201

De la Cruz N, Giménez J, Esplugas S, Grandjean D, de Alencastro LF, Pulgarín C (2012) Degradation of 32 emergent contaminants by UV and neutral photofenton in domestic wastewater effluent previously treated by activated sludge. Water Res 46:1947–1957

DellaGreca M, Iesce MR, Isidori M, Nardelli A, Previtera L, Rubino M (2007) Phototransformation products of tamoxifen by sunlight in water. Toxicity of the drug and its derivatives on aquatic organisms. Chemosphere 67(10):1933–1939

Dorr RT, VonHoff DD (1994) Cancer chemotherapy handbook, 2nd edn. Norwalk, Appleton & Lange

Eitel A, Scherrer M, Kümmerer K (1999) Handling cytostatic drugs. Kehrer Digital & Print, Freiburg

Ferrando-Climent L, Rodriguez-Mozaz S, Barceló D (2013) Development of a UPLC-MS/MS method for the determination of ten anticancer drugs in hospital and urban wastewaters, and its application for the screening of human metabolites assisted by information-dependent acquisition tool (IDA) in sewage samples. Anal Bioanal Chem 405(18):5937–5952

Ferrando-Climent L, Cruz-Morató C, Marco-Urrea E, Vicent T, Sarrà M, Rodriguez-Mozaz S, Barceló D (2015) Non conventional biological treatment based on Trametes versicolor for the elimination of recalcitrant anticancer drugs in hospital wastewater. Chemosphere 136:9–19

Ferrando-Climent L, Gonzalez-Olmos R, Anfruns A, Aymerich I, Corominas L, Barceló D, Rodriguez-Mozaz S (2017) Elimination study of the chemotherapy drug tamoxifen by different advanced oxidation processes: Transformation products and toxicity assessment. Chemosphere 168:284–292

Ferre-Aracil J, Valcárcel Y, Negreira N, de Alda ML, Barceló D, Cardona SC, Navarro-Laboulais J (2016) Ozonation of hospital raw wastewaters for cytostatic compounds removal. Kinetic modelling and economic assessment of the process. Sci Total Environ 556:70–79

Froehner K, Backhaus T, Grimme LH (2000) Bioassays with Vibrio fischeri for the assessment of delayed toxicity. Chemosphere 40:821–828

Garcia-Ac A, Broseus R, Vincent S, Barbeau B, Prevost M, Sauve S (2010) Oxidation kinetics of cyclophosphamide and methotrexate by ozone in drinking water. Chemosphere 79(11):1056–1063

Glaze WH, Kang J-W, Chapin DH (1987) The chemistry of water treatment processes involving ozone, hydrogen peroxide and ultraviolet radiation. Ozone Sci Eng 9(4):335–352

Gómez-Canela C, Cortés-Francisco N, Ventura F, Caixach J, Lacorte S (2013) Liquid chromatography coupled to tandem mass spectrometry and high resolution mass spectrometry as analytical tools to characterize multi-class cytostatic compounds. J Chromatogr A 1276(0):78–94

Governo M, Santos MSF, Alves A, Madeira LM (2017) Degradation of the cytostatic 5-Fluorouracil in water by Fenton and photo-assisted oxidation processes. Environ Sci Pollut Res 24(1):844–854

Haddad T, Baginska E, Kümmerer K (2015) Transformation products of antibiotic and cytostatic drugs in the aquatic cycle that result from effluent treatment and abiotic/biotic reactions in the environment: An increasing challenge calling for higher emphasis on measures at the beginning of the pipe. Water Res 72(Supplement C):75–126

Halling-Sørensen B, Nors Nielsen S, Lanzky PF, Ingerslev F, Holten Lützhøft HC, Jørgensen SE (1998) Occurrence, fate and effects of pharmaceutical substances in the environment- a review. Chemosphere 36(2):357–393

Henschel KP, Wenzel A, Diedrich M, Fliedner A (1997) Environmental hazard assessment of pharmaceuticals. Regul Toxicol Pharmacol 25(3):220–225

Johnson AC, Jürgens MD, Williams RJ, Kümmerer K, Kortenkamp A, Sumpter JP (2008) Do cytotoxic chemotherapy drugs discharged into rivers pose a risk to the environment and human health? An overview and UK case study. J Hydrol 348(1–2):167–175

Kiffmeyer T, Götze H-S, Jursch M, Lüders U (1997) Trace enrichment, chromatographic separation and biodegradation of cytostatic compounds in surface water. Fresenius J Anal Chem 361:185–191

Koltsakidou A, Antonopoulou M, Evgenidou E, Konstantinou I, Lambropoulou DA (2017a) Cytarabine degradation by simulated solar assisted photocatalysis using TiO_2. Chem Eng J 316:823–831

Koltsakidou A, Antonopoulou M, Sykiotou M, Evgenidou E, Konstantinou I, Lambropoulou DA (2017b) Photo-Fenton and Fenton-like processes for the treatment of the antineoplastic drug 5-fluorouracil under simulated solar radiation. Environ Sci Pollut Res 24(5):4791–4800

Kosjek T, Heath E (2011) Occurrence, fate and determination of cytostatic pharmaceuticals in the environment. TrAC Trends Anal Chem 30(7):1065–1087

Kosjek T, Negreira N, de Alda ML, Barceló D (2015) Aerobic activated sludge transformation of methotrexate: identification of biotransformation products. Chemosphere 119(Supplement): S42–S50

Kovalova L, McArdell CS, Hollender J (2009) Challenge of high polarity and low concentrations in analysis of cytostatics and metabolites in wastewater by hydrophilic interaction chromatography/tandem mass spectrometry. J Chromatogr A 1216(7):1100–1108

Kümmerer K, Steger-Hartmann T, Baranyai A, Burhaus I (1996) Tests of biodegradation of cytostatics cyclophosphamide and ifosfamide using the Closed Bottle Test (OECD 301 D). Zentralbl Hyg Umweltmed 198:215–225

Kümmerer K, Steger-Hartmann T, Meyer M (1997) Biodegradability of the anti-tumour agent ifosfamide and its occurrence in hospital effluents and communal sewage. Water Res 31(11):2705–2710

Kümmerer K, Alexy R, Hüttig J, Schöll A (2004) Standardized tests fail to assess the effects of antibiotics on environmental bacteria. Water Res 38(8):2111–2116

Kümmerer K, Haiß A, Schuster A, Hein A, Ebert I (2016) Antineoplastic compounds in the environment—substances of special concern. Environ Sci Pollut Res 23(15):14791–14804

Lai WW-P, Lin HH-H, Lin AY-C (2015) TiO_2 photocatalytic degradation and transformation of oxazaphosphorine drugs in an aqueous environment. J Hazard Mater 287:133–141

Lee Y, Kovalova L, McArdell CS, von Gunten U (2014) Prediction of micropollutant elimination during ozonation of a hospital wastewater effluent. Water Res 64:134–148

Lenz K, Koellensperger G, Hann S, Weissenbacher N, Mahnik SN, Fuerhacker M (2007) Fate of cancerostatic platinum compounds in biological wastewater treatment of hospital effluents. Chemosphere 69(11):1765–1774

Li W, Nanaboina V, Chen F, Korshin GV (2016) Removal of polycyclic synthetic musks and antineoplastic drugs in ozonated wastewater: quantitation based on the data of differential spectroscopy. J Hazard Mater 304:242–250

Lin HH-H, Lin AY-C (2014) Photocatalytic oxidation of 5-fluorouracil and cyclophosphamide via UV/TiO_2 in an aqueous environment. Water Res 48(0):559–568

Lin AY-C, Hsueh JH-F, Hong PKA (2015) Removal of antineoplastic drugs cyclophosphamide, ifosfamide, and 5-fluorouracil and a vasodilator drug pentoxifylline from wastewaters by ozonation. Environ Sci Pollut Res 22(1):508–515

Lutterbeck C, Kern D, Machado Ê, Kümmerer K (2014) 'Ecotoxicological evaluation of selected anticancer drugs. Toxicol Lett 229(Supplement):S71–S72

Lutterbeck CA, Baginska E, Machado ÊL, Kümmerer K (2015a) Removal of the anti-cancer drug methotrexate from water by advanced oxidation processes: Aerobic biodegradation and toxicity studies after treatment. Chemosphere 141:290–296

Lutterbeck CA, Machado ÊL, Kümmerer K (2015b) Photo degradation of the antineoplastic cyclophosphamide: a comparative study of the efficiencies of UV/H_2O_2, UV/Fe^{2+}/H_2O_2 and UV/TiO_2 processes. Chemosphere 120(0):538–546

Lutterbeck CA, Wilde ML, Baginska E, Leder C, Machado ÊL, Kümmerer K (2015c) Degradation of 5-FU by means of advanced (photo)oxidation processes: UV/H_2O_2, UV/Fe^{2+}/H_2O_2 and UV/TiO_2 — comparison of transformation products, ready biodegradability and toxicity. Sci Total Environ 527:232–245

Lutterbeck CA, Wilde ML, Baginska E, Leder C, Machado ÊL, Kümmerer K (2016) Degradation of cyclophosphamide and 5-fluorouracil by UV and simulated sunlight treatments: Assessment of the enhancement of the biodegradability and toxicity. Environ Pollut 208:467–476

Mahmoud WM, Kümmerer K (2012) Captopril and its dimer captopril disulfide: photo degradation, aerobic biodegradation and identification of transformation products by HPLC-UV and LC-ion trap-MS(n). Chemosphere 88(10):1170–1177

Mahmoud WMM, Trautwein C, Leder C, Kümmerer K (2013) Aquatic photochemistry, abiotic and aerobic biodegradability of thalidomide: identification of stable transformation products by LC–UV–MSn. Sci Total Environ 463–464(0):140–150

Mahnik SN, Rizovski B, Fuerhacker M, Mader RM (2004) Determination of 5-fluorouracil in hospital effluents. Anal Bioanal Chem 380(1):31–35

Mahnik SN, Rizovski B, Fuerhacker M, Mader RM (2006) Development of an analytical method for the determination of anthracyclines in hospital effluents. Chemosphere 65(8):1419–1425

Mahnik SN, Lenz K, Weissenbacher N, Mader RM, Fuerhacker M (2007) Fate of 5-fluorouracil, doxorubicin, epirubicin, and daunorubicin in hospital wastewater and their elimination by activated sludge and treatment in a membrane-bio-reactor system. Chemosphere 66(1):30–37

Margot J, Kienle C, Magnet A, Weil M, Rossi L, de Alencastro LF, Abegglen C, Thonney D, Chèvre N, Schärer M, Barry DA (2013) Treatment of micropollutants in municipal wastewater: ozone or powdered activated carbon? Sci Total Environ 461–462:480–498

Martín J, Camacho-Muñoz D, Santos JL, Aparicio I, Alonso E (2011) Simultaneous determination of a selected group of cytostatic drugs in water using high-performance liquid chromatography–triple-quadrupole mass spectrometry. J Sep Sci 34(22):3166–3177

Mioduszewska K, Maszkowska J, Białk-Bielińska A, Krüger O, Kalbe U, Liberek B, Łukaszewicz P, Stepnowski P (2016) The leaching behavior of cyclophosphamide and ifosfamide from soil in the presence of co-contaminant — mixture sorption approach. Sci Total Environ 542:915–922

Mompelat S, Le Bot B, Thomas O (2009) Occurrence and fate of pharmaceutical products and by-products, from resource to drinking water. Environ Int 35(5):803–814

Negreira N, Regueiro J, López de Alda M, Barceló D (2015a) Transformation of tamoxifen and its major metabolites during water chlorination: identification and in silico toxicity assessment of their disinfection by-products. Water Res 85:199–207

Negreira N, Regueiro J, López de Alda M, Barceló D (2015b) Degradation of the anticancer drug erlotinib during water chlorination: non-targeted approach for the identification of transformation products. Water Res 85:103–113

Norton-Brandão D, Scherrenberg SM, van Lier JB (2013) Reclamation of used urban waters for irrigation purposes – a review of treatment technologies. J Environ Manag 122:85–98

Ocampo-Pérez R, Sánchez-Poloa M, Rivera-Utrilla J, Leyva-Ramos R (2010) Degradation of antineoplastic cytarabine in aqueous phase by advanced oxidation processes based on ultraviolet radiation. Chem Eng J 165(2):581–588

Ofiarska A, Pieczyńska A, Fiszka Borzyszkowska A, Stepnowski P, Siedlecka EM (2016) 'Pt–TiO2-assisted photocatalytic degradation of the cytostatic drugs ifosfamide and cyclophosphamide under artificial sunlight. Chem Eng J 285(Supplement C):417–427

Olalla A, Negreira N, López de Alda M, Barceló D, Valcárcel Y (2018) A case study to identify priority cytostatic contaminants in hospital effluents. Chemosphere 190(Supplement C):417–430

Pedrero F, Kalavrouziotis I, Alarcón JJ, Koukoulakis P, Asano T (2010) Use of treated municipal wastewater in irrigated agriculture—review of some practices in Spain and Greece. Agric Water Manag 97(9):1233–1241

Pérez-Estrada LA, Agüera A, Hernando MD, Malato S, Fernández-Alba AR (2008) Photo degradation of malachite green under natural sunlight irradiation: kinetic and toxicity of the transformation products. Chemosphere 70(11):2068–2075

Pignatello JJ, Oliveros E, MacKay A (2006) Advanced oxidation processes for organic contaminant destruction based on the Fenton reaction and related chemistry. Crit Rev Environ Sci Technol 36:1–84

Prasse C, Stalter D, Schulte-Oehlmann U, Oehlmann J, Ternes TA (2015) 'Spoilt for choice: a critical review on the chemical and biological assessment of current wastewater treatment technologies. Water Res 87:237–270

Richardson ML, Bowron JM (1985) The fate of pharmaceutical chemicals in the aquatic environment. J Pharm Pharmacol 37(1):1–12

Roig B, Marquenet B, Delpla I, Bessonneau V, Sellier A, Leder C, Thomas O, Bolek R, Kummerer K (2014) Monitoring of methotrexate chlorination in water. Water Res 57:67–75

Rowney NC, Johnson AC, Williams RJ (2009) Cytotoxic drugs in drinking water: a prediction and risk assessment exercise for the thames catchment in the United Kingdom. Environ Toxicol Chem 28(12):2733–2743

Rubino FM (2001) Separation methods for methotrexate, its structural analogues and metabolites. J Chromatogr B Biomed Sci Appl 764(1–2):217–254

Santana-Viera S, Montesdeoca-Esponda S, Sosa-Ferrera Z, Santana-Rodríguez JJ (2016) 'Cytostatic drugs in environmental samples: an update on the extraction and determination procedures. TrAC Trends Anal Chem 80(Supplement C):373–386

Santana-Viera S, Guedes-Alonso R, Sosa-Ferrera Z, Santana-Rodríguez JJ, Kabir A, Furton KG (2017) 'Optimization and application of fabric phase sorptive extraction coupled to ultra-high performance liquid chromatography tandem mass spectrometry for the determination of cytostatic drug residues in environmental waters. J Chromatogr A 1529(Supplement C):39–49

Santos MSF, Franquet-Griell H, Lacorte S, Madeira LM, Alves A (2017) Anticancer drugs in Portuguese surface waters – estimation of concentrations and identification of potentially priority drugs. Chemosphere 184:1250–1260

Somensi CA, Simionatto EL, Dalmarco JB, Gaspareto P, Radetski CM (2012) 'A comparison between ozonolysis and sonolysis/ozonolysis treatments for the degradation of the cytostatic drugs methotrexate and doxorubicin: kinetic and efficiency approaches. J Environ Sci Health A Tox Hazard Subst Environ Eng 47(11):1543–1550

Steger-Hartmann T, Kümmerer K, Schecker J (1996) Trace analysis of the antineoplastics ifosfamide and cyclophosphamide in sewage water by twostep solid-phase extraction and gas chromatography-mass spectrometry. J Chromatogr A 726(1):179–184

Ternes TA (1998) Occurrence of drugs in German sewage treatment plants and rivers. Water Res 32(11):3245–3260

Thomas KV, Hilton MJ (2004) The occurrence of selected human pharmaceutical compounds in UK estuaries. Mar Pollut Bull 49(5):436–444

Toolaram AP, Kümmerer K, Schneider M (2014) Environmental risk assessment of anti-cancer drugs and their transformation products: a focus on their genotoxicity characterization-state of knowledge and short comings. Mutat Res Rev Mutat Res 760(0):18–35

Wang X-H, Lin AY-C (2014) Is the phototransformation of pharmaceuticals a natural purification process that decreases ecological and human health risks? Environ Pollut 186(0):203–215

Weissbrodt D, Kovalova L, Ort C, Pazhepurackel V, Moser R, Hollender J, Siegrist H, McArdell CS (2009) Mass flows of X-ray contrast media and cytostatics in hospital wastewater. Environ Sci Technol 43(13):4810–4817

WHO/UNICEF (2015) 25 Years – Progress on Drinking Water and Sanitation – 2015 Update and MDG Assessement, Geneva, Switzerland. Available at: http://files.unicef.org/publications/files/Progress_on_Sanitation_and_Drinking_Water_2015_Update_.pdf

Yin J, Shao B, Zhang J, Li K (2010) A preliminary study on the occurrence of cytostatic drugs in hospital effluents in Beijing, China. Bull Environ Contam Toxicol 84(1):39–45

Yin J, Niu Y, Shao B (2017) Products of methotrexate during chlorination. J Environ Sci 55:100–108

Yu JT, Bouwer EJ, Coelhan M (2006) Occurrence and biodegradability studies of selected pharmaceuticals and personal care products in sewage effluent. Agric Water Manag 86:72–80

Zhang J, Chang VWC, Giannis A, Wang J-Y (2013) Removal of cytostatic drugs from aquatic environment: a review. Sci Total Environ 445–446(0):281–298

Chapter 8
Analytical Methodologies for the Determination of Cytostatic Compounds in Environmental Matrices

Sergio Santana-Viera, Sarah Montesdeoca-Esponda, Zoraida Sosa-Ferrera, and José Juan Santana-Rodríguez

Abstract The presence of cytostatic drugs in the environment is mainly due to the entry of nonmetabolized compounds prescribed in anticancer therapies that are excreted by patients. These compounds have been detected in wastewaters in very low concentrations. The analysis of cytostatic drugs in environmental samples is characterized by difficulty in the quantification of these low concentrations in complex matrices. Therefore, to be able to detect them, first an extraction process and preconcentration are necessary for their subsequent determination by the correct sensitive analytical techniques. Data about the mostly used techniques, their optimal conditions, the predominant compounds, and the results obtained will contribute to the knowledge of these emerging pollutants.

In this chapter, a comprehensive description of methodologies used for the determination of cytostatic drugs in environmental samples is presented in terms of selecting suitable extraction and clean-up procedures. The most commonly used extraction/preconcentration techniques are solid-phase extraction (SPE) for liquid samples and pressurized liquid extraction (PLE) as well as ultrasound-assisted extraction (UAE) for sludge samples. The detection is carried out mainly by liquid chromatography along with mass spectrometry detection.

Keywords Cytostatic compounds · Environmental samples · Sewage · Sludges · Chromatography

S. Santana-Viera · S. Montesdeoca-Esponda · Z. Sosa-Ferrera · J. J. Santana-Rodríguez (✉)
Instituto Universitario de Estudios Ambientales y Recursos Naturales (i-UNAT), Universidad de Las Palmas de Gran Canaria, Las Palmas de Gran Canaria, Spain
e-mail: josejuan.santana@ulpgc.es

E. Heath et al. (eds.), *Fate and Effects of Anticancer Drugs in the Environment*,
https://doi.org/10.1007/978-3-030-21048-9_8

Abbreviations

4-OHTAM	4-Hydroxytamoxifen
5-FU	5-Fluorouracil
A-GLU	Amino glutethimide
APCI	Atmospheric pressure chemical ionization
APPI	Atmospheric pressure photoionization
araU	1-β-d-Arabinofuranoside
AZA	Azathioprine
BP1	3'-O-Desmethyl etoposide
CAP	Capecitabine
Car-Pt	Carboplatin
CE	Capillary electrophoresis
CHLO	Chlorambucil
CIPRO	Ciprofloxacin
Cis-Pt	Cisplatin
CP	Cyclophosphamide
CYP	Cyproterone
CYT	Cytarabine
DAU	Daunorubicin
dFdU	2′,2′-Difluorodeoxyuridine
Di-Pt	Diaquacisplatin
DOC	Docetaxel
DOX	Doxorubicin
EI	Electron impact
ENDO	Endoxifen
EPI	Epirubicin
ERLO	Erlotinib
ESI	Electrospray ionization
ETO	Etoposide
FLU	Fludarabine
GAC	Green analytical chemistry
GC	Gas chromatography
GEM	Gemcitabine
GOS	Goserelin
HILIC	Hydrophilic interaction chromatography
ICP-MS	Inductively coupled plasma mass spectrometry
IDL	Instrumental detection limits
IF	Ifosfamide
IMA	Imatinib
IRI	Irinotecan
LC	High-performance liquid chromatography

LC	Liquid chromatography
LEU	Leuprolide
LIT	Linear ion trap
LLE	Liquid-liquid extraction
LOD	Limit of detection
LOQ	Limit of quantification
MAE	Microwave-assisted extraction
MEG	Megestrol acetate
MEL	Melphalan
MET	Methotrexate
MI	Mitotic indexes
MISPE	Molecularly imprinted solid-phase extraction
MIT-C	Mitomycin C
Mono-Pt	Monoaquacisplatin
MS	Mass spectrometry
MS/MS	Mass Spectrometry in tandem
MTBSTFA	N-Methyl-N-[tert-butyldimethylsilyl]-trifluoroacetamide
N.D.	Not detectable
OH-D-TAM	4-Hydroxy-Ndesmethyl- tamoxifen
OH-MET	Hydroxymethotrexate
OH-PAC	6(α)-Hydroxypaclitaxel
OH-TAM	Hydroxytamoxifen
Oxa-Pt	Oxaliplatin
PAC	Paclitaxel
PGC	Porous graphitic carbon
PLE	Pressurized liquid extraction
PRED	Prednisone
QqLit	Triple quadrupole-linear ion trap
SPE	Solid-phase extraction
TAM	Tamoxifen
TCB	1,2,3,5-Tetrachloro-benzene
TEM	Temozolomide
TF	Trofosfamide
TIS	Turbo ion spray
ToF	Time of Flight
TP	Transformation products
UAE	Ultrasound-assisted extraction
UHPLC	Ultrahigh-performance liquid chromatography
VINB	Vinblastine
VINC	Vincristine
VINO	Vinorelbine
WWTP	Wastewater treatment plant

8.1 Introduction

Cytostatic drugs are compounds used in chemotherapy to fight cancer. Although the current trend is toward the nonhospitalization of patients, hospitals remain the main sources of anticancer residues (Zhang et al. 2013). Hospital effluents are normally connected to the sewage system without any pretreatment, so once administered, cytostatic compounds excreted in the urine or feces as metabolized or original compounds eventually reach the wastewater treatment plants (WWTPs). Since WWTPs are designed for the removal of organic load, they are not specifically designed for the treatment of particular chemical compounds, and their effluents could be considered as the main source of these compounds and their metabolites in the aquatic environment. An important feature of cytostatic compounds is that they have low biodegradability in conventional treatments and are considered recalcitrant compounds (Kümmerer et al. 1997, 2000; Mahnik et al. 2007; Lenz et al. 2007; Kazner et al. 2008). Similar to other families of pharmaceutical compounds, though their concentrations in wastewater are very low (in the $ng{\cdot}L^{-1}$ range), the consequences of their continued input are unknown (Buchberger 2007).

These compounds have been designed to disrupt or prevent cell proliferation, usually by interfering with DNA synthesis. Since they do not act selectively on the growth of cancer cells but act on all fast-dividing cells, it is important to establish the risk from the presence of these compounds in the environment (Toolaram et al. 2014). The risk of anticancer drugs to the environmental organisms is not very clear, mainly because of the lack of standardized toxicity approaches. Therefore, residues of cytostatic compounds are considered to be emerging pollutants, although many of them have already been described as genotoxic, and they will cause adverse effects on the aquatic ecosystems (Kovács et al. 2015). To determine the negative impact of these pollutants, qualitative and quantitative analyses of the aquatic systems must be conducted.(Ferrando-Climent et al. 2013). However, despite the need to obtain data that help us understand their hazard in the environment (Zuloaga et al. 2012), only a small number of studies exist regarding cytostatic compounds, with a limited amount of experimental data on ecotoxicity (Zounkova et al. 2010). Some of them are sometimes included in multiresidual studies, but only a few papers are completely devoted to the determination of this family of pollutants. Therefore, one of the challenges in the analytical chemistry field is the development of fast and efficient procedures for the analysis of these slowly discovered compounds and their transformation products, which are compounds that are formed by reactions of cytostatic compounds that give rise to new unknown compounds and that can be even more dangerous than the original ones. For example, when the elimination of 5-fluorouracil (5-FU) by photo-oxidation processes such as UV/H_2O_2, $UV/Fe^{2+}/H_2O_2$ and UV/TiO_2 was attempted, parent compounds can be eliminated, but the transformation products formed can still be toxic. Therefore, toxicity screening after advanced treatment is recommended (Lutterbeck et al. 2015b).

Studies with certain cytostatic compounds demonstrate its potential persistence in the environment and its danger. Cyclophosphamide (CP) and ifosfamide (IF) are persistent in the aquatic environment, and they can reach the drinking water through the surface water. Though they can react with the DNA and that the risk is higher for newborns and children than for adults, a safe threshold concentration with regard to health effects could not be given (Kuemmerer and Al-Ahmad 2010). Assays performed for MET presented a high number of cell deaths, indicating that these compounds may affect the growth and normal development of these plants (Lutterbeck et al. 2015a). Compounds like 5-FU, etoposide (ETO), cisplatin (Cis-Pt), carboplatin (Car-Pt), vincristine (VINC), and CP lead to genotoxic effects invariably (Mišík et al. 2014).

The presence of different cytostatic drugs in organisms of higher trophic level has also been studied to investigate their possible adverse effects. Studies on fish demonstrate that exposure to 5-FU at high concentrations (0.01, 1, and 100 $\mu g\ L^{-1}$) can damage the integrity of their DNA and induce massive whole-transcriptome changes, which might affect fish populations over the long-term exposure of several generations (Kovács et al. 2015). At concentrations detected in hospital effluents, in the range of $\mu g{\cdot}L^{-1}$, a decrease in the reproductive capacity has been observed in organisms exposed to Cis-Pt, 5-FU (Parrella et al. 2014), tamoxifen (TAM), 4-hydroxytamoxifen (4-OHTAM), and endoxifen (ENDO) (Borgatta et al. 2015, 2016).

Among the rare references regarding the presence of these compounds in environmental liquid samples, most of the relevant works are focused on the analysis of sewage waters from wastewater treatment plants or hospital effluents, and only a few papers consider river water or groundwater samples for the analysis of cytostatic compounds. In 2011, Kosjek and Heath 2011 performed a review of the occurrence, fate, and determination of cytostatic drugs in the environment, examining a total of 17 papers published until 2009 (Kosjek and Heath 2011). Subsequently, since the number of papers had increased and different techniques of determination had been used, sometimes focused on a different objective such as the detection of metabolites, another overview was performed updating the existing information (Santana-Viera et al. 2016). This time, 42 papers in the techniques used for the determination of cytostatic compounds in environmental samples were reviewed. In these last years, the use of new detection technologies along with the greater use of the online SPE has assumed a main role, focusing on faster multiresidue methods. Since the number of publications about cytostatic compounds is scarce, the number of studies where these compounds have been detected is very low and almost exclusively devoted to Europe (Fig. 8.1).

Cytostatic drugs can also be accumulated in solid matrices, mainly in the sludge from the WWTPs; however, the available information about this accumulation is even scarcer than the information on aqueous samples. The extraction of analytes from solid samples is even more tedious than from liquid samples because of the difficulty in predicting and overcoming the solute-matrix interactions.

It is important to know which cytostatic compounds can be adsorbed into the sludge depending on their hydrophobicity and the concentrations in which they can be found, especially when such sludge is used as fertilizer.

Sensitive analytical methods capable of detecting these contaminants at low concentrations in the aquatic environment and sludge are essential. Furthermore, a

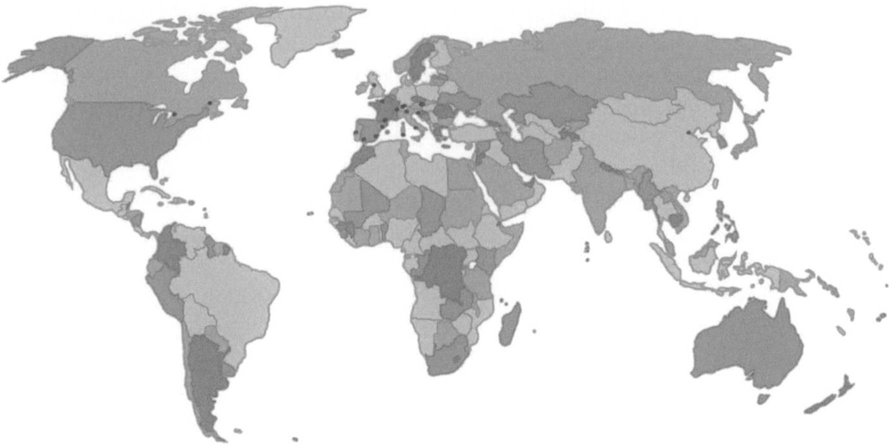

Fig. 8.1 Occurrence of cytostatic compounds until 2016

high selectivity is required to avoid interference due to complex matrix components (Garcia-Ac et al. 2009a, b). Besides sophisticated detection systems, which are imperative in this type of determination, the previous incorporation of isolation and purification steps is mandatory in order to obtain high sensitivity. Therefore, sample preparation often represents the most tedious and time-consuming part of the analytical process.

For liquid samples, both off-line and online solid-phase extractions (SPE) are the preparation techniques that have been mainly used for the analysis of these drugs (Mahugo-Santana et al. 2011), including the use of new materials such as molecularly imprinted polymers. However, microextraction techniques, like solid-phase microextraction, stir bar sorptive extraction, single-drop microextraction, etc., have not been implemented. Similar to the techniques used in the last decade to analyze environmental samples, gas chromatography (GC) and liquid chromatography (LC) separation combined with mass spectrometry (MS) detection are the mostly used techniques for the analysis of these compounds (Buchberger 2007). However, determination techniques like inductively coupled plasma mass spectrometry (ICP-MS) (Vyas et al. 2014) or capillary electrophoresis (CE) (Mahnik et al. 2004) have been also applied.

8.2 Extraction Techniques

Conventional techniques for the extraction from liquid samples such as liquid-liquid extraction (LLE) have been replaced by alternative methodologies. Though LLE provides high recoveries and good repeatability, it is relatively time consuming, harmful (due to the use of large volumes of organic solvents), and quite expensive. Thus, some alternative techniques have been developed: SPE is the main technique

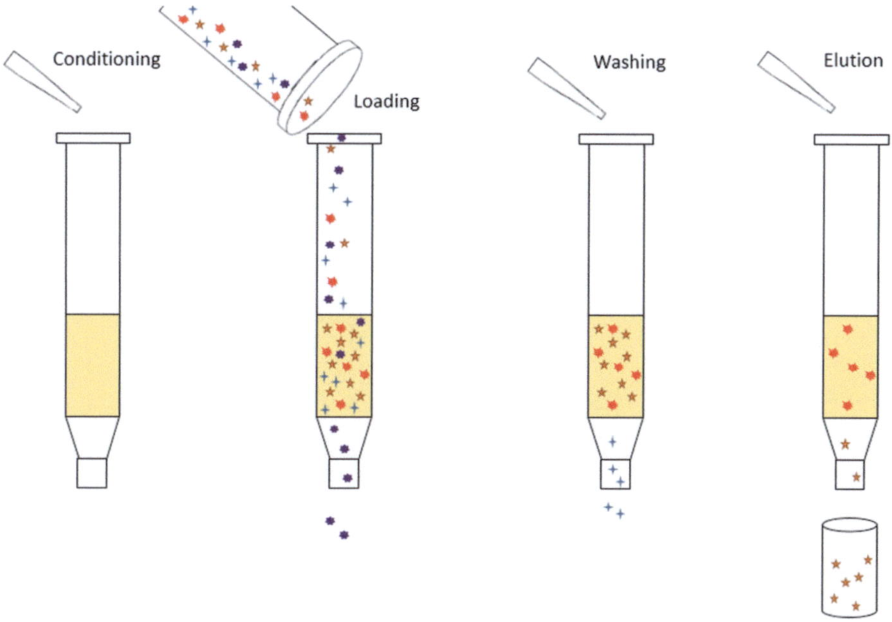

Fig. 8.2 Scheme of the solid-phase extraction technique

used for the extraction of cytotoxic compounds in liquid environmental samples. Only Tauxe-Wuersch et al. used the LLE technique for the extraction of cytostatic compounds as a previous step for purification before using SPE (Tauxe-Wuersch et al. 2006).

SPE is a sample treatment technique that enables the concentration and purification of analytes from solution via adsorption onto a solid sorbent (Camel 2003), and it is widely used in the environmental analytical field, because it extracts and preconcentrates the sample in a single step. In Fig. 8.2, a scheme is shown, in which each step of the SPE technique can be seen.

SPE is undoubtedly the most common extraction technique used for the extraction of cytostatic compounds from environmental liquid samples (Tables 8.1, 8.2 and 8.3). Different cartridges have been used depending on the cytostatics intended to be determined. For example, Biotage Isolute ENV+ cartridges have been used for the extraction of pyrimidine analogues, while Phenomenex Strata-X are preferred when we tried to extract cytostatics from a mixture of different pharmaceutical compounds and Waters Oasis HLB cartridges to extract a larger group of them from different families. Kosjek et al. (2013) studied 5-FU and its prodrug, capecitabine (CAP), and tested different cartridge mechanisms, concluding that ion pair and anion exchange retention mechanisms are not viable for complex matrices (Kosjek et al. 2013).

This technique can be used in a conventional way (off-line) as described in Fig. 8.2, automated in workstations, or coupled online to the chromatographic column. Only two works have been published regarding the determination of

Table 8.1 Methodologies for the determination of cytostatic compounds in liquid samples from WWTPs

Compounds	Extraction	Determination technique	Recovery (%)	LOD ($ng \cdot L^{-1}$)	Found concentration ($ng \cdot L^{-1}$)	References
IF, CP	Off-line SPE	GC-MS	39 (CP)	7 (IF)	24 (IF)	Steger-Hartmann et al. (1996)
			30 (IF)	6 (CP)	146 (CP)	
CP, IF	Off-line SPE	LC-ESI-MS/MS	51–57	10		Ternes et al. (1998)
IF, CP	Same as reference Ternes et al. (1998)	Same as reference Ternes et al. (1998)	Not provided	10	N.D.	Ternes (1998)
CP	Off-line SPE	LC-ESI-MS/MS	Not provided	5–20	4–8	Metcalfe et al. (2003b)
CP	Off-line SPE	LC-TIS-MS/MS	106 (CP)	1.9 (CP)	2.1–9 (CP)	Castiglioni et al. (2005)
			76 (MET)	0.83 (MET)	12.6 (MET)	
5-FU	Off-line SPE	GC-MS			N.D.	Yu et al. (2006)
IF, CP	Off-line SPE	Technique: LC-TIS-MS	74–93	0.2–2	1.4–11	Buerge et al. (2006)
TAM, 5-FU	Off-line SPE as purification step after LLE	GC-MS	81 (TAM)	1 (TAM)	1–4 (TAM)	Tauxe-Wuersch et al. (2006)
			73 (5-FU)	15–30 (5-FU)	N.D. (5-FU)	
TAM	Off-line SPE	LC-ESI-MS/MS	19	0.03	N.D.	Nebot et al. (2007)

5-FU, GEM, dFdU	Same as reference Kovalova et al. (2009)	LC-ESI-MS/MS	54–118	0.9–9[a]	N.D.	Weissbrodt et al. (2009)
IF, CP	Off-line SPE	Technique: LC-ESI-MS/MS	167–96 (IF)	5–25 (R.O.)[d]	< L.D.	Busetti et al. (2009)
			106–98 (CP)	(100)[d]		
CP, MET	Online SPE	SPE-ESI-LC-MS/MS	148 (CP)	5–11	9 (CP)	Garcia-Ac et al. (2009a)
			55 (MET)		59 (MET)	
TAM	Online SPE	SPE-LC-ESI-MS/MS	ABS: 10.36–14.34	0.63–5.41	N.D.	López-Serna et al. (2010)
			REL: 74.16–131.3			
CP, IF	Off-line SPE	LC-ESI-MS	57–70	0.03–0.12 (CP)	4–11 (CP)	Llewellyn et al. (2011)
				0.05–0.09 (IF)	2 (IF)	
5-FU, GEM, CP, IF, CYT, MET, PAC, ETO, IRI, DOC, EPI, DOX, VINO, MIT-C	Off-line SPE	LC-ESI-QqQ-MS	11–105	0.1–38	1.2–15	Martín et al. (2011)
CP, MET	Online SPE	SPE-LC-ESI-MS/MS	Not provided	5–38		Garcia-Ac et al. (2011)
CP, EPI	Off-line SPE	LC-ESI-Orbitrap-MS	37–107	3.1–85	13.1 (CP)[c]	Gomez-Canela et al. (2012)
5-FU, CAP	Off-line SPE	GC-MS		0.48 (5-FU)	4.7–14	Kosjek et al. (2013)
GEM, TEM, MET, IRI, IMA, IF, CP, ERLO, ETO, DOX, CAP, TAM, PAC, OH-MET, OH-D-TAM, OH-TAM, OH-PAC	Online SPE	SPE-LC-ESI-MS/MS	72–119[d]	0.3–36	2.1–29.7	Negreira et al. (2013)

(continued)

Table 8.1 (continued)

Compounds	Extraction	Determination technique	Recovery (%)	LOD ($ng \cdot L^{-1}$)	Found concentration ($ng \cdot L^{-1}$)	References
CP, IF, DOC, PAC, ETO, VINC, TAM, MET, AZA, CIPRO	Off-line SPE	UHPLC-ISS-QqLit	47–115	0.8–24	18.2–32,000	Ferrando-Climent et al. (2013)
CP, GEM, IF, MET, IRI, EPI	Online SPE	SPE-ESI-LC-MS	> 70 (except GEM)	4–20	13 (CP)	Rabii et al. (2014)
					60 (MET)	
CP, CHLO, MEL, IF, FLU, CYT, GEM, CAP, VINB, VINC, ETO, PAC, DOC, DOX, DAU, EPI, IMA, ERLO, IRI, LEU, GOS, TAM, A-GLU, MEG, CYP, PRED	Off-line	LC-HESI-Orbitrap-MS	6–110	0.7–356	0.003–0.22[c]	Gómez-Canela et al. (2014)

[a] $\mu g \cdot L^{-1}$
[b] Limit of quantification (LOQ)
[c] $\mu g \cdot L^{-1}$
[d] Instrumental detection limits (IDL)
[e] Relative recoveries
N.D. = Not detectable

Table 8.2 Methodologies for the determination of cytostatic compounds in liquid samples from hospital effluents

Compounds	Extraction	Determination techniques	Recovery (%)	LOD ($ng \cdot L^{-1}$)	Found concentration ($ng \cdot L^{-1}$)	References
Cis-Pt, car-Pt, Oxa-Pt, mono-Pt, Di-Pt	Same as reference Lenz et al. (2005)	LC-ICP-MS	90–116	0.09–0.15[a]	0.3–1.7[a]	Hann et al. (2005)
EPI, DOX, DAU	Off-line SPE	RP-LC-FD	80–90	0.26–0.29[c]	0.1–1.4 (EPI)[a]	Mahnik et al. (2006)
					0.1–0.5 (DOX)[a]	
5-FU	Off-line SPE	GC-MS	93–101	12	0.09–4[a]	Mullot et al. (2009)
5-FU, GEM, dFdU	Same as reference Kovalova et al. (2009)	LC-ESI-MS/MS	54–118	0.9–9[b]	0.9–839	Weissbrodt et al. (2009)
5-FU, CYT, GEM, araU, dFdU	Off-line SPE	LC-ESI-MS/MS (−)	72–127	0.9–9[b]	27 (5-FU)	Kovalova et al. (2009)
					38 (GEM)	
					840 (dFdU)	
CP, EPI	Off-line SPE	LC-ESI-Orbitrap-MS	37–107	3.1–85	5.73[a]	Gomez-Canela et al. (2012)
CIPRO, CP, IF	Online SPE	SPE-LC-ESI-MS	90–149 (CIPRO)	16–3500 (CIPRO)	0.151–0.958[a]	Kovalova et al. (2012)
			103–134 (CP)	9–110 (CP)		
			101–152 (IF)	2–26 (IF)		
5-FU, CAP	Off-line SPE	GC-MS		0.16	35–92	Kosjek et al. (2013)

(continued)

Table 8.2 (continued)

Compounds	Extraction	Determination techniques	Recovery (%)	LOD ($ng{\cdot}L^{-1}$)	Found concentration ($ng{\cdot}L^{-1}$)	References
CP, IF, DOC, PAC, ETO, VINC, TAM, MET, AZA, CIPRO	Off-line SPE	UHPLC-ISS-QqLit	46–106	0.8–24	14.5–14.725[a]	Ferrando-Climent et al. (2013)
CP, CHLO, MEL, IF, FLU, CYT, GEM, CAP, VINB, VINC, ETO, PAC, DOC, DOX, DAU, EPI, IMA, ERLO, IRI, LEU, GOS, TAM, A-GLU, MEG, CYP, PRED	Off-line	LC-HESI-Orbitrap-MS	6–110	0.7–356	0.02–12.3[a]	Gómez-Canela et al. (2014)

[a] $\mu g{\cdot}L^{-1}$
[b] Limit of quantification (LOQ)
[c] $\mu g{\cdot}L^{-1}$

Table 8.3 Methodologies for the determination of cytostatic compounds in other sources of liquid samples

Matrix	Compounds	Extraction condition	Determination	Recovery (%)	LOD ($ng \cdot L^{-1}$)	Found concentration ($ng \cdot L^{-1}$)	References
River water Tap water	CP, IF	Off-line SPE	LC-ESI-MS/MS	51–57	10		Ternes et al. (1998)
Groundwater	IF, CP	Off-line SPE	LC-ESI-MS/MS	71–102 (CP) 73–87 (IF)	32 (CP) 14 (IF)	N.D.	Sacher et al. (2001)
River water	CP	Off-line SPE:	LC-ESI-MS/MS		1–10	5	Metcalfe et al. (2003b)
River water	CP, MET	Online SPE	SPE-ESI-LC-MS/MS	120 (CP) 125 (MET)	3–6	N.D.	Garcia-Ac et al. (2009a)
Tap water Surface water	CP, MET	Online SPE	SPE-LC-ESI-ToF-MS/MS	60–109	1	N.D.	Garcia-Ac et al. (2009a, b)
River water	5-FU, GEM, CP, IF, CYT, MET, PAC, ETO, IRI, DOC, EPI, DOX, VINO, MIT-C	Off-line SPE	LC-ESI-QqQ-MS	21–91	0.1–34	2.4–13	Martín et al. (2011)
Surface water	5-FU, CAP	Off-line SPE	GC-MS	Not provided	0.16 (5-FU)	N. D.	Kosjek et al. (2013)

(continued)

Table 8.3 (continued)

Matrix	Compounds	Extraction condition	Determination	Recovery (%)	LOD ($ng \cdot L^{-1}$)	Found concentration ($ng \cdot L^{-1}$)	References
Groundwater	TAM	Same as reference López-Serna et al. (2010)	Same as reference (López-Serna et al. 2010)	ABS: 74.16–131.33		26.9–72. 7	López-Serna et al. (2013)
				REL: 10.36–10.7			
Superficial water	GEM, TEM, MET, IRI, IMA, IF, CP, ERLO, ETO, DOX, CAP, TAM, PAC, OH-MET, OH-D-TAM, OH-TAM, OH-PAC	Online SPE	SPE-LC-ESI-MS/MS	76–113 (SW)	0.4–45 (SW)	N.D.	Negreira et al. (2013)
Groundwater				72–119 (GW)	0.1–24 (GW)		

cytostatics using automatic off-line SPE (Mullot et al. 2009; Ferrando-Climent et al. 2013). In the first of these works, Oasis HLB was used for the extraction and preconcentration of CP, IF, docetaxel (DOC), paclitaxel (PAC), ETO, VINC, TAM, MET, azathioprine (AZA), and ciprofloxacin (CIPRO), while in the second work, Isolute ENV+ was used for the study of 5-FU. However, SPE coupled to LC-MS is much more common (9.1 1–9.3). The use of automatic SPE can carry out both extraction and desorption processes without an analyst, which implies numerous advantages including the following: reducing the consumption of organic solvents, needing a shorter extraction time, and minimizing errors related to the manipulation of the sample. All of these advantages are even improved when the elution can be done with the same mobile phase that is further used in the LC system. For example, Negreira et al. (2013) developed a multiresidue method for the extraction and determination of 13 cytostatics (gemcitabine (GEM), temozolomide (TEM), MET, iIrinotecan (IRI), IMA, IF, CP, erlotinib (ERLO), ETO, doxorubicin (DOX), CAP, TAM, and PAC) and 4 metabolites (hydroxymethotrexate (OH-MET), 4-hydroxy-N-desmethyl-tamoxifen (OH-D-TAM), hydroxytamoxifen (OH-TAM), and 6(α)-hydroxypaclitaxel (OH-PAC)) in wastewater samples (Negreira et al. 2013).

Regarding environmental solid matrices, as indicated above, the cytostatic compounds can be adsorbed into the sludge of the purifiers due to their K_{ow} constants, and the extraction and purification of these compounds are limited by the large number of interferences and strong interactions between the analytes and sample. The extraction of cytostatics in solid matrices, concretely in sludge, has been carried out using two methods: PLE and UAE (Table 8.4). They replace conventional extraction techniques of pollutants from solids, such as Soxhlet, which consume large volumes of toxic organic solvents and have a long working time.

UAE can also be denominated as a classical technique (Zuloaga et al. 2012), but both the extraction time, due to the cavitation process by the ultrasound bath, and the use of organic solvents decrease in comparison with Soxhlet (Luque-García and Luque de Castro 2003). Although in this technique, the diffusion of analytes from the solid sample to the solvent is facilitated by ultrasonic energy, the lack of uniformity in their distributions generates poor reproducibility as well as low selectivity and limited sample-enrichment capabilities.

In PLE, a relatively small amount of solvent can be used at temperatures above its boiling temperature, because it is under pressure (1500–2000 psi). In this manner, it is possible to increase the solubility of the analyte in the solvent, the transfer rate, and the extraction rate (Ramos 2012). PLE also reduces the required extraction time and has a high level of automation. Its main limitation is low selectivity and coextraction of interferences, especially in complex samples such as sludge. The acquisition cost of the equipment and the dilution of the analytes when a large number of cycles are used imply other disadvantages that must be considered. Other techniques, such as microwave-assisted extraction (MAE), which has not been applied to cytostatic compounds till now, could be considered. MAE is based on the application of the energy of the microwaves to heat the sample. This technique offers advantages such as short extraction times, small volumes of solvents, and the possibility of extracting

Table 8.4 Methodologies for the determination methodologies of cytostatic compounds in sludge

Compounds	Extraction	Determination technique	Recovery (%)	LOD ($ng \cdot g^{-1}$)	Found concentration ($ng \cdot g^{-1}$)	References
CP, IF	UAE: 4 × 5 min	LC-ESI-MS	52–78	20		Ternes et al. (2005)
CP, CIPRO	PLE (used for CP)	LC-MS/MS; UHPLC-MS/MS	85 (CIPRO)		20–500 (CIPRO)	Okuda et al. (2009)
	UAE (used for CIPRO)		75 (CP)		30–5 (CP)	
IF, CP, TAM	PLE	UHPLC-ESI-MS/MS	95–108 (IF)	3.5–74 $ng \cdot L^{-1}$ (IF)	11.4–42.5 (IF)	Seira et al. (2013)
			94–100 (CP)	2.5–51 $ng \cdot L^{-1}$ (CP)	12.6 (CP)	

[a] $\mu g \cdot L^{-1}$
[b] Limit of quantification (LOQ)
[c] $\mu g \cdot L^{-1}$
[d] Instrumental detection limits (IDL)
[e] Relative recoveries
N.D. Not detectable

many samples at the same time. However, this technique cannot be automated and needs an additional sample clean-up step prior to its injection (Montesdeoca-Esponda et al. 2014).

8.3 Determination Techniques

The first determinations of cytostatic agents in wastewater were performed in the 1990s (Steger-Hartmann et al. 1996; Kümmerer et al. 1997) using gas chromatography coupled to mass spectrometry (GC-MS) as a detection technique after the SPE. GC-MS is the combination of two techniques, a separation technique (GC) and an identification technique (MS) (Watson and Sparkman 2007).

In the first of these reports (Steger-Hartmann et al. 1996), IF and CP were studied in wastewater from Germany by the SPE technique. The limits of detection (LODs) obtained were approximately 7 ng L^{-1} for IF and 6 ng L^{-1} for CP, with the measured concentrations ranging from 24 ng L^{-1} (IF) to 146 ng L^{-1} (CP). 1,2,3,5-Tetrachlorobenzene (TCB) and trofosfamide (TF) were used as internal standards for the underivatized and derivatized samples, respectively. In each series of samples, a standard and a calibration curve was employed, since the authors indicated that the compounds might have undergone decomposition on the column, resulting in a loss of efficiency, because the response of the standards rapidly changed from one analysis to another for underivatized samples. Kümmerer et al. (1997) used the same procedure, and they achieved 6 ng L^{-1} as the LOD for IF, detecting concentrations of 109 ng L^{-1} in the hospital's effluent and between 6.2 and 9.3 ng L^{-1} in the wastewater (Kümmerer et al. 1997).

Though these authors achieved satisfactory LODs (6–146 ng L^{-1}) using GC, almost all their subsequent analyses have been performed using LC. Gas chromatography has some disadvantages such as the need of a derivatization step of low volatile compounds (generally with high-molecular-weight compounds) and drawbacks related with the time of sample pretreatment and incapability of determining very polar compounds or transformation products (TPs) (Hernández et al. 2004).

However, certain cytostatic compounds, specifically pyrimidine analogues, are often analyzed using GC-MS. For example, TAM and 5-FU were analyzed using GC-MS, where TAM was detected in the water samples from the hospital, residential areas, and WWTPs (Tauxe-Wuersch et al. 2006). However, the concentrations of this drug were between the limit of quantification and limit of detection (1 and 4 ng L^{-1}) (Table 8.1). The authors conclude that since the use of cytostatic drugs is less than other drugs such as anti-inflammatory drugs, it is necessary to use more powerful techniques to detect them.

A GC-MS method was used for the determination of 5-FU because of its poor retention in reversed-phase materials and the difficulty of its determination by reversed-phase LC. An automated SPE workstation was used for the extraction, detecting concentrations between 0.09 and 4 μg·L^{-1} in hospital wastewater (Mullot et al. 2009).

Kosjek et al. (2013) studied 5-FU and its prodrug capecitabine (CAP) (Kosjek et al. 2013). Three silylation reagents were tested for the derivatization of 5-FU in GC-MS with electron impact (EI) ionization. Finally, they used N-methyl-N-[tert-butyldimethylsilyl]-trifluoroacetamide (MTBSTFA) as it has the best hydrolytic stability of the derivatives and the most favorable fragmentation as well as a superior chromatographic response. Finally, concentrations of 5-FU 4.7 and 14 ng L^{-1} and of 35 and 92 ng L^{-1} were measured in wastewater and hospital wastewater samples, respectively.

LC coupled to MS/MS with electrospray ionization (ESI) in the positive mode is undoubtedly the mostly used technique in the determination of cytostatic compounds. In relation to the ionization mode, it has been found that ESI is the best ionization mode for cytostatic compounds (Garcia-Ac et al. 2011).

The first papers that used LC-ESI-MS/MS for the determination of cytostatics in environmental samples were carried out in 1998 (Ternes 1998) (Ternes et al. 1998). In both papers, the extraction of IF and CP from wastewater, river water, and tap water was performed by SPE. In the latter work, the authors compared the detection levels between GC-MS and LC-MS. They achieved LODs of 100–250 ng L^{-1} in wastewater and 50 ng L^{-1} in drinking water using GC-MS and LODs of 10 ng L^{-1} by LC-MS for both matrices using dihydrocarbamazepine as the surrogate standard.

Some works aimed to identify cytostatics such as CP, IF, or MET besides other pharmaceuticals. Sacher et al. (2001) analyzed IF and CP among 60 pharmaceutical compounds in groundwater using 2,3-dichlorophenoxyacetic acid as the internal standard for the overall procedure. SPE followed by LC-ESI-MS/MS was chosen for IF and CP, but these cytostatic compounds were not detected in real samples (Sacher et al. 2001).

Metcalfe et al. analyzed different pharmaceuticals, including CP in river water (Metcalfe et al. 2003b) and CP and IF in effluents of WWTPs (Metcalfe et al. 2003a). The objective of the first work was to evaluate the occurrence and distribution of the drugs being studied (Metcalfe et al. 2003b) in the lower Great Lakes region (Canada), since this region has a large number of municipalities discharging treated sewage. In this case, an internal standard (dihydrocarbamazepine) for neutral drugs was added before the evaporation of samples to check the precision of the procedure. They detected CP at concentrations of 4–8 ng L^{-1} in the Otonabee River, Little River, and Detroit River (Metcalfe et al. 2003b). The objective of the second work was to evaluate their presence in 18 effluents from different wastewater treatment plants in Canada (Metcalfe et al. 2003a), which had different treatment processes varying from the primary treatment to the tertiary treatment. The LODs ranged between 0.1 and 0.5 μg L^{-1}, but the compounds were not detected.

Some authors (Castiglioni et al. 2005; Castiglioni et al. 2006) investigated MET and CP among other compounds by SPE and LC-turbo ion spray(TIS)-MS/MS in urban wastewater from Italy using salbutamol-d_3 as an internal standard. In the first of the papers (Castiglioni et al. 2005), they detected concentrations of 2.1 (CP) and 12.6 ng L^{-1}(MET). In a subsequent paper, the authors applied this optimized procedure for the determination of the same compounds in other WWTPs, but cytostatics were not found (Castiglioni et al. 2006).

Buerge et al. (Buerge et al. 2006) analyzed IF and CP in wastewater in Switzerland using SPE and LC-TIS-MS. The samples were fortified with an internal standard, $^{13}C_3$-caffeine (0.052 ng·L^{-1}), achieving LODs in the range of 0.2–0.3 ng L^{-1}. The authors managed to detect these compounds in untreated and treated wastewaters, and they carried out a degradation study of CP and IF in activated sludge, finding no degradation after 24 h. To make sure that the sludge was biologically activated, they used caffeine as a reference compound, which was degraded in 1 h, concluding that CP and IF are persistent compounds. After studying surface waters, they found very low concentrations of CP, but it was present in all of the samples analyzed from Lake Zurich and the Limmat River.

It seems that the main entry of cytostatic compounds into the environment is through the wastewater of hospitals, so their study would yield important information. With this aim, Weissbrodt et al. (2009) and Kovalova et al. (2009) studied different cytostatics in wastewater from hospitals (Weissbrodt et al. 2009) (Kovalova et al. 2009). Authors used LC-ESI-MS/MS analyzing several compounds, with half of them being the same in both works. However, the first work was carried out with ESI in the positive mode, as is usual for cytostatics, and the second was carried out in the negative mode. Kovalova et al. (2009) suggested that retaining such polar compounds in reversed-phase LC can be very complicated, and thus, the authors studied retention mechanisms on the hydrophilic interaction chromatography (HILIC) stationary phase. Finally, both papers analyzed hospital wastewater from Switzerland, reaching the same range of LODs (0.9–9 ng·L^{-1}) measuring the same cytostatic drugs (5-FU, GEM, and 2′,2′-difluorodeoxyuridine (dFdU)) at the same concentrations.

Multiresidue methods for different types of pharmaceutical compounds that include different cytostatics have been reported. Martín et al. (2011) investigated 14 cytostatic drugs (5-FU, GEM, CP, IF, cytarabine (CYT), MET, PAC, ETO, IRI, DOC, epirubicin (EPI), DOX, vinorelbine (VINO), and mitomycin C (MIT-C)) in wastewater and river water from Spain. These authors pointed out that the lack of works determining a group of cytostatics could be due to the absence of methods for their simultaneous determination. They used SPE followed by LC-ESI-QqQ-MS, measuring concentrations in the range from 1.2 to 15 ng L^{-1}. Only CYT and GEM were detected in river water, and almost all the compounds detected in the influent of the WWTP were detected also in the effluent (Martín et al. 2011).

Regarding the extraction of cytostatic compounds in sludge, only three papers have been published describing the determination of these compounds in sludge samples, with ultrasound-assisted extraction (UAE) (Ternes et al. 2005; Okuda et al. 2009) and pressurized liquid extraction (PLE) (Okuda et al. 2009; Seira et al. 2013) being the selected techniques. Seira et al. 2013, studied CP, IF, and TAM: they obtained recoveries close to 100% and were able to detect CP and IF in the sludge. Okuda et al. 2009 developed a procedure for the extraction of 66 compounds, including CP and CIPRO, which were detected at concentrations of 5–30 ng g^{-1} (CP) and 20–500 ng g^{-1} (CIPRO).

Tables 8.1, 8.2 and 8.3 show analytical methodologies employed to determine cytostatic compounds from different liquid samples, while Table 8.4 summarizes the methodologies used in solid samples.

8.4 Current Analytical Procedures to Determine Cytostatic Compounds in Environmental Samples

Green analytical chemistry (GAC) is a movement within analytical chemistry that seeks to reduce the consumption of solvents, to replace toxic solvents with others that are not so toxic, and to minimize and automate the methods (Armenta et al. 2015). Thus, it is intended to alleviate the side effects of the chemistry. The development of online SPE methods coupled to LC-MS/MS has been shown to improve method sensitivity, reduce sample pretreatment and analysis time, and increase the number of samples that can be analyzed simultaneously (Pan et al. 2014). This is the goal of most environmental monitoring programs, to analyze the maximum number of compounds spending minimal resources. Garcia-Ac et al. (2009a, b) determined CP and MET, among five other pharmaceuticals, in residual and surface water using a completely automated method (Garcia-Ac et al. 2009a). For this purpose, online SPE-LC-ESI-MS/MS was used, obtaining LODs in the range of 9–20 ng·L^{-1} using the standard addition method. Concentrations of 9 ng·L^{-1} (CP) and 59 (MET) ng·L^{-1} were measured.

Although the online SPE process is one of the most promising techniques, reversed-phase sorbents show some limitations, such as poor extraction of quite polar compounds (Castiglioni et al. 2006). Alternative phases may be used in online SPE to obtain the desired selectivity (e.g., HILIC, porous graphitic carbon (PGC), or molecularly imprinted solid-phase extraction (MISPE)) (Rogeberg et al. 2014), but difficulties can appear when combining different analytes with a wide range of hydrophobicities. As a result, low recovery and poor resolution might occur.

With regard to the determination techniques, high-performance liquid chromatography (LC) has been giving way to ultrahigh-performance liquid chromatography (UHPLC), which decreases the analysis time, since it works at a higher pressure. It also decreases solvent consumption when working at lower flow rates. The use of the reverse phase is the most common in the determination of cytostatic compounds, as only one paper has been published with HILIC for the determination of 5-FU (Kovalova et al. 2009).

Concerning detection techniques, mass spectrometry in tandem presents the disadvantage of only being able to focus on known analytes. This situation changed radically with the high resolution of time-of-flight (ToF) systems, where not only the target compounds but also the non-targets are analyzed (Eichhorn et al. 2016). The principal operation of this detector is to measure the time it takes for an ion to travel from the source to the detector, which is located at a known distance. The ions receive the same acceleration, but, due to their different m/z values, they acquire different speeds. This method would have no upper limits of m/z (Watson and Sparkman 2007). Garcia-Ac et al. (2009a, b) also attempted to determine the cytostatics CP and MET among 14 pharmaceuticals compounds in drinking and surface water using SPE-LC-ToF-MS (Garcia-Ac et al. 2009a, b).

Gomez-Canela et al. identified CP and EPI using LC-Orbitrap-MS, revealing their presence in WWTP influents as well as hospital and urban effluents in Catalonia

(Spain) at concentration levels ranging from 5.73 to 24.8 μg L^{-1}, which are higher than those levels determined by other authors in wastewaters, which rarely reach the hundreds of ng L^{-1} (Gomez-Canela et al. 2012) (Tables 8.1 and 8.2). The advantages of this technique allow other studies beyond the determination of known compounds to be performed, as demonstrated by the work of Negreira et al. 2015 who studied metabolites of anticancer compounds (Negreira et al. 2015). Sometimes cytostatic compounds are detected in sewage treatment plants but not in effluents from hospitals due to degradation reactions that could take place in the presence of disinfectant. As a result, the UHPLC coupled to quadrupole-orbitrap mass spectrometry (Qq-Orbitrap-MS/MS) technique has been used to elucidate the degradation of ETO and the determination of its byproducts in chlorinated water. This work showed that ETO degrades in few seconds into two products, with concentrations in the range of 14–33 ng L^{-1} for one of these byproducts, the 3'-O-desmethyl etoposide (BP1), and without traces of the parent compound.

Ferrando-Climent et al. developed a method for the determination of ten cytostatic agents and their metabolites in hospital effluents and wastewater treatment plant influents (Ferrando-Climent et al. 2013). They used automated off-line SPE and UHPLC coupled to a triple quadrupole-linear ion trap (QqLit)-MS using isotopically labelled compounds as internal standards. The authors obtained concentrations of up to 14 μg L^{-1} of CIPRO in the hospital effluent, but they also suggest that the concentrations did not differ too much between hospital effluents and wastewater treatment plant influents, and thus, the hospital cannot be considered to be the main source of these drugs.

8.5 Conclusions and Future Trends

Determining the concentrations at which cytostatic compounds are found in the environment is essential to place them under surveillance taking into account their mutagenic and genotoxic potentials. CP, IF, TAM, 5-FU, EPI, MET, and DOX are the most commonly detected compounds in sewage, given that they are the most frequently used anticancer drugs.

Though the main sources of contamination of these compounds are considered to be hospital effluents by the majority of the scientific community, some authors do not agree with this affirmation, and they claim that the levels of cytostatic agents in effluents from hospitals and in the influent of wastewater treatment plants are similar. However, the differences in concentrations between hospital effluents and wastewater are remarkable. In hospitals, the effluent concentrations are measured at the μg L^{-1} level, whereas in wastewater, surface water, and river water, the concentrations are at the ng L^{-1} level.

The combination most frequently used method for the analyses of these compounds in liquid samples is SPE followed by LC-MS/MS. Nevertheless, due to the low concentrations at which cytostatic compounds are found, the use of more selective techniques for purification and preconcentration as well as sensible detection methods is required.

Although online strategies such as online SPE have been implemented in recent years because of their undoubted advantages in terms of repeatability, efficiency, and speed, one of their weak points is that there is not a sufficient variety of sorbents. Future trends should be focused on developing new extraction materials capable of extracting polar analytes suitable for multiresidue analysis of compounds with a wide range of hydrophobicity levels.

Continuing with the philosophy of green chemistry, the development of microextraction techniques has been boosted, which should be attempted to be implemented for the determination of this group of compounds. These techniques of microextraction have, among their objectives, the miniaturization, diminution of extraction time, automation, and onsite analysis. These methods are also characterized by the use of small amounts of sample, small volumes of organic solvents, etc., thus generating less waste. Concerning the use of these principles in the extraction of cytostatic compounds, microextraction techniques have not been used, and only the extraction is carried out by online or off-line SPE. Due to the variety of structures, molecular and physicochemical properties of the family of cytostatic compounds can complicate its application.

However, the most important challenge regarding cytostatic analysis is probably related to their determination in solid samples. The literature for this type of sample is very scarce, and it is only devoted to sludge samples. Thus, it is mandatory to implement new approaches for the analysis of these compounds in other solid samples such as marine sediments or organisms close to the marine outfalls of wastewaters. In addition, new procedures for the extraction of analytes in solid matrices such as MAE should be implemented.

Allowing policies to be established for the control of these drugs in the near future only will be possible through the wide monitoring of cytostatic drugs, including all of the environmental compartments.

Acknowledgments This work was supported by funds provided by the Spanish Ministry of Economy and Competitiveness [Research Project CTM2015-66095-C2-1-R].

References

Armenta S, Garrigues S, de la Guardia M (2015) The role of green extraction techniques in green analytical chemistry. TrAC Trends Anal Chem 71:2–8. https://doi.org/10.1016/j.trac.2014.12.011

Borgatta M, Decosterd L-A, Waridel P, Buclin T, Chèvre N (2015) The anticancer drug metabolites endoxifen and 4-hydroxy-tamoxifen induce toxic effects on Daphnia pulex in a two-generation study. Sci Total Environ 520:232–240. https://doi.org/10.1016/j.scitotenv.2015.03.040

Borgatta M, Waridel P, Decosterd L-A, Buclin T, Chèvre N (2016) Multigenerational effects of the anticancer drug tamoxifen and its metabolite 4-hydroxy-tamoxifen on Daphnia pulex. Sci Total Environ 545–546:21–29. https://doi.org/10.1016/j.scitotenv.2015.11.155

Buchberger WW (2007) Novel analytical procedures for screening of drug residues in water, waste water, sediment and sludge. Anal Chim Acta 593:129–139. https://doi.org/10.1016/j.aca.2007.05.006

Buerge IJ, Buser H-R, Poiger T, Müller MD (2006) Occurrence and fate of the cytostatic drugs cyclophosphamide and Ifosfamide in wastewater and surface waters†. Environ Sci Technol 40:7242–7250. https://doi.org/10.1021/es0609405

Busetti F, Linge KL, Heitz A (2009) Analysis of pharmaceuticals in indirect potable reuse systems using solid-phase extraction and liquid chromatography–tandem mass spectrometry. J Chromatogr A 1216:5807–5818. https://doi.org/10.1016/j.chroma.2009.06.001

Camel V (2003) Solid phase extraction of trace elements. Spectrochim Acta Part B At Spectrosc 58:1177–1233. https://doi.org/10.1016/S0584-8547(03)00072-7

Castiglioni S, Bagnati R, Calamari D, Fanelli R, Zuccato E (2005) A multiresidue analytical method using solid-phase extraction and high-pressure liquid chromatography tandem mass spectrometry to measure pharmaceuticals of different therapeutic classes in urban wastewaters. J Chromatogr A 1092:206–215. https://doi.org/10.1016/j.chroma.2005.07.012

Castiglioni S, Bagnati R, Fanelli R, Pomati F, Calamari D, Zuccato E (2006) Removal of Pharmaceuticals in Sewage Treatment Plants in Italy. Environ Sci Technol 40:357–363. https://doi.org/10.1021/es050991m

Eichhorn P, Pérez S, Barceló D (2016) Applications of time-of-flight and Orbitrap mass spectrometry in environmental, food, doping, and forensic analysis. Elsevier. ISBN: 9780444635730

Ferrando-Climent L, Rodriguez-Mozaz S, Barceló D (2013) Development of a UPLC-MS/MS method for the determination of ten anticancer drugs in hospital and urban wastewaters, and its application for the screening of human metabolites assisted by information-dependent acquisition tool (IDA) in sewage samples. Anal Bioanal Chem 405:5937–5952. https://doi.org/10.1007/s00216-013-6794-4

Garcia-Ac A, Segura PA, Gagnon C, Sauvé S (2009a) Determination of bezafibrate, methotrexate, cyclophosphamide, orlistat and enalapril in waste and surface waters using on-line solid-phase extraction liquid chromatography coupled to polarity-switching electrospray tandem mass spectrometry. J Environ Monit 11:830–838. https://doi.org/10.1039/B817570E

Garcia-Ac A, Segura PA, Viglino L, Fürtös A, Gagnon C, Prévost M, Sauvé S (2009b) On-line solid-phase extraction of large-volume injections coupled to liquid chromatography-tandem mass spectrometry for the quantitation and confirmation of 14 selected trace organic contaminants in drinking and surface water. J Chromatogr A 1216:8518–8527. https://doi.org/10.1016/j.chroma.2009.10.015

Garcia-Ac A, Segura PA, Viglino L, Gagnon C, Sauvé S (2011) Comparison of APPI, APCI and ESI for the LC-MS/MS analysis of bezafibrate, cyclophosphamide, enalapril, methotrexate and orlistat in municipal wastewater. J Mass Spectrom 46:383–390. https://doi.org/10.1002/jms.1904

Gomez-Canela C, Cortes-Francisco N, Oliva X, Pujol C, Ventura F, Lacorte S, Caixach J (2012) Occurrence of cyclophosphamide and epirubicin in wastewaters by direct injection analysis-liquid chromatography-high-resolution mass spectrometry. Environ Sci Pollut Res Int 19:3210–3218. https://doi.org/10.1007/s11356-012-0826-z

Gómez-Canela C, Ventura F, Caixach J, Lacorte S (2014) Occurrence of cytostatic compounds in hospital effluents and wastewaters, determined by liquid chromatography coupled to high-resolution mass spectrometry. Anal Bioanal Chem 406:3801–3814. https://doi.org/10.1007/s00216-014-7805-9

Hann S, Stefánka Z, Lenz K, Stingeder G (2005) Novel separation method for highly sensitive speciation of cancerostatic platinum compounds by HPLC-ICP-MS. Anal Bioanal Chem 381:405–412. https://doi.org/10.1007/s00216-004-2839-z

Hernández F, Ibáñez M, Sancho JV, Pozo ÓJ (2004) Comparison of different mass spectrometric techniques combined with liquid chromatography for confirmation of pesticides in environmental water based on the use of identification points. Anal Chem 76:4349–4357. https://doi.org/10.1021/ac049768i

Kazner C, Lehnberg K, Kovalova L, Wintgens T, Melin T, Hollender J, Dott W (2008) Removal of endocrine disruptors and cytostatics from effluent by nanofiltration in combination with adsorption on powdered activated carbon. Water Sci Technol 58(8):1699–1706

Kosjek T, Heath E (2011) Occurrence, fate and determination of cytostatic pharmaceuticals in the environment. TrAC Trends Anal Chem 30:1065–1087. https://doi.org/10.1016/j.trac.2011.04.007

Kosjek T, Perko S, Žigon D, Heath E (2013) Fluorouracil in the environment: analysis, occurrence, degradation and transformation. J Chromatogr A 1290:62–72. https://doi.org/10.1016/j.chroma.2013.03.046

Kovács R, Csenki Z, Bakos K, Urbányi B, Horváth Á, Garaj-Vrhovac V, Gajski G, Gerić M, Negreira N, López de Alda M, Barceló D, Heath E, Kosjek T, Žegura B, Novak M, Zajc I, Baebler Š, Rotter A, Ramšak Ž, Filipič M (2015) Assessment of toxicity and genotoxicity of low doses of 5-fluorouracil in zebrafish (Danio rerio) two-generation study. Water Res 77:201–212. https://doi.org/10.1016/j.watres.2015.03.025

Kovalova L, McArdell CS, Hollender J (2009) Challenge of high polarity and low concentrations in analysis of cytostatics and metabolites in wastewater by hydrophilic interaction chromatography/tandem mass spectrometry. J Chromatogr A 1216:1100–1108. https://doi.org/10.1016/j.chroma.2008.12.028

Kovalova L, Siegrist H, Singer H, Wittmer A, McArdell CS (2012) Hospital wastewater treatment by membrane bioreactor: performance and efficiency for organic micropollutant elimination. Environ Sci Technol 46:1536–1545. https://doi.org/10.1021/es203495d

Kuemmerer K, Al-Ahmad A (2010) Estimation of the cancer risk to humans resulting from the presence of cyclophosphamide and ifosfamide in surface water. Environ Sci Pollut Res Int 17:486–496. https://doi.org/10.1007/s11356-009-0195-4

Kümmerer K, Steger-Hartmann T, Meyer M (1997) Biodegradability of the anti-tumour agent ifosfamide and its occurrence in hospital effluents and communal sewage. Water Res 31:2705–2710. https://doi.org/10.1016/S0043-1354(97)00121-8

Kümmerer K, Al-Ahmad A, Bertram B, Wießler M (2000) Biodegradability of cytostatic compounds in screening tests: influence of glucosidation and of stereochemistry. Chemosphere 40:767–773. https://doi.org/10.1016/S0045-6535(99)00451-8

Lenz K, Hann S, Koellensperger G, Stefanka Z, Stingeder G, Weissenbacher N, Mahnik SN, Fuerhacker M (2005) Presence of cancerostatic platinum compounds in hospital wastewater and possible elimination by adsorption to activated sludge. Sci Total Environ 345:141–152. https://doi.org/10.1016/j.scitotenv.2004.11.007

Lenz K, Mahnik SN, Weissenbacher N, Mader RM, Krenn P, Hann S, Koellensperger G, Uhl M, Knasmüller S, Ferk F, Bursch W, Fuerhacker M (2007) Monitoring, removal and risk assessment of cytostatic drugs in hospital wastewater. Water Sci Technol 56:141–149

Llewellyn N, Lloyd P, Jürgens MD, Johnson AC (2011) Determination of cyclophosphamide and ifosfamide in sewage effluent by stable isotope-dilution liquid chromatography–tandem mass spectrometry. J Chromatogr A 1218:8519–8528. https://doi.org/10.1016/j.chroma.2011.09.061

López-Serna R, Pérez S, Ginebreda A, Petrović M, Barceló D (2010) Fully automated determination of 74 pharmaceuticals in environmental and waste waters by online solid phase extraction-liquid chromatography- electrospray-tandem mass spectrometry. Talanta 83:410–424. https://doi.org/10.1016/j.talanta.2010.09.046

López-Serna R, Jurado A, Vázquez-Suñé E, Carrera J, Petrović M, Barceló D (2013) Occurrence of 95 pharmaceuticals and transformation products in urban groundwaters underlying the metropolis of Barcelona, Spain. Environ Pollut 174:305–315. https://doi.org/10.1016/j.envpol.2012.11.022

Luque-García JL, Luque de Castro MD (2003) Ultrasound: a powerful tool for leaching. TrAC Trends Anal Chem 22:41–47. https://doi.org/10.1016/S0165-9936(03)00102-X

Lutterbeck CA, Kern DI, Machado ÊL, Kümmerer K (2015a) Evaluation of the toxic effects of four anti-cancer drugs in plant bioassays and its potency for screening in the context of waste water reuse for irrigation. Chemosphere 135:403–410. https://doi.org/10.1016/j.chemosphere.2015.05.019

Lutterbeck CA, Wilde ML, Baginska E, Leder C, Machado ÊL, Kümmerer K (2015b) Degradation of 5-FU by means of advanced (photo)oxidation processes: UV/H2O2, UV/Fe2 +/H2O2 and UV/TiO2 — comparison of transformation products, ready biodegradability and toxicity. Sci Total Environ 527–528:232–245. https://doi.org/10.1016/j.scitotenv.2015.04.111

Mahnik SN, Rizovski B, Fuerhacker M, Mader RM (2004) Determination of 5-fluorouracil in hospital effluents. Anal Bioanal Chem 380:31–35. https://doi.org/10.1007/s00216-004-2727-6

Mahnik SN, Rizovski B, Fuerhacker M, Mader RM (2006) Development of an analytical method for the determination of anthracyclines in hospital effluents. Chemosphere 65:1419–1425. https://doi.org/10.1016/j.chemosphere.2006.03.069

Mahnik SN, Lenz K, Weissenbacher N, Mader RM, Fuerhacker M (2007) Fate of 5-fluorouracil, doxorubicin, epirubicin, and daunorubicin in hospital wastewater and their elimination by activated sludge and treatment in a membrane-bio-reactor system. Chemosphere 66:30–37. https://doi.org/10.1016/j.chemosphere.2006.05.051

Mahugo-Santana C, Sosa-Ferrera Z, Torres-Padrón ME, Santana-Rodríguez JJ (2011) Application of new approaches to liquid-phase microextraction for the determination of emerging pollutants. TrAC Trends Anal Chem 30:731–748. https://doi.org/10.1016/j.trac.2011.01.011

Martín J, Camacho-Muñoz D, Santos JL, Aparicio I, Alonso E (2011) Simultaneous determination of a selected group of cytostatic drugs in water using high-performance liquid chromatography–triple-quadrupole mass spectrometry. J Sep Sci 34:3166–3177. https://doi.org/10.1002/jssc.201100461

Metcalfe CD, Koenig BG, Bennie DT, Servos M, Ternes TA, Hirsch R (2003a) Occurrence of neutral and acidic drugs in the effluents of Canadian sewage treatment plants. Environ Toxicol Chem 22:2872–2880. https://doi.org/10.1897/02-469

Metcalfe CD, Miao X-S, Koenig BG, Struger J (2003b) Distribution of acidic and neutral drugs in surface waters near sewage treatment plants in the lower Great Lakes, Canada. Environ Toxicol Chem 22:2881–2889. https://doi.org/10.1897/02-627

Mišík M, Pichler C, Rainer B, Filipic M, Nersesyan A, Knasmueller S (2014) Acute toxic and genotoxic activities of widely used cytostatic drugs in higher plants: possible impact on the environment. Environ Res 135:196–203. https://doi.org/10.1016/j.envres.2014.09.012

Montesdeoca-Esponda S, Vega-Morales T, Sosa-Ferrera Z, Santana-Rodríguez JJ (2014) Analysis of personal care products in sediments and soils. In: Personal care products in the aquatic environment. Springer, Cham, pp 231–262

Mullot J-U, Karolak S, Fontova A, Huart B, Levi Y (2009) Development and validation of a sensitive and selective method using GC/MS-MS for quantification of 5-fluorouracil in hospital wastewater. Anal Bioanal Chem 394:2203–2212. https://doi.org/10.1007/s00216-009-2902-x

Nebot C, Gibb SW, Boyd KG (2007) Quantification of human pharmaceuticals in water samples by high performance liquid chromatography–tandem mass spectrometry. Anal Chim Acta 598:87–94. https://doi.org/10.1016/j.aca.2007.07.029

Negreira N, López de Alda M, Barceló D (2013) On-line solid phase extraction-liquid chromatography-tandem mass spectrometry for the determination of 17 cytostatics and metabolites in waste, surface and ground water samples. J Chromatogr A 1280:64–74. https://doi.org/10.1016/j.chroma.2013.01.031

Negreira N, López de Alda M, Barceló D (2015) Degradation of the cytostatic etoposide in chlorinated water by liquid chromatography coupled to quadrupole-Orbitrap mass spectrometry: identification and quantification of by-products in real water samples. Sci Total Environ 506–507:36–45. https://doi.org/10.1016/j.scitotenv.2014.10.097

Okuda T, Yamashita N, Tanaka H, Matsukawa H, Tanabe K (2009) Development of extraction method of pharmaceuticals and their occurrences found in Japanese wastewater treatment plants. Environ Int 35:815–820. https://doi.org/10.1016/j.envint.2009.01.006

Pan J, Zhang C, Zhang Z, Li G (2014) Review of online coupling of sample preparation techniques with liquid chromatography. Anal Chim Acta 815:1–15. https://doi.org/10.1016/j.aca.2014.01.017

Parrella A, Lavorgna M, Criscuolo E, Russo C, Fiumano V, Isidori M (2014) Acute and chronic toxicity of six anticancer drugs on rotifers and crustaceans. Chemosphere 115:59–66. https://doi.org/10.1016/j.chemosphere.2014.01.013

Rabii FW, Segura PA, Fayad PB, Sauvé S (2014) Determination of six chemotherapeutic agents in municipal wastewater using online solid-phase extraction coupled to liquid chromatography-tandem mass spectrometry. Sci Total Environ 487:792–800. https://doi.org/10.1016/j.scitotenv.2013.12.050

Ramos L (2012) Critical overview of selected contemporary sample preparation techniques. J Chromatogr A 1221:84–98. https://doi.org/10.1016/j.chroma.2011.11.011

Rogeberg M, Malerod H, Roberg-Larsen H, Aass C, Wilson SR (2014) On-line solid phase extraction–liquid chromatography, with emphasis on modern bioanalysis and miniaturized systems. J Pharm Biomed Anal 87:120–129. https://doi.org/10.1016/j.jpba.2013.05.006

Sacher F, Lange FT, Brauch H-J, Blankenhorn I (2001) pharmaceuticals in groundwaters: analytical methods and results of a monitoring program in baden-württemberg, germany. J Chromatogr A 938:199–210. https://doi.org/10.1016/S0021-9673(01)01266-3

Santana-Viera S, Montesdeoca-Esponda S, Sosa-Ferrera Z, Santana-Rodríguez JJ (2016) Cytostatic drugs in environmental samples: an update on the extraction and determination procedures. TrAC Trends Anal Chem 80:373–386. https://doi.org/10.1016/j.trac.2015.08.016

Seira J, Claparols C, Joannis-Cassan C, Albasi C, Montréjaud-Vignoles M, Sablayrolles C (2013) Optimization of pressurized liquid extraction using a multivariate chemometric approach for the determination of anticancer drugs in sludge by ultra high performance liquid chromatography–tandem mass spectrometry. J Chromatogr A 1283:27–38. https://doi.org/10.1016/j.chroma.2013.01.114

Steger-Hartmann T, Kümmerer K, Schecker J (1996) Trace analysis of the cytostatics ifosfamide and cyclophosphamide in sewage water by twostep solid-phase extraction and gas chromatography-mass spectrometry. J Chromatogr A 726:179–184. https://doi.org/10.1016/0021-9673(95)01063-7

Tauxe-Wuersch A, De Alencastro LF, Grandjean D, Tarradellas J (2006) Trace determination of tamoxifen and 5-fluorouracil in hospital and urban wastewaters. Int J Environ Anal Chem 86:473–485. https://doi.org/10.1080/03067310500291502

Ternes TA (1998) Occurrence of drugs in German sewage treatment plants and rivers. Water Res 32:3245–3260. https://doi.org/10.1016/S0043-1354(98)00099-2

Ternes TA, Hirsch R, Mueller J, Haberer K (1998) Methods for the determination of neutral drugs as well as betablockers and β2-sympathomimetics in aqueous matrices using GC/MS and LC/MS/MS. Fresenius J Anal Chem 362:329–340. https://doi.org/10.1007/s002160051083

Ternes TA, Bonerz M, Herrmann N, Löffler D, Keller E, Lacida BB, Alder AC (2005) Determination of pharmaceuticals, iodinated contrast media and musk fragrances in sludge by LC tandem MS and GC/MS. J Chromatogr A 1067:213–223. https://doi.org/10.1016/j.chroma.2004.10.096

Toolaram AP, Kümmerer K, Schneider M (2014) Environmental risk assessment of anti-cancer drugs and their transformation products: a focus on their genotoxicity characterization-state of knowledge and short comings. Mutat Res Rev Mutat Res 760:18–35. https://doi.org/10.1016/j.mrrev.2014.02.001

Vyas N, Turner A, Sewell G (2014) Platinum-based anticancer drugs in waste waters of a major UK hospital and predicted concentrations in recipient surface waters. Sci Total Environ 493:324–329. https://doi.org/10.1016/j.scitotenv.2014.05.127

Watson JT, Sparkman OD (2007) Introduction to mass spectrometry: instrumentation, applications, and strategies for data interpretation. Wiley, Hoboken

Weissbrodt D, Kovalova L, Ort C, Pazhepurackel V, Moser R, Hollender J, Siegrist H, McArdell CS (2009) Mass flows of X-ray contrast media and cytostatics in hospital wastewater. Environ Sci Technol 43:4810–4817. https://doi.org/10.1021/es8036725

Yu JT, Bouwer EJ, Coelhan M (2006) Occurrence and biodegradability studies of selected pharmaceuticals and personal care products in sewage effluent. Agric Water Manag 86:72–80. https://doi.org/10.1016/j.agwat.2006.06.015

Zhang J, Chang VWC, Giannis A, Wang J-Y (2013) Removal of cytostatic drugs from aquatic environment: a review. Sci Total Environ 445–446:281–298. https://doi.org/10.1016/j.scitotenv.2012.12.061

Zounkova R, Kovalova L, Blaha L, Dott W (2010) Ecotoxicity and genotoxicity assessment of cytotoxic cytostatic drugs and their metabolites. Chemosphere 81:253–260. https://doi.org/10.1016/j.chemosphere.2010.06.029

Zuloaga O, Navarro P, Bizkarguenaga E, Iparraguirre A, Vallejo A, Olivares M, Prieto A (2012) Overview of extraction, clean-up and detection techniques for the determination of organic pollutants in sewage sludge: a review. Anal Chim Acta 736:7–29. https://doi.org/10.1016/j.aca.2012.05.016

Chapter 9
Removal of Cytostatic Drugs from Water and Wastewater: Progress in the Development of Advanced Treatment Methods

Ewa Maria Siedlecka

Abstract Cytostatic drugs and other pharmaceuticals are newly recognized classes of environmental pollutants that receive considerable attention because of their categorization as carcinogenic, mutagenic, and teratogenic compounds. Although the cytostatics belong to currently unregulated trace-level contaminants, this situation can change in the near future. Due to poor biodegradability and low removal efficiency by conventional wastewater treatments, the alternative methods for a successful elimination of these drugs have been developed and investigated. This chapter provides a review of recent scientific research on the elimination of cytostatic drugs from water and wastewater by advanced physicochemical, chemical and biological methods, and advanced oxidation processes. The advanced treatments including membrane filtration, adsorption, ozonation, biomembrane filtration and advanced oxidation processes (AOPs), such as the Fenton reaction, photodegradation under solar and UV irradiation, and electrochemical oxidation, constitute a promising technology for the treatment of water and wastewater containing cytostatic drugs. In the presented review, data published on the degradation of cytostatic drugs by the aforementioned alternative methods have been evaluated for the period 2007–2017. Additional aspects of the problem such as the operating conditions, influence of the aqueous matrix quality on the removal efficacy, decomposition mechanism of cytostatic drugs based on the identified intermediates, and the advantages and disadvantages of the applied processes are also discussed.

Keywords Wastewater treatment · AOPs · Filtration · Adsorption · Cytostatics removal

---

E. M. Siedlecka (✉)
Department of Chemistry, University of Gdansk, Gdansk, Poland
e-mail: ewa.siedlecka@ug.edu.pl

E. Heath et al. (eds.), *Fate and Effects of Anticancer Drugs in the Environment*,
https://doi.org/10.1007/978-3-030-21048-9_9

9.1 Introduction

Cytostatic compounds, partially transformed or even unchanged, are directly discharged into the sewer system mainly via urinary and/or fecal excretions of patients undergoing medical treatment (Zhang et al. 2013). The residuals of these pharmaceutically active compounds are highly dangerous to human health and the aquatic environment due to their cytotoxicity, genotoxicity, mutagenicity, and teratogenicity (Besse et al. 2012; Zhang et al. 2013). Moreover, rather limited knowledge is available about the chronic health effects related to the consumption of drinking water containing trace amounts of pharmaceuticals, including cytostatic drugs or their metabolites (Johnson et al. 2008).

The municipal and hospital wastewaters are the main sources of cytostatic compounds discharged into wastewater treatment plants (WWTPs). Studies conducted in various countries indicate that hospitals account for 5.5–17% (Weissbrodt et al. 2009) of the total discharged pharmaceuticals (Franquet-Griell et al. 2015). In general, cytostatic compounds have complex structures and are often recalcitrant to biodegradation (Kosjek and Heath 2011), hydrolysis, and/or photolysis (Fiszka Borzyszkowska et al. 2016). Due to a very poor removal efficiency of conventional activated sludge systems, cytostatic drugs are frequently detected in the WWTP effluents, and subsequently in surface waters. The efficiency with which cytostatic drugs are treated largely depends on their physicochemical properties and the operating parameters of a certain WWTP (Seira et al. 2016).

It is well known that conventional methods of water and wastewater treatment are not adequate for the effective removal of many trace organic contaminants, including pharmaceuticals (Ternes et al. 2004). Although drugs belong to currently unregulated trace level pollutants, there is a growing consensus among the scientific community and water authorities that their optimized removal during wastewater treatment is a justifiably prudent approach to environmental protection. Therefore, the wastewater treatment methods should be specifically adapted to the local pollution sources and, if necessary, the tertiary treatment of wastewater should be considered. Currently, chemical, physicochemical, and biological advanced treatment processes are under discussion in order to improve the removal rates and performance of methods for the elimination of anticancer drugs from treated wastewater.

9.2 Physicochemical and Biological Properties of Cytostatic Drugs

The physicochemical properties of pollutants can be used to identify physical hazards and to understand or predict a chemical's environmental fate, or to design the effective removal process of pollutants from water and wastewater. The physicochemical

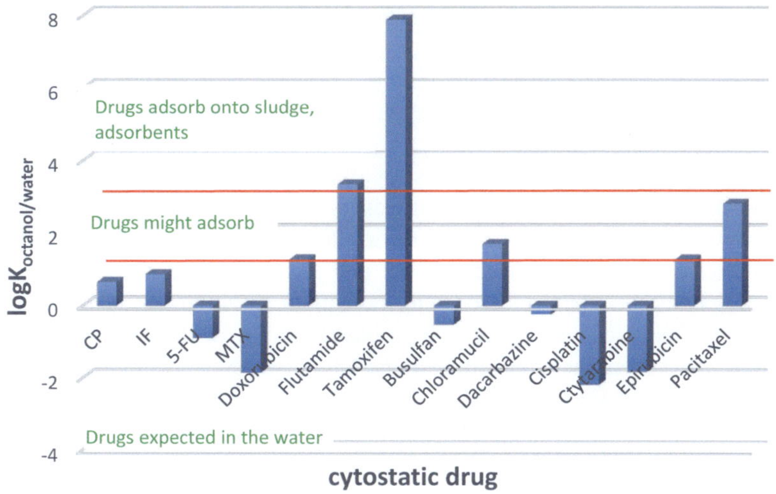

Fig. 9.1 Log $K_{o/w}$ coefficient of selected cytostatic drugs

properties having particular meaning in wastewater treatment are octanol–water partition coefficient (log $K_{o/w}$), optical properties, constant of ionization (pKa or pKb), and solubility in water.

Molecular hydrophobicity (or lipophilicity) is expressed as log $K_{o/w}$ and is one of the most studied physicochemical properties in environmental chemistry. Log $K_{o/w}$ is defined as the ratio between the concentration of compound in the hydrophobic phase (octanol simulates the affinity to lipids, biomass, carbon, etc.) and the concentration of all species (ionized and unionized) in an aqueous phase at a given pH.

Figure 9.1 shows log $K_{o/w}$ of selected cytostatic drugs. Many cytostatic drugs, particularly those commonly used (e.g., CP, IF, and 5-FU), are polar (log $K_{o/w} < 1.5$) and characterized by high water solubility. Therefore, after being discharged to wastewater, these compounds mainly exist in the aqueous phase. Due to the low log$K_{o/w}$, the low adsorption potential of cytostatic drugs on activated sludge is also to be expected. On the one hand, flutamide, chloramucil, and pacitaxel, all of which display log$K_{o/w}$ values higher than 1.5 but lower than 3, can partially adsorb on activated sludge, natural suspended solids or adsorbents, for example, activated carbon, while on the other hand, finding tamoxifen (log$K_{o/w} > 3$) is more likely to be detected in sediments than in water.

Optical properties define the material response to the incident radiation, and absorption of photons is one such property of these chemical compounds. The photolytic removal efficiency of pollutants under sunlight or UV-light applied in disinfection process depends on their optical properties. The values of maximum absorption wavelength for cytostatic drugs, with an exception of doxorubicin and epirubicin, are lower than 290 nm, indicating that some of these chemicals are poorly decomposed by photolysis under sunlight irradiation ($\lambda > 290$ nm).

On the other hand, the susceptibility to biodegradation of pollutants is a parameter which determines the type of wastewater treatment technology. 5-FU, Methotrexate (MTX), Busulfan (BSF), and Cytarabine (CTB) have been described as relatively biodegradable compounds, while the remaining cytostatics are considered as recalcitrant to biological treatment.

Due to the aforementioned reasons, cytostatic drugs are often resistant to the physical and biological degradation processes that are employed in conventional wastewater treatment systems (Johnson et al. 2008). The high resistance to biodegradation and low adsorption ability of CP and IF indicate that these drugs will be extremely persistent when released into the aquatic environment. Thus both compounds are often used in the application-based studies of the alternative removal processes. 5-FU is another frequently studied drug because of its high consumption. The advanced removal methods are mainly based on the oxidation and/or separation processes.

9.3 Advanced Treatment Processes

Since the removal of cytostatic drugs and other pharmaceuticals by conventional wastewater treatment is often incomplete and inefficient (Zhang et al. 2013), other alternatives need to be investigated. In addition, the efficiency of cytostatic drug removal via conventional drinking water treatment processes is limited because the treatments in use were not designed to control these emerging micropollutants (Ternes et al. 2004). Advanced treatment processes employing the membrane techniques and AOPs have been proven to successfully eliminate pharmaceuticals at the laboratory and pilot-plant scale (Wang et al. 2009; Lester et al. 2011; Köhler et al. 2012; Ofiarska et al. 2016; Seira et al. 2016). Until now, the studies focusing on the removal of cytostatics have included, inter alia, NF; RO; membrane bioreactor (MBR); and chemical, photocatalytic, and electrochemical oxidation (Fig. 9.2). The application of AOPs and MBR leads to the partial or total decomposition of cytostatic drugs, which, in many cases, eliminates their mutagenicity or toxicity. However, the oxidation process may sometimes result in the formation of by-products whose effects can be of potential concern in relation to ecotoxicity. For example, a serious problem with applying AOPs to treat hospital wastewater is the presence of pollutants containing iodine or bromine atoms, which will become oxidized to bromates or iodates during the treatment. The presence of these ions in water is associated with the negative influence on the environment and human health. On the other hand, the physical treatment processes only transfer the compounds from the aqueous phase to the solid phase (e.g., adsorption on activated carbon), or concentrate the compounds in another stream, for example, in the retentate in the case of NF/RO. The concentrates (retentates) in the form of a liquid (NF/RO) or in the solid state

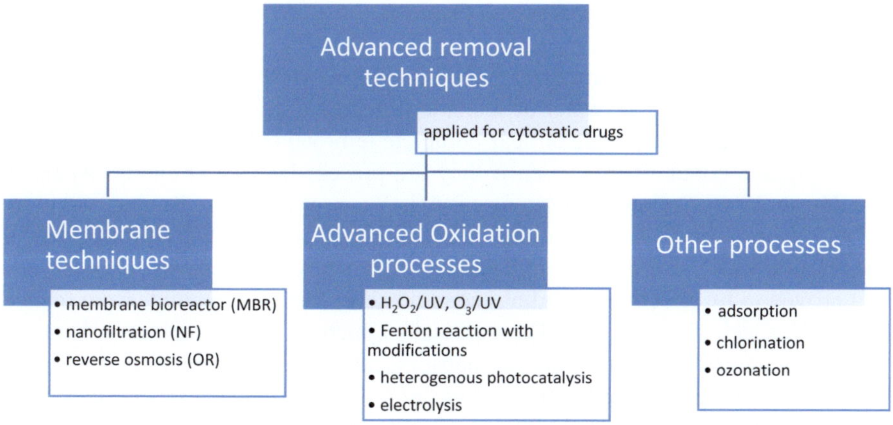

Fig. 9.2 Advanced removal techniques used for cytostatic drugs elimination from aqueous environment

(adsorption process) are hazardous wastes and need further disposal. This will invariably lead to the additional treatment or disposal of the residues of the process and, subsequently, an increase in costs of the whole operation.

9.4 Alternative Treatment Processes

The methods that can be applied to remove the residues of cytostatic drugs from effluents are advanced treatment processes, including RO, NF, MBR, and AOPs.

9.4.1 Membrane Filtration Processes

The membrane filtration processes are increasingly used in advanced water and wastewater treatment for the removal of bacteria, activated sludge, particulates, and natural and anthropogenic organic materials. A membrane is a thin layer of semipermeable material that separates substances when a driving force is applied across the membrane. As more and more advancements are made in the field of membrane production and membrane-based module design, the capital and operating costs associated with the application of membrane filtration continue to decline. The membrane processes investigated with regard to the elimination of cytostatic drugs are NF, RO, and microfiltration/biofiltration in MBR.

9.4.1.1 Nanofiltration and Reverse Osmosis

The membrane filtration processes have been successfully used to produce high-quality water from the seawater or secondary treated effluent, because the applied procedures secure a complete or near-complete removal of a wide range of organic and inorganic contaminants, even if they are present in trace amounts (Agenson and Urase 2007; Bellona and Drewes 2007: Yoon et al. 2007; Radjenović et al. 2008; Verliefde et al. 2007a, b, 2009; Yangali-Quintanilla et al. 2009). Because of the relatively low size and low molecular weight of cytotoxic drugs, only NF and RO proved to be effective regarding the removal of these compounds from treated water (Verlicchi et al. 2015). The parameters such as the membrane's filtration properties, chemistry of the filtered matrices, and physicochemical properties of investigated drugs can substantially alter the removal efficiency of filtration (Bellona et al. 2004; Nghiem et al. 2004; Verliefde et al. 2008). Researchers trying to explain the mechanisms of the removal of trace organic contaminants by NF/RO (Kiso et al. 2001; Nghiem et al. 2005) have demonstrated that the separation is based on (i) size exclusion or steric hindrance effect, (ii) electrostatic interactions between the charged trace organic compounds and the negatively charged surface of filtration membrane (Bellona et al. 2004), and (iii) adsorption of hydrophobic trace organics to membrane surfaces and their subsequent retention on RO and NF membranes. The latter phenomenon results in lower rejection than would be expected based solely on the size exclusion mechanisms (Nghiem et al. 2004; Wang et al. 2009). The NF and RO systems were used to remove trace amounts of CP (between 1 and 600 μg/L) from water during the tertiary treatment of raw water and a posttreatment process of MBR in wastewater treatment. The CP rejection from ultrapure water by NF membrane was relatively low, ranging from 20% to 40%, while RO membrane provided excellent rejection (>90%) under the applied operating conditions (Wang et al. 2009). These results confirmed the hypothesis proposed by Kiso et al. (2001): in the case of hydrophilic solutes, steric hindrance is the most important controlling factor for the molecule rejection by NF membrane. In addition, the adsorption onto the organic membrane material has also been recognized as an important mechanism of rejecting trace contaminants by the NF/RO processes, especially when the feed concentration reaches the ng/L level (Kimura et al. 2003).

Wang et al. (2009) studied the influence of aqueous matrix on the removal of CP residuals during NF. When the MBR effluent was used as the background solution, the CP rejection rate was much higher (60%) than that measured in ultrapure water. The authors explained that the higher performance of filtration was connected to the membrane fouling and CP interactions with the aqueous matrix. In conclusion, NF is not efficient enough in terms of CP removal to be considered as a tertiary treatment for CP-containing raw water, but it could be adopted as a posttreatment unit, located after an MBR unit, in a wastewater treatment system.

An MBR–RO system is an ideal choice for treating hospital effluents since it can be expected to totally remove many pharmaceutical compounds present at quite high concentrations in this type of wastewater (Wang et al. 2009). Nevertheless, further studies are needed to develop suitable procedures for the disposal of drug concentrates in the obtained NF/OR retentate. Moreover, currently the price of NF/OR techniques application for hospital WWTP is also limiting factor.

9.4.1.2 Membrane Bioreactor (MBR)

Recent developments in the MBR techniques have resulted in the availability of MBR-based systems that are seriously considered as the next-generation WWTPs and an alternative to conventional activated sludge treatment. A MBR combines biological degradation and microfiltration in one unit, where the microfiltration membranes have a nominal pore size between 0.1 and 0.4 μm. In comparison to conventional activated sludge process, the MBR appears to be more robust, with a much smaller carbon footprint and improved effluent quality (Judd 2004, 2008). In some cases, MBRs have shown relatively better removal of contaminants with special characteristics, such as low biodegradability and low concentration, compared to conventional activated sludge treatment (Quintana et al. 2005; Bernhard et al. 2006; Weiss and Reemtsma 2008). This is related to the MBR operating parameters, that is, high sludge retention time (SRT) and high concentration of biomass. These operating parameters allow for the intensification of biological processes by the inclusion of resistant and low-growth microorganisms in the biomass (De Wever et al. 2007). Moreover, the hydrophobic organic contaminants present in trace amounts can adsorb to the mixed liquor suspended solids (MLSS), resulting in a longer actual retention time in the bioreactor and, in turn, in the enhanced removal efficiency.

The improved removal efficiencies were observed in the case of micropollutants existing in the hydrophobic neutral forms compared to their negatively charged species present at higher pH values; this phenomenon is due to the fact that the negative forms cannot adsorb on the negatively charged suspended solids in the reactor (Urase et al. 2005; Tadkaew et al. 2010). Delgado et al. (2011) suggested that molecules exhibiting toxic effects may be either adsorbed or entrapped in humic acids, which increases their rejection in a MBR.

The MBR technology applied to a hospital WWTP in order to remove CP exhibited low degradation (< 20%) for a SRT of up to 50 days (Verliefde et al. 2007a, b), while a similar investigation on the presence of MTX in domestic wastewater-treated via inoculated activated sludge showed the removal values of up to 80%. Despite the high efficiency of drug removal, the ecotoxicity assessment of permeate revealed residual ecotoxicity, which was probably due to the presence of newly formed bioproducts (Delgado et al. 2009). According to more recent reports, CP removal of up to 80% was achieved for a hydraulic retention time (HRT) of 48 h and a SRT of 50 days (Delgado et al. 2010). Mahnik et al. (2007) employed a pilot-scale MBR to treat effluents from an oncology ward. It was found that 5-FU had rapidly degraded as anthracyclines were removed (removal efficiency >90%) by adsorption onto the biomass. A high efficiency of CP removal was also demonstrated for a SRT of 20 days irrespective of the variations in COD, N, or CP concentrations in the inlet flow. It is noteworthy that the removal via sorption on sludge was neglected and biodegradation was confirmed as the predominant mechanism (Ternes et al. 2004; Seira et al. 2016). The cometabolism mechanisms by ammonia oxidizers were involved in the CP elimination (Seira et al. 2016). The probable role of AMO enzyme, originating from the autotrophic ammonia oxidizers, in the simultaneous degradation of ammonium and some micropollutants was suggested also by some other authors (Clara et al. 2005; Helbling et al. 2012; Tran et al. 2013).

Nevertheless, the main problem with the application of membrane in a MBR is a rapid decline in the permeation flux due to membrane fouling, which requires frequent membrane cleaning/replacement, thus increasing the running costs (Judd 2004). This point still needs investigation, especially when toxic compounds are present in treated wastewater. Currently, available research data suggest that despite the low concentrations of cytostatic drugs present in wastewater, the toxicity of these compounds modifies the activated sludge behavior (Delgado et al. 2009, 2010). The effect of CP and its metabolites, that is, 4-hydroxycyclophophosphamide, aldophosphamide, acrolein, and phosphoramide mustard, on the formation of extracellular polymeric substances (EPS) induced an increase in the production of polysaccharides and protein–polysaccharides complexes. The EPS form a complex polymeric network with a large surface area adsorbing pollutants, nutrients, and minerals. The accumulation of EPS in the MBR sludge mixture would facilitate the formation of an EPS-fouling gel layer on the membrane surface and eventually pore narrowing/blocking, thus potentially causing a serious fouling problem (Delgado et al. 2010).

9.4.2 Advanced Oxidation Processes

AOPs represent an interesting alternative, since they can be employed in association with biological treatments for wastewater remediation, as a pretreatment, increasing the biodegradability through partial oxidation, or as a posttreatment for the degradation of persistent compounds (Klavarioti et al. 2009; De la Cruz et al. 2012). The stages of cytostatic drug oxidation via AOPs are shown in Fig. 9.3.

Advanced oxidation processes (AOPs) are based on a series of the oxidative reactions of organic matter in the aqueous phase by means of in situ generated highly reactive oxidants such as the hydroxyl radicals •OH and other chemical species, leading to the decomposition of pollutants or their mineralization. The •OH radicals and other oxidants can be generated by chemical, photocatalytic, or electrochemical methods. The AOPs used for the elimination of cytostatic drugs from the aquatic environment are presented in Fig. 9.4.

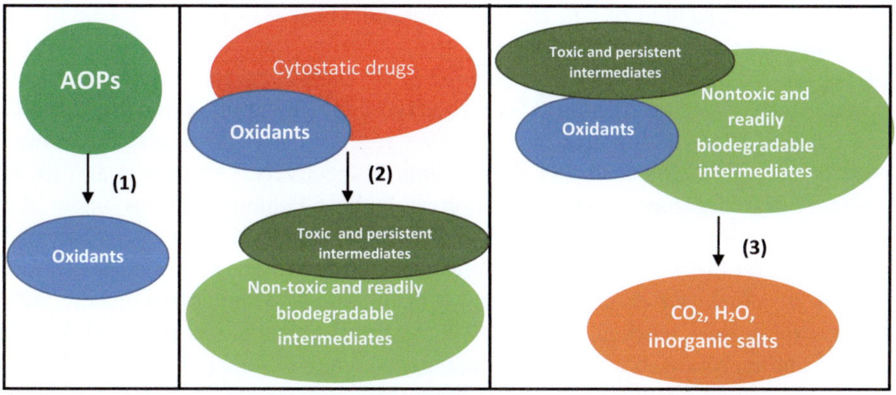

Fig. 9.3 Organic matter removal by advanced oxidation processes (AOPs); (1) generation of oxidants, (2) initial oxidation of organic matter to intermediates, and (3) oxidation of intermediates to inorganic products

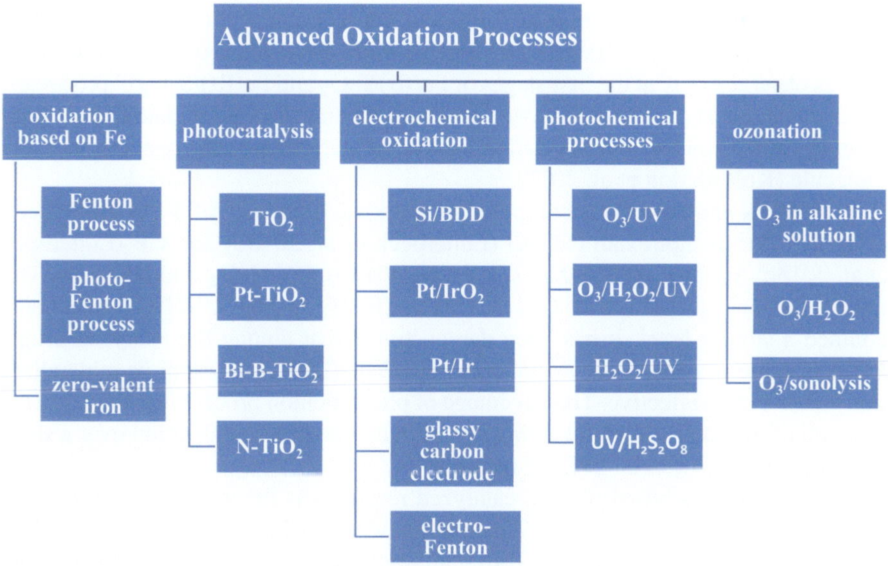

Fig. 9.4 Classification of AOPs applied in the removal of cytostatic drugs

9.4.2.1 AOPs Based on Iron-Catalysis

The Fenton's reagent and zero-valent iron in the presence of oxygen have been applied to eliminate the cytostatic drugs. The classical Fenton reaction (Fe(II)/H_2O_2) was used by Barek et al. (1998) to decompose Amsacrine, Azathioprine, Asparaginase, and Thiotepa. This simple method has also been reported as a useful tool for the degradation of CP, IF, and melphalan (Hansel et al. 1997). Due to the possibility of solar light application to accelerate the slow reaction of Fe (II) regeneration in the Fenton reaction, the photoassisted Fenton reaction has been investigated. The photo-Fenton (UV/Fe(II)/H_2O_2) process applied to degrade 5-FU and CP was faster than conventional Fenton reaction, and the investigated drugs were completely eliminated after 4 and 2 min, respectively. Governo et al. (2017) studied 5-FU decomposition via Fenton and photo-Fenton reactions under the optimal operating conditions (T = 30 °C, $[Fe(II)]_0 = 0.5$ mM; $[H_2O_2]_0 = 240$ mM and at pH 3 for $[5\text{-FU}]_0 = 0.38$ mM, simulated solar light). The authors reported that 5-FU was completely eliminated after 2 h of treatment, while approximately 50% mineralization was reached after 8 h. The best performance was achieved for the photo-Fenton process in which the mineralization level of 5-FU reached 67% for the iron dose remaining within the legal limits required for direct water discharge. Moreover, it was found that the degradation products generated during the photo-assisted Fenton process were less toxic toward *Vibro fisheri* than the parent compound, fully supporting the relevance of such technologies in the degradation of cytostatics such as 5-FU (Governo et al. 2017). In other studies, low toxicity was recorded for the initial solution of 5-FU and the solution at the end of the photo-Fenton-like (sunlight/Fe(III)/H_2O_2) treatment, while an increase in the overall toxicity was observed only at the first stages of Fe(III)/H_2O_2 and Fe(III)/$S_2O_8^{2-}$ processes. All iron-catalyzed processes followed quite similar transformation pathways, which included defluorination-hydroxylation as well as pyrimidine ring opening in 5-FU molecule (Koltsakidou et al. 2017).

Lutterbeck and coworkers compared the removal efficiency of 5-FU (Lutterbeck et al. 2015a), methotrexate (MTX) (Lutterbeck et al. 2015b) and CP (Lutterbeck et al. 2015c) for three AOPs, that is, UV/Fe(II)/H_2O_2, UV/H_2O_2, and UV/TiO_2. The mineralization of 5-FU and CP via UV/Fe(II)/H_2O_2 and UV/TiO_2 treatments attained a similar level. It has been demonstrated that among the analyzed drugs, MTX is the least persistent cytostatic and in UV/Fe(II)/H_2O_2 system its mineralization was the most effective. The advantage of photo-Fenton process was the effective removal of cytostatic drugs, resulting in a high degree of mineralization over a short time period (i.e., 1.6 min. of IMA removal).

Another process based on iron-catalyzed reactions is the zero-valent iron/water system under aerobic conditions. This method was considered as a suitable process for the degradation of cytostatic drugs in wastewater and source-separated urine originating from hospitals. The results reported by Stieber et al. (2011) indicate that IF in this system gets transformed by both reductive and oxidative reactions. The important species participating in the decay of cytostatic drugs are the hydroxyl radicals, ferryl ions, superoxide radicals, and hydrogen. Due to the fact that hydrogen peroxide is an intermediate product produced in the system, acidic conditions promote the production of highly reactive •OH in the Fenton reaction, whereas neutral and basic conditions lead to the formation of ferryl ions. On the other hand, the reductive dehalogenation of IF can be achieved by catalytic hydrogenation, because the atomic hydrogen is formed through the reduction of H_2O or acidic iron corrosion. This hypothesis has been confirmed by the presence of m/z = 227 intermediate identified in the posttreatment solution. The detected compound corresponds to the IF molecule with one chlorine atom substituted by a hydrogen atom (Stieber et al. 2011). The authors proposed that the separation and treatment of pharmaceutical-containing urine by the aforementioned method could be a reasonable alternative to conventional wastewater treatment.

A next-generation "green treatment" is represented by the ferrate(VI) process, which is not an AOP due to the lack of the •OH radical generation. Because of its strong oxidizing capacity, ferrate(VI) has been employed as a potential chemical that is suitable for the degradation of the variety of inorganic and organic contaminants. Moreover, ferrate(VI) is an emerging water-treatment disinfectant and coagulant, which can address the stringent water standards maintained by the agencies. The reaction mechanism proposed for Fe(VI) process is based on the one-electron and two-electron transfer reactions that are associated with the degradation of organics. The conducted studies showed that Fe(VI) can be reduced to Fe(V) through one-electron transfer to organic matter, generating organic radical compounds, which are well-known one-electron reductants (Rush and Bielski 1986; Bielski 1991; Barisci and Dimoglo 2016). Ferrate(VI) was explored in the context of MTX removal. The ferrate ions were formed by electrolysis at an iron anode under alkaline conditions. More than 99% degradation efficiency of MTX was provided by Fe (VI) at pH values equal to or higher than 7. Only the acidic pH values were found to be less efficient, with 70% of MTX degraded. Besides, COD removal efficiency was almost 80% for real wastewater samples. Ferrate (VI) has several advantages because of its dual functions of oxidation and coagulation. Among the identified transformation products, $C_{15}H_{16}N_8O$ was the most abundant compound that was formed by the C-N bond cleavage in the hydrolyzed glutamic acid (Barisci et al. 2015). Other transformation products resulted from the cleavage of C-N bond, oxidative cleavage of amine group, hydrolysis of glutamic acid, and decarboxylation (Barisci et al. 2015).

9.4.2.2 Photocatalysis

Photocatalytic oxidation has been frequently considered as a promising technique in the removal of pharmaceuticals, including cytostatic drugs. The photocatalytic oxidation mediated by TiO_2 is easy to set up, it runs under ambient conditions, and employs the most effective technologies. Moreover, the TiO_2 semiconductor is nontoxic, inexpensive, and has a higher chemical and physical stability than other photocatalysts. Degussa P25 showed the best results with regard to degrading 5-FU and CP present at low (μg/L) as well as high concentrations (mg/L). Many researchers indicated that the photocatalyst loading, target compound concentration, and initial pH are important parameters of the photocatalytic degradation process (Lin and Lin 2014; Lai et al. 2015; Ofiarska et al. 2016). Under optimal conditions (C_{drug} = 100 μg/L, P25 = 20 mg/l, pH 5.8), CF and 5-FU were almost decomposed within 2 h. The TiO_2/UV system was also effective in the case of MTX and IF. Despite the rapid removal of parent compounds, more than 24 h was required to reach the complete mineralization of the studied drugs. Lutterbeck et al. (2015b) showed that the mineralization of CP at a concentration as high as 20 mg/l by means of the UV/TiO_2 system reached 89.6% after about 4 h. The photocatalytic degradation pathways of IF, CP, and trofosfamide were described as the bond-breaking processes, that is, breaking of the C-Cl, P-N, and C-N bonds (N-dechloroethylation). The chloride (Cl^-) release was recognized as the likely primary step of the decomposition process (Lai et al. 2015). However, ecotoxicity testing by Microtox bioassay identified an alarming trend. Despite the fact that CP was rapidly removed from the solution, the toxicity increased during this particular treatment. Such phenomenon, which appears to be compound-dependent (for CP, chlorinated by-products were a probable source of toxicity), was not observed in the case of 5-FU (Lin and Lin 2014).

Despite the aforementioned advantages, the photocatalytic process in the presence of TiO_2 has the following drawbacks: (i) UVA light–initiated photocatalysis represents only 3–5% of sunlight, which is the free energy, and (ii) there is the electron–hole recombination step, which results in the low photocatalytic efficiency of TiO_2 semiconductor (Wu et al. 2010; Zhang et al. 2010). The efforts to enhance the photocatalytic efficiency of TiO_2 can be divided into two methodological categories: (i) the enhancement by increasing the adsorption of organic pollutants on the surface of TiO_2 particles, and (ii) the enhancement by improving the separation of electron–hole pairs in TiO_2 (Cui et al. 2009; Hafez 2009; Zhang et al. 2009; Ide et al. 2011). The second category of methods was used to enhance the photocatalytic efficiency of TiO_2 during the degradation of cytostatic drugs via noble metal deposition and ion-doping, as shown in Fig. 9.4 (Lin and Lin 2014; Ofiarska et al. 2016). Ofiarska et al. (2016) reported that in the photocatalytic oxidation of CP and IF under simulated sunlight, Pt-doped TiO_2 catalyzed faster and more effectively than undoped TiO_2. Moreover, it was observed that phosphates present in the solution were adsorbed onto the photocatalyst, which accelerated the separation of photogenerated holes and electrons. This was possible because the negative electrostatic field, formed by the surface anion forcing, helped in the formation of free hydroxyl radicals. In another study, the Bi and B ion-doping method was used to inhibit the electron–hole recombination process in TiO_2. The rates and efficiencies of the photocatalytic decomposition of four selected cytostatic drugs by applying 3%Bi-2%B-TiO_2 (Fiszka Borzyszkowska et al. 2016) were high and increasing in the following order: CP $\approx$ IF (30% removal efficiency after 45 min of irradiation) < 5-FU (38%) < imatinib (100%). However, the mineralization of IF in the presence of modified catalyst was lower than that obtained by using unmodified TiO_2. This phenomenon was explained by the inhibited generation of •OH free radicals after B and Bi codoping. The postulated hypothesis has been confirmed by other researchers who used N-doped TiO_2 to photocatalytically oxidize CP and 5-FU (Lin and Lin 2014). The authors suggested that a lower concentration of •OH was generated on the synthesized N-doped TiO_2 compared to pure TiO_2 due to the lower basicity of the active sites at the photocatalyst surface. In this study, both drugs were more effectively removed in the presence of microwave-treated N-doped TiO_2 under visible light than in the presence of Degussa P25. After

20 h of treatment, the removal efficiency values for 5-FU and CP reached 88.8% and 59.3%, respectively. All results clearly indicate that the cytostatic drugs such as CP, IF, and 5-FU are resistant to oxidation, and only the powerful quasifree •OH can efficiently remove the parent compounds and their intermediates from water.

Photocatalytic oxidation was shown to be effective in decomposing 5-FU, CP, and other cytostatics. This technology has the potential to be widely applied in the hospital wastewater treatment plants to further eliminate recalcitrant compounds that are known to resist conventional wastewater treatment processes and to remain persistent in surface waters after their release. However, if the total mineralization of cytostatics becomes a target, an extended operation time will be necessary.

9.4.2.3 Electrochemical Oxidation

A relatively new group of AOPs is constituted by electrochemical methods (EAOPs) such as (i) electro-oxidation, (ii) Fenton-based electrochemical AOP, and (iii) photoelectrocatalysis. Currently, a successful treatment of the residual anti-inflammatory, analgesic, and cytostatic drugs, and sulfonamides by these methods has been reported (Feng et al. 2013; Fabiańska et al. 2014, 2015; Brillas and Sires 2015). The above-mentioned electrochemical processes are classified as AOPs because of their ability to generate reactive oxygen species, as either superoxides (MOx, where M is metal) or the hydroxyl radicals (M(•OH)), adsorbed on the anode (M) surface due to the oxidation of water or OH^- ions. Active anodes, such as Pt/Ir or Pt/IrO_2, promote the MOx formation, whereas nonactive anodes, for example, boron-doped diamond (BDD), foster the generation of quasifree •OH radicals. In the Fenton-based electrochemical AOP, namely, the electro-Fenton process, the •OH radicals are mainly produced in the homogeneous Fenton reaction. This is possible thanks to the in situ H_2O_2 generation from the cathodic O_2 reduction, while Fe (II) ions are simultaneously regenerated at the cathode under acidic conditions.

The elimination of cytotoxicity, mutagenicity, and antibacterial activity in clinical wastewater containing anticancer drugs was the objective of the investigations conducted by Kobayashi et al. (2008) and Hirose et al. (2005). The reactive anodes, such as Pt/Ir, Pt/IrO_2, were used to electrochemically oxidize cytostatic drugs in various matrices, that is, in urine samples, and clinical wastewater. The Pt/Ir and Pt/IrO_2 electrodes were chosen because of their immortality and the ability to generate Cl_2/HOCl, the latter being strong and relatively long-lived oxidants. Nearly 100% of cytostatic compounds were removed after the electrochemical treatment of wastewater. The advantage of this method is that it enables the reduction of toxicity and the disinfection of wastewater in a one-step process. The decomposition of MTX in the urine samples showed increased oxidation efficiency in the presence of Cl^-, and a decrease in efficiency in the presence of urine. The boron-doped diamond (BDD) thin film electrode is an exemplary nonactive electrode. Such electrode was used by Fabiańska et al. (2015) to study the oxidation of IF and CP. The complete oxidation of both compounds was observed after 4 h of electrolysis. Moreover, the influence of naturally occurring ions in water and wastewater (Cl^-, PO_4^{3-}, NH_4^+,

NO_3^-) on the electrochemical decomposition of cytostatic drugs at the BDD electrode was investigated for the first time. The presence of Cl^- and PO_4^{3-} ions increased the rate of this process, while other ions had no significant effect on the drug decomposition. The main advantage of electrochemical decomposition is high mineralization of organic matter, while its drawback is the possible oxidation of chloride to ClO_3^-, which can have a negative influence on the aquatic biota and human health (Fabiańska et al. 2015).

Last but not least, the electro-Fenton process is an interesting alternative with regard to the degradation of cytostatic drugs. The galvanostatic electrolysis was investigated with a boron-doped diamond anode and a carbon felt cathode in an undivided laboratory-scale cell. In the case of 5-FU, the fastest drug decomposition (6 min at I = 300 mA) accompanied by a high degree of mineralization was reached at the optimal Fe(II) concentration of 0.2 mM. The main identified intermediates were oxalic and acetic acids and short-chain aliphatic by-products, which were completely degraded to inorganic ions (NH_4^+, NO_3^-, and F^-) after 6 h of electrolysis. The final solution contained less than 10% of residual organics (Ganzenko et al. 2017).

9.4.2.4 Ozonation

Ozonation continues to attract wide interest because of its important role in disinfecting drinking water or treating hospital wastewater. In general, molecular ozone reacts selectively with unsaturated bonds, and aromatic and amino groups in organic compounds, whereas •OH reacts much more indiscriminately (von Gunten 2003). It is known that the dominant pathway involved in a specific ozonation process largely depends on the characteristics of aqueous matrix (i.e., pH, alkalinity, redox conditions, natural and anthropogenic organic matter). The oxidation of cytostatic drugs by ozonation can occur via (i) ozone, (ii) •OH generated by the decomposition of ozone, or (iii) their combination. Not only alkalinity and pH can change the ratio of ozone to •OH by directly affecting the decomposition rate of ozone. Bicarbonate can also hamper ozonation by reacting with •OH to form carbonate ions, while bromates/iodates may be formed during the ozonation of hospital wastewater containing bromide/iodide (Kovalova et al. 2013). Furthermore, natural dissolved organic matter can act as •OH scavenger, which will significantly decrease the ozone treatment efficiency of target compounds (Kim et al. 2009). Therefore, the water quality must be considered in order to ensure the efficient removal of potential chemicals via ozonation. In addition, the pertinent ozone reaction mechanism must be investigated to develop and implement the specific plan to reduce the negative effects of ozonated hospital wastewater.

MTX and CP can be effectively removed from drinking water via ozonation, although the removal of CP required a significant treatment time (Garcia-Ac et al. 2010). Pharmaceuticals such as CP, IF, and 5-FU were rapidly removed by ozonation under alkaline conditions (pH 11) from the distilled water, pharmaceutical wastewater, and hospital effluent. However, at low pH values, their removal was less effective. The pH during treatment plays an important role, because it determines the attainable concentrations of dissolved ozone (high concentrations at low pH) and hydroxyl radicals (high concentrations at high pH). Under an alkaline pH of 11, all of the target compounds rapidly degraded through the attack of hydroxyl radicals, which resulted in the complete removal of drugs within 5 min at an ozone supply rate of 3 g O_3/h. Under acidic conditions (pH 5.6), CP and IF exhibited slower removal rates, while compounds with unsaturated C-C bonds, such as 5-FU, were still removed at rapid rates (Lin et al. 2015). Hernández et al. (2008) reported that cisplatin was totally removed by ozonation after a 2-min reaction at pH 9, while Ferre-Aracil et al. (2016) showed that CP degraded after 10 min at pH 8.5. The ozonation of MTX and doxorubicin was also found to be a pH-dependent process, with alkaline conditions being the favorable ones (Somensi et al. 2012). Despite the complete removal of the parent compounds, TOC removal was incomplete (50%). This fact indicated that by-products were still in the solution, and the biological activity of some of them resulted in the acute toxicity of ozonated effluent. (Lin et al. 2015). The results of another study showed that the ozonation efficiency can be enhanced by the application of ultrasound (Somensi et al. 2012) or hydrogen peroxide (Lester et al. 2011). The combined sonolysis/ozonation process was more efficient than ozonolysis alone in relation to the degradation of doxorubicin, while MTX was easily decomposed by both methods. Yet another conclusions from the investigations of cytostatic drugs during ozonation were that CP, IF (Česen et al. 2015), and BSF (Li et al. 2016), which belong to nitrogen mustard derivatives, are mainly eliminated via free radical-mediated oxidation mechanism. The molecular mechanism of ozone attack on the drugs is also possible in the presence of •OH scavenger in the water. In this case, however, the ozonation effectiveness is low in relation to the removal of CP (42%) and IF (36%) even in ultrapure and natural waters (Česen et al. 2015). Other studied cytostatic drugs, that is, MTX (Garcia-Ac et al. 2010), TMX, and Irinotecan (Chen et al. 2008), were less persistent during ozonolysis. MTX and TMX are large-molecular-size aromatic compounds, while IF, BSF, and CP are saturated aliphatic heteroatomic substances of low molecular size. Due to these characteristics, the decomposition of CP and IF by the •OH radical attack is favored, while MTX and TMX are sensitive to a variety of oxidants. Ferre-Aracil et al. (2016) showed that direct ozonation is technically and economically feasible as a treatment process for the chemical abatement of particular cytostatic compounds (IF, CP, and CAP at concentrations ranging from 273 to 1139 ng/L) and other dissolved organics present in raw wastewater from hospitals. In the case of CP, the elimination efficiency reached 97%. The corresponding values for other organic compounds were higher, and had been achieved over a relatively short reaction time (10 min). The concentration of ozone gas for the described treatment was ca. 43.9 g/m^3 (Ferre-Aracil et al. 2016).

The aqueous matrix composition is the key factor in reference to residual toxicity of water or ozonated wastewater. In the distilled water, the resulting ozonation products exhibited minimal mineralization but high acute toxicity, whereas in naturally buffered pharmaceutical and hospital effluents, the byproducts were more amenable to removal and detoxification. The studies conducted have also demonstrated that ozone in the presence of urine can only partially degrade recalcitrant pharmaceutical compounds; thus, the complete mineralization is not attained (Escher et al. 2006; Dodd et al. 2008). The direct reactions with ozone play an important role in governing oxidation rates during urine ozonation. Compared to the highly diluted influent of a wastewater treatment plant, source-separated urine contains relatively low concentrations of organic substrate but much higher concentrations (100–500 times) of pharmaceuticals (Larsen et al. 2004). This fact can help to achieve more cost-effective removal of various pharmaceutical residues from human urine by ozonation (Ferre-Aracil et al. 2016; Li et al. 2016). The researchers predicted that the pilot-scale treatment of hospital wastewater with ozone, following the treatment in a membrane biological reactor, would cost 2.4–2.7 €/m^3. In comparison, the reports on the exploitation of municipal wastewater treatment plants in Switzerland, which use ozonation as an additional stage after conventional biological treatment, stated the cost of ca.1.8 €/m^3 (Kovalova et al. 2013).

Until now, the available data have been inadequate to conclude whether current water treatment processes are successful in preventing the various polar residues of cytostatics from entering tap water and threatening public health. Based on present knowledge, we can propose that the by-products generated during the water treatment processes (e.g., ozonation, chlorination) might be even more toxic than their parent compounds.

9.4.3 Chlorination

It was determined that the neutral and deprotonated 5-FU interacts readily with free chlorine, HOCl, and OCl^-, while the protonated and deprotonated forms of the drug dominate the bromination process. In the presence of chloramine, 5-FU is stable. However, the drug degrades relatively rapidly in the pH range between 7 and 8, especially in the presence of bromide, which forms more kinetically active species of bromine when bromide is oxidized by chlorine. However, under real conditions, this effect can be insignificant due to potentially fast consumption of bromine by organic matter present in wastewater effluents (Tanumihardj 2013; Li et al. 2015). In addition, increased NaCl concentration results in an increase in the chlorination rate. The chlorination of 5-FU proceeds via chlorine incorporation at the sixth carbon in the heterocyclic pyrimidine ring of this drug followed by the formation of mono- and dihalogenated 5-FU products. Next, the halogenated products are decomposed into simple organic and inorganic compounds. The authors suggested that the obtained results can be used for predicting the behavior of 5-FU during the treatment of wastewater and drinking water (Tanumihardj 2013). The fate of 5-FU in chlorinated environment was currently studied by Hok et al. (2018). EC_{50} value obtained for 5-FU in test with *Daphnia magna* was 50% higher than that after the chlorination process, in which a chlorohydrin 5-chloro-5-fluoro-6-hydroxy-5,6-dihydrouracil was formed. 2-chloro-2-fluoro-3,3-dihydroxypropanoic acid was identified as a final product, with toxicity parameter EC_{50} more than twice lower compared to the parent 5-FU (Hok et al., 2018).

The study of the fate of other cytostatic drugs MTX and tamoxifen during chlorination was investigated, respectively, by Roig et al. (2014) and by Negreira et al. (2015). Methotrexate was rapidly eliminated ($t_{1/2}$ of 21 min) from water and the formation of mono-chlorinated by-products of parent drug was identified. But tamoxifen was fairly stable in chlorination conditions. On the other hand, its metabolites 4-hydroxy-tamoxifen and 4-hydroxy-N-desmethyl-tamoxifen were quickly degraded, and thirteen mono- or di- chlorinated by-products were tentatively identified (Negreira et al. 2015).

9.4.4 Adsorption

During the water treatment, adsorption to activated carbon (AC) can be an important sink of contaminating pharmaceuticals. For this purpose, the activated carbon is used in the form of powder activated carbon (PAC) and granulated activated carbon (GAC). The removal via AC adsorption largely depends on the adsorbent dosage and the value of octanol–water partition coefficient ($logK_{o/w}$) of the adsorbates being separated, especially in the case of hydrophobic species (see substances with log $K_{o/w} > 2$ in Fig. 9.2) (Westerhoff et al. 2005). Chen et al. (2008) demonstrated that the PAC adsorption capacity for TMX ($logK_{o/w} = 7.88$) was better than for other cytostatic drugs, such as irinotecan ($logK_{o/w} = 3.73$) and CP ($logK_{o/w} = 0.63$), but it also depended on the initial dose of PAC. In the literature review published by Verlicchi et al. (2015), it was stressed that the removal of hydrophobic CP and IF from hospital WWTPs by means of PAC was only possible for high doses of adsorbent (150–450 mg/L). Capecitabine ($logK_{o/w} = 0.56$), the prodrug of 5-FU, also showed low removal efficiency even at high doses of adsorbent (<NORIT SEA of 40%), while the batch-scale experiments confirmed the poor sorption of 5-FU and CTB to Norit SAE Super in a municipal WWTP effluent. A 30 mg/L dose of the adsorbent, sufficient to completely eliminate hydrophobic micropollutants (bisphenol A with $logK_{o/w} = 3.32$, and 17-α-ehinyl-estradiol with $logK_{o/w} = 3.67$), removed only 25% and 65% of CT and 5-FU, respectively (Kazner et al. 2008). GAC seems to be a more effective adsorbent than PAC. Lenz et al. (2007) coupled an MBR with a GAC adsorption column to treat the effluent from the oncological ward of a hospital in Vienna, managing to remove cancerostatic platinum compounds (CPCs) and 5-FU, and thus reaching the high elimination efficiency of the process. After the MBR step, 5-FU was removed below the detection limit, while the GAC column allowed for the removal of CPCs (Lenz et al. 2007). The GAC filtration was also successfully employed to almost completely remove the neutral form of CP from pretreated surface water (Verliefde et al. 2007a, b). On the other hand, the results of the study on 5-FU adsorption on montmorillonite and saponite indicate that the interaction between 5-FU and the clay fraction is possible by direct or indirect coordination (via water molecules) to the Lewis acidic centers

through water bridges (Akalin et al. 2007). However, since many water-soluble cytostatic drugs are readily protonated or deprotonated in water at a neutral pH and are characterized by the low $\log K_{o/w}$, AC adsorption as a stand-alone unit does not seem suitable and economically justified for a satisfactory removal of these drugs from water or industrial wastewater.

9.5 Conclusions

The existing WWTPs are not able to efficiently remove pharmaceuticals, including the cytostatic drugs, as they have been principally designed to eliminate nutrients and macropollutants present in water at the mg/L levels. For this reason, some countries started to raise awareness about the environmental danger due to micropollutants in water. For example, in Switzerland, a new water treatment step aimed at the removal of micropollutants, consisting of ozonation and/or adsorption on activated carbon, has become mandatory for the WWTPs in 2016.

Cytotoxic compounds diluted in wastewater and surface water seem to be successfully eliminated by the advanced treatment processes such as AOPs, NF/RO, MBR, adsorption on activated carbon and their combination.

Photocatalytic processes are used in the most advanced technologies, and it is likely that they will be applied on the industrial scale in the future. Electrochemical processes are a promising method for treating pharmaceutical wastewater, especially when such wastewater contains relatively high concentrations of salts. The application of AOPs, such as the photo-Fenton process or photocatalytic oxidation, frequently improves the biodegradability of effluent and reduces its ecotoxicity and/or mutagenicity.

Until now, the combination of advanced and conventional treatments for water reuse and wastewater treatment applications remains very uncommon in practice. This situation might be connected to the high investment and exploitation costs of such ventures.

Acknowledgment The authors would like to acknowledge the financial support of the Polish Ministry of Research and Higher Education, Poland under the Grant DS530-8626-D596-18.

References

Agenson KO, Urase T (2007) Change in membrane performance due to organic fouling in nanofiltration (NF)/reverse osmosis (RO) applications. Sep Purif Technol 55(2):147–156

Akalin E, Akyuz S, Akyuz T (2007) Adsorption and interaction of 5-Xuorouracil with montmorillonite and saponite by FT-IR spectroscopy. J Mol Struct 834–836:477–481

Barek J, Cvacka J, Zima J, Meot MDE, Lagett M, Michelon J, Castegnarot M (1998) Chemical degradation of wastes of antineoplastic agents Amsacrine, azathioprine, Asparaginase and Thiotepa. Ann Occup Hyg 42:259–266

Barisci S, Dimoglo A (2016) Chapter 12: Review on the stability of ferrate (VI) species in aqueous medium and oxidation of Pharmaceuticals and Personal Care Products (PPCPs) by ferrate (VI): identification of transformation by-products. In: Ferrites and ferrates: chemistry and applications in sustainable energy and environmental remediation. American Chemical Society, pp 287–335. https://doi.org/10.1021/bk-2016-1238.ch012

Barisci S, Turkay O, Ulusoy E, Dimoglo A (2015) Degradation of the cytostatic methotrexate by electrosynthesized ferrate (VI): identification of oxidation by-products. Conference paper, https://www.researchgate.net/publication/288416292

Bellona C, Drewes JE (2007) Viability of a low-pressure nanofilter in treating recycled water for water reuse applications: a pilot-scale study. Water Res 41(17):3948–3958

Bellona C, Drewes JE, Xu P, Amy G (2004) Factors affecting the rejection of organic solutes during NF/RO treatment—a literature review. Water Res 38(12):2795–2809

Bernhard M, Müller J, Knepper TP (2006) Biodegradation of persistent polar pollutants in wastewater: comparison of an optimised lab-scale membrane bioreactor and activated sludge treatment. Water Res 40:3419–3428

Besse JP, Latour JF, Garric J (2012) Anticancer drugs in surf waters: what can we say about the occurrence and environmental significance of cytotoxic, cytostatic and endocrine therapy drugs. Environ Int 39:73–86

Bielski BHJ (1991) Studies of hypervalent iron. Free Radic Res Commun 12:469–477

Brillas E, Sires I (2015) Electrochemical removal of pharmaceuticals from water streams: reactivity elucidation by mass spectrometry. Trends Anal Chem 70:112–121

Česen M, Kosjek T, Laimou-Geraniou M, Kompare B, Širok B, Lambropolou D, Heath E (2015) Occurrence of cyclophosphamide and ifosfamide in aqueous environment and their removal by biological and abiotic wastewater treatment processes. Sci Total Environ 527–528:465–473

Chen Z, Park G, Herckes P, Westerhoff P (2008) Physicochemical treatment of three chemotherapy drugs: irinotecan, tamoxifen, and cyclophosphamide. J Adv Oxid Technol 11:254–260

Clara M, Strenn B, Gans O, Martinez E, Kreuzinger N, Kroiss H (2005) Removal of selected pharmaceuticals, fragrances and endocrine disrupting compounds in a membrane bioreactor and conventional wastewater treatment plants. Water Res 39:4797–4807

Cui L, Wang Y, Niu M, Chen G, Cheng Y (2009) Synthesis and visible light photocatalysis of Fe-doped TiO_2 mesoporous layer deposited on hollow glass microbeads. J Solid Sate Chem 182:2785–2790

de la Cruz N, Giménez J, Esplugas S, Grandjean D, de Alencastro LF, Pulgarín C (2012) Degradation of 32 emergent contaminants by UV and neutral photo-Fenton in domestic wastewater effluent previously treated by activated sludge. Water Res 46:1947–1957

de Wever H, Weiss S, Reemtsma T, Vereecken J, Müller J, Knepper T, Rörden O, Gonzalez S, Barcelo D, Hernando MD (2007) Comparison of sulfonated and other micropollutants removal in membrane bioreactor and conventional wastewater treatment. Water Res 41(4):935–945

Delgado LF, Dorandeu C, Marion B, Gonzalez C, Faucet-Marquis V, Schetrite S, Albasi C (2009) Removal of a cytostatic drug by a membrane bioreactor. Desalin Water Treat 9(1–3):112–118

Delgado LF, Faucet-Marquis V, Schetrite S, Pfohl-Leszkowicz A, Paranthoen S, Albasi C (2010) Effect of cytostatic drugs on the sludge and on the mixed liquor characteristics of a cross-flow membrane bioreactor: consequence on the process. J Membr Sci 347:165–173

Delgado LF, Faucet-Marquis V, Pfohl-Leszkowicz A, Dorandeu C, Marion B, Schetrite S, Albasi C (2011) Cytotoxicity micropollutant removal in a crossflow membrane bioreactor. Bioresour Technol 102:4395–4401

Dodd MC, Zuleeg S, von Gunten U, Pronk W (2008) Ozonation of source-separated urine for resource recovery and waste minimization: process modeling, reaction chemistry, and operational considerations. Ozonation source separated urine resource. Recovery waste minimization process model. React Chem Oper Consider 42:9329–9337

Escher BI, Pronk W, Suter MJF, Maurer M (2006) Monitoring the removal efficiency of pharmaceuticals and hormones in different treatment processes of source-separated urine with bioassays. Environ Sci Technol 40:5095–5101

Fabiańska A, Białk-Bielińska A, Stepnowski P, Stolte S, Siedlecka EM (2014) Electrochemical degradation of sulfonamides at BDD electrode: kinetics, reaction pathway and eco-toxicity evaluation. J Hazard Mater 280:579–587

Fabiańska A, Ofiarska A, Fiszka-Borzyszkowska A, Stepnowski P, Siedlecka EM (2015) Electrodegradation of ifosfamide and cyclophosphamide at BDD electrode: decomposition pathway and its kinetics. Chem Eng J 276:274–282

Feng L, van Hullebusch ED, Rodrigo MA, Esposito G, Oturan MA (2013) Removal of residual anti-inflammatory and analgesic pharmaceuticals from aqueous systems by electrochemical advanced oxidation processes. A review. Chem Eng J 228:944–964

Ferre-Aracil J, Valcárcel Y, Negreira N, de Alda ML, Barceló D, Cardona SC, Navarro-Laboulais J (2016) Ozonation of hospital raw wastewaters for cytostatic compounds removal. Kinetic modelling and economic assessment of the process. Sci Total Environ 556:70–79

Fiszka Borzyszkowska A, Pieczyńska A, Ofiarska A, Nikiforow K, Stepnowski P, Siedlecka EM (2016) Bi-B-TiO_2-based photocatalytic decomposition of cytostatic drugs under simulated sunlight treatments. Sep Purif Technol 169:113–120

Franquet-Griell H, Gómez-Canela C, Ventura F, Lacorte S (2015) Predicting concentrations of cytostatic drugs in sewage effluents and surface waters of Catalonia (NE Spain). Environ Res 138:161–172

Ganzenko O, Oturan N, Sires I, Huguenot D, van Hullebusch ED, Esposito G, Oturan MA (2017) Fast and complete removal of the 5-fluorouracil drug from water by electro-Fenton oxidation. Environ Chem Lett. https://doi.org/10.1007/s10311-017-0659-6

Garcia-Ac A, Broséus R, Vincent S, Barbeau B, Prévost M, Sauvé S (2010) Oxidation kinetics of cyclophosphamide and methotrexate by ozone in drinking water. Chemosphere 79:1056–1063

Governo M, Santos MSF, Alves A, Madeira LM (2017) Degradation of the cytostatic 5-fluorouracil in water by Fenton and photoassisted oxidation processes. Environ Sci Pollut Res 24:844–854

Hafez HS (2009) Synthesis of highly-acitve single-crystalline TiO_2 nanorods and its application in environmental photocatalysis. Mater Lett 63:1471–1474

Hansel S, Castegnaro M, Sportouch MH, De Meo M, Milhavet JC, Laget M, Dumenil G (1997) Chemical degradation of wastes of antineoplastic agents: cyclophosphamide, ifosfamide and melphalan. Int Arch Occup Environ Health 69:109–114

Helbling DE, Johnson DR, Honti M, Fenner K (2012) Micropollutant biotransformation kinetics associate with WWTP process parameters and microbial community characteristics. Environ Sci Technol 46:10579–10588

Hernández C, Ramos Y, Fernández LA, Ledea O (2008) Ozonation of cisplatin in aqueous solution at pH 9. Ozone Sci Eng 30:189–196

Hirose J, Kondo F, Nakano T, Kobayashi T, Hiro N, Ando Y, Sano K (2005) Inactivation of antineoplastics in clinical wastewater by electrolysis. Chemosphere 60:1018–1024

Hok L, Ulm L, Tandari T, Krivohlavek A, Šakič D, Vrček V (2018) Chlorination of 5-fluorouracil: reaction mechanism and ecotoxicity assessment of chlorinated products. Chemosphere 207:612–619

Ide Y, Koike Y, Ogawa M (2011) Molecular selective photocatalysis by TiO_2/nanoporous silica core/shell particulates. J Colloid Interface Sci 358(1):245–251

Johnson AC, Jurgens MD, Williams RJ, Kummerer K, Kortenkamp A, Sumpter JP (2008) Do cytotoxic chemotherapy drugs discharged into rivers pose a risk to the environment and human health ? An overview and UK case study. J Hydrol 44:167–175

Judd SJ (2004) A review of fouling of membrane bioreactors in sewage treatment. Water Sci Technol 49(2):229–235

Judd SJ (2008) The status of membrane bioreactor technology. Trends Biotechnol 26(2):109–112

Kazner C, Lehnberg K, Kovalova L, Wintgens T, Melin T, Hollender J, Dott W (2008) Removal of endocrine disruptors and cytostatics from effluent by Nanofiltration in combination with adsorption on powdered activated carbon, 5-th IWA leading-edge Conference on Water and Wastewater Technologies , 1–4 June 2008 Zurich, Switzerland

Kim I, Yamashita N, Tanaka H (2009) Chemosphere Photodegradation of pharmaceuticals and personal care products during UV and UV/H2O2 treatments. Chemosphere 77:518–525

Kimura K, Amy G, Drewes JE, Heberer T, Kim TU, Watanabe Y (2003) Rejection of organic micropollutants (disinfection by-products, endocrine disrupting compounds, and pharmaceutically active compounds) by NF/RO membranes. J Membr Sci 227:113–121

Kiso Y, Sugiura Y, Kitao T, Nishimura K (2001) Effects of hydrophobicity and molecular size on rejection of aromatic pesticides with nanofiltration membranes. J Membr Sci 192(1):1–10

Klavarioti M, Mantzavinos D, Kassinos D (2009) Removal of residual pharmaceuticals from aqueous systems by advanced oxidation processes. Environ Int 35:402–417

Kobayashi T, Hirose J, Sano K, Hiro N, Ijiri Y, Takiuchi H, Nakano T (2008) Evaluation of an electrolysis apparatus for inactivating antineoplastics in clinical wastewater. Chemosphere 72:659–665

Köhler C, Venditti S, Igos E, Klepiszewski K, Benetto E, Cornelissen A (2012) Elimination of pharmaceutical residues in biologically pre-treated hospital wastewater using advanced UV irradiation technology: a comparative assessment. J Hazard Mater 239–240:70–77

Koltsakidou A, Antonopoulou M, Sykiotou M, Evgenidou E, Konstantinou I, Lambropoulou DA (2017) Photo-Fenton and Fenton-like processes for the treatment of the antineoplastic drug 5-fluorouracil under simulated solar irradiation. Environ Sci Pollut Res 24:4791–4800

Kosjek T, Heath E (2011) Occurrence, fate and determination of cytostatic pharmaceuticals in the environment. Trends Anal Chem 30:1065–1087

Kovalova L, Siegrist H, von Gunten U, Eugster J, Hagenbuch M, Wittmer A, Moser R, McArdell CS (2013) Elimination of micropollutants during post-treatment of hospital wastewater with powdered activated carbon, ozone, and UV. Environ Sci Technol 47(14):7899–7908

Lai WW-P, Lin HH-H, Lin AY-C (2015) TiO2 photocatalytic degradation and transformation of oxazaphosphorine drugs in an aqueous environment. J Hazard Mater 287C:133–141

Larsen TA, Lienert J, Joss A, Siegrist H (2004) How to avoid pharmaceuticals in the aquatic environment. J Biotechnol 113:295–304

Lenz K, Mahnik SN, Weissenbacher N, Mader RM, Krenn P, Hann S, Koellensperger G, Uhl M, Knasmu¨ lS, Ferk F, Bursch W, Fuerhacker M (2007) Monitoring, removal and risk assessment of cytostatic drugs in hospital wastewater. Water Sci Technol 56:141–149

Lester Y, Avisar D, Gozlan I, Mamane H (2011) Removal of pharmaceuticals using combination of UV/H2O2/O3 advanced oxidation process. Water Sci Technol J Int Assoc Water Pollut Res 64:2230–2238

Li W, Tanumihardj J, Masuyama T, Korshin G (2015) Examination of the kinetics of degradation of the antineoplastic drug 5-fluorouracil by chlorine and bromine. J Hazard Mater 282:125–132

Li W, Nanaboina V, Chen F, Korshin GV (2016) Removal of polycyclic synthetic musks and antineoplastic drugs in ozonated wastewater: quantitation based on the data of differential spectroscopy. J Hazard Mater 304:242–250

Lin HH-H, Lin AY-C (2014) Photocatalytic oxidation of 5-fluorouracil and cyclophosphamide via UV/TiO2 in an aqueous environment. Water Res 48:559–568

Lin AY-C, Hsueh JH-F, Hong PKA (2015) Removal of antineoplastic drugs cyclophosphamide, ifosfamide, and 5-fluorouracil and a vasodilator drug pentoxifylline from wastewaters by ozonation. Environ Sci Pollut Res 22:508–515

Lutterbeck CA, Baginska E, Machado EL, Kümmerer K (2015a) Removal of the anti-cancer drug methotrexate from water by advanced oxidation processes: aerobic biodegradation and toxicity studies after treatment. Chemosphere 141:290–296

Lutterbeck CA, Machado EL, Kümmerer K (2015b) Photodegradation of the antineoplastic cyclophosphamide: a comparative study of the efficiencies of UV/H2O2, UV/Fe^{2+}/H2O2 and UV/TiO2 processes. Chemosphere 120:538–546

Lutterbeck CA, Wilde ML, Baginska E, Leder C, Machado EL, Kümmerer K (2015c) Degradation of 5-FU by means of advanced (photo)oxidation processes: UV/H2O2, UV/Fe^{2+}/H2O2 and UV/TiO2 – comparison of transformation products, ready biodegradability and toxicity. Sci Total Environ 527–528:232–245

Mahnik SN, Lenz K, Weissenbacher N, Mader RM, Fuerhacker M (2007) Fate of 5-fluorouracil, doxorubicin, epirubicin, and daunorubicin in hospital wastewater and their elimination by activated sludge and treatment in a membrane-bio-reactor system. Chemosphere 66(1):30–37

Negreira N, Regueiro J, Lopez de Alda M, Barcelo D (2015) Transformation of tamoxifen and its major metabolites during water chlorination: identification and in silico toxicity assessment of their disinfection byproducts. Water Res 85:199–207

Nghiem LD, Schäfer AI, Elimelech M (2004) Removal of natural hormones by nanofiltration membranes: measurement, modeling, and mechanisms. Environ Sci Technol 38:1888–1896

Nghiem LD, Schafer AI, Elimelech M (2005) Pharmaceutical retention mechanisms by nanofiltration membranes. Environ Sci Technol 39(19):7698–7705

Ofiarska A, Pieczyńska A, Fiszka Borzyszkowska A, Stepnowski P, Siedlecka EM (2016) Pt–TiO_2-assisted photocatalytic degradation of the cytostatic drugs ifosfamide and cyclophosphamide under artificial sunlight. Chem Eng J 285:417–427

Quintana J, Weiss S, Reemtsma T (2005) Pathways and metabolites of microbial degradation of selected acidic pharmaceutical and their occurrence in municipal wastewater treated by a membrane bioreactor. Water Res 39:1664–2654

Radjenović J, Petrović M, Ventura F, Barceló D (2008) Rejection of pharmaceuticals in nanofiltration and reverse osmosis membrane drinking water treatment. Water Res 42(14):3601–3361

Roig B, Marquenet B, Delpla I, Bessonneau V, Sellier A, Leder C, Thomas O, Bolek R, Kummerer K (2014) Monitoring of methotrexate chlorination in water. Water Res 57:67–75

Rush JD, Bielski BHJ (1986) Pulse radiolysis studies of alkaline Fe(III) and Fe(VI) solutions. Observation of transient ion complexes with intermediate oxidation states. Am Chem Soc 108:523–525

Seiraa J, Sablayrolles C, Montréjaud-Vignoles M, Albasi C, Joannis-Cassan C (2016) Elimination of an anticancer drug (cyclophosphamide) by amembrane bioreactor: comprehensive study of mechanismsJordan. Biochem Eng J 114:155–163

Somensi CA, Simionatto EL, Dalmarco JB, Gaspareto P, Radetski CMA (2012) A comparison between ozonolysis and sonolysis/ozonolysis treatments for the degradation of the cytostatic drugs methotrexate and doxorubicin: kinetic and efficiency approaches. J Environ Sci Health A Tox Hazard Subst Environ Eng 47:1543–1550

Stieber M, Putschew A, Jekel M (2011) Treatment of pharmaceuticals and diagnostic agents using zero-Valent Iron – kinetic studies and assessment of transformation products assay. Environ Sci Technol 45:4944–4950

Tadkaew N, Sivakumar M, Khan SJ, McDonald JA, Nghiem LD (2010) Effect of mixed liquor pH on the removal of trace organic contaminants in a membrane bioreactor. Bioresour Technol 101(5):1494–1500

Tanumihardj J (2013) Examination of the degradation of the antineoplastic drug 5-fluorouracil by chlorine at varying treatment conditions. Msc thesis, University of Washington

Ternes TA, Joss A, Siegrist H (2004) Scrutinizing pharmaceuticals and personal care products in wastewater treatment. Environ Sci Technol 38(20):392A–399A

Tran NH, Urase T, Ngo HH, Hu J, Ong SL (2013) Insight into metabolic and co-metabolic activities of autotrophic and heterotrophic microorganisms in the biodegradation of emerging trace organic contaminants. Bioresour Technol 146:721–731

Urase T, Kagawa C, Kikuta T (2005) Factors affecting removal of pharmaceutical substances and estrogens in membrane separation bioreactors. Desalination 178(1–3):107–113

Verlicchi P, Al Aukidy M, Zambello E (2015) What have we learned from worldwide experiences on the management and treatment of hospital effluent? – an overview and a discussion on perspectives. Sci Total Environ 514:467–491

Verliefde A, Cornelissen E, Amy G, Van der Bruggen B, van Dijk H (2007a) Priority organic micropollutants in water sources in Flanders and the Netherlands and assessment of removal possibilities with nanofiltration. Environ Pollut 146(1):281–289

Verliefde ARD, Heijman SGJ, Cornelissen ER, Amy G, Van der Bruggen B, van Dijk JC (2007b) Influence of electrostatic interactions on the rejection with NF and assessment of the removal efficiency during NF/GAC treatment of pharmaceutically active compounds in surface water. Water Res 41(15):3227–3240

Verliefde ARD, Cornelissen ER, Heijman SGJ, Verberk JQJC, Amy GL, Van der Bruggen B, van Dijk JC (2008) The role of electrostatic interactions on the rejection of organic solutes in aqueous solutions with nanofiltration. J Membr Sci 322(1):52–66

Verliefde ARD, Cornelissen ER, Heijman SGJ, Petrinic I, Luxbacher T, Amy GL, Van der Bruggen B, van Dijk JC (2009) Influence of membrane fouling by (pretreated) surface water on rejection of pharmaceutically active compounds (PhACs) by nanofiltration membranes. J Membr Sci 330(1–2):90–103

von Gunten U (2003) Ozonation of drinking water: part I. Oxidation kinetics and product formation. Water Res 37(7):1443–1467

Wang L, Albasi C, Faucet-Marquis V, Pfohl-Leszkowicz A, Dorandeu C, Marion B, Causserand C (2009) Cyclophosphamide removal from water by nanofiltration and reverse osmosis membrane. Water Res 43(17):4115–4122

Weiss S, Reemtsma T (2008) Membrane bioreactors for municipal wastewater treatment—a viable option to reduce the amount of polar pollutants discharged into surface waters? Water Res 42:3837–3847

Weissbrodt D, Kovalova L, Ort C, Pazhepurackel V, Moser R, Hollender J, Siegrist H, McArdell CS (2009) Mass flows of X-ray contrast media and cytostatics in hospital wastewater. Environ Sci Technol 43:4810–4817

Westerhoff P, Yoon Y, Snyder S, Wert E (2005) Fate of endocrine-disruptor, pharmaceutical, and personal care product chemicals during simulated drinking water treatment processes. Environ Sci Technol 39:6649–6663

Wu R-J, Chen C-C, Lu C-S, Hsu P-Y, Chen M-H (2010) Phorate degradation by TiO_2 photocatalysis: parameter and reaction pathway investigations. Desalination 250:869–875

Yangali-Quintanilla V, Sadmani A, McConville M, Kennedy M, Amy G (2009) Rejection of pharmaceutically active compounds and endocrine disrupting compounds by clean and fouled nanofiltration membranes. Water Res 43(9):2349–2362

Yoon Y, Westerhoff SA, Snyder EC, Wert J, Yoon J (2007) Removal of endocrine disrupting compounds and pharmaceuticals by nanofiltration and ultrafiltration membranes. Desalination 202(1–3):16–23

Zhang C, Gu L, Lin Y, Wang Y, Fu D, Gu Z (2009) Degradation of X-3B dye by immobilized TiO_2 photocatalysis coupling anodic oxidation on BDD electrode. J Photochem Photobiol A Chem 207:66–72

Zhang X, Wu F, Deng N (2010) Degradation of paracetamol in self assembly β-cyclodextrin/TiO_2 suspension under visible irradiation. Catal Commun 11:422–425

Zhang J, Chang VWC, Giannis A, Wang J-Y (2013) Removal of cytostatic drugs from aquatic environment: a review. Sci Total Environ 445–446:281–298

Chapter 10
Occurrence of Cytostatics in Different Water Compartments

Paola Verlicchi, Aina Campos Garrigós, and Mustafa Al Aukidy

Abstract The chapter deals with the occurrence of a selection of anticancer drugs in different water environments: hospital wastewater, wastewater treatment plant influents and effluents, surface water, sea water, and drinking water. Unfortunately, no data are available for groundwater up to now. The chapter presents and discusses measured environmental concentrations of anticancer drugs collected in 56 peer-reviewed papers referring to investigations carried out in 18 countries all over the world. It focuses on the variability of observed concentrations in the different environments, and it highlights the importance of planning efficient sampling strategies in order to obtain representative water samples.

The highest concentrations in hospital effluents were found for platinum-based compounds and 5-fluorouracil (> 10^5 ng L^{-1}), in the influent for ciprofloxacin (> 10^3 ng L^{-1}), in the effluent for platinum-based compounds, ifosfamide and bicalutamide (> 10^3 ng L^{-1}), and in surface water for cyclophosphamide, tamoxifen, ciprofloxacin, and bicalutamide (> 10^2 ng L^{-1}). In addition, a comparison is provided between measured and predicted concentrations of some anticancer drugs and a brief discussion of the strengths and weaknesses of the two approaches is reported.

Keywords Cytostatic · Hospital wastewater · Occurrence · Raw wastewater · Surface water · Treated effluent

---

P. Verlicchi (✉)
Department of Engineering, University of Ferrara, Ferrara, Italy

Terra&Acqua Tech, Technopole of the University of Ferrara, Ferrara, Italy
e-mail: paola.verlicchi@unife.it

A. Campos Garrigós
Department of Engineering, University of Ferrara, Ferrara, Italy

Faculty of Earth Science, University of Barcelona, Barcelona, Spain

M. Al Aukidy
Department of Engineering, University of Ferrara, Ferrara, Italy

E. Heath et al. (eds.), *Fate and Effects of Anticancer Drugs in the Environment*,
https://doi.org/10.1007/978-3-030-21048-9_10

10.1 Introduction

In the last few years, researchers and scientists have paid increasing attention to investigating the occurrence, fate, and distribution of *new* pollutants as well as their cocktails in the different water matrices: ground, surface and rainwater, municipal and industrial wastewater, treatment plant effluents, and drinking water. The development of new analytical techniques and the possibility to detect compounds at the ng/L level permitted the search for contaminants of *emerging interest* in water occurring at very low concentrations, so-called *micropollutants*. This group includes a multitude of substances of different natures and uses (flame retardants, pharmaceuticals and personal care products, endocrine disruptor compounds, etc.). Among these, diagnostic agents (iodinated contrast media), disinfectants, and pharmaceuticals (PhACs) have been the object of many studies due to increasing concern regarding their disposal and potential ecotoxic and human health impact. These substances are worthy of attention, although they are not yet included in any list of compounds subjected to periodic monitoring and control in water due to their adverse effects on the environment, with the exception of the substances included in a *watch list* (for instance, that set by the European Community, according to the recent Decision (EU) 2018/840). Each of these drugs was conceived of and then administered in order to exert a specific function: as a disinfectant, an antibiotic, a hormone, an antimetabolite, etc. When administered, its action takes place within the human body. A fraction of the compound is excreted via urine or feces as an unchanged compound or/and as its metabolites and released into the sewer, thus entering the water cycle. Its activity continues in both the public sewer and the receiving water compartment, and aquatic organisms might be susceptible to its toxicity.

Among pharmaceuticals, the class of drugs used in cancer treatment is the object of worldwide investigations. There are over 100 different compounds regularly used in developed countries; new drugs are entering the market and some others are administered differently based on changes in medical practices, and therefore some compounds are less used while others are increasingly used (Kümmerer et al. 2016). The demand for chemotherapy treatment in developed countries is increasing at around 10% per year (Besse et al. 2012). The growing interest is related to the awareness that most of them (the antineoplastic agents) exert cytotoxic, genotoxic, mutagenic, carcinogenic, or teratogenic effects on aquatic life. In addition, as they often interact with the structure and functions of DNA, all eukaryote organisms might be the target of their toxicity (Besse et al. 2012). Other anticancer drugs have hormonal and antihormonal properties, and because of this they require adequate consideration. The term "anticancer drugs" is used when accounting for the following types of molecules: cytostatics, cytotoxic, and hormonally active drugs.

Due to the wide spectrum of compounds belonging to this group, attempts to classify them have been made in recent years. Among them, Espinosa et al. (2003),

Table 10.1 Classification of anticancer drugs

ATC	Class	Mechanism of action
	Cytotoxic drugs that directly interact with DNA	Alkylating agents
		Pt-based compounds
		Intercalating agents
ATC L01: Antineoplastic agents	Cytotoxic drugs with indirect interaction with DNA	Antimetabolites
		Antitumor antibiotics
		Topoisomerase inhibitors
		Mitotic inhibitors
	Cytostatics	Kinase inhibitors
		Monoclonal antibodies
	Hormones	Steroids
		LH–RH analogs
ATC L02 Endocrine therapy	Hormone inhibitors	Estrogen inhibitors
		Androgen inhibitors
		Aromatase agent inhibitors

Besse et al. (2012), and Kümmerer et al. (2016) proposed a scheme starting from the Anatomical Therapeutic Chemical (ATC) classification and developing into a tree (Table 10.1).

The main sources of antineoplastics in the environment are households and hospitals, depending on the compound and on its administration mode.

Consumption of these compounds may vary from one country to another and from one year to another. For instance, the specific consumption of cyclophosphamide, a cytotoxic alkylating agent, was 0.002 mg/(d person) in Denmark, 0.003 mg/(d person) in Austria, 0.012 mg/(d person) in Germany, 0.013 mg/(d person) in France, and 0.056 mg/(d person) in Switzerland (Kümmerer et al. 2016). Moreover, in Germany, the consumption of all the anticancer drugs (class L) was about 22,000 kg in 2001, 42,000 kg in 2008, and 50,000 kg in 2012.

In order to provide a snapshot of the observed concentration ranges of selected anticancer drugs in water cycle segments, this chapter deals with an overview of literature data on measured environmental concentrations (*MECs*) in the following different water environments: hospital effluents, wastewater treatment plant (WWTP) influents and effluents, surface water, marine water, groundwater, and tap water. Data is provided from 56 peer-reviewed papers referring to investigations carried out between 1990 and 2017 in 18 different countries all over the world. The chapter also includes a section dealing with studies that evaluate the predicted environmental concentrations (*PECs*) of selected anticancer drugs and presents a comparison between *PECs* and *MECs* in the different compartments for the compounds of interest.

10.2 Occurrence of Cytostatic Drugs in Aquatic Media

The occurrence of cytostatic drugs in different water compartments has been investigated by different authors in the last two decades. Table 10.2 reports them together with the monitored compartment: surface water (SW) (including lakes, rivers, and channels), hospital wastewater (HWW), municipal wastewater treatment plant (WWTP) influent and effluent, drinking water (DW), and seawater (SeaW). Twenty-three studies have dealt with cytostatic drugs in hospital effluents, 22 in WWTP influents, 31 in WWTP effluents, 21 in SW, 5 in DW, and 3 in SeaW.

Most of the studies were carried out in European countries: fifteen in Spain, six in the United Kingdom, five in Italy, four in Austria and Germany, three in Switzerland, two in France and Slovenia, and one in the Netherlands, Norway, Romania, and Serbia; few studies were carried out in other countries: three in China and Japan, two in the United States, and one in Australia, Canada, and Thailand. The investigations dealing with monitoring of groundwater compartments did not address any of the anticancer drugs.

10.2.1 The Studied Compounds in Published Papers

Investigations reported in Table 10.2 refer to the compounds reported in Table 10.3, where they are grouped into classes according to their mechanism of action. Some compounds (in italics in Table 10.3) were always found below the corresponding limit of detection (LOD) in the studies included in this review. This is the case for mitomycin in WWTP influent and effluents (Martín et al. 2014) and imatinib in drinking water (Mendoza et al. 2016).

10.3 Excretion factor

The occurrence of pharmaceuticals in the water cycle is primarily related to the amount administered in a specific place and to excretion via urine or feces as unchanged form or as their metabolites. Other parameters contributing to the definition of the concentration level, as discussed in the final section of this chapter, are treatment removal efficiency, dilution in the SW, and attenuation processes within the receiving water body. Excretion rates vary widely among compounds. As an example, it was found that the excretion rate is less than 2% for erlotinib and nearly 100% for carboplatin. Moreover, for most of the compounds, the excretion data, reported in the literature, varies greatly. This is because it depends on different factors such as patient health, age and gender, and

Table 10.2 Studies referring to the occurrence of cytostatic drugs in different countries and investigated water environments (●)

Country	References	HWW	Influent	Effluent	SW	DW	SeaW
Australia	Busetti et al. (2009)			●			
Austria	Lenz et al. (2007)	●		●			
Austria	Mahnik et al. (2004)	●					
Austria	Mahnik et al. (2006)	●					
Austria	Mahnik et al. (2007)	●					
Canada	Rabii et al. (2014)		●	●			
China	Liu et al. (2010)	●	●	●			
China	Yang et al. (2011)						●
China	Yin et al. (2010)	●					
France	Catastini et al. (2008)	●	●	●			
France	Coetsier et al. (2009)			●	●		
Germany	Kümmerer et al. (1997)	●	●	●			
Germany	Steger-Hartmann et al. (1996)	●					
Germany	Steger-Hartmann et al. (1997)	●	●	●			
Germany	Ternes (1998)			●	●		
Italy	Al Aukidy et al. (2012)			●			
Italy	Castiglioni et al. (2005)			●			
Italy	Verlicchi et al. (2012)	●	●	●	●		
Italy	Zuccato et al. (2000)			●	●	●	
Italy	Zuccato et al. (2005)			●			
Japan	Azuma et al. (2015)	●	●	●	●		
Japan	Azuma et al. (2016)		●				
Japan	Matsuo et al. (2011)			●	●		
Norway	Thomas et al. (2007)	●	●	●			
Romania	Moldovan (2006)				●		
Serbia	Petrović et al. (2014)			●	●		
Slovenia	Isidori et al. (2016)	●	●	●			
Slovenia	Kosjek et al. (2013)	●	●				
Spain	Ferrando-Climent et al. (2013)	●	●	●			
Spain	Ferrando-Climent et al. (2014)	●	●				
Spain	Franquet-Griell et al. (2017)	●	●	●	●		
Spain	Gómez-Canela et al. (2014)				●		
Spain	Isidori et al. (2016)	●					
Spain	López-Serna et al. (2010)			●	●	●	
Spain	López-Serna et al. (2011)				●		

(continued)

Table 10.2 (continued)

Country	References	HWW	Influent	Effluent	SW	DW	SeaW
Spain	López-Serna et al. (2012a)				●		
Spain	López-Serna et al. (2012b)				●		
Spain	Martín et al. (2011)		●	●	●		
Spain	Martín et al. (2014)		●	●			
Spain	Mendoza et al. (2016)					●	
Spain	Negreira et al. (2013)		●				
Spain	Negreira et al. (2014)	●	●	●			
Spain	Valcárcel et al. (2011)				●		
Switzerland	Buerge et al. (2006)		●	●	●		
Switzerland	Kovalova (2009)	●					
Switzerland	Tauxe-Wuersch et al. (2005)	●	●				
Thailand	Usawanuwat et al. (2014)				●		
The Netherlands	Houtman et al. (2013)				●		
UK	Aherne et al. (1990)			●	●	●	
UK	Ashton et al. (2004)			●			
UK	Llewellyn et al. (2011)			●			
UK	Roberts and Thomas (2006)		●	●	●		●
UK	Thomas and Hilton (2004)						●
UK	Vyas et al. (2014)	●					
USA	Furlong et al. (2017)					●	
USA	Yu et al. (2012)		●	●			
Number of studies	56	23	22	31	21	5	3

administration mode. Table 10.4 compiles the values available in the literature and adopted by some research groups for 35 different substances.

Most cytostatic drugs are administered to outpatients in a hospital and more than 70% of the administered amount is expected to be excreted out of the hospitals. For example, Weissbrodt et al. (2009) found that only 5.5% of the expected excreted amount of 5-fluorouracil was recovered in the hospital sewer. Its half-life in plasma varies in the range of 5–20 min, whereas for gemcitabine, it is only 8 min. With regard to this compound, 5% of the total administered amount is excreted, unchanged in urine during the first 6 h after treatment and 60% of the administered substance is excreted as a metabolite within 24 h. The authors also reported that for 5-fluorouracil the duration of full excretion is around 6 h.

Table 10.3 The investigated compounds included in this overview and the corresponding class

Class	Compounds
Alkylating agents (A)	Carboxy-cyclophosphamide, Chlorambucil, Cyclophosphamide (CP), Ifosfamide, Keto-Cyclophosphamide (Keto-CP), N-Dechloroethylcyclophosphamide (N-declCP), Pt-based drugs (sum of Cisplatin, Carboplatin, Oxaliplatin)
Antimetabolites (B)	5-fluorouracil, 2′,2′-difluorodeoxyuridine (dFdU), Azathioprine, Capecitabine, Cytarabine, Gemcitabine, Hydroxycarbamide (Hydroxy-urea), Hydroxy-methotrexate, Methotrexate, Tegafur
Antitumor antibiotics (C)	Bleomycin, Ciprofloxacin, Doxorubicin, Epirubicin, *Mitomycin C,* Mycophenolic acid
Topoisomerase inhibitors (D)	Etoposide, Irinotecan
Mitotic inhibitors (E)	Docetaxel, Paclitaxel, OH-Paclitaxel (OH-PAC), Vincristine, Vinorelbine
Hormones (F)	Anastrozole, Bicalutamide, Endoxifen (OH-D-TAM), Letrozole, Megestrol, OH-tamoxifen, Tamoxifen,
Kinase inhibitors (G)	Erlotinib, *imatinib*
Total compounds	38

Table 10.4 Excretion rates of cytostatic drugs found in the literature

Class	Compound	References	Excretion rate [%]	Range
A	Carboplatin	Besse et al. (2012)	100	32–100
		Booker et al. (2014)	54	
		Kovalova (2009)	45–75	
		Rowney et al. (2009)	32–65	
		Weissbrodt et al. (2009)	70	
	Cisplatin	Booker et al. (2014)	33	25–45
		Franquet-Griell et al. (2015)	60	
		Kovalova (2009)	25–45	
		Rowney et al. (2009)	27–65	
		Weissbrodt et al. (2009)	40	
	Cyclophosphamide	Besse et al. (2012)	> 25	15–68
		Booker et al. (2014)	21	
		Kovalova (2009)	15–25	
		Kümmerer et al. (2016)	< 25	
		Ortiz de García et al. (2013)	15	
		Rowney et al. (2009)	30–68; < 20	
		Tauxe-Wuersch et al. (2005)	50	
		Usawanuwat et al. (2014)	25	
		Weissbrodt et al. (2009)	20 (urine)	

(continued)

Table 10.4 (continued)

Class	Compound	References	Excretion rate [%]	Range
	Ifosfamide	Besse et al. (2012)	50	12–90
		Booker et al. (2014)	26	
		Kovalova (2009)	15–45	
		Kümmerer et al. (2016)	< 50	
		Ortiz de García et al. (2013)	61	
		Tauxe-Wuersch et al. (2005)	12–90	
		Weissbrodt et al. (2009)	34	
	Oxaliplatin	Booker et al. (2014)	40	5–75
		Franquet-Griell et al. (2015)	50	
		Kovalova (2009)	45–75	
		Rowney et al. (2009)	5–50	
B	5-fluorouracil	Besse et al. (2012)	20	5–20
		Booker et al. (2014)	18	
		Kovalova (2009)	5–15	
		Kümmerer et al. (2016)	10	
		Rowney et al. (2009)	7–11	
		Tauxe-Wuersch et al. (2005)	< 20	
		Usawanuwat et al. (2014)	20	
		Weissbrodt et al. (2009)	10	
	Azathioprine	Franquet-Griell et al. (2015)	2	2
	Capecitabine	Besse et al. (2012)	3	0.5–84
		Booker et al. (2014)	3	
		Kümmerer et al. (2016)	84	
		Rowney et al. (2009)	7–11	
		Tauxe-Wuersch et al. (2005)	0.5	
	Cytarabine	Besse et al. (2012)	< 10	5–15
		Booker et al. (2014)	10	
		Kovalova (2009)	5–15	
	Gemcitabine	Besse et al. (2012)	10	< 5–15
		Booker et al. (2014)	8	
		Kovalova (2009)	5–15	
		Ortiz de García et al. (2013)	10	
		Rowney et al. (2009)	< 10 (urine); < 5 (feces)	
		Weissbrodt et al. (2009)	5	

(continued)

Table 10.4 (continued)

Class	Compound	References	Excretion rate [%]	Range
	Hydroxycarbamide (hydroxyurea)	Besse et al. (2012)	50	30–60
		Booker et al. (2014)	58	
		Tauxe-Wuersch et al. (2005)	30–60	
		Usawanuwat et al. (2014)	50	
	Hydroxy-methotrexate	Besse et al. (2012)	15	15
	Methotrexate	Besse et al. (2012)	90	50–90
		Booker et al. (2014)	83	
		Kovalova (2009)	< 75	
		Lienert et al. (2007)	81 (urine); 15 (feces)	
		Tauxe-Wuersch et al. (2005)	50–80	
		Weissbrodt et al. (2009)	81 (urine); 15 (feces)	
	Tegafur	Booker et al. (2014)	20	5–20
		Franquet-Griell et al. (2015)	5	
C	Bleomycin	Booker et al. (2014)	62	15–62
		Kovalova (2009)	15–25	
	Ciprofloxacin	Ortiz de García et al. (2013)	70	55–70
		Pal et al. (2010)	70	
		Ternes and Joss (2006)	55	
	Doxorubicin	Besse et al. (2012)	50	5–50
		Booker et al. (2014)	14	
		Kovalova (2009)	5–15	
		Rowney et al. (2009)	12–45	
		Weissbrodt et al. (2009)	10 (urine); 45 (feces)	
	Epirubicin	Booker et al. (2014)	11	< 5–20
		Kovalova (2009)	< 5	
		Rowney et al. (2009)	6–20	
	Mitomycin C	Besse et al. (2012)	10	10–25
		Booker et al. (2014)	10	
		Kovalova (2009)	15–25	
		Ortiz de García et al. (2013)	10	
	Mycophenolic acid	Franquet-Griell et al. (2015)	63	63
D	Etoposide	Besse et al. (2012)	93	25–93
		Booker et al. (2014)	43	
		Kovalova (2009)	25–45	

(continued)

Table 10.4 (continued)

Class	Compound	References	Excretion rate [%]	Range
		Weissbrodt et al. (2009)	50	
	Irinotecan	Besse et al. (2012)	> 50	15–> 50
		Booker et al. (2014)	16	
		Kovalova (2009)	15–25	
		Weissbrodt et al. (2009)	22 (urine); 33 (feces)	
E	Docetaxel	Besse et al. (2012)	< 8	< 5–75
		Booker et al. (2014)	7	
		Ferrando-Climent et al. (2014)	6 (urine); 75 (feces)	
		Franquet-Griell et al. (2015)	14	
		Kovalova (2009)	< 5	
	Paclitaxel	Besse et al. (2012)	18	< 5–18
		Booker et al. (2014)	7	
		Kovalova (2009)	< 5	
	Vincristine	Kovalova (2009)	25–45	25–45
	Vinorelbine	Besse et al. (2012)	11	11–25
		Booker et al. (2014)	13	
		Kovalova (2009)	15–25	
F	Anastrozole	Franquet-Griell et al. (2015)	10	10
	Bicalutamide	Besse et al. (2012)	55	55
	Letrozole	Franquet-Griell et al. (2015)	6	6
	Tamoxifen	Coetsier et al. (2009)	30	13–50
		Franquet-Griell et al. (2015)	13	
		Ortiz de García et al. (2013)	50	
		Tauxe-Wuersch et al. (2005)	20	
	Megestrol	Franquet-Griell et al. (2015)	78	78–92
		Ortiz de García et al. (2013)	92	
G	Erlotinib	Besse et al. (2012)	< 2	< 2–6
		Booker et al. (2014)	6	
	Imatinib	Besse et al. (2012)	25	9–25
		Booker et al. (2014)	9	

10.4 Occurrence of Cytostatics in Hospital Effluents

Great attention has been paid in recent years to the monitoring of hospital effluents as they were considered the major source of cytostatic drugs released into the environment, whereas in more recent years this fact has been questioned (Kümmerer et al. 2016). The first study included in this overview dates back 1990 and was carried out by the biochemistry department of the University of Surrey in United Kingdom, which sampled water from activated sludge effluents, rivers, and DW to analyze the occurrence of bleomycin (Aherne et al. 1990). In 1996, a research group of the Institute of Environmental Medicine and Hospital Epidemiology in Freiburg, Germany, proposed a method of detecting concentrations of ifosfamide and cyclophosphamide at a ppt level in a hospital effluent (Steger-Hartmann et al. 1996). In the following years, other research groups mainly from Switzerland, Austria, Spain, Japan, and Italy carried out monitoring investigations on hospital effluents. Their collected results are reported in Fig. 10.1 as a box-plot graph with the compounds reported according to their therapeutic classes (A–G). Sampling mode and frequency varied in the different studies and measured concentrations refer to both grab samples or composite samples. In total, 28 compounds were detected in HWWs in a wide range of concentrations: from 0.2 to 266,000 ng/L. The ranges of variability were 0.85–266,000 ng/L for alkylating

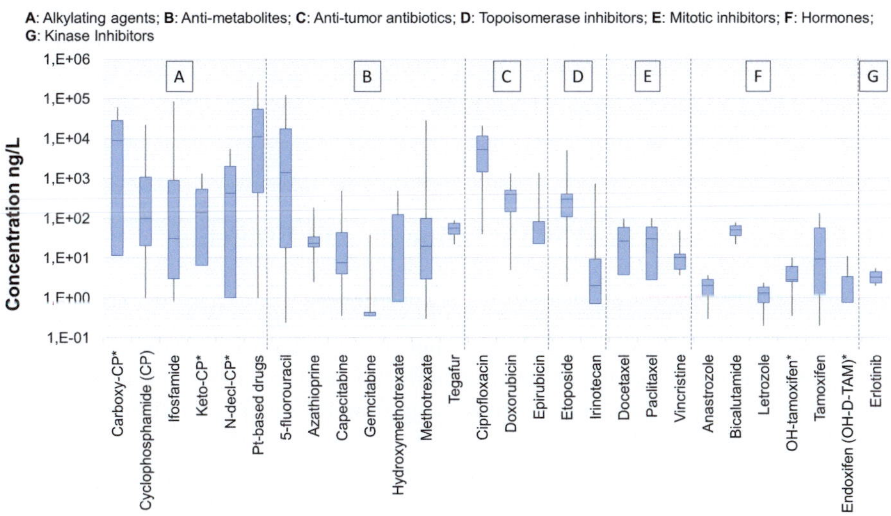

Fig. 10.1 Occurrence of cytostatic drugs in hospital effluents
Data from (Steger-Hartmann et al. 1996,1997; Kümmerer et al. 1997; Tauxe-Wuersch et al. 2005; Mahnik et al. 2004, 2006, 2007; Lenz et al. 2007; Catastini et al. 2008; Weissbrodt et al. 2009; Liu et al. 2010; Yin et al. 2010; Verlicchi et al. 2012; Ferrando-Climent et al. 2013, 2014; Kosjek et al. 2013; Negreira et al. 2014; Gómez-Canela et al. 2014; Vyas et al. 2014; Azuma et al. 2016; Isidori et al. 2016)

agents, 0.24–124,000 ng/L for anti-metabolites, 5–21,000 ng/L for antitumor antibiotics, 0.70–5000 ng/L for topoisomerase inhibitors, 2.75–99.70 ng/L for mitotic inhibitors, and 0.2–133.40 for hormones.

The most frequently investigated compounds were cyclophosphamide, ifosfamide, methotrexate, 5-fluorouracil, platinum-based drugs (cisplatin + carboplatin + oxaliplatin), tamoxifen, ciprofloxacin, and etoposide.

The highest concentrations in HWWs were found for platinum-based drugs (266,000 ng/L), 5-fluorouracil (124,000 ng/L), ifosfamide (86,200 ng/L), carboxy-cyclophosphamide (60,600 ng/L), methotrexate (28,700 ng/L), cyclophosphamide (22,100 ng/L), and ciprofloxacin (21,000 ng/L).

It is important to underline that micropollutant concentration may vary widely on a daily basis and also on a weekly basis. In planning sampling mode and frequency, these aspects must be taken into account. In this context, the profile provided by Weissbrodt et al. (2009) of the mass flow (i.e., concentration x flow rate) of 5-fluorouracil and gemcitabine in the hospital effluent versus daily hours and versus the days in a typical week may be quite useful. It was shown that peak mass load occurred at 1 pm for 5-fluorouracil and at 4 pm for gemcitabine. During the night, their mass flow was always very low, mainly due to very low concentrations. The authors also found differences in the measured mass flow over the observation periods (8–9 consecutive days), depending on the administration protocols and on the number of treated patients. Due to the pulse concentrations of these micropollutants, selection of the sampling mode (grab samples or composite samples and, in this case, flow-, time-, or proportional composite samples) and frequency (weekdays and weekend) is crucial in order to reduce the uncertainty associated with direct measurements. In this context, Weissbrodt et al. (2009) found that in their investigation, the overall uncertainty of the measured load was in the range of 74–123%.

10.5 Occurrence in WWTP Influents and Effluents

Many studies investigated the occurrence of cytostatic drugs in WWTP influents and effluents (Table 10.2), where grab or composite samples were analyzed. In total, 25 compounds were detected in the urban influents and 22 compounds in the treated effluent with a range of concentrations varying from 0.12 to 3700 ng/L for the urban influents and from < LOD to 144,000 ng/L for the effluents (Figs. 10.2 and 10.3).

The ranges in WWTP influents were 0.30–400 ng/L for alkylating agents, 0.25–366 ng/L for antimetabolites, 2.1–3700 ng/L for antitumor antibiotics, 0.6–83 ng/L for topoisomerase inhibitors, 0.13–219 ng/L for mitotic inhibitors, and 0.12–663 ng/L for hormones.

Figure 10.3 shows the range of variability observed in the effluent of different WWTPs, which was 0.10–144,000 ng/L for alkylating agents, 0.04–75.70 ng/L for antimetabolites, 0.05–1100 ng/L for antitumor antibiotics, 0.2–55 ng/L for topoisomerase inhibitors, 0.05–2014 ng/L for mitotic inhibitors, and 0.05–694 ng/L for

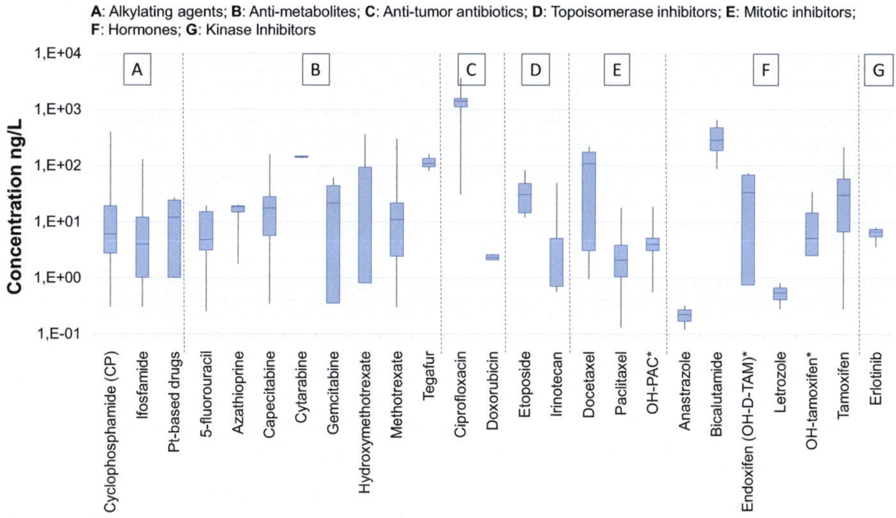

Fig. 10.2 Occurrence of cytostatic drugs in WWTP influents
Data from (Kümmerer et al. 1997; Steger-Hartmann et al. 1997; Tauxe-Wuersch et al. 2005; Buerge et al. 2006; Roberts and Thomas 2006; Catastini et al. 2008; Liu et al. 2010; Matsuo et al. 2011; Verlicchi et al. 2012; Ferrando-Climent et al. 2013, 2014; Kosjek et al. 2013; Martín et al. 2014; Negreira et al. 2013, 2014; Rabii et al. 2014; Azuma et al. 2016; Isidori et al. 2016)

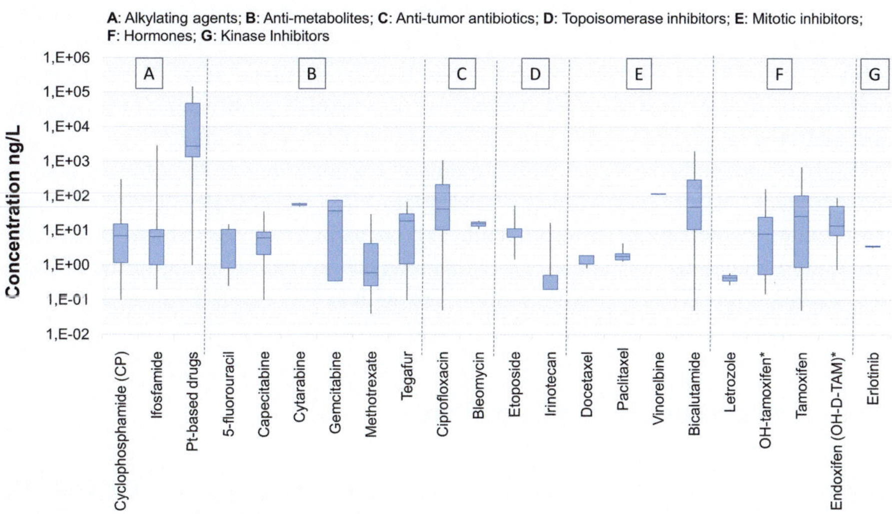

Fig. 10.3 Occurrence of cytostatic drugs in WWTP effluents
Data from (Aherne et al. 1990; Kümmerer et al. 1997; Steger-Hartmann et al. 1997; Ternes 1998; Ashton et al. 2004; Castiglioni et al. 2005; Zuccato et al. 2005; Buerge et al. 2006; Roberts and Thomas 2006; Lenz et al. 2007; Catastini et al. 2008; Coetsier et al. 2009; López-Serna et al. 2010; Liu et al. 2010; Llewellyn et al. 2011; Al Aukidy et al. 2012; Verlicchi et al. 2012; Ferrando-Climent et al., 2014; Martín et al. 2014; Rabii et al. 2014; Azuma et al. 2015, 2016; Isidori et al. 2016)

hormones. The most commonly studied compounds were capecitabine, ciprofloxacin, cyclophosphamide, ifosfamide, methotrexate, and tamoxifen.

Based on the collected data and considering 1000 ng/L as a threshold value, ciprofloxacin had the highest concentration in the urban influent (3700 ng/L), which was expected due to its wide use as an antibiotic, and, concerning the WWTP effluents, the highest concentrations were found for Pt-based drugs (144,000 ng/L), ifosfamide (2900 ng/L), bicalutamide (2010 ng/L), and ciprofloxacin (1100 ng/L).

The high concentrations of Pt-based drugs in the effluent are due to the fact that they also include values of the effluent of a specific treatment plant (an MBR) installed within a hospital in Austria, which is fed with wastewater from the oncological ward (Lenz et al. 2007). If we only consider municipal WWTPs, the concentration for platinum-based drugs was always found to be below the limit of detection (Isidori et al. 2016).

A comparison between HWW and WWTP influents highlights that 28 cytostatics were detected in both matrices, and for 19 substances median concentrations in HWW were higher than those found in WWTP influents. In particular, the ratio between median concentrations in HWW and the corresponding WWTP influents was in the range of 1–10 for 10 compounds, in the range of 11–100 for 4 compounds, and greater than 100 for 5 substances (5-fluorouracil, carboxy-CP, doxorubicin, N-decl-CP, and Pt-based substances). For 9 compounds, the median concentrations in HWWs were lower than the concentrations detected in WWTP influents.

With regard to WWTP influent and effluent concentrations, the comparison refers to 32 substances detected in both water compartments. The median concentration of 22 compounds was higher in the influent than in the effluent and the ratio between influent and effluent median concentrations was in the range of 1–10 for 15 compounds and greater than 10 for 4 substances: docetaxel, ciprofloxacin, azathioprine, and methotrexate.

Eight compounds exhibited higher median concentrations in the effluent than in the influent: cyclophosphamide, mitomycin C, anastrazole, OH-tamoxifen, ifosfamide, gemcitabine, 5-fluoroucail, doxorubicin, and vinrelbine. Platinum-based compounds were excluded from this analysis as data referring to the effluent also included the WWTP (MBR) effluent fed with HWW (Lenz et al. 2007).

10.6 Occurrence in Surface Water

Measured concentrations refer to grab samples with one exception—the study by Roberts and Thomas (2006), which refers to a final representative composite sample obtained as a mixture of samples withdrawn every 2 h over a 12-h period.

The occurrence and fate of cytostatic drugs released into SW have been investigated by different authors (Table 10.2). A limited number of compounds was detected and the observed variability ranges were < LOD–1907 ng/L for alkylating agents, 0.10–56 ng/L for antimetabolites, 0.05–224 ng/L for antitumor antibiotics, and 0.05–533 ng/L for hormones. Detected compounds with only one or two data are

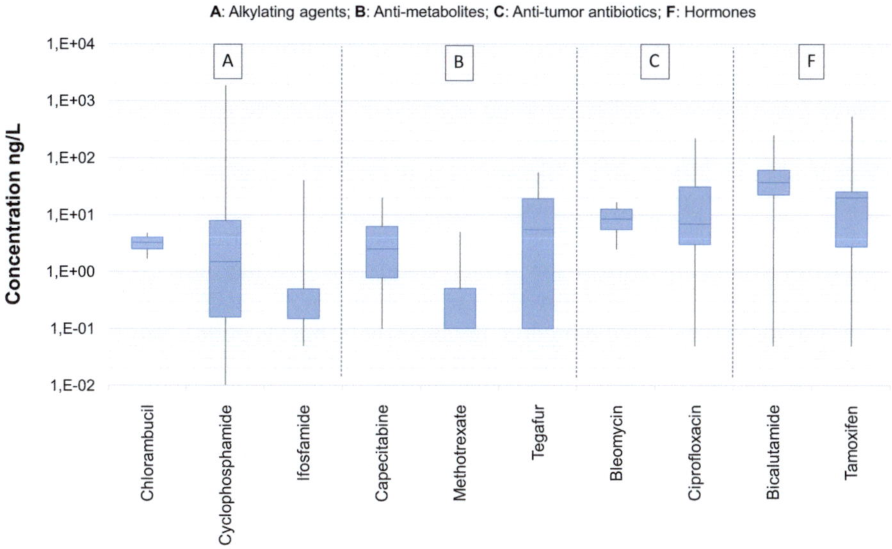

Fig. 10.4 Occurrence of cytostatic drugs in surface water
Data from (Aherne et al. 1990; Ternes 1998; Zuccato et al. 2000; Buerge et al. 2006; Moldovan 2006; Roberts and Thomas 2006; Coetsier et al. 2009; Valcárcel et al. 2011; López-Serna et al. 2010, 2011, 2012a, b; Houtman et al., 2013; Ferrando-Climent et al. 2014; Usawanuwat et al. 2014; Verlicchi et al. 2014; Azuma et al. 2015, 2016; Franquet-Griell et al. 2017)

not reported in Fig. 10.4 (this was the case of 5-fluorouracil (578 ng/L), hydroxycarbamide (788 ng/L), megestrol (6 ng/L), cytarabine (13 ng/L), gemcitabine (2.4 ng/L), and mycophenolic acid (8.6 and 656 ng/L) (Usawanuwat et al. 2014; Franquet-Griell et al. 2017; Martín et al. 2011).

As previously mentioned, WWTP effluents represent the main sources of cytostatic drugs in the environment, but in some cases farms and livestock activities might consistently contribute to river contamination. In this context, two anticancer drugs were detected in SW in Girona (Spain), *prior* to the local WWTP discharge point and downstream of a farm location (Ferrando-Climent et al. 2014). The drugs in question were ciprofloxacin, which is used as an antibiotic in farms and tamoxifen, also used for veterinary treatment, such as reproduction control or hormonal treatment.

10.7 Occurrence in Other Water Compartments

Very few studies have investigated the occurrence of cytostatic drugs in DW. Recently, Furlong et al. (2017) examined source and treated waters from 25 drinking-water treatment plants from across the United States. None of the selected cytostatic drugs (namely, methotrexate, prednisone, prednisolone, fadrozolef) were detected.

Mendoza et al. (2016) investigated the occurrence of 17 compounds in DW from the Madrid region (Central Spain). Their results show that all the selected cytostatic drugs were found at levels below their corresponding LOD. Similar results were obtained in previous studies referring to cyclophosphamide and ifosfamide in tap water in Italy (Zuccato et al. 2000), in France (Mompelat et al., 2011) and in Spain (Valcárcel et al. 2011, 2013). Bleomycin was detected in tap water at a maximum concentration of 13 ng/L in United Kingdom (Aherne et al. 1990).

Fabbri and Franzellitti (2016) reviewed the presence of pharmaceuticals in marine water from different parts of the world. Tamoxifen was the only cytostatic drug included in this review. Its observed range was between 27 and 212 ng/L in the Tyne river estuary in the United Kingdom (Roberts and Thomas 2006), between 13 and 71 ng/L in five estuaries from the United Kingdom (Thomas and Hilton 2004), and between 120 and 224 ng/L in the Yangtse estuary and coastal area in China (Yang et al. 2011).

With regard to groundwater, the concentration values of cytostatic drugs are not available in the literature.

10.8 Considerations About Water Sampling

It is important to underline that the measured concentrations are affected by uncertainties related to water sampling, sample conservation, and analytical methods (Weissbrodt et al. 2009; Kovalova 2009; Ort et al. 2010a, b; Verlicchi et al. 2014). In the case of compounds administered to outpatients in hospitals, a part of their excretion takes place once they have left the hospital and are at home. In the case of HWW, Weissbrodt et al. (2009) found that all the mass load of 5-fluorouracil and gemcitabine was between 11 a.m. and 10 p.m., whereas during the night the mass load was equal to 0. This leads to the suggestion of monitoring cytostatics in this daily period. In monitoring WWTP influents, the adopted sampling mode should "catch" the flushing containing the cytostatics excreted at home by the patients and, in this case, should also cover the night due to the travelling time necessary for the released drugs to reach the WWTP through the sewer network. With regard to SW, grab samples were generally analyzed and monitoring campaigns included different days of the week.

10.9 Predicted Environmental Concentrations

Another approach that can be used to estimate the concentrations of cytostatics in the different water compartments is based on the adoption of predictive models. They require knowledge of different parameters, in particular the annual consumption of the specific drug *Cons* (g/year); the excretion factor; the daily water demand per

$$PEC = \frac{Cons\ x\ E}{WW_{inh}\ x\ P\ x\ 365}\ x\ (1-R)\ x\ \frac{1}{D}$$

PEC_{inf} $(R = 0, D = 0)$

PEC_{eff} $(D = 0)$

PEC_{SW}

Fig. 10.5 Models used for the estimation of the predicted concentrations
PEC_{inf} predicted concentration in the influent, PEC_{eff} predicted concentration in the effluent, PEC_{SW} predicted concentration in surface water, *Cons* drug consumption (g/year), *P* population, WW_{inh} water demand per day per inhabitant (L/inh d), *E* excretion factor, *R* removal efficiency at the WWTP, *D* dilution factor *D*

inhabitant WW_{inh} (L/inh d); the removal efficiency *R* at the WWTP, and the dilution factor *D* once the treated effluent is released into the SW body.

Figure 10.5 represents the most adopted model in its complete expression (Verlicchi et al. 2014) in estimating the concentration in raw wastewater, treated effluents, and SW. Figure 10.5 shows the parameters that are necessary to estimate the concentration in the influent (PEC_{inf}), in the effluent (PEC_{eff}), and in surface water (PEC_{SW}).

As deduced from Eq. 10.1, PEC_{inf} depends on drug consumption, population, water demand, and excretion factor; PEC_{eff} also requires knowledge of the removal efficiency *R* at the WWTP and for PEC_{SW} it is necessary to know the dilution factor *D*, that is the ratio between effluent flow rate and surface water flow rate, following mixing. With regard to drug consumption, data is quite often available in terms of costs on an annual basis but not in terms of amounts or daily defined doses of the specific active principle. Moreover, it generally refers to the whole population in a country and not in a specific area or region.

Table 10.5 reports *PECs* for some of the studied compounds together with direct measured concentrations (*MECs*) in the investigated water environment.

It emerges that:

- Studies dealing with a comparison between the *PEC* and *MEC* of cytostatic drugs were carried out in a few countries: in Spain (the majority), Italy, France, United Kingdom and Thailand.
- The comparisons were more frequently evaluated for influent concentrations (15 cases), followed by surface water concentrations (12 cases) and only in three cases for effluent concentrations.
- Consumption of the same compound may differ from one country to another (e.g., tamoxifen) or even within the same country in different years (e.g., cyclophosphamide).
- The assumed value of excretion factor for a compound may vary from study to study. Some of them assumed total excretion ($E = 100\%$), others a value provided by previous studies.

Table 10.5 *PECs* of cytostatics included in this review and comparison with the corresponding *MEC* in the WWTP influent, effluent, and surface water (/ =data not available)

ATC	Compound	Country	*Specific Cons* mg/(inh year)	*E* %	*WW* L/(inh d)	PEC_{inf} ng/L	MEC_{inf} ng/L	*R* %	PEC_{eff} ng/L	MEC_{eff} ng/L	*D*	PEC_{SW} ng/L	MEC_{SW} ng/L	References
A	Chlorambucil	Spain	0.112	1	130	/	/	2	/	/	1.2	0.02	**1.7–4.8**	Franquet-Griell et al. (2017)
	Cyclophosphamide	Spain	1.714	100	103	**456**	< LOD-2.13		/	/	/	/	/	Martín et al. (2014)
	Cyclophosphamide	Spain	0.598	25	130	/	/	0	/	/	1.2	2.57	5–13.7	Franquet-Griell et al. (2017)
	Cyclophosphamide	Thailand	1.7e-5	25	250	/	/	0	/	/	140	5750	1907	Usawanuwat et al. (2014)
	Ifosfamide	Spain	1.30	100	103	34.5	12.3		/	/	/	/	/	Martin et al. (2014)
	Ifosfamide	Spain	0.997	50	130	/	/	0	/	/	1.2	8.48	10.1–13.9	Franquet-Griell et al. (2017)
B	5-fluorouracil	Spain	16.57	100	103	**4408**	< LOD-38.4		/	/	/	/	/	Martín et al. (2014)
	5-fluorouracil	Thailand	3.7e-5	20	250	/	/	0	/	/	140	**7890**	578	Usawanuwat et al. (2014)
	Capecitabine	Spain	78.07	11	130	/	/	15	/	/	1.2	**126**	< LOD-2.9	Franquet-Griell et al. (2017)
	Cytarabine	Spain	0.816	100	103	21.7	**151**		/	/	/	/	/	Martín et al. (2014)
	Gemcitabine	Spain	2.43	100	103	**646**	44.2		/	/	/	/	/	Martín et al. (2014)

	Methotrexate	Spain	0.557	100	103	14.8	21.8		/	/	/	/	/	Martín et al. (2014)
	Hydroxycarbamide	Thailand	2.26e-5	50	250	/	/	0	/	/	140	**3564**	788	Usawanuwat et al. (2014)
C	Doxorubicin	Spain	0.059	100	103	1.56	< LOD-4.15		/	/	/	/	/	Martín et al. (2014)
	Epirubicin	Spain	0.0929	100	103	2.47	< LOD-3.7		/	/	/	/	/	Martín et al. (2014)
	Mytomicin C	Spain	0.0343	100	103	0.91	< LOD-1.66		/	/	/	/	/	Martín et al. (2014)
	Ciprofloxacin	Italy	370	55	200	2780	2200	71	810	630	/	5.35	25	Verlicchi et al. (2014)
	Mycophenolic acid	Spain	313	63	130	/	/	41	/	/	1.2	**2018**	8.6–656	Franquet-Griell et al. (2017)
D	Etoposide	Spain	0.144	100	103	3.84	**30.5**	/	/	/	/	/	/	Martín et al. (2014)
	Irinotecan	Spain	0.124	100	103	3.31	< LOD-1.1	/	/	/	/	/	/	Martín et al. (2014)
E	Docetaxel	Spain	0.106	100	103	2.81	< LOD-1.89	/	/	/	/	/	/	Martín et al. (2014)
	Paclitaxel	Spain	0.291	100	103	7.75	< LOD-0.2	/	/	/	/	/	/	Martín et al. (2014)
	Vinorelbine	Spain	0.0443	100	103	1.18	< LOD-LOD5.18	/	/	/	/	/	/	Martín et al. (2014)
F	Tamoxifen	France	5.325	30	200	/	/	0	22	26.5 (med)	/	7 (med)	11 (med)	Coetsier et al. (2009)
	Tamoxifen	UK	/	100	150	/	/	0	/	/	10	63	< LOD-10	Ashton et al. (2004)
	Tamoxifen	Spain	7.3	13	130	/	/	93	/	/	1.2	1.13	0.2–25.1	Franquet-Griell et al. (2017)

- Water consumption varies from country to country: Table 10.5 reports values in the range of 100–250 L/inh d.
- With regard to removal efficiency, in some studies, no removal was assumed, even though literature data suggested other values.
- Dilution factor varies from 1.2 to 140.

In the influent, predicted concentrations for cyclophosphamide, 5-fluorouracil, and gemcitabine were found to be 1 or 2 orders of magnitude higher than the corresponding *MECs*, whereas for cytarabine and etoposide, *PECs* were 1 order lower than *MECs* (see characters in bold in Table 10.5).

With regard to effluent concentration, there was quite good agreement in both cases. Finally, in surface water, *PECs* were of the same order of magnitude in all cases with the exception of 5-fluorouracil, capecitabine, hydroxycarbamide, and mycophenolic acid, for which *PECs* were found to be 1–2 orders of magnitude higher than *MECs*.

As discussed in the literature, both approaches present strengths and weaknesses (see, for instance, Verlicchi et al. 2014; Verlicchi and Zambello 2016) leading to different levels of uncertainties.

Unless *MEC* values can be extrapolated to a longer period characterized by identical PhAC consumption patterns and environmental conditions (including flow rate, constant consumption and meteorological conditions, and constant removal efficiencies at the WWTP), they can only be considered valid for a particular observation period. However, to obtain annual data, monitoring campaigns would become even more complex and expensive. On the other hand, irrespective of the model used, *PEC* values extrapolated from annual data (annual values assumed for each variable in the commonly used models) can only be considered theoretical values. Predictive models might include terms accounting for possible degradation processes occurring in the receiving water body (mainly photocatalytic and biological reactions) or to "generation" mechanisms (mainly due to desorption of the PhAC or reactions among its metabolites leading to the parent compound itself).

In the case of new compounds or for those for which analytical techniques are not available or too expensive and time-consuming, predictions represent the only way to estimate concentrations.

References

Aherne GW, Hardcastle A, Nield AH (1990) Cytotoxic drugs and the aquatic environment: estimation of bleomycin in river and water samples. J Pharm Pharmacol 42(10):741–742

Al Aukidy M, Verlicchi P, Jelic A, Petrovic M, Barcelò D (2012) Monitoring release of pharmaceutical compounds: occurrence and environmental risk assessment of two WWTP effluents and their receiving rivers in the Po Valley, Italy. Sci Total Environ 438:15–25

Ashton D, Hilton M, Thomas KV (2004) Investigating the environmental transport of human pharmaceuticals to streams in the United Kingdom. Sci Total Environ 333:167–184

Azuma T, Ishiuchi H, Inoyama T, Teranishi Y, Yamaoka M, Sato T, Mino Y (2015) Occurrence and fate of selected anticancer, antimicrobial, and psychotropic pharmaceuticals in an urban river in a subcatchment of the Yodo river basin, Japan. Environ Sci Pollut Res 22(23):18676–18686

Azuma T, Arima N, Tsukada A, Hirami S, Matsuoka R, Moriwake R, Ishiuchi H, Inoyama T, Teranishi Y, Yamaoka M, Mino Y, Hayashi T, Fujita Y, Masada M (2016) Detection of pharmaceuticals and phytochemicals together with their metabolites in hospital effluents in Japan, and their contribution to sewage treatment plant influents. Sci Total Environ 548–549:189–197

Besse JP, Latour JF, Garric J (2012) Anticancer drugs in surface waters: what can we say about the occurrence and environmental significance of cytotoxic, cytostatic and endocrine therapy drugs? Environ Int 39:73–86

Booker V, Halsall C, Llewellyn N, Johnson A, Williams R (2014) Prioritising anticancer drugs for environmental monitoring and risk assessment purposes. Sci Total Environ 473–474:159–170

Buerge IJ, Buser H, Poiger T, Müller MD (2006) Occurrence and fate of the cytostatic drugs cyclophosphamide and ifosfamide in wastewater and surface waters. Environ Sci Technol 40(23):7242–7250

Busetti F, Linge KL, Heitz A (2009) Analysis of pharmaceuticals in indirect potable reuse systems using solid-phase extraction and liquid chromatography-tandem mass spectrometry. J Chromatogr A1216(31):5807–5818.

Castiglioni S, Bagnati R, Calamari D, Fanelli R, Zuccato E (2005) A multiresidue analytical method using solid-phase extraction and high-pressure liquid chromatography tandem mass spectrometry to measure pharmaceuticals of different therapeutic classes in urban wastewaters. J Chromatogr A 1092:206–215

Catastini C, Mullot J, Boukari S, Mazellier P, Levi Y, Cervantes P, Ormsby J (2008) Assessment of antineoplastic drugs in effluents of two hospitals. J Eur Hydrol 39(2):171–180

Coetsier CM, Spinelli S, Lin L, Roig B, Touraud E (2009) Discharge of pharmaceutical products (PPs) through a conventional biological sewage treatment plant: MECs vs PECs? Environ Int 35(5):787–792

Commission Implementing Decision (EU) 2018/840 of 5 June 2018 establishing a watch list of substances for Union-wide monitoring in the field of water policy pursuant to Directive 2008/105/EC of the European Parliament and of the Council and repealing Commission Implementing Decision (EU) 2015/495

Espinosa E, Zamora P, Feliu J, González Barón M (2003) Classification of anticancer drugs – a new system based on therapeutic targets. Cancer Treat Rev 29(6):515–523

Fabbri E, Franzellitti S (2016) Human pharmaceuticals in the marine environment: focus on exposure and biological effects in animal species. Environ Toxicol Chem 35(4):799–812

Ferrando-Climent L, Rodriguez-Mozaz S, Barceló D (2013) Development of a UPLC-MS/MS method for the determination of ten anticancer drugs in hospital and urban wastewaters, and its application for the screening of human metabolites assisted by information-dependent acquisition tool (IDA) in sewage samples. Anal Bioanal Chem 405(18):5937–5952

Ferrando-Climent L, Rodriguez-Mozaz S, Barceló D (2014) Incidence of anticancer drugs in an aquatic urban system: from hospital effluents through urban wastewater to natural environment. Environ Pollut 193:216–223

Franquet-Griell H, Gómez-Canela C, Ventura F, Lacorte S (2015) Predicting concentrations of cytostatic drugs in sewage effluents and surface waters of Catalonia (NE Spain). Environ Res 138:161–172

Franquet-Griell H, Cornadó D, Caixach J, Ventura F, Lacorte S (2017) Determination of cytostatic drugs in Besòs river (NE Spain) and comparison with predicted environmental concentrations. Environ Sci Pollut Res 24(7):6492–6503

Furlong ET, Batt AL, Glassmeyer ST, Noriega MC, Kolpin DW, Mash H, Schenck KM (2017) Nationwide reconnaissance of contaminants of emerging concern in source and treated drinking waters of the united states: pharmaceuticals. Sci Total Environ 579:1629–1642

Gómez-Canela C, Ventura F, Caixach J, Lacorte S (2014) Occurrence of cytostatic compounds in hospital effluents and wastewaters, determined by liquid chromatography coupled to high-resolution mass spectrometry. Anal Bioanal Chem 406(16):3801–3814

Houtman CJ, ten Broek R, de Jong K, Pieterse B, Kroesbergen J (2013) A multicomponent snapshot of pharmaceuticals and pesticides in the river Meuse basin. Environ Toxicol Chem 32(11):2449–2459

Isidori M, Lavorgna M, Russo C, Kundi M, Žegura B, Novak M, Filipič M, Mišík M, Knasmueller S, de Alda ML, Barceló D, Žonja B, Česen M, Ščančar J, Kosjek T, Heath E (2016) Chemical and toxicological characterisation of anticancer drugs in hospital and municipal wastewaters from Slovenia and Spain. Environ Pollut 219:275–287

Kosjek T, Perko S, Žigon D, Heath E (2013) Fluorouracil in the environment: analysis, occurrence, degradation and transformation. J Chromatogr A 1290:62–72

Kovalova L (2009) Cytostatics in the aquatic environment: analysis, occurrence, and possibilities for removal. PhD Thesis University of Aachen

Kümmerer K, Steger-Hartmann T, Meyer M (1997) Biodegradability of the anti-tumour agent ifosfamide and its occurrence in hospital effluents and communal sewage. Water Res 31(11):2705–2710

Kümmerer K, Haiß A, Schuster A, Hein A, Ebert I (2016) Antineoplastic compounds in the environment- substances of special concern. Environ Sci Pollut Res 223:14791–14804

Lenz K, Mahnik SN, Weissenbacher N, Mader RM, Krenn P, Hann S, Koellensperger G, Uhl M, Knasmüller S, Ferk F, Bursch W, Fuerhacker M (2007) Monitoring, removal and risk assessment of cytostatic drugs in hospital wastewater. Water Sci Technol 56(12):141–149

Lienert J, Bürki T, Escher BI (2007) Reducing micropollutants with source control: substance flow analysis of 212 pharmaceuticals in faeces and urine. Water Sci Technol 56(5):87–96

Liu X, Zhang J, Yin J, Duan H, Wu Y, Shao B (2010) Analysis of hormone antagonists in clinical and municipal wastewater by isotopic dilution liquid chromatography tandem mass spectrometry. Anal Bioanal Chem 396(8):2977–2985

Llewellyn N, Lloyd P, Jürgens MD, Johnson AC (2011) Determination of cyclophosphamide and ifosfamide in sewage effluent by stable isotope-dilution liquid chromatography-tandem mass spectrometry. J Chromatogr A 1218(47):8519–8528

López-Serna R, Pérez S, Ginebreda A, Petrović M, Barceló D (2010) Fully automated determination of 74 pharmaceuticals in environmental and waste waters by online solid phase extraction-liquid chromatography- electrospray-tandem mass spectrometry. Talanta 83(2):410–424

López-Serna R, Petrović M, Barceló D (2011) Development of a fast instrumental method for the analysis of pharmaceuticals in environmental and wastewaters based on ultra high performance liquid chromatography (UHPLC)-tandem mass spectrometry (MS/MS). Chemosphere 85(8):1390–1399

López-Serna R, Petrović M, Barceló D (2012a) Occurrence and distribution of multi-class pharmaceuticals and their active metabolites and transformation products in the Ebro river basin (NE spain). Sci Total Environ 440:280–289

López-Serna R, Postigo C, Blanco J, Pérez S, Ginebreda A, de Alda ML, Petrović M, Munné A, Barceló D (2012b) Assessing the effects of tertiary treated wastewater reuse on the presence emerging contaminants in a mediterranean river (llobregat, NE spain). Environ Sci Pollut Res 19(4):1000–1012

Mahnik SN, Rizovski B, Fuerhacker M, Mader RM (2004) Determination of 5-fluorouracil in hospital effluents. Anal Bioanal Chem 380(1):31–35

Mahnik SN, Rizovski B, Fuerhacker M, Mader RM (2006) Development of an analytical method for the determination of anthracyclines in hospital effluents. Chemosphere 65(8):1419–1425

Mahnik S, Lenz K, Weissenbacher N, Mader R, Fuerhacker M (2007) Fate of 5-fluorouracil, doxorubicin, epirubicin, and daunorubicin in hospital wastewater and their elimination by activated sludge and treatment in a membrane-bioreactor system. Chemosphere 6:30–37

Martín J, Camacho-Muñoz D, Santos JL, Aparicio I, Alonso E (2011) Simultaneous determination of a selected group of cytostatic drugs in water using high-performance liquid chromatography-triple-quadrupole mass spectrometry. J Sep Sci 34(22):3166–3177

Martín J, Camacho-Munõz D, Santos JL, Aparicio I, Alonso E (2014) Occurrence and ecotoxicological risk assessment of 14 cytostatic drugs in wastewater. Water Air Soil Pollut 225(3):1–10

Matsuo H, Sakamoto H, Arizono K, Shinohara R (2011) Behavior of pharmaceuticals in wastewater treatment plant in Japan. Bull Environ Contam Toxicol 87(1):31–35

Mendoza A, Zonja B, Mastroianni N, Negreira N, López de Alda M, Pérez S, Barceló D, Gil A, Valcárcel Y (2016) Drugs of abuse, cytostatic drugs and iodinated contrast media in tap water from the Madrid region (central Spain): a case study to analyse their occurrence and human health risk characterization. Environ Int 86:107–118

Moldovan Z (2006) Occurrences of pharmaceutical and personal care products as micropollutants in rivers from Romania. Chemosphere 64(11):1808–1817

Mompelat S, Thomas O, Le Bot B (2011) Contamination levels of human pharmaceutical compounds in French surface and drinking water. J Environ Monit 13(10):2929–2939

Negreira N, López de Alda M, Barceló D (2013) On-line solid phase extraction-liquid chromatography-tandem mass spectrometry for the determination of 17 cytostatics and metabolites in waste, surface and ground water samples. J Chromatogr A 1280:64–74

Negreira N, de Alda ML, Barcelo´ D (2014) Cytostatic drugs and metabolites in municipal and hospital wastewaters in Spain: filtration, occurrence, and environmental risk. Sci Total Environ 497–498:68–77

Ort C, Lawrence MG, Rieckermann J, Joss A (2010a) Sampling for pharmaceuticals and personal care products (PPCPs) and Illicit drugs in wastewater systems: are your conclusions valid? A critical review. Environ Sci Technol 44:6024–6035

Ort C, Lawrence MG, Reungoat J, Mueller JF (2010b) Sampling for PPCPs in wastewater systems: comparison of different sampling modes and optimization strategies. Environ Sci Technol 44(16):6289–6296

Ortiz de García S, Pinto Pinto G, GarcíaEncina P, Irusta Mata R (2013) Consumption and occurrence of pharmaceutical and personal care products in the aquatic environment in Spain. Sci Total Environ 444:451–465

Pal A, Gin K, Lin A, Reinhard M (2010) Impacts of emerging organic contaminants on freshwaterresources: review of recent occurrences, sources, fate and effects. Sci Total Environ 408:6062–6069

Petrović M, Škrbić B, Živančev J, Ferrando-Climent L, Barcelo D (2014) Determination of 81 pharmaceutical drugs by high performance liquid chromatography coupled to mass spectrometry with hybrid triple quadrupole-linear ion trap in different types of water in Serbia. Sci Total Environ 468–469:415–428

Rabii FW, Segura PA, Fayad PB, Sauvé S (2014) Determination of six chemotherapeutic agents in municipal wastewater using online solid-phase extraction coupled to liquid chromatography-tandem mass spectrometry. Sci Total Environ 487(1):792–800

Roberts PH, Thomas KV (2006) The occurrence of selected pharmaceuticals in wastewater effluent and surface waters of the lower Tyne catchment. Sci Total Environ 356(1–3):143–153

Rowney NC, Johnson AC, Williams RJ (2009) Cytotoxic drugs in drinking water: a prediction and risk assessment exercise for the Thames catchment in the United Kingdom. Environ Toxicol Chem 28:2733–2743

Steger-Hartmann T, Kümmerer K, Schecker J (1996) Trace analysis of the antineoplastics ifosfamide and cyclophosphamide in sewage water by two-step solid-phase extraction and gas chromatography-mass spectrometry. J Chromatogr A 726(1–2):179–184

Steger-Hartmann T, Kümmerer K, Hartmann A (1997) Biological degradation of cyclophosphamide and its occurrence in sewage water. Ecotoxicol Environ Saf 36(2):174–179

Tauxe-Wuersch A, De Alencastro LF, Grandjean D, Tarradellas J (2005) Occurrence of several acidic drugs in sewage treatment plants in Switzerland and risk assessment. Water Res 39:1761–1772

Ternes TA (1998) Occurrence of drugs in German sewage treatment plants and rivers. Water Res 32:3245–3260

Ternes TA, Joss A (2006) Human Pharmaceuticals, hormones and fragrances, The challenge of micropollutants in urban water management. IWA Publishing, London

Thomas KV, Hilton MJ (2004) The occurrence of selected human pharmaceutical compounds in UK estuaries. Mar Pollut Bull 49:436–444

Thomas KV, Dye C, Schlabach M, Langford KH (2007) Source to sink tracking of selected human pharmaceuticals from two Oslo city hospitals and a wastewater treatment works. J Environ Monit 9(12):1410–1418

Usawanuwat J, Boontanon N, Kitpati Boontanon S (2014) Analysis of three anticancer drugs (5-Fluorouracil, cyclophosphamide and hydroxyurea) in water samples by HPLC-MS/MS. Int J Adv Agric Environ Eng 1:72–76

Valcárcel Y, González Alonso S, Rodríguez-Gil JL, Gil A, Catalá M (2011) Detection of pharmaceutically active compounds in the rivers and tap water of the Madrid region (Spain) and potential ecotoxicological risk. Chemosphere 84(10):1336–1348

Valcárcel Y, Alonso SG, Rodríguez-Gil JL, Castaño A, Montero JC, Criado-Alvarez JJ, Mirón IJ, Catalá M (2013) Seasonal variation of pharmaceutically active compounds in surface (Tagus river) and tap water (central Spain). Environ Sci Pollut Res 20(3):1396–1412

Verlicchi P, Zambello E (2016) Predicted and measured concentrations of pharmaceuticals in hospital effluents. Examination of the strengths and weaknesses of the two approaches through the analysis of a case study. Sci Total Environ 565:82–94. https://doi.org/10.1016/j.SciTotEnviron2016.04.165

Verlicchi P, Al Aukidy M, Galletti A, Petrovic M, Barceló D (2012) Hospital effluent: investigation of the concentrations and distribution of pharmaceuticals and environmental risk assessment. Sci Total Environ 430:109–118

Verlicchi P, Al Aukidy M, Jelic A, Petrovic M, Barceló D (2014) Comparison of measured and predicted concentrations of selected pharmaceuticals in wastewater and surface water: a case study of a catchment area in the Po Valley (Italy). Sci Total Environ 470–471:844–854

Vyas N, Turner A, Sewell G (2014) Platinum-based anticancer drugs in waste waters of a major UK hospital and predicted concentrations in recipient surface waters. Sci Total Environ 493:324–329

Weissbrodt D, Kovalova L, Ort C, Pazhepurackel V, Moser R, Hollender J, Siegrist H, Mcardell CS (2009) Mass flows of x-ray contrast media and cytostatics in hospital wastewater. Environ Sci Technol 43(13):4810–4817

Yang Y, Fu J, Peng H, Hou L, Liu M, Zhou J (2011) Occurrence and phase distribution of selected pharmaceuticals in the Yangtze Estuary and its coastal zone. J Hazard Mater 190:588–596

Yin J, Shao B, Zhang J, Li K (2010) A preliminary study on the occurrence of cytostatic drugs in hospital effluents in Beijing, china. Bull Environ Contam Toxicol 84(1):39–45

Yu JT, Bisceglia KJ, Bouwer EJ, Roberts AL, Coelhan M (2012) Determination of pharmaceuticals and antiseptics in water by solid-phase extraction and gas chromatography/mass spectrometry: analysis via pentafluorobenzylation and stable isotope dilution. Anal Bioanal Chem 403(2):583–591

Zuccato E, Calamari D, Natangelo M, Fanelli R (2000) Presence of therapeutic drugs in the environment. Lancet 355(9217):1789–1790

Zuccato E, Castiglioni S, Fanelli R (2005) Identification of the pharmaceuticals for human use contaminating the Italian aquatic environment. J Hazard Mater 122:205–209

Chapter 11
Photodegradation of Cytostatic Drugs in Low-Pressure UV Photoreactor Through Direct and Indirect Pathways

Yiqing Zhang and Teik-Thye Lim

Abstract Various advanced technologies have been suggested to remove the biorefractive and recalcitrant cytostatic drugs (CSDs) in aquatic environment, such as membrane biofiltration, electrolysis, ozonation, and UV-based technologies. This chapter focuses on the photodegradation of CSDs in low-pressure UV photoreactor through direct and indirect pathways. UV photolysis was applied to investigate the degradation effect of eight CSDs, including azathioprine (AZA), cyclophosphamide (CP), cytarabine (CYT), doxorubicin (DOX), methotrexate (MET), 5-fluorouracil (5FU), flutamide (FLU), and mitotane (MIT). Six CSDs were degraded by less than 10% using UV dose of 400 mJ·cm^{-2}, indicating the ineffectiveness of UV photolysis toward most of the studied CSDs. UV/H_2O_2 and UV/persulfate (PS) processes were then evaluated to compare the removal of CSDs. With the addition of H_2O_2 or PS, the degradation rates of CSDs increased significantly. The main degradation pathways of AZA were proposed based on its identified intermediates. The operating costs of different processes for AZA degradation were also calculated.

Keywords UV photolysis · UV/H_2O_2 · UV/persulfate · Degradation pathway

11.1 Overview of Advanced Wastewater Treatment Technologies for Cytostatic Drugs

A large number of cytostatic drugs (CSDs) have been found at ng·L^{-1} to μg·L^{-1} levels in surface water bodies worldwide, exhibiting potential threats to both human health and aquatic ecosystem. Conventional wastewater treatment processes may exhibit low removal efficiency toward CSDs (Kosjek and Heath 2011). Therefore, developing effective technologies to remove these biorefractive and recalcitrant pharmaceuticals from contaminated water is urgently required. Advanced

Y. Zhang · T.-T. Lim (✉)
School of Civil and Environmental Engineering, Nanyang Technological University, Singapore, Singapore
e-mail: cttlim@ntu.edu.sg

E. Heath et al. (eds.), *Fate and Effects of Anticancer Drugs in the Environment*,
https://doi.org/10.1007/978-3-030-21048-9_11

wastewater treatment technologies have been investigated for CSDs removal as shown in Table 11.1, such as membrane bioreactor (MBR), electrolysis, ozonation, and UV-based technologies (Zhang et al. 2013).

MBR technology combines membrane filtration with a conventional biological-activated sludge process as a unified system for secondary wastewater treatment. Mahnik et al. (2007) indicated that fluorouracil, doxorubicin, epirubicin, and daunorubicin in a hospital effluent could be removed effectively using a pilot-scale MBR system through biodegradation and/or adsorption. However, MBR technology becomes inefficient toward less biodegradable CSDs. For example, Kovalova et al. (2012) reported a low removal rate of merely 20% for cyclophosphamide degradation in hospital effluent using MBR technology.

Due to the generation of hypochlorite from sodium chloride, electrolysis has been applied to inactivate microorganisms and degrade chemicals. Thus, the cytotoxicity, mutagenicity, and antibacterial activity of epirubicin hydrochloride in hospital wastewater could be removed completely within 6 h by electrolysis process (Hirose et al. 2005). However, human carcinogens such as trihalomethanes could be accumulated during the electrolysis process (Pérez et al. 2010).

Ozonation process includes both direct and indirect reactions with the compounds present in a certain aquatic environment. Direct ozonation process can degrade organic compounds efficiently by reacting with their unsaturated bonds. Somensi et al. (2012) indicated that direct ozonation process could degrade methotrexate and doxorubicin, with the second-order rate constants of 0.0267 and 0.3373 min^{-1}, respectively. Garcia-Ac et al. (2010) also demonstrated that direct ozonation process could remove methotrexate from contaminated water efficiently with second-order rate constant of more than $3.6 \times 10^3\ M^{-1} \cdot s^{-1}$, while higher ozone dose and longer reaction time were required for cyclophosphamide degradation with second-order rate constant of $3.3 \pm 0.2\ M^{-1} \cdot s^{-1}$. Recently, indirect ozonation processes have been investigated due to the highly effective radicals generated. Free $HO^{\bullet}$ can be generated through the process and react with compounds effectively, but ozonation is highly affected by the pH and NOM of the water matrix. In addition, the disinfection by-products can also be generated during ozonation process, such as bromate, haloacetaldehydes, and halonitromethanes (Li and Mitch 2018).

11.2 UV Photolysis

UV photolysis using electromagnetic radiation with wavelengths of 10–400 nm has been recognized as a cost-effective water treatment process for pathogen disinfection, by, e.g., damaging their nucleic acids and disabling their cellular functions. As a chemical-free technology, UV technology shows environmental-friendliness when compared with other chemical-based disinfection technologies. In addition, UV irradiation does not influence the aesthetic quality (e.g., taste and odor) of the treated water, as compared with other treatment processes such as chlorination or ozonation

Table 11.1 Removal of cytostatic drugs using advanced wastewater treatment technologies

Methods	Target CSDs	Matrices	Water treatment/ mechanism	References
MBR	Fluorouracil, doxorubicin, epirubicin, daunorubicin, cisplatin, carboplatin, cyclophosphamide	Oncologic wastewater, domestic wastewater, hospital wastewater	Combination of biological treatment and membrane filtration	Mahnik et al. (2007), Lenz et al. (2007), Delgado et al. (2011) and Kovalova et al. (2012)
Electrolysis	Epirubicin, irinotecan, vincristine, mitomycin C, paclitaxel, methotrexate, cisplatin	Clinic wastewater, urine	Generation of hypochlorite	Hirose et al. (2005) and Kobayashi et al. (2012)
Ozonation	Cyclophosphamide, irinotecan, tamoxifen, methotrexate, doxorubicin	Pure water, biologically treated water, drinking water	Selective reaction with unsaturated bonds	Kim et al. (2009), Garcia-Ac et al. (2010) and Somensi et al. (2012)
UV photolysis	Azathioprine, cyclophosphamide, cytarabine, doxorubicin, fluorouracil, flutamide, methotrexate, mitotane	DI water	Homolytic bond cleavage	Zhang et al. (2017), Miolo et al. (2011) and Lutterbeck et al. (2016)
UV/H_2O_2	Cyclophosphamide	MilliQ water, tap water, pre-treated water	Hydrogen abstraction, electron transfer, and electrophilic addition	Wols et al. (2013) and Lutterbeck et al. (2015)
$UV/H_2O_2/O_3$	Cyclophosphamide, ifosfamide	DI water, hospital wastewater	Hydrogen abstraction, electron transfer, and electrophilic addition	Lester et al. (2011) and Česen et al. (2015)
$UV/Fe^{2+}/H_2O_2$	Cyclophosphamide	Ultrapure water	Hydrogen abstraction, electron transfer, and electrophilic addition	Lutterbeck et al. (2015)
UV/TiO_2	Cyclophosphamide, fluorouracil	MilliQ water	Hydrogen abstraction, electron transfer, and electrophilic addition	Lin and Lin (2014) and Lutterbeck et al. (2015)
UV/ persulfate	Azathioprine	DI water	Electron transfer	Zhang et al. (2016)

(Oppenländer 2003). UV technology has also been applied as an effective purification method for certain organic compound degradation in water.

11.2.1 Kinetics Determination

The photolysis of CSDs can be described using a pseudo-first-order kinetic model:

$$\ln(C_0/C) = k_t \cdot t \tag{11.1}$$

where C_0 and C are the CSDs concentrations before and after UV photolysis treatment ($\mu g \cdot L^{-1}$), k_t is the time-based pseudo-first-order rate constant for direct UV photolysis (min^{-1}), and t is the reaction time (min).

The photolysis rates of CSDs are significantly influenced by the molar absorption coefficient and quantum yield. Molar absorption coefficient (ε_λ, $M^{-1} \cdot cm^{-1}$) measures the probability that a chemical absorbs photon of λ wavelength, while quantum yield (Φ_λ, $mol \cdot E^{-1}$) indicates the ratio of photons that destroy the chemical normalized by the amount of photons absorbed.

Zhang et al. (2017) compared the UV photolysis effect of eight frequently used CSDs, including azathioprine (AZA), cyclophosphamide (CP), cytarabine (CYT), doxorubicin (DOX), fluorouracil (5FU), flutamide (FLU), methotrexate (MET), and mitotane (MIT). The physicochemical properties are summarized in Table 11.2. The direct photolytic rates of the eight CSDs followed the order of 5FU > DOX > AZA > MET > CYT > MIT ≈ FLU > CP. Their degradation rates varied significantly from 0.31×10^{-2} min^{-1} (CP) to 3.32×10^{-2} min^{-1} (5FU). Removal rates R (%) of the CSDs by direct UV photolysis were also calculated. According to Table 11.2, ε_{254} of the eight CSDs vary significantly from 7 to 10,037 $M^{-1} \cdot cm^{-1}$. The insignificant removal of CP and MIT can be explained by their extremely low ε_{254}, which are 7 and 300 $M^{-1} \cdot cm^{-1}$, respectively. In addition, the negligible degradability of FLU can be ascribed to its low Φ_{254} of 0.19×10^{-3} $mol \cdot E^{-1}$. Applying a UV dose of 400 $mJ \cdot cm^{-2}$ (ten times higher than the typically used UV dose for disinfection), only DOX and 5FU was degraded by more than 10%, while the removal of the other six CSDs was negligible. This indicates that most CSDs are resistant toward direct UV photolysis.

11.2.2 Proposed Pathways

Table 11.3 shows the hypothetical photolytic degradation pathways of the CSDs based on their detected by-products. It can be concluded that most of the studied chemicals are degraded by aromatic ring cleavage under direct UV photolysis (Zhang et al. 2017).

Table 11.2 Physicochemical properties of eight CSDs (Zhang et al. 2017)

Compound	Molecule	MW ($g \cdot mol^{-1}$)	Structure	$k_t \times 10^2$ (min^{-1})[a]	R (%) (400 $mJ \cdot cm^{-2}$)	ε_{254} ($M^{-1} \cdot cm^{-1}$)	$\Phi_{254} \times 10^3$ ($mol \cdot E^{-1}$)[a]
Azathioprine (AZA)	$C_9H_7N_7O_2S$	277	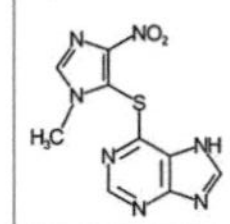	3.32 (±0.27)	8.10	5369	0.80 (±0.07)
Cyclophosphamide (CP)	$C_7H_{15}Cl_2N_2O_2P$	260	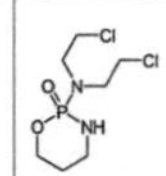	0.31 (±0.03)	0.79	7	54.83 (±5.31)
Cytarabine (CYT)	$C_9H_{13}N_3O_5$	243	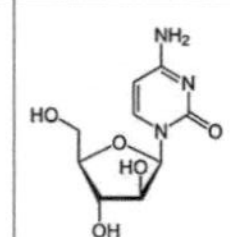	1.78 (±0.20)	4.43	4337	0.53 (±0.06)
Doxorubicin (DOX)	$C_{27}H_{29}NO_{11}$	544	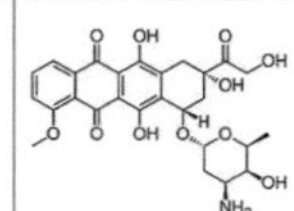	7.51 (±0.45)	17.40	10,037	0.97 (±0.06)
Methotrexate (MET)	$C_{20}H_{22}N_8O_5$	454	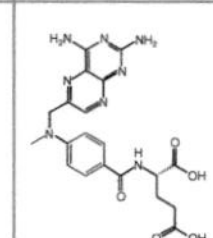	1.95 (±0.21)	4.84	1695	1.59 (±0.16)
Fluorouracil (5FU)	$C_4H_3FN_2O_2$	130	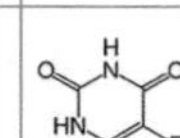	10.9 (±0.37)	24.32	3506	4.06 (±0.14)

(continued)

Table 11.2 (continued)

Compound	Molecule	MW ($g \cdot mol^{-1}$)	Structure	$k_t \times 10^2$ (min^{-1})[a]	R (%) ($400\ mJ \cdot cm^{-2}$)	ε_{254} ($M^{-1} \cdot cm^{-1}$)	$\Phi_{254} \times 10^3$ ($mol \cdot E^{-1}$)[a]
Flutamide (FLU)	$C_{11}H_{11}F_3N_2O_3$	276	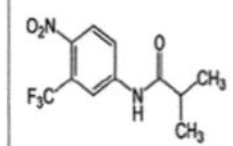	0.47 (±0.02)	1.19	3244	0.19 (±0.008)
Mitotane (MIT)	$C_{14}H_{10}Cl_4$	322	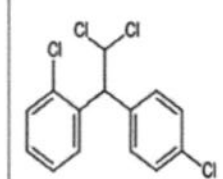	0.50 (±0.03)	1.26	300	2.17 (±0.13)

[a]Presented in mean values ± standard deviation

Table 11.3 Proposed degradation pathways of the CSDs during UV photolysis (Zhang et al. 2017)

Compound	Proposed degradation pathway
AZA	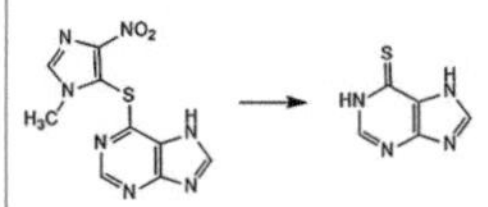
CP	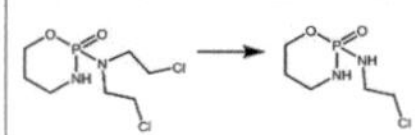
CYT	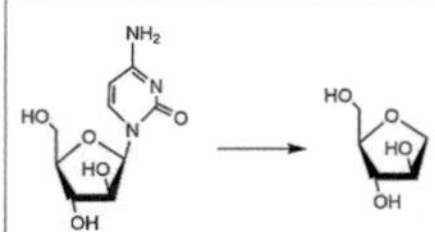
DOX	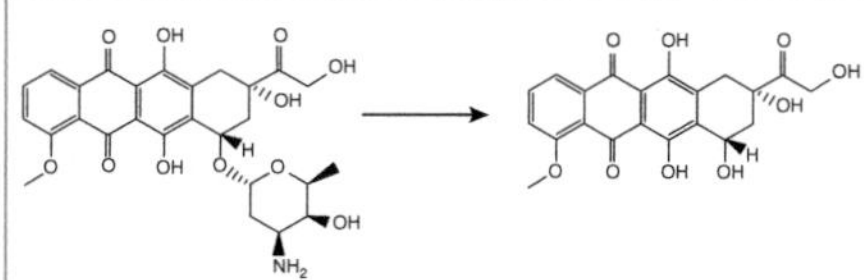
MET	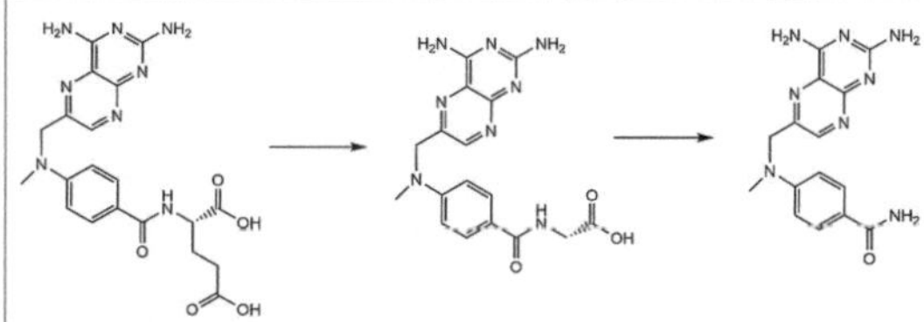
5FU*	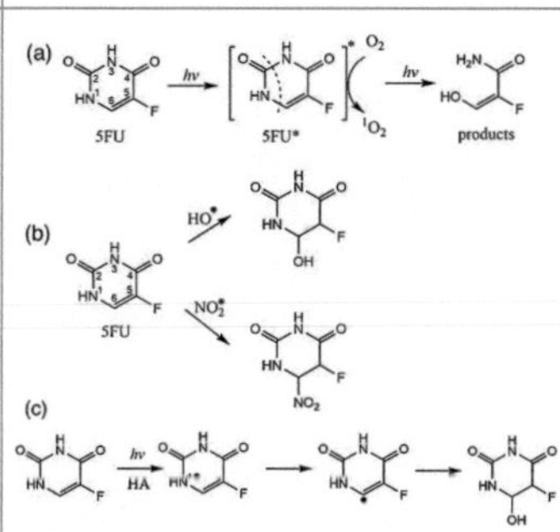
FLU	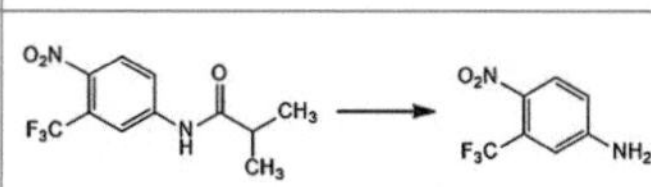
MIT	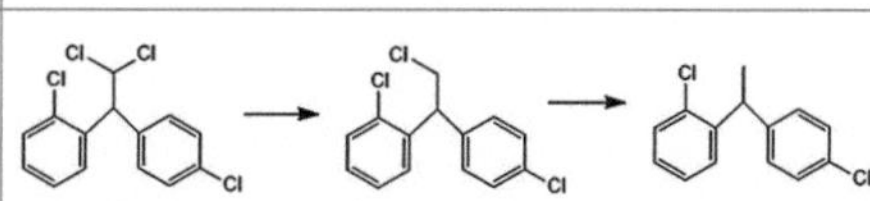

*(a) direct UV photolysis, (b) NO_3^--induced irradiation and (c) DOM-induced irradiation

The indirect photolytic degradation pathways with the addition of important natural water components NO_3^- or dissolved organic matter (DOM) were also investigated. Various radical species, such as $HO^\bullet$ and $NO_2^\bullet$, can be formed in the presence of NO_3^- according to Eqs. (11.2, 11.3, 11.4, 11.5 and 11.6). The single state ($^1DOM^*$) and triplet state ($^3DOM^*$) of sensitizer molecule (DOM) can be converted from its ground state during the irradiation according to Eq. (11.7). Singlet molecular oxygen ($^1O_2$) can be formed from O_2 by the deactivation of reactive species $^3DOM^*$ Eq. (11.8). As a high-energy form of O_2, $^1O_2$ can react efficiently with various organic chemicals:

$$NO_3^- + h\nu \rightarrow NO_2^- + O \tag{11.2}$$

$$NO_3^- + h\nu \rightarrow O^{\bullet -} + NO_2^\bullet \tag{11.3}$$

$$2NO_2^\bullet + H_2O \rightarrow NO_2^- + NO_3^- + 2H^+ \tag{11.4}$$

$$O + H_2O \rightarrow 2HO^\bullet \tag{11.5}$$

$$O^{\bullet -} + H_2O \rightarrow HO^\bullet + HO^- \tag{11.6}$$

$$DOM + h\nu \rightarrow {}^1DOM^* \rightarrow {}^3DOM^* \tag{11.7}$$

$${}^3DOM^* + O_2 \rightarrow DOM + {}^1O_2 \tag{11.8}$$

The direct and indirect photolysis pathways of 5FU were compared, due to its high photolytic degradation efficiency. In the direct UV photolysis, the reaction opens the nucleophilic ring of 5FU. The carbon-centered radical $5FU^*$ and the corresponding final product are formed as a result of bond cleavage. However, in the NO_3^--induced photolysis, an additional hydroxylation step occurs on the unsaturated site of 5FU. $NO_2^\bullet$ can also react with 5FU to form nitrated derivative compound. In addition, in the DOM-induced photolysis, a charge-transfer interaction takes place between excited HA and the nitrogen atom of 5FU, followed by the transition of hydrogen atom to form a carbon radical intermediate, and the addition of HO group.

11.3 UV-Based AOPs

11.3.1 Degradation Effect

Considering the low removal rate of CSDs (Sect. 11.2.1), UV photolysis is mostly insufficient to be an effective removal technology. Therefore, UV-based advanced oxidation processes, including UV/H_2O_2 and UV/persulfate (PS, $S_2O_8^{2-}$), were applied to remove CSDs. UV/H_2O_2 has been widely studied in micropollutant photodegradation and microorganism inactivation due to the formation of highly reactive species $HO^\bullet$ Eq. (11.9). It can react with various pollutants through e^- transfer, H abstraction, or electrophilic addition reactions with non-selectivity (Baxendale and Wilson 1957).

$$H - O - O - H + h\nu \rightarrow 2HO^{\bullet} \; \left(\Phi = 1.0 \; \text{mol} \cdot E^{-1}\right) \tag{11.9}$$

UV/PS has also been investigated recently in pharmaceutical degradation due to the generation of $SO_4^{\bullet-}$ Eq. (11.10). The degradation efficiency of UV/PS process seems to be higher than that of UV/H_2O_2 for chemicals such as azathioprine (Zhang et al. 2016).

$$^{-}SO_3 - O - O - SO_3^{-} + h\nu \rightarrow 2SO_4^{\bullet-} \; \left(\Phi = 1.4 - 1.8 \; \text{mol} \cdot E^{-1}\right) \tag{11.10}$$

As shown in Fig. 11.1, the overall degradation rates of CSDs increase with addition of 0.1 mM H_2O_2 or PS, indicating the significant contribution of the oxidant. For four CSDs, including AZA, CYT, DOX, and MET, their degradation rates in UV/PS process are much higher than that in UV/H_2O_2 process. The reason may be due to the electron-donor functional groups contained in these chemicals. $SO_4^{\bullet-}$ can attack the electron-donor groups of CSDs selectively via electron transfer. In contrast, the degradation efficiency of CP in UV/PS process is lower than that in UV/H_2O_2 process. The $^{+}NR_3$ and -Cl groups in CP are both electron withdrawing, leading to the slow reaction between CP and $SO_4^{\bullet-}$. The degradation rates of other three CSDs (5FU, FLU, and MIT) in the two UV-AOPs systems are comparable, due to their mixture composition of electron-donor groups and electron-withdrawing groups.

11.3.2 Proposed Pathways

Due to its high photoresistance and also toxicity, AZA was chosen as a model but representative CSD for the mechanistic and economic comparison of UV/H_2O_2 and UV/PS processes. The degradation pathways of AZA in the two UV-AOP processes were proposed according to its identified by-products (Fig. 11.2a). The degradation product $C_9H_7N_7O_3S$ (m/z 294) of AZA as detected in both UV-AOP systems

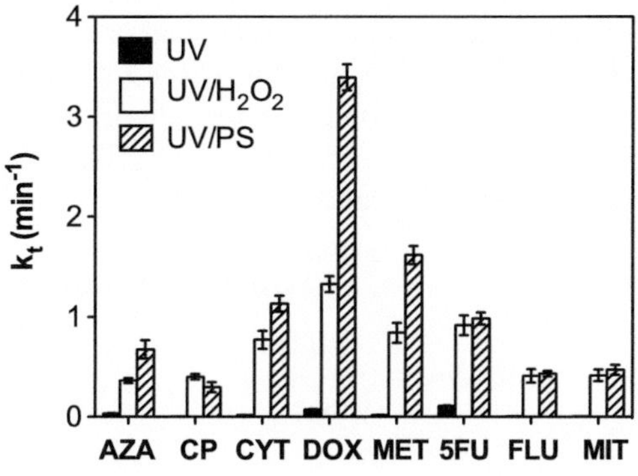

Fig. 11.1 First-order rate constants comparison of various CSDs by UV, UV/H_2O_2, and UV/PS processes

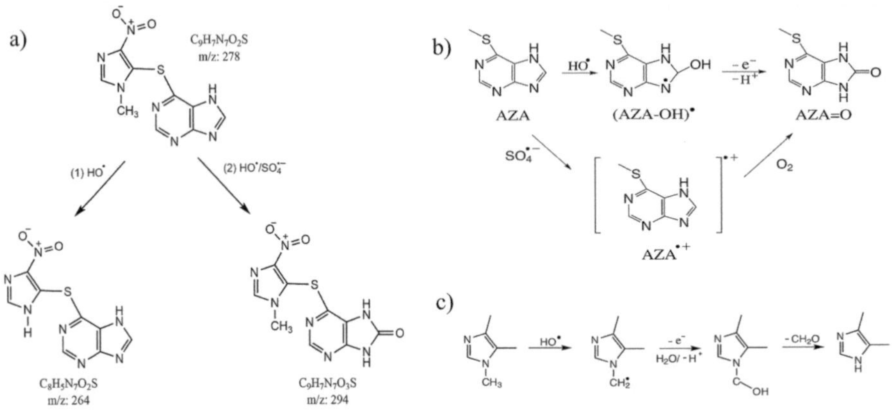

Fig. 11.2 (**a**) Proposed degradation pathways of AZA, and mechanism of the (**b**) hydroxylation and (**c**) demethylation

indicates a possible pathway of hydroxylation (+16 Da). During a HO•-induced oxidation, carbon atom at the purine ring is typically the major target, leading to the generation of hydroxyl adduct (AZA-OH)• and AZA=O. In the UV/PS system, electron transfer is the main mechanism. A radical cation AZA•+ is converted from AZA, followed by the deprotonation and oxidization to form AZA=O (Fig. 11.2b).

The degradation product $C_8H_5N_7O_2S$ (m/z 264) of AZA identified only in UV/H_2O_2 system suggests another possible degradation pathway of demethylation (−14 Da). As shown in Fig. 11.2c, a carbon-centered radical is initially generated by hydrogen abstraction of the methyl chain. After electron release and hydrolysis, the carbon-centered radical can be converted into an AZA hydroxymethyl intermediate compound, followed by the demethylation of the compound (Zhang et al. 2016).

11.3.3 Economic Comparison

To compare the cost-efficiency of the three UV-based processes and to determine the optimal oxidant concentration, total cost per order ($Cost/O_{total}$) was evaluated based on the 90% removal of AZA. $Cost/O_{total}$ refers to the total cost required to degrade a pollutant by one order of magnitude in 1 m^3 of contaminated water, which is determined by the oxidant dose, oxidant price, and electricity price through Eqs. (11.11, 11.12 and 11.13):

$$Cost/O_{total} = Cost/O_{UV} + Cost/O_{ox} \quad (11.11)$$

$$EE/O_{UV} = \frac{1.45Pt}{V\log\left(\frac{C_i}{C_f}\right)} = \frac{5.57 \times 10^{-2}P}{Vk}\left(\mathrm{kWh \cdot m^{-3} \cdot order^{-1}}\right) \quad (11.12)$$

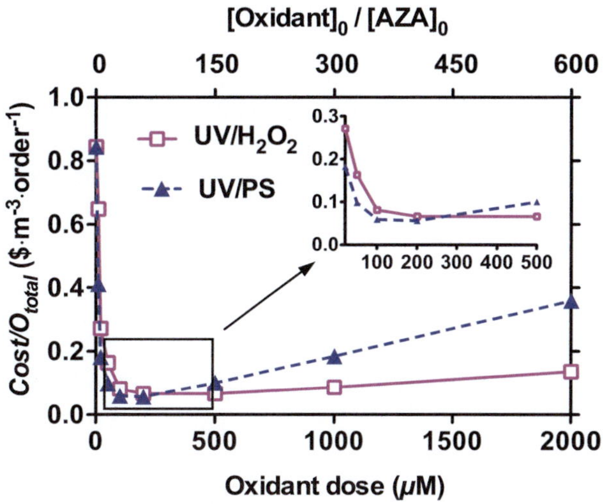

Fig. 11.3 *Cost/O*$_{total}$ comparison of AZA removal by UV, UV/H_2O_2, and UV/PS

$$Cost/O_{UV}\left(\$\cdot \mathrm{m}^{-3}\cdot \mathrm{order}^{-1}\right) = EE/O_{UV}\left(\mathrm{kWh}\cdot \mathrm{m}^{-3}\cdot \mathrm{order}^{-1}\right) \times \text{electricity cost}\left(\$\cdot \mathrm{kWh}^{-1}\right) \tag{11.13}$$

where $Cost/O_{UV}$ and $Cost/O_{ox}$ are the costs of electricity and oxidants ($\$\cdot m^{-3}\cdot order^{-1}$), respectively, EE/O_{UV} is the electrical energy per order ($kWh\cdot m^{-3}\cdot order^{-1}$), P is the UV lamp output power (kW), t is the degradation time (h), V is the volume (m^3) of treated water, k is the first-order rate constant (min^{-1}); C_i and C_f are the initial and final concentrations ($mg\cdot L^{-1}$) of AZA, respectively. This study considered following values for calculations: US\$0.10 kWh^{-1} for electricity price, US\$1.5 kg^{-1} for H_2O_2, and US\$0.74 kg^{-1} for PS.

Figure 11.3 shows that the $Cost/O_{total}$ decreases significantly with the increasing oxidant concentration from 0 to 200 μM and increases slightly with excess oxidant due to self-scavenging and high chemical cost. At low oxidant concentration (less than 200 μM), the $Cost/O_{total}$ of UV/PS process is lower than that of UV/H_2O_2 process. Since a low PS dose is sufficient for complete AZA removal, UV/PS process is the most cost-effective in AZA degradation compared with UV photolysis and UV/H_2O_2 (Zhang et al. 2016).

11.4 Conclusions

UV photolysis generally exhibits insignificant degradation effect toward most CSDs, i.e., CSDs were degraded by less than 10% at UV dose of 400 $mJ\cdot cm^{-2}$. Direct UV photolysis degrades most of the studied CSDs by cleaving their aromatic ring, while NO_3^--induced and DOM-induced photolysis eliminated the compounds by radical addition and electron transfer reaction, respectively. The overall degradation rates of

CSDs increased significantly with addition of H_2O_2 or PS. For CSDs with electron-donor groups, their degradation rates in UV/PS process were higher than that in UV/H_2O_2. Two main photodegradation pathways were proposed for AZA degradation, including hydroxylation and demethylation. The treatment costs of UV/H_2O_2 and UV/PS process decrease significantly compared with that of UV photolysis, where UV/PS process is more cost-effective than UV/H_2O_2 for AZA degradation.

References

Baxendale JH, Wilson JA (1957) The photolysis of hydrogen peroxide at high light intensities. Trans Faraday Soc 53:344–356

Česen M, Kosjek T, Laimou-Geraniou M et al (2015) Occurrence of cyclophosphamide and ifosfamide in aqueous environment and their removal by biological and abiotic wastewater treatment processes. Sci Total Environ 527–528:465–473

Delgado LF, Faucet-Marquis V, Pfohl-Leszkowicz A et al (2011) Cytotoxicity micropollutant removal in a crossflow membrane bioreactor. Bioresour Technol 102:4395–4401

Garcia-Ac A, Broseus R, Vincent S et al (2010) Oxidation kinetics of cyclophosphamide and methotrexate by ozone in drinking water. Chemosphere 79:1056–1063

Hirose J, Kondo F, Nakano T et al (2005) Inactivation of antineoplastics in clinical wastewater by electrolysis. Chemosphere 60:1018–1024

Kim I, Yamashita N, Tanaka H (2009) Photodegradation of pharmaceuticals and personal care products during UV and UV/H_2O_2 treatments. Chemosphere 77:518–525

Kobayashi T, Hirose J, Sano K et al (2012) Application of electrolysis for detoxification of an antineoplastic in urine. Ecotoxicol Environ Saf 78:123–127

Kosjek T, Heath E (2011) Occurrence, fate and determination of cytostatic pharmaceuticals in the environment. TrAC Trends Anal Chem 30:1065–1087

Kovalova L, Siegrist H, Singer H et al (2012) Hospital wastewater treatment by membrane bioreactor: performance and efficiency for organic micropollutant elimination. Environ Sci Technol 46:1536–1545

Lenz K, Koellensperger G, Hann S et al (2007) Fate of cancerostatic platinum compounds in biological wastewater treatment of hospital effluents. Chemosphere 69:1765–1774

Lester Y, Avisar D, Gozlan I et al (2011) Removal of pharmaceuticals using combination of UV/H_2O_2/O_3 advanced oxidation process. Water Sci Technol 64:2230–2238

Li XF, Mitch WA (2018) Drinking Water Disinfection Byproducts (DBPs) and human health effects: multidisciplinary challenges and opportunities. Environ Sci Technol 52:1681–1689

Lin H, Lin AC (2014) Photocatalytic oxidation of 5-fluorouracil and cyclophosphamide via UV/TiO_2 in an aqueous environment. Water Res 48:559–568

Lutterbeck CA, Machado EL, Kammerer K (2015) Photodegradation of the antineoplastic cyclophosphamide: a comparative study of the efficiencies of UV/H_2O_2, UV/Fe^{2+}/H_2O_2 and UV/TiO_2 processes. Chemosphere 120:538–546

Lutterbeck CA, Wilde ML, Baginska E et al (2016) Degradation of cyclophosphamide and 5-fluorouracil by UV and simulated sunlight treatments: assessment of the enhancement of the biodegradability and toxicity. Environ Pollut 208:467–476

Mahnik SN, Lenz K, Weissenbacher N et al (2007) Fate of 5-fluorouracil, doxorubicin, epirubicin, and daunorubicin in hospital wastewater and their elimination by activated sludge and treatment in a membrane-bio-reactor system. Chemosphere 66:30–37

Miolo G, Marzano C, Gandin V et al (2011) Photoreactivity of 5-fluorouracil under UVB light: photolysis and cytotoxicity studies. Chem Res Toxicol 24:1319–1326

Oppenländer T (2003) Photochemical purification of water and air: advanced oxidation processes (AOPs) – Principles, reaction mechanisms, reactor concepts. Wiley, Weinheim

Pérez G, Fernández-Alba AR, Urtiaga AM et al (2010) Electro-oxidation of reverse osmosis concentrates generated in tertiary water treatment. Water Res 44:2763–2772

Somensi CA, Simionatto EL, Dalmarco JB et al (2012) A comparison between ozonolysis and sonolysis/ozonolysis treatments for the degradation of the cytostatic drugs methotrexate and doxorubicin: kinetic and efficiency approaches. J Environ Sci Health A Tox Hazard Subst Environ Eng 47:1543–1550

Wols BA, Hofman-Caris CHM, Harmsen DJH et al (2013) Degradation of 40 selected pharmaceuticals by UV/H_2O_2. Water Res 47(15):5876–5888

Zhang J, Chang VWC, Giannis A et al (2013) Removal of cytostatic drugs from aquatic environment: a review. Sci Total Environ 445:281–298

Zhang YQ, Zhang JF, Xiao YJ et al (2016) Kinetic and mechanistic investigation of azathioprine degradation in water by UV, UV/H_2O_2 and UV/persulfate. Chem Eng J 302:526–534

Zhang YQ, Zhang JF, Xiao YJ et al (2017) Direct and indirect photodegradation pathways of cytostatic drugs under UV germicidal irradiation: process kinetics and influences of water matrix species and oxidant dosing. J Hazard Mater 324:481–488

Chapter 12
Analysis, Occurrence, and Fate of Cyclophosphamide and Ifosfamide in Aqueous Environment

Marjeta Česen, Tina Kosjek, and Ester Heath

Abstract Among numerous active pharmaceutical ingredients registered for chemotherapy, two of the oldest, cyclophosphamide (CP) and ifosfamide (IF), are still widely prescribed. Their administration can result in side effects such as cytotoxicity, genotoxicity, mutagenicity, and teratogenicity, which might affect aqueous biota once introduced into the environment. These compounds, which are excreted from the human body as parent compounds and metabolites, find their way into the environment via the sewerage system from hospitals and from homes, where cancer outpatients live. Concentrations of CP and IF in hospital wastewaters (WW), wastewater treatment plant (WWTP) influents and effluents, and surface waters (SW) range from ng L^{-1} to μg L^{-1}. To reduce the burden of CP and IF residues in wastewater and consequently surface and drinking water (DW), the development and optimization of biological and abiotic water treatment technologies is essential, especially since both compounds are recalcitrant. Studies report complete removal of CP and IF during certain advanced oxidation processes; however, these treatments are still not available due to the high costs involved. In addition, understanding the degradation pathways of these compounds is important, since their transformation products (TPs) could exhibit higher toxicity toward aquatic ecosystems than the parent compounds. Finally, several studies describing the analysis, occurrence, and formation of CP and IF transformation products during various water treatments are discussed in this chapter.

Keywords Cyclophosphamide · Ifosfamide · Occurrence · Analysis · Removal · Transformation products

M. Česen · T. Kosjek · E. Heath (✉)
Department of Environmental Sciences, Jožef Stefan Institute, Ljubljana, Slovenia

Jožef Stefan International Postgraduate School, Ljubljana, Slovenia
e-mail: Ester.Heath@ijs.si

E. Heath et al. (eds.), *Fate and Effects of Anticancer Drugs in the Environment*,
https://doi.org/10.1007/978-3-030-21048-9_12

12.1 Methodology for Determining CP and IF in Water Samples

Cyclophosphamide (CP) and ifosfamide (IF) are two cytostatic agents used to treat cancer patients. In particular, CP is used to treat different types of leukemia, malignant lymphoma, some malignant solid tumors with or without metastases, Ewings' sarcoma, for various progressive autoimmune diseases (e.g., rheumatoid arthritis, erythematosus lupus, and myasthenia gravis) and as immunosuppressive therapy after organ transplantations. Ifosfamide is used to treat bronchial carcinoma, ovarian cancer, some testicular cancer tumors, soft tissue sarcomas, breast cancer, pancreatic carcinoma, renal cell cancer, carcinoma of the endometrium, and malignant lymphomas. Once excreted from our bodies, CP and IF residues reach SW and ground waters via treated WW. For quantitative analysis of cytostatic residues in aqueous samples, analytical methods typically employ solid-phase extraction (SPE) as sample preparation step followed by either gas chromatography (GC) or liquid chromatography (LC) coupled to mass spectrometry (MS). In the case of GC-MS, derivatization step is also applied, which was in the case of CP and IF successfully achieved by acylation with trifluoroacetic anhydride (Momerency et al. 1994; Steger-Hartmann et al. 1996; Česen et al. 2015).

Sample preconcentration for trace analysis of CP and IF is typically performed with N-vinylpyrrolidone and divinylbenzene (Oasis HLB™) copolymers (Ferrando-Climent et al. 2013, 2015; Gómez-Canela et al. 2012; Kovalova et al. 2012; Köhler et al. 2012; Martín et al. 2011; Moldovan 2006; Valcárcel et al. 2011) or surface-modified styrene-divinyl benzene (Strata X™) cartridges (Buerge et al. 2006; Busetti et al. 2009; Delgado et al. 2010; Garcia-Ac et al. 2010; Llewellyn et al. 2011). Several studies have extracted CP and IF using "on-line" SPE also with N-vinylpyrrolidone and divinylbenzene copolymer sorbent, proving that this technique is highly applicable for routine analysis of water samples (Garcia-Ac et al. 2009; Kovalova et al. 2012; Negreira et al. 2013). In these studies, multianalyte analysis was performed and the optimal conditions were determined for all investigated compounds.

Several studies report the use of GC or LC coupled to MS for determining the occurrence of CP and IF in aqueous environment. Among them, only two studies use GC-MS technique for their quantification (Table 12.1). Despite different instrumentation, the limits of detection (LODs) and quantification (LOQs) are comparable (all in low ng L^{-1} range), suggesting the adequate sensitivity of these methods for trace analysis, with the only exception being a study by Kiffmeyer et al. (1998), who used an UV detector (Table 12.1). In the case of GC-MS analysis, HP-5MS (5% diphenyl/95% dimethylpolysiloxane) and Permabond SE-52-DF (5% phenyl/95% methylpolysiloxan) columns were used for separation (Moldovan 2006; Steger-Hartmann et al. 1996). In both cases, ionization and mass analysis were based on EI and single quadrupole (Q; Table 12.1). Studies based on LC-MS used mainly reversed phase (RP) C18 columns and water in combination with either methanol (MeOH) or acetonitrile (ACN) as mobile phases (MPs). In addition, acidification of

Table 12.1 Studies reporting quantitative analysis of CP and IF in aqueous samples

		Quantitative analysis		LOQ/LOD	References
Comp.	Matrix	Separation	Detection		
CP, IF	WW	GC: on-column injection (Permabond SE-52-DF); carrier gas: helium	Single Q; electron ionization (EI) at 70eV; detection mode: selected ion monitoring (SIM)	LODs: 6 ng L^{-1} (CP) 7 ng L^{-1} (IF)	Steger-Hartmann et al. (1996)
CP, IF	River water, WW	RP HPLC: C18 column; MP: water/ACN with addition of 10 mM ammonium acetate (pH = 5.7)	QqQ-MS; ESI(+); detection mode: multiple reaction monitoring (MRM)	LODs: 10 ng L^{-1} (CP and IF)	Ternes (1998)
CP	SW	RP HPLC: C18 column; MP: phosphate buffer (pH = 3)/MeOH	UV detector (200 nm)	LOD: 200 μg L^{-1}	Kiffmeyer et al. (1998)
CP	WW	RP HPLC: C8 column. MP: 0.1% formic acid (pH = 2)/ACN	QqQ MS; ESI(+); detection mode: multiple reaction monitoring (MRM)	LOQ: 1.9 ng L^{-1}	Castiglioni et al. (2005)
CP	WW, SW	RP HPLC: C8 column MP: 0.1% formic acid (pH = 2)/ACN	QqQ MS; ESI(+) detection mode: MRM	n.a.	Zuccato et al. (2005)
CP, IF	WW, SW	RP HPLC: C18 column. MP: 0.1% formic acid/0.1% formic acid in MeOH	QqQ MS; ESI(+); detection mode: MRM	LODs: 0.02 ng L^{-1} (SW; CP and IF) 0.3 ng L^{-1} (WW; CP and IF)	Buerge et al. (2006)
CP	SW	GC: HP-5MS column; carrier gas: n.a.	Single Q; EI mode at 70 eV; detection mode: SIM	LOQ: 30 ng L^{-1}	Moldovan (2006)
CP, IF	WW	RP UPLC: C18 column; MP: 0.1% formic acid/ACN	QqQ MS; ESI(+); detection mode: MRM	LODs: 2 ng L^{-1} (CP and IF)	Yin et al. (2010b)
CP, IF	WW	RP HPLC: C18 column; MP: 0.1% formic acid/0.1% formic acid in MeOH	QqQ MS; ESI(+) and APCI; detection mode: selected reaction monitoring (SRM)	LOQs: 0.11–0.4 ng L^{-1} (CP) 0.16–0.24 ng L^{-1} (IF)	Llewellyn et al. (2011)
CP, IF	WW	RP UPLC: C18 column; MP: 0.1% formic acid/ACN	QqQ-LIT; ESI(+); detection mode: 2 SRMs for each compound	LOQs: 3.6 ng L^{-1} (CP) 5.8 ng L^{-1} (IF)	Ferrando-Climent et al. (2013)

(continued)

Table 12.1 (continued)

		Quantitative analysis		LOQ/LOD	References
Comp.	Matrix	Separation	Detection		
CP, IF	WW	RP UPLC: C18 column; MP: 0.1% formic acid/ACN	QqQ-LIT; ESI(+); detection mode: MRM	LOQs: 1.3 ng L^{-1} (CP) 1.7 ng L^{-1} (IF)	Ferrando-Climent et al. (2015)
CP	WW	RP HPLC: C18 column; MP: 0.1% formic acid/0.1% (v/v) formic acid in MeOH	Orbitrap ESI(+); detection mode: full scan and HRMS (resolving power = 50,000)	LOQ: 0.35 ng L^{-1}	Gómez-Canela et al. (2012)
CP, IF	WW	On-line solid-phase extraction (SPE)-RP HPLC: C18 column; MP: MeOH, ACN, 0.1% formic acid	QqQ ESI(+); detection mode: SRM	LOQs: 10–84 ng L^{-1} (CP) 2–17 ng L^{-1} (IF)	Kovalova et al. (2012)
CP, IF	SW, WW	RP HPLC: C18 column; MP: 0.1% formic acid in ACN/15 mM ammonium formate containing 0.1% formic acid	QqQ; ESI(+); detection mode: 2 MRMs	LOQs: 1.7–2.3 ng L^{-1} (CP) 1.1–1.7 ng L^{-1} (IF)	Martín et al. (2011)
CP, IF	SW, tap water	RP HPLC: C18 column; MP: ACN/0.1% formic acid	QqQ-LIT; ESI(+); detection mode: two SRMs	LOQs: 4 ng L^{-1} (CP) 1 ng L^{-1} (IF)	Valcárcel et al. (2011)
CP, IF	WW	RP HPLC: C18 column; MP: 0.4% formic acid/1% formic acid in MeOH	QqQ; ESI(+); detection mode: MRM	LOQs: 310 pg on column (CP) 454 pg on column (IF)	Busetti et al. (2009)
CP	Drinking water	On-line SPE-RP HPLC: C18 column; MP: 0.2% acetic acid/ACN	QqQ; ESI(+); detection mode: SRM	LOQ: 3.2 ng L^{-1}	Garcia-Ac et al. (2010)

MP with formic acid was often applied and the ionization was operated in electrospray ionization (ESI) positive mode with triple quadrupole (QqQ) being the most commonly used mass analyzer, followed by either QqQ-LIT (triple quadrupole Linear Ion-Trap) or Orbitrap (Table 12.1).

There are only few published studies concerning the formation of CP and IF TPs (Table 12.2). Separation of TPs was achieved in all cases using an RP C18 column. For ionization, ESI was used, while the applied mass analyzers differed (Table 12.2).

Table 12.2 Qualitative analysis for identification of CP and/or IF TPs

Comp.	Qualitative analysis		Reference
	Separation	Detection	
CP	RP LC: C18 column MP: water/ACN	Q-TOF; ESI(+)	Fernández et al. (2010) and Venta et al. (2005)
CP	RP LC: C18 column MP: 0.1% formic acid/ACN	IT; ESI(+)	Lutterbeck et al. (2015)
CP, IF	RP LC: C18 column MP: 0.1% formic acid for positive mode and 5 mM ammonium acetate for negative mode(A); 0.1% formic acid in MeOH for positive mode and 5 mM ammonium acetate in MeOH for negative mode (B)	QqQ; ESI (+/−)	Lai et al. (2015)
CP, IF	RP LC: C18 column MP: water/ACN	HCT Ultra IT (+/−)	Ofiarska et al. (2016)
CP, IF	RP LC: C18 column MP: 0.1% formic acid/0.1% formic acid in ACN	LTQ Orbitrap-XL	Česen et al. (2016)
CP	RP LC: C18 column MP: 0.1% formic acid/0.1% formic acid in MeOH	QqQ; ESI (+)	Zhang et al. (2017)

The suitability of these analyzers (QTOF, IT, QqQ, and Orbitrap) for the identification of unknown TPs is discussed in a review paper by Kosjek et al. (2007). Interestingly, only Česen et al. (2016), Fernández et al. (2010), and Venta et al. (2005) used hyphenated techniques enabling both, MSn experiments and HRMS.

12.2 Environmental Occurrence and Transformations

12.2.1 Sources and Physicochemical Parameters of Cyclophosphamide and Ifosfamide

The current trend in chemotherapy is toward outpatient treatment, that is, patients go home once they have received their therapy at the hospital. This reduces the cost of cancer therapy and increases patient comfort. These patients may excrete cytostatic residues including CP and IF in the hospital, since intravenous treatment can last several hours or at home due to their long half-lives in the body (Kosjek and Heath 2011). In addition, there is still a number of hospitalized patients receiving chemotherapy with CP and IF, which makes hospitals an important source of anticancer drug residues that end up in WW (Kümmerer 2001). There have been several attempts to reduce pollution from hospitals by separating urine, but the emergence of outpatient therapies has meant that this strategy has not been implemented to any significant degree (Janssens et al. 2017).

Once in the environment, physicochemical properties, namely, solubility, dissociation constant (pK_a), bioconcentration factor (BCF), sorption constant (K_d), octanol–water (K_{ow}) and organic carbon–water (K_{oc}) partition coefficients, and Henry's law constant (HLC), will dictate distribution and fate of a certain compound. The solubility of CP and IF is significantly higher than their environmental concentrations; hence, it does not limit their occurrence in the aquatic compartment (Table 12.3). Based on their pK_a values, both compounds act as weak acids and are partially dissociated in neutral environment suggesting low sorption to organic matter. This agrees with their K_{oc} values that also indicate only partial adsorption onto organic matter in the soil and sediment compartments, for example, humus (Table 12.3). Moreover, the log K_{ow} value determines the distribution of a compound between water and organic matter, in particular, lipids and fats. In the case of CP and IF, their log K_{ow} values are <1, indicating their high polarity and, consequently, a tendency to distribute into the water phase (Table 12.3). In addition, the bioconcentration factor (BCF) predicts the potential of a compound to accumulate in aquatic organisms. For CP and IF, their BCF values (Table 12.3) indicate low potential for bioaccumulation. The data for sorption of CP and IF on solids like sludge, sediment, and soil are very scarce. Mioduszewska et al. (2016) report the low sorption potential of CP and IF onto soil and rapid leaching from soils once exposed to aqueous environment. However, the authors do not give the K_d values of CP and IF. It is known that CP and IF do not sorb onto activated sludge at wastewater treatment plants (WWTPs), suggesting limited elimination from WW by this mechanism (Kümmerer et al. 1997). Finally, reported HLC values (Table 12.3) suggest CP and IF have low volatility.

Table 12.3 Physicochemical characteristics of the investigated compounds

Structure	Solubility ($g\ L^{-1}$)	pKa	log K_{ow} and K_{oc}	BCF	K_d	HLC ($atm \times m^3\ mole^{-1}$)	References
CP 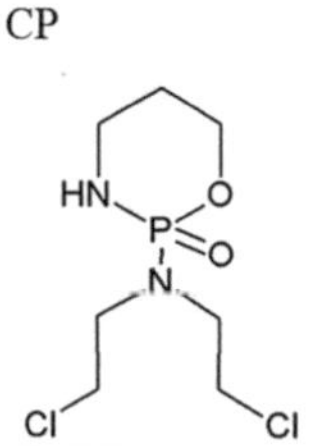	40	6.00	0.63 52	3	n.a.[a]	1.4×10^{-11}	Mahoney et al. (2003) and Kosjek and Heath (2011)
IF 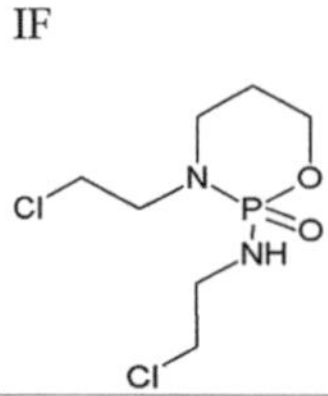	38	3.75	0.86 70	3	n.a.	1.36×10^{-11}	Mahoney et al. (2003) andKosjek and Heath (2011)

[a]*n.a. not available*

12.2.2 Occurrence of Cyclophosphamide and Ifosfamide in Wastewaters and Surface Waters

Physicochemical properties of CP and IF suggest that they will occur mainly in the aqueous environment; however, a number of additional factors are also important for quantifying their presence in the environment. They include their consumption, disposal, pharmacokinetics, and fate during WW treatment. Table 12.4 gives the detected concentrations of CP and IF in various WWs (hospital WW and WWTP influents and effluents) and SWs as determined concentration ranges or, where these data was not available, as the mean value ± SD (standard deviation). The first studies, reporting the levels of CP and IF in SW and WW, were published 20 years ago (Ternes 1998; Steger-Hartmann et al. 1996, 1997). The presence of CP and IF in either ground water or tap water remains to be evaluated.

The highest concentrations of CP and IF are in hospital WWs, followed by WWTP influents and effluents (< LODs or LOQs to μg L^{-1}) and the lowest in SWs (Table 12.4). The low concentrations in SWs (< LODs or LOQs to ng L^{-1}) can be attributed to effluent dilution once it is introduced into the receiving SW. Except for Gómez-Canela et al. (2012) and Ternes (1998), who reported levels of CP $\leq$ 13,100 ng L^{-1} and of IF $\leq$2900 ng L^{-1} in WWTP effluent, respectively, the reported concentrations in influents and effluents ranged from below the LOD to ng L^{-1} (Table 12.4). In addition, several studies report comparable concentrations of CP and IF in pairs of WWTP influents and effluents, suggesting only limited biodegradation of these compounds (Buerge et al. 2006; Česen et al. 2015; Negreira et al. 2014; Franquet-Griell et al. 2017b). Recently, Franquet-Griell et al. (2017b) reported the occurrence of CP in WW effluent using novel macroporous ceramic passive samplers. The authors report comparable concentrations of CP in effluent using either passive or grab sampling approach, confirming the former as a useful tool for monitoring time-weighted average concentrations of CP in WWs (Table 12.4).

12.2.3 Environmental Transformations

The rate at which chemical (hydrolysis, oxidation), microbiological, and/or physicochemical (photodegradation) degradation occurs depends on many factors, including ambient temperature, the amount of solar irradiation, pH, the presence of other species, and the nature of the compound of interest. For example, Khetan (2007) found that seasonal variations in temperature and light intensity affect the fate of pharmaceutical residues in SW.

The environmental fate of CP and IF has been rarely reported. Haddad et al. (2015) reviewed all the available data on transformation products (TPs) of cytostatics, but no CP and IF TPs, formed under environmental conditions, are

Table 12.4 The occurrence of CP and IF in WW and SW

	Type of water	Sampling (flow-proportional, time-proportional or grab and number of samplings)	Concentration (ng L^{-1})	References
CP	Hospital WW	24 h time-proportional $n = 7$	19–4500	Steger-Hartmann et al. (1997)
		24 h time-proportional $n = 1$	146	Steger-Hartmann et al. (1996)
		Grab $n = 12$	< LOD (2) – 21	Thomas et al. (2007)
		Grab $n = 65$ (21 hospitals)	6–2000	Yin et al. (2010a)
		24 h time-proportional $n = 1$	5730	Gómez-Canela et al. (2012)
		Grab n = 1 (4 hospitals)	< LOQ (3.6) – 200.7	Ferrando-Climent et al. (2013)
		24 h time-proportional $n = 7$	< LOQ (3.0) – 100.0	Negreira et al. (2014)
		Grab $n = 1$ (5 hospitals)	< LOD (0.78) – 22,000	Česen et al. (2015)
		Grab $n = 7$	76–2680	Česen et al. (2016)
		3 grab samples/day – mixed together	Effluent 1: 114–1187	Olalla et al. (2018)
		$n = 1$ (2 effluents from one hospital, 5 days in a row)	Effluent 2: 46–3000	
CP	WWTP influent	8 h time-proportional	< LOD (6) – 143	Steger-Hartmann et al. (1997)
		$n = 2$	< LOD (6) – 8	
		Flow-proportional (24 h) $n = 5$ (3 WWTPs)	2–11	Buerge et al. (2006)
		24 h time-proportional $n = 2$	< LOD (2)	Thomas et al. (2007)
		24 h time-proportional $n = 1$	< LOQ (7.1)	Martín et al. (2011)
		24 h time-proportional $n = 2$ (3 WWTPs)	< LOD (0.35) – 13,100	Gómez-Canela et al. (2012)
		Grab $n = 2$ (3 WWTPs)	< LOQ (3.6)–25.5	Ferrando-Climent et al. (2013)

(continued)

Table 12.4 (continued)

	Type of water	Sampling (flow-proportional, time-proportional or grab and number of samplings)	Concentration (ng L^{-1})	References
		24 h time-proportional	< LOQ (3.0) – 43.8	Negreira et al. (2014)
		$n = 1$ (12 WWTPs)		
		24 h time-proportional	< LOD (0.55) – 27	Česen et al. (2015)
		$n = 1$ (3 WWTPs)		
		24 h time-proportional	< LOD (2.3)	Česen et al. (2016)
		$n = 1$		
		Grab	15 ± 9	Franquet-Griell et al. (2017b)
		$n = 4$		
	WWTP effluent	8 h time-proportional	6–17	Steger-Hartmann et al. (1997)
		$n = 2$	8–15	
		Grab	< LOD (10) – 20	Ternes (1998)
		$n = 1$		
		24 h time-proportional	< LOQ (1.9) – 9	Castiglioni et al. (2005)
		$n = 9$ (different WWTPs)		
		Flow-proportional (24 h)	2–10	Buerge et al. (2006)
		$n = 5$ (3 WWTPs)		
		24 h time-proportional	Median: 0.6	Zuccato et al. (2005)
		$n = 1$ (8 WWTPs)		
		24 h time-proportional	< LOD (2)	Thomas et al. (2007)
		$n = 2$		
		24 h time-proportional and grab	< LOQ (5)	Busetti et al. (2009)
		$n = 3$ (2 WWTPs)		
		24 h time-proportional	< LOQ (7.7)	Martín et al. (2011)
		$n = 1$		
		Grab	0.19–3.7	Llewellyn et al. (2011)
		$n = 3$ (2 WWTPs)		
		24 h time-proportional	< LOD (0.35)	Gómez-Canela et al. (2012)
		$n = 2$ (3 WWTPs)		
		24 h time-proportional	< LOQ (SM) – 25.0	Negreira et al. (2014)
		$n = 1$ (12 WWTPs)		
		24 h time-proportional	< LOD (0.55) – 17	Česen et al. (2015)
		$n = 1$ (3 WWTPs)		
		24 h time-proportional	< LOD (2.3)	Česen et al. (2016)
		$n = 1$		
		Grab	17 ± 4	Franquet-Griell et al. (2017b)
		$n = 4$		
		Passive sampling with macroporous ceramic passive sampler	19 ± 3	
		$n = 3$		

(continued)

Table 12.4 (continued)

	Type of water	Sampling (flow-proportional, time-proportional or grab and number of samplings)	Concentration (ng L^{-1})	References
	SW	Grab $n = 1$	< LOD (10)	Ternes (1998)
		Grab $n = 5$ (3 SWs)	0.05–0.17	Buerge et al. (2006)
		2.5 h time-proportional $n = 1$ (2 SWs)	< LOD (not available)	Zuccato et al. (2005)
		Grab $n = 2$ (4 SWs)	< LOQ (30) – 65	Moldovan (2006)
		Grab $n = 1$	< LOQ (5.5)	Martín et al. (2011)
		Grab $n = 5$ (5 rivers)	< LOD (3)	Valcárcel et al. (2011)
		Grab $n = 7$ (7 rivers)	< LOQ (10)	de Jongh et al. (2012)
IF	Hospital WW	Grab	< LOD (6) – 1914	Kümmerer et al. (1997)
		24 h time-proportional $n = 1$	24	Steger-Hartmann et al. (1996)
		Grab $n = 12$	< LOD (2) – 338	Thomas et al. (2007)
		Grab $n = 65$ (21 hospitals)	4–10,647	Yin et al. (2010a)
		Grab $n = 1$ (4 hospitals)	< LOQ (5.8) – 227.9	Ferrando-Climent et al. (2013)
		24 h time-proportional $n = 7$	< LOQ (2.0) – 19.4	Negreira et al. (2014)
		Grab $n = 1$ (5 hospitals)	< LOD (2.8) – 6800	Česen et al. (2015)
		Grab $n = 7$	26–47	Česen et al. (2016)
		3 grab samples/day – mixed together	Effluent 1: < LOD (0.2) – 31	Olalla et al. (2018)
		$n = 1$ (2 effluents from one hospital, 5 days in a row)	Effluent 2: 58–4761	
	WWTP influent	6 h time-proportional	7–29	Kümmerer et al. (1997)
		$n = 2$	< LOD (6) – 29	
		Flow-proportional (24 h) $n = 5$ (3 WWTPs)	< LOD (0.3) – 15	Buerge et al. (2006)
		24 h time-proportional $n = 2$	< LOD (2)	Thomas et al. (2007)

(continued)

Table 12.4 (continued)

	Type of water	Sampling (flow-proportional, time-proportional or grab and number of samplings)	Concentration (ng L^{-1})	References
		24 h time-proportional	3.5 ± 0.1	Martín et al. (2011)
		$n = 1$	(mean ± SD)	
		Grab	< LOQ (5.8) – 130.1	Ferrando-Climent et al. (2013)
		$n = 2$ (3 WWTPs)		
		24 h time-proportional	< LOQ (2.0) – 27.9	Negreira et al. (2014)
		$n = 1$ (12 WWTPs)		
		24 h time-proportional	< LOD (0.36)	Česen et al. (2015)
		$n = 1$ (3 WWTPs)		
		24 h time-proportional	< LOD (4.8)	Česen et al. (2016)
		$n = 1$		
		Grab	< IDL[a] (0.009 ng)	Franquet-Griell et al. (2017b)
		$n = 4$		
	WWTP effluent	6 h time-proportional	10–40	Kümmerer et al. (1997)
		$n = 2$	< LOD (6) – 43	
		Grab	< LOD (10) – 2900	Ternes (1998)
		$n = 1$		
		Flow-proportional (24 h)	1.7–6	Buerge et al. (2006)
		$n = 5$ (3 WWTPs)		
		24 h time-proportional	< LOD (2) – 71	Thomas et al. (2007)
		$n = 2$		
		24 h time-proportional and grab	< LOQ (25)	Busetti et al. (2009)
		$n = 3$ (2 WWTPs)		
		24 h time-proportional	1.2 ± 0.1	Martín et al. (2011)
		$n = 1$	(mean ± SD)	
		Grab	< LOQ (0.24)	Llewellyn et al. (2011)
		$n = 3$ (2 WWTPs)		
		24 h time-proportional	< LOQ (2.0) – 15.9	Negreira et al. (2014)
		$n = 1$ (12 WWTPs)		
		24 h time-proportional	< LOD (0.36)	Česen et al. (2015)
		$n = 1$ (3 WWTPs)		
		24 h time-proportional	< LOD (4.8)	Česen et al. (2016)
		$n = 1$		
		Grab	< IDL (0.009 ng)	Franquet-Griell et al. (2017b)
		$n = 4$		
		Passive sampling with macroporous ceramic passive sampler	< IDL (0.009 ng)	
		$n = 3$		
	SW	Grab	< LOD (10)	Ternes (1998)
		$n = 1$		

(continued)

Table 12.4 (continued)

	Type of water	Sampling (flow-proportional, time-proportional or grab and number of samplings)	Concentration (ng L^{-1})	References
		Grab	0.05–0.14	Buerge et al. (2006)
		$n = 5$ (3 SWs)		
		Grab	< LOQ (4.4)	Martín et al. (2011)
		$n = 1$		
		Grab	< LOD (1) – 41	Valcárcel et al. (2011)
		$n = 5$ (5 rivers)		
		Grab	< LOQ (10)	de Jongh et al. (2012)
		$n = 7$ (7 rivers)		

[a]*IDL* instrumental detection limit

reported. To the author's knowledge, only two studies address the environmental degradation of CP and/or IF in Switzerland and Taiwan, both in synthetic and natural SWs (Buerge et al. 2006; Lin et al. 2013). Lin et al. (2013) investigated the degradation of CP, while Buerge et al. (2006) investigated the fate of both compounds. Both studies suggest limited environmental biodegradation and that direct photodegradation plays only a minor (if any) role in the degradation of CP and/or IF in the environment. This agrees with the findings from a recent study by Franquet-Griell et al. (2017a), who also report low degradation (< 20%) during artificial solar irradiation experiments for both compounds. However, the authors report an increase in photochemical degradation, which correlates to an increase in •OH formation in the presence of NO_3-N, a naturally present photosensitizer. They conclude that the highest degradation of CP and/or IF occurs in shallow, clear, NO_3-N-rich natural waters (Buerge et al. 2006; Lin et al. 2013).

12.3 Removal and Transformation During Various Water Treatments

Various WW treatment technologies exist, which are designed to remove compounds, particles, dissolved gasses, and pathogens from WW (Jjemba 2008). Certain compounds that are resistant to biodegradation, including CP and IF, can pass through the WWTPs either partially or completely unchanged (Eggen et al. 2015). The research toward upgrading existing conventional biological treatment has led to the development of new treatment technologies. The efficiency of conventional and advanced treatment techniques in terms of removal of CP and IF is discussed in the following paragraphs.

12.3.1 Biological Treatment

The results of published studies concerning the removal of CP and IF during biological treatment are given in Table 12.5. In general, both compounds show limited removal under experimental conditions with either suspended biomass or fungi. Despite different concentrations of CP and IF applied in the studies (ng L^{-1} to mg L^{-1} range), their highest removal efficiency was reported for conventional treatment, that is, 17% and 15%, respectively. In addition, these tests, lasting days to months, revealed no improvement in removal efficiency with prolonged time (Table 12.5). Four studies report the removal efficiency for CP using the MBR with inconsistent results (Delgado et al. 2011; Kovalova et al. 2012; Köhler et al. 2012; Seira et al. 2016). Delgado et al. (2011) and Seira et al. (2016) reported significant removal ($\leq$ 80% and 60%, respectively), while Kovalova et al. (2012) and Köhler et al. (2012) reported lower removals (< 20%). One reason for this discrepancy could be the use of different matrices, that is, hospital WW with varying amounts of contaminants that could affect biomass activity (real situation) versus artificial/semiartificial WW, that is less contaminated and has a constant composition to which biomass adapts. On the contrary, Česen et al. (2015) reports higher removal using attached growth biomass in the case of hospital WW compared to an artificial WW matrix (Table 12.5). However, the duration of experiments described by Česen et al. (2015) differs significantly (artificial WW: 120 days and hospital WW: 2 days). Higher removal (35%) in this study was observed also for IF, when hospital WW was introduced into bioreactors. To the author's knowledge, this the highest reported IF removal during biological WW treatment.

12.3.2 Abiotic Treatment

Various abiotic treatment technologies like UV irradiation, ozonation, advanced oxidation processes (AOPs), and physical treatment can be used to disinfect and/or remove not readily biodegradable compounds like CP and IF from water (Glaze et al. 1987; Huber et al. 2005; Legrini et al. 1993). A review of such treatments is given in the following paragraphs.

12.3.2.1 UV Irradiation

UV irradiation can be used for disinfection and removal (complete or partial degradation) of organic compounds in water. The latter can be achieved by direct and indirect photolysis (Klavarioti et al. 2009; Legrini et al. 1993). A review of the literature reveals four studies on the removal of CP and IF by UV irradiation. All four studies report similar results (Table 12.6). These compounds do not absorb

Table 12.5 The removal efficiency for CP and IF during various biological treatments

	Treatment type	Type of water	Conc.	Duration	Removal	References
CP	Modified Zahn-Wellens test (OECD 302 B)	OECD medium + activated sludge (AS) from WWTP	160 mg L^{-1}	28 days	None	Steger-Hartmann et al. (1997)
	Simulated WWTP	Synthetic WW + AS from WWTP	10 μg L^{-1}	42 days	Poor (≈ 17%)	Steger-Hartmann et al. (1997)
	OECD Confirmatory test (Degradation and Accumulation, 1992)	Synthetic WW + AS from WWTP	375 mg L^{-1} 750 mg L^{-1}	10 days	None (0 ± 5%)	Kiffmeyer et al. (1998)
			150 mg L^{-1}	14 days		
	Simulated WWTP	Influent + AS from WWTP	90 ng L^{-1} 900 ng L^{-1}	24 h	None	Buerge et al. (2006)
	Membrane bioreactor (MBR)	Synthetic WW + AS from WWTP	5 μg L^{-1}	139 days 115 days	≤ 80%	Delgado et al. (2011)
	MBR	Hospital WW	161 ng L^{-1}	1 year	< 20%	Kovalova et al. (2012)
	MBR	Hospital WW	Data not provided	5 days	≈ 12%	Köhler et al. (2012)
	Biological treatment with fungi *Trametes versicolor*	Hospital WW	10 mg L^{-1} 100 μg L^{-1}	8 days	None	Ferrando-Climent et al. (2015)
	Bioreactors with attached biomass on Mutag™ carriers	Artificial WW + AS from WWTP	10 μg L^{-1}	120 days	42 ± 12%	Česen et al. (2015)
		Hospital WW + AS from WWTP	5.3 μg L^{-1}	2 days	59 ± 15%	Česen et al. (2015)
	MBR	Semi-synthetic WW	5 μg L^{-1}	77 days	60%	Seira et al. (2016)
	Sequential batch reactors	WW effluent + AS from WWTP	50 μg L^{-1}	2 days	≈ 15%	Franquet-Griell et al. (2017a)
IF	Modified Zahn-Wellens test (OECD 302 B)	DW + AS from WWTP	160 mg L^{-1}	42 days	None	Kümmerer et al. (1997)
		Hospital WW + AS from WWTP	4.3 mg L^{-1}			

(continued)

Table 12.5 (continued)

Treatment type	Type of water	Conc.	Duration	Removal	References
Simulated WWTP	Effluent + AS from WWTP	11.4 μg L^{-1}	56 days	< 3%	Kümmerer et al. (1997)
Simulated WWTP	Influent + AS from WWTP	120 ng L^{-1} 1200 ng L^{-1}	24 h	None	Buerge et al. (2006)
Biological treatment with fungi *Trametes versicolor*	Hospital WW	10 mg L^{-1} 100 μg L^{-1}	8 days	None	Ferrando-Climent et al. (2015)
Bioreactors with attached biomass on Mutag™ carriers	Artificial WW + AS from WWTP	10 μg L^{-1}	120 days	18 ± 11%	Česen et al. (2015)
	Hospital WW + AS from WWTP	6.8 μg L^{-1}	2 days	35 ± 9.3%	Česen et al. (2015)
Sequential batch reactors	WW effluent + AS from WWTP	50 μg L^{-1}	2 days	≈ 15%	Franquet-Griell et al. (2017a)

photons under UV irradiation (due to the lack of aromatic rings or C = C bonds), which means that removal is poor regardless of the experimental conditions applied (Russo et al. 2017).

12.3.2.2 Ozonation

Ozonation is a treatment process, where ozone (O_3) is introduced into water. Similar to UV irradiation, it can be used for disinfecting and/or removing compounds from water via direct or indirect degradation processes.

Seven studies report the removal efficiency of CP and IF by ozonation using varying O_3 concentrations (Table 12.7). In general, removal efficiencies >60% can be achieved in up to 30 min regardless of the matrix type (deionized water or hospital WW) and initial CP or IF concentration. Only Česen et al. (2015) and Li et al. (2016) report lower removal, which can be related to the lower O_3 concentration used in their experiments (10 mg L^{-1} and 0.25–5 mg L^{-1}, respectively) compared to other studies. Table 12.7 also shows how pH plays an important role in removal. For example, Venta et al. (2005) report 20% removal of CP at pH 7 and 60% at pH 9. The crucial role played by pH in the removal is described also by Fernandez et al. (2010) and Lin et al. (2015) for both compounds (Table 12.7). These outcomes suggest that ozonation is a promising technique, especially for highly contaminated hospital WWs; however, installation and maintenance costs are high and further detailed operational costs of this treatment are needed (Ferre-Aracil et al. 2016).

Table 12.6 The removal efficiency for CP and IF during sole UV irradiation

	Type of water	UV lamp	Conc.	Duration	Removal	References
CP	Pure water	Low pressure (UV_{dose} = 230.4 mJ cm^{-2})	10 μg L^{-1}	60 min	< 20%	Kim and Tanaka (2009)
	Artificial WW	Low pressure (UV_{dose} = 44 mJ cm^{-2})	10 μg L^{-1}	120 min	21%	Česen et al. (2015)
	Pure water	Low pressure (UV_{dose} = 14,472 mJ cm^{-2})	50 μg L^{-1}	90 min	Negligible	Franquet-Griell et al. (2017a)
	Pure water	Low pressure (UV_{dose} = 400 mJ cm^{-2})	261 μg L^{-1}	3 min	Negligible	Zhang et al. (2017)
IF	Artificial WW	Low pressure (UV_{dose} = 44 mJ cm^{-2})	10 μg L^{-1}	120 min	16%	Česen et al. (2015)
	Pure water	Low pressure (UV_{dose} = 14,472 mJ cm^{-2})	50 μg L^{-1}	90 min	Negligible	Franquet-Griell et al. (2017a)

Table 12.7 The removal efficiency for CP and IF during ozonation treatment experiments

	O_3 concentration	Type of water	pH	Conc.	Removal	References
CP	45 mg L^{-1}	Pure water	7 9	261 mg L^{-1}	≈ 20% (pH = 7, after 12 min); ≈ 60% (pH = 9, after 12 min)	Venta et al. (2005)
	32 mg L^{-1}	Pure water	5.6 9 11	5 mg L^{-1} 20 mg L^{-1}	61% (pH = 5.6; after 30 min)–100% (pH = 11; after 5 min) not concentration dependent	Lin et al. (2015)
		Hospital WW	7.8	20 mg L^{-1}	100% after 20 min	
	6–15 mg L^{-1}	Pure water	8.1	100 ng L^{-1}	87% after 2 min 100% after 30 min	Garcia-Ac et al. (2010)
		DW	Ambient		96% after ≈ 5 min	
	30 mg L^{-1} 45 mg L^{-1}	Buffered water	7 9 11	130.5 mg L^{-1} 261 mg L^{-1}	75% (pH = 7) and 90% (pH = 9 or 11) after 40 min not concentration dependent	Fernández et al. (2010)
	10 mg L^{-1}	Artificial WW	7	10 μg L^{-1}	42% after 120 min	Česen et al. (2015)
	0.25–5 mg L^{-1}	Diluted treated WW with ultrapure water	7.2	5 μg L^{-1}	≈ 10–70% after 30 min (O_3 dose dependent)	Li et al. (2016)
	60 mg L^{-1}	Hospital WW	8.9	0.14–1187 μg L^{-1} (native concentrations)	97–100% after 10 min (O_3 dose dependent)	Ferre-Aracil et al. (2016)
IF	3 g O_3 h^{-1}	Deionized water	5.6 9 11	5 mg L^{-1} 20 mg L^{-1}	79% (pH = 5.6; after 30 min) – 100% (pH = 11; after 5 min) not concentration dependent	Lin et al. (2015)
		Hospital WW	7.8	20 mg L^{-1}	100% after 20 min	
	10 mg L^{-1}	Artificial WW	7	10 μg L^{-1}	36% after 120 min	Česen et al. (2015)

(continued)

Table 12.7 (continued)

O_3 concentration	Type of water	pH	Conc.	Removal	References
0.25–5 mg L^{-1}	Diluted treated WW with ultrapure water	7.2	5 μg L^{-1}	≈ 10–70% after 30 min (O_3 dose dependent)	Li et al. (2016)
60 mg L^{-1}	Hospital WW	8.9	0.016–0.031 μg L^{-1} (native concentrations)	100% after 10 min (regardless of the O_3 dose)	Ferre-Aracil et al. (2016)

12.3.2.3 Advanced Oxidation Processes

Glaze et al. (1987) defined advanced oxidation processes as "those which involve the generation of hydroxyl radicals (•OH) in sufficient quantity to affect water purification." They described only O_3/H_2O_2, UV/O_3, and UV/H_2O_2 as AOPs. Nowadays, also other AOPs such as UV/TiO_2, Fe^{2+}/H_2O_2 (Fenton), UV/Fe^{2+}/H_2O_2 (photoassisted Fenton), and UV/O_3/H_2O_2 represent efficient DW and WW treatment technologies (Linden and Mohseni 2014; Saharan et al. 2014; Fabiańska et al. 2015). During AOP, the formation of •OH is followed by their reaction with the organic compounds present. These interactions lead to a series of complex oxidation reactions, which results in either their partial or complete degradation (Saharan et al. 2014). The high costs involved means that AOPs as WW treatment technologies can be applied as a tertiary treatment for WW containing high amounts of proteins or sugars, which are degraded during biological treatment, while the remaining biorecalcitrant organic matter can be degraded by an AOP (Oller et al. 2011).

The formation of •OH is common to all AOPs; however, the mechanism of their "synthesis" differs. For example, in the case of the Fenton process, •OH are formed due to the oxidation of Fe^{2+} to Fe^{3+}. This is a metal-catalyzed oxidation, in which iron acts as a catalyst (Saharan et al. 2014). A number of photoassisted AOP treatments also exist, such as UV/TiO_2, UV/H_2O_2, UV/O_3, UV/O_3/H_2O_2, and UV/Fe^{2+}/H_2O_2 (photoassisted Fenton AOP), which is an advanced version of Fe^{2+}/H_2O_2 with a higher •OH formation rate (Legrini et al. 1993; Saharan et al. 2014; Glaze et al. 1987; Andreozzi et al. 1999). Except for UV/TiO_2, which is a photocatalytic process, others can be described as photoactivated chemical reactions, where interactions between photons with sufficient energy levels and H_2O_2 or O_3 result in the formation of free radicals (mostly •OH), which react with the compounds present in water (Saharan et al. 2014). To achieve homolytic cleavage of H_2O_2, an UV irradiation (254 nm) is usually applied. When UV is used in combination with O_3, it is also recommended to use UV light with a wavelength of 254 nm (Andreozzi et al. 1999). An alternative way to produce •OH is by photo-catalytic oxidation with UV/TiO_2, where •OH are formed on the surface of a semiconductor catalyst, for example, titanium dioxide (TiO_2). The absorption of UV irradiation and consequent formation of electron–hole pairs on the catalyst's surface reduces the dissolved O_2 to the superoxide radical (O_2^{-}) ion and H_2O and OH^- to •OH (Saharan et al. 2014).

Besides UV/O_3, there are also other O_3-based AOPs: O_3/H_2O_2 and UV/O_3/H_2O_2. It is known that decomposition of O_3 in an aqueous solution is accompanied by the formation of both H_2O_2 and •OH (Legrini et al. 1993). The rate of •OH formation can be increased by adding H_2O_2 and by applying UV irradiation (Legrini et al. 1993).

To the author's knowledge, there are 15 AOP-based studies (Table 12.8), within which four report low removal efficiency of CP and/or IF (Wols et al. 2013; Lai et al. 2015; Zhang et al. 2017; Česen et al. 2015). Wols et al. (2013) report 10–15% (CP) and 10–30% (IF) removal efficiency during UV/H_2O_2 treatment in tap water and WWTP effluent (Table 12.8). This contradicts Kim et al. (2009a), who used similar experimental conditions, that is, WWTP effluent as a matrix, similar initial H_2O_2 concentration and UV dose, and reported $\leq$90% CP removal (Table 12.8). However, the initial CP concentration reported by Kim et al. (2009a), 3 ng L^{-1}, is far less than what is reported in the other cases. In addition, this value was below the LOD, which was determined using standard solutions, directly analyzed by LC-MS/MS without taking into account the concentration factor of SPE. This represents an additional ambiguity in their determination of CP removal. The two studies that report high CP removal efficiency ($\approx$ 90%) from WWTP effluent with similar initial CP concentrations used considerably higher UV and H_2O_2 doses (Kim et al. 2009b; Köhler et al. 2012).

Another study reporting low IF removal was described by Lai et al. (2015), who investigated the removal efficiency of IF during UV/TiO_2 treatment in one hospital WW, whereas higher removal was achieved in another hospital WW, deionized water and two WWs coming from pharmaceutical industry. The authors report a DOC-dependent removal efficiency, resulting from 10% for hospital WW with highest DOC value (29 mg L^{-1}) to 100% removal efficiency in deionized water with the lowest DOC value (data not provided) within 120 min of treatment (Lai et al. 2015). Although significantly shorter UV/H_2O_2 treatment (3 min) was performed by Zhang et al. (2017), the authors also report matrix-dependent removal efficiency with the lowest CP removal from treated WW ($\approx$ 45%). Interestingly, in Lai et al. (2015)'s study, who addressed IF, removal can be compared to that of Hui-Hsiang et al., (2013), who investigated CP removal using UV/TiO_2. Similar matrices (purified water) and initial CP/IF concentrations were applied in both cases (Table 12.8). The only difference was the TiO_2 concentration (20 and 100 mg L^{-1}), which accounts for the decrease in the time needed to remove 100% of either CP (2 h) or IF (10 min; Table 12.8).

Česen et al. (2015) also report low CP and IF removal during O_3/H_2O_2 treatment, that is, 30–40% and 26–39% after 120 min of treatment, respectively (Table 12.8). The authors report comparable or even decreased removal with an increased amount of H_2O_2. On the contrary, Ferre-Aracil et al. (2016) achieved complete CP removal using the same treatment of hospital WW in only 20 min for similar CP concentrations, but with higher O_3 and a significantly lower H_2O_2 concentration (Table 12.8). It can be assumed that in the first study, the high amount of H_2O_2 acted as scavenger of •OH produced by ozonation, which resulted in low CP and IF removal.

Table 12.8 The removal efficiency of CP and IF under various AOPs

	AOP	Conc.	$Fe^{2+}/TiO_2/H_2O_2/O_3$ conc.	Conditions	Matrix	Removal	References
CP	O_3/H_2O_2	261 mg L^{-1}	45 mg L^{-1} O_3 4.7–12.5 mg L^{-1} H_2O_2	pH = 7	Pure water	100% (15 min, all conditions)	Venta et al. (2005)
				pH = 9			
				pH = 11			
	UV/ H_2O_2	µg L^{-1} (not reported)	6 mg L^{-1}	8 W lamp	Pure water	≈ 90% (75 min; UV dose = 1695 mJ cm^{-2})	Kim et al. (2009b)
			8.2 mg L^{-1}		Effluent		
	UV/ H_2O_2	≈ 3 ng L^{-1}	7.8 mg L^{-1}	65 W lamp	Effluent	≈ 90% (5 min; UV dose = 923 mJ cm^{-2})	Kim et al. (2009a)
	O_3/H_2O_2	100 ng L^{-1}	10 mg L^{-1} O_3 2.5 mg L^{-1} H_2O_2	pH = 8	DW	100% (2 min)	Garcia-Ac et al. (2010)
	O_3/H_2O_2	261 mg L^{-1}	30–45 mg L^{-1} O_3 34 mg L^{-1} H_2O_2	pH = 7	Pure water	70–95% (15 min)	Fernández et al. (2010)
				pH = 9		60–80% (15 min)	
	UV/ H_2O_2	1 µg L^{-1}	0.83–1.1 g L^{-1}	2–10 W lamp 250 W lamp	Effluent	≤ 90% (16 h with 4 kWh m^{-3} energy input)	Köhler et al. (2012)
	UV/ H_2O_2	1 µg L^{-1}	5–20 mg L^{-1}	60 W lamp 2000 W lamp	Pure water	80–100% (500 mJ cm^{-2}; 5–20 mg L^{-1} H_2O_2	Wols et al. (2013)
					DW	10–15% (1000 mJ cm^{-2}; 10 mg L^{-1} H_2O_2)	
					Effluent		
	UV/ TiO_2	100 µg L^{-1}	5–3000 mg L^{-1}	8 W; pH = 3–10	Pure water	100% (2 h, pH = 5.8, 20 mg L^{-1} TiO_2)	Hui-Hsiang Lin and Yu-Chen Lin (2013)
		27.6 mg L^{-1}	20 mg L^{-1} 300 mg L^{-1}	8 W; pH not reported		100% (4 h with 300 and 5 h with 20 mg L^{-1} TiO_2)	
	UV/ H_2O_2	20 mg L^{-1}	333 mg L^{-1} 500 mg L^{-1} 666 mg L^{-1}	150 W	Pure water	100% (8 min, 333 mg L^{-1})	Lutterbeck et al. (2015)
	UV/ Fe^{2+}/ H_2O_2		434 mg L^{-1} Fe^{2+} 333 mg L^{-1} H_2O_2			100% (< 2 min)	
	UV/ TiO_2		100–1000 mg L^{-1}			100% (32 min, 500 mg L^{-1})	

	UV/ H_2O_2	10 μg L^{-1}	2.5–5 g L^{-1}	12 W	Artificial WW	65% (120 min or 44 mJ cm^{-2}; 5 g L^{-1})	Česen et al. (2015)
	UV/O_3		10 mg L^{-1}			59% (120 min or 44 mJ cm^{-2})	
	O_3/ H_2O_2		10 mg L^{-1} O_3 2.5–5 g L^{-1} H_2O_2			30–40% (120 min)	
	UV/ O_3/ H_2O_2		10 mg L^{-1} O_3 2.5–5 g L^{-1} H_2O_2			99% (120 min or 44 mJ cm^{-2}; 5 g L^{-1} H_2O_2)	
	O_3/ H_2O_2	0.14–1187 μg L^{-1}	60 mg L^{-1} O_3 21–128 mg L^{-1} H_2O_2	pH = 8.1–8.5	Hospital WW	100% (20 min; 21 mg L^{-1} H_2O_2)	Ferre-Aracil et al. (2016)
	UV/ TiO_2	50 mg L^{-1}	5 g L^{-1}	Solar simulator (550 W m^{-2}); pH = 5.5	Pure water	≈ 100% (240 min)	Ofiarska et al. (2016)
	UV/Pt/ TiO_2		5 g L^{-1}			≈ 100% (60 min)	
	UV/ H_2O_2	261 μg L^{-1}	0.34–170 mg L^{-1}	5 W; pH = 7	Pure water	64% (3 min or 510 mJ cm^{-2}; 3.4 mg L^{-1}) ≈ 83% (3 min, 6.8 mg L^{-1})	Zhang et al. (2017)
			6.8 mg L^{-1}		DW	≈ 79% (3 min)	
			6.8 mg L^{-1}		Treated WW	≈ 45% (3 min)	
	UV/ H_2O_2	50 μg L^{-1}	15 mg L^{-1}	Solar simulator; pH = 7.6	Pure water	100% (4 min or 643 mJ cm^{-2})	Franquet-Griell et al. (2017a)
IF	UV/ H_2O_2	1 μg L^{-1}	5–20 mg L^{-1}	60 W lamp 2000 W lamp	Pure water	90–100% (500 mJ cm^{-2}; 5–20 mg L^{-1} H_2O_2)	Wols et al. (2013)
					DW	10–30% (1000 mJ cm^{-2}; 10 mg L^{-1} H_2O_2)	
					Effluent		
	UV/ TiO_2			8 W	Pure water	100% after 10 min (100 mg L^{-1} TiO_2)	Lai et al. (2015)

(continued)

Table 12.8 (continued)

AOP	Conc.	$Fe^{2+}/TiO_2/H_2O_2/$ O_3 conc.	Conditions	Matrix	Removal	References
	100–20 mg L^{-1} (optimal: 100 μg L^{-1})	2–1000 mg L^{-1} (optimal: 100 mg L^{-1})		Hospital WW 1 and 2	1: 100% after 60 min 2: 10% after 120 min (DOC dependent)	
				Pharmaceutical industry WW 1 and 2	1 and 2: 40% after 120 min	
UV/ H_2O_2	10 μg L^{-1}	2.5–5 g L^{-1}	12 W	Artificial WW	70% (120 min or 44 mJ cm^{-2}; 5 g L^{-1})	Česen et al. (2015)
UV/O_3		10 mg L^{-1}			49% (120 min or 44 mJ cm^{-2})	
$O_3/$ H_2O_2		10 mg L^{-1} O_3 2.5–5 g L^{-1} H_2O_2			26–39% (120 min)	
UV/ $O_3/$ H_2O_2		10 mg L^{-1} O_3 2.5–5 g L^{-1} H_2O_2			94% (120 min or 44 mJ cm^{-2}; 5 g L^{-1})	
UV/ TiO_2	50 mg L^{-1}	5 g L^{-1}	Solar simulator (550 W m^{-2}); pH = 5.5	Pure water	≈ 100% (240 min)	Ofiarska et al. (2016)
UV/Pt/ TiO_2	5 mg L^{-1}	0.05–0.5 g L^{-1}	Solar simulator (550 W m^{-2}); pH = 5.5–9.5		≈ 100% (60 min)	
	50 mg L^{-1}	1.25–5 g L^{-1}	Solar simulator (550 W m^{-2}); pH = 5.5		≈ 100% (60 min; 5 g L^{-1})	
UV/ H_2O_2	50 μg L^{-1}	15 mg L^{-1}	Solar simulator; pH = 7.6	Pure water	100% (4 min or 643 mJ cm^{-2})	Franquet-Griell et al. (2017a)

Fernandez et al. (2010), who also investigated CP removal during O_3/H_2O_2, observed a decrease in removal efficiency at elevated pH values. This differs from ozonation treatment, where higher pH values result in more •OH being produced and consequently enhanced degradation (von Gunten and von Sonntag 2012). The authors explain the reverse phenomenon observed within the experiments, where the added H_2O_2 acts as a scavenger of the •OH produced at higher pH values. The same observation was reported by Venta et al. (2005), who reports complete CP removal within 15 min, but with a lower amount of H_2O_2 compared to Fernandez et al. (2010).

Within the UV-based AOPs (Table 12.8), the most efficient is photo-Fenton ($UV/Fe^{2+}/H_2O_2$), where CP was completely degraded in less than 2 min (Lutterbeck et al. 2015). This is comparable to O_3-based AOP, that is, O_3/H_2O_2, conducted at an environmentally relevant initial CP concentration, 100 ng L^{-1} (Garcia-Ac et al. 2010). In the latter study, the amount of H_2O_2 used is relatively small (2.5 mg L^{-1} compared to 333 mg L^{-1}); however, O_3-based treatment technologies are more costly compared to UV-based AOPs (von Gunten and von Sonntag 2012; Saharan et al. 2014). Wols et al. (2013), Zhang et al. (2017), and Franquet-Griell et al. (2017a) also achieved 100% CP and IF removal within only few min of UV/H_2O_2 treatment (comparable UV doses; Table 12.8), where low amounts of H_2O_2 (20, 6.8, and 15 mg L^{-1}, respectively) were applied. In all studies, environmentally relevant concentrations of CP and IF in pure water were used. In addition, the authors report a drop in removal efficiency with increased matrix complexity (Table 12.8). This can be explained by CP/IF competition with other species present in WW for reaction with •OH (Zhang et al. 2017; Wols et al. 2013).

A direct comparison among the different studies (Table 12.8) in terms of cost-efficiency for real-world applications is not possible at this point since the described experimental conditions vary significantly. For example, studies were performed in different matrices and volumes of samples (laboratory to pilot-scale experiments) using varying instrumentation and were conducted at different initial concentrations of CP and IF.

12.3.3 Physical Treatment

Adsorption on activated carbon (AC), nanofiltration (NF), and reverse osmosis (RO) are common physical treatment technologies, which can improve the quality of WW (Jjemba 2008). The main disadvantage of these techniques is that retained compounds are not degraded and require further treatment (Rakić et al. 2015).

The data on CP and IF removal using physical treatment are scarce (Table 12.9). A study by Chen et al. (2008) reports a carbon dose–dependent removal efficiency of CP (AC dose of 100 mg L^{-1} resulted in $\approx$ 90% removal). In addition, a correlation

Table 12.9 Removal efficiency of CP during various physical treatments (data for IF is unavailable)

Treatment	CP conc.	Matrix type	Removal	References
AC (0.1–100 mg L^{-1})	10 μg L^{-1}	Pure water	≈ 1–90%	Chen et al. (2008)
AC (22 mg L^{-1})	2 μg L^{-1}	Pure water	70%	de Ridder et al. (2009)
		SW	55%	
		Effluent	28%	
NF	1–10 μg L^{-1}	Pure water	20–40%	Wang et al. (2009)
		Effluent	60%	
RO	1–10 μg L^{-1}	Pure water	> 90%	
		Effluent	> 90%	

between matrix complexity and removal efficiency was also reported, with CP removal between 28% and 70% depending on tested matrix (de Ridder et al. 2009).

Nanofiltration and RO can be also used to treat WW by physically removing the dissolved compounds. In case of NF, particles with a diameter > 1 nm are retained, whereas in RO, only particles <0.1 nm in diameter can pass through the membrane. Pretreatment is also necessary to remove any solid particles that could affect the rejection efficiency of NF and RO (Ravikumar et al. 2014; Radjenović et al. 2008; von Gunten et al. 2006). Wang et al. (2009) studied the rejection efficiency of CP in pure water and treated WW by NF and RO. For NF, the rejection efficiency was matrix dependent (Table 12.9), where the lower rejection efficiency for untreated WW was correlated to membrane fouling by the organic matter present. The authors report over 90% rejection efficiency of CP by RO regardless of the matrix type (Table 12.9).

12.3.4 Transformations

Compounds undergo similar transformation reactions during water treatment as in the environment, that is, chemical, physicochemical, and/or microbiological transformations. However, these processes are typically more intense during treatment, where degradation and formation of TPs strongly depend on the applied conditions (Mompelat et al. 2009; Saharan et al. 2014). The transformations of CP and IF during biological treatment have not been studied yet, most likely due to their poor biodegradability, whereas TPs formed during abiotic treatments have been extensively investigated (Table 12.10). Seven studies have looked at CP degradation and identified 16 different TPs, whereas three studies report 17 different IF TPs (Tables 12.10 and 12.11). O_3-based treatments of CP produced one TP, a keto-CP. Ketonization was the most common reaction also during UV treatment and UV-based AOPs. Apart from keto-CP, there are several other reports of TPs that share the same molecular structure as known CP and IF human metabolites,

Table 12.10 Reported TPs of CP during various AOPs

AOP	Identified TPs			Reference
O_3/H_2O_2	Keto-CP 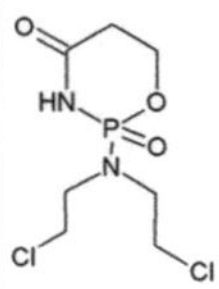			Venta et al. (2005)
O_3, pH = 9	Keto-CP 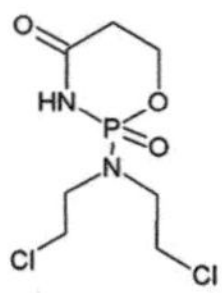			Fernández et al. (2010)
UV/H_2O_2 UV/TiO_2	3-dechloroethyl-CP (only UV/H_2O_2) 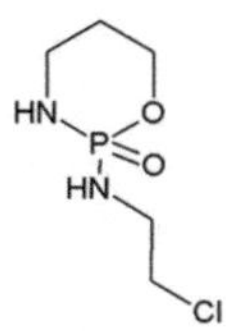	Keto-CP 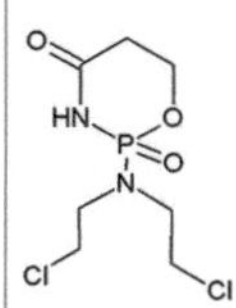	Imino-phosphamide 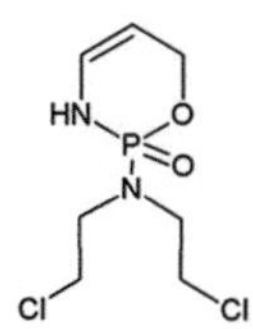	Lutterbeck et al. (2015)
	CP-TP1 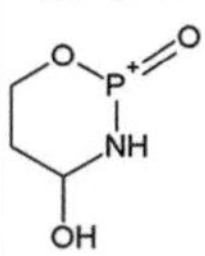			
UV/TiO_2	Keto-CP 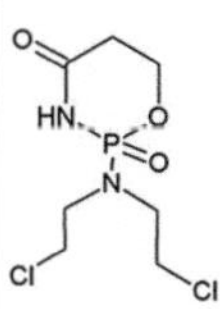	CP-TP2 (loss of Cl) 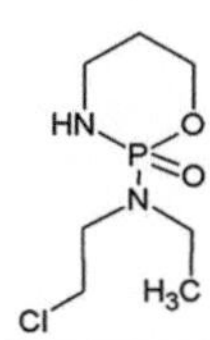	bis(2-chloroethyl)amine 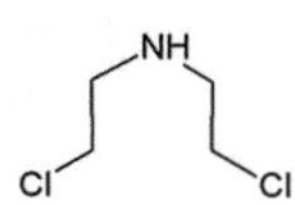	Lai et al. (2015)
	3-dechloroethyl-CP 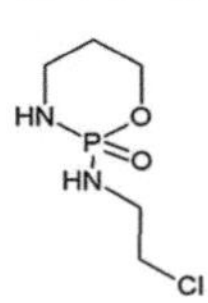	CP-TP3 (loss CH_2) 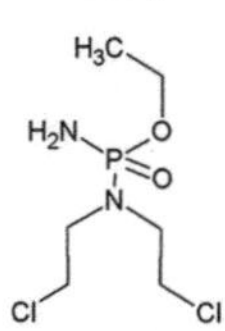	2-chloroethylamine 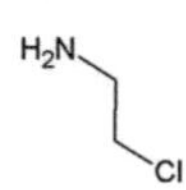	

(continued)

Table 12.10 (continued)

AOP	Identified TPs		Reference	
	CP-TP4 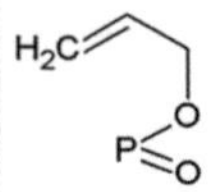			
UV and UV/H_2O_2	3-dechloroethyl-CP 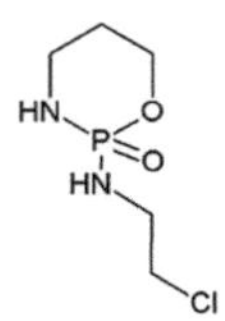	Keto-CP 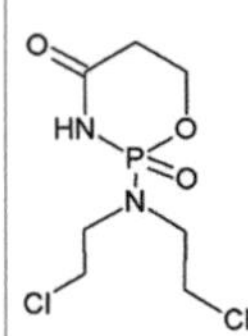	Imino-phosph-amide (only UV/H_2O_2) 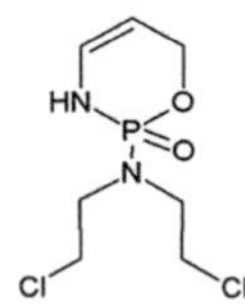	Česen et al. (2016)
	CP-TP5 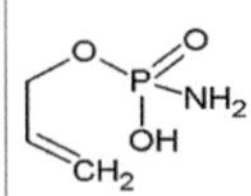	CP-TP6 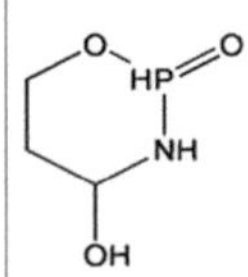		
UV/TIO_2 and $UV/Pt\text{-}TiO_2$	Inorganic species: NH^{4+} PO_4^{3-} Cl^-	Keto-CP 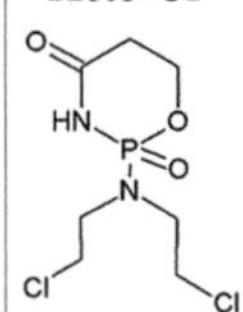	Unidentified TP $[M + H]^+ = 213$	Ofiarska et al. (2016)
UV and UV/H_2O_2	CP-TP7 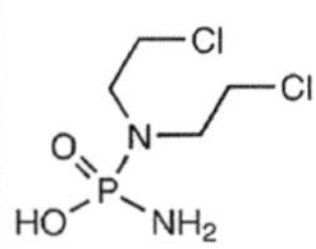	CP-TP8 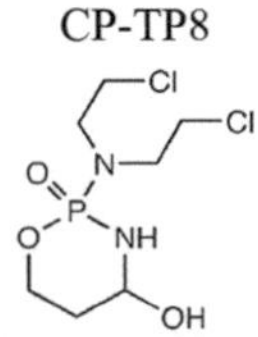		Zhang et al. (2017)

Table 12.11 Reported TPs of CP during various AOPs

AOP	Identified TPs			Reference
UV/TiO_2	Keto-IF 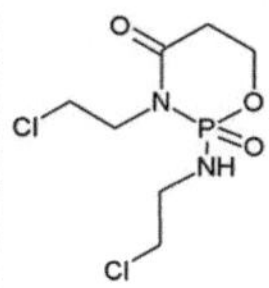	IF-TP1 (loss 1 Cl) 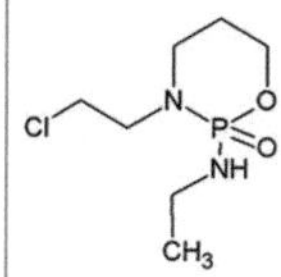	IF-TP2 (loss CH_2) 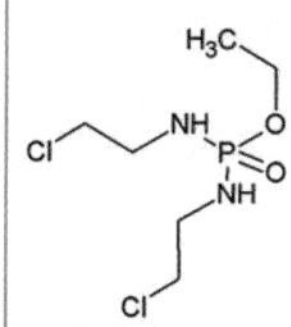	Lai et al. (2015)
	3-dechloroethyl-ifosfamide 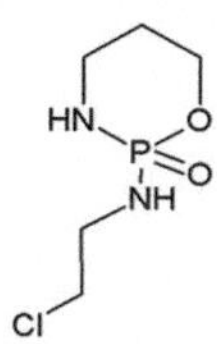	IF-TP3 (loss 1 Cl) 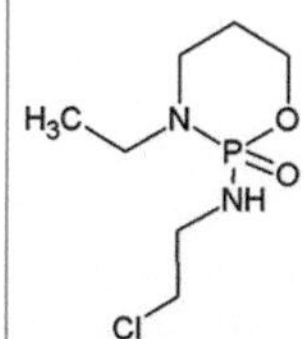	IF-TP4 (loss CH_3-Cl) 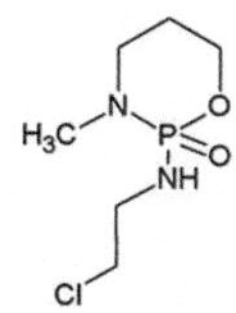	
	IF-TP5 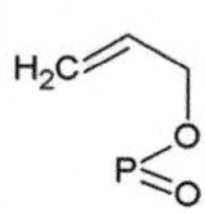	IF-TP6 (loss 2 Cl) 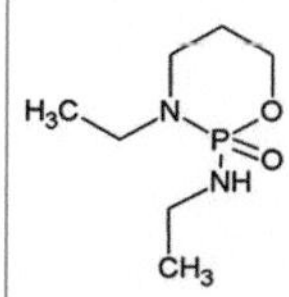	2-chloroethylamine 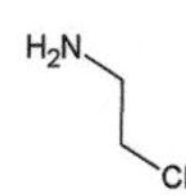	
UV and UV/H_2O_2	3-dechloro-ethyl-IF 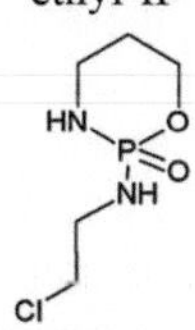	Keto-IF 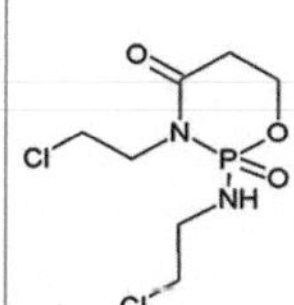	Imino-IF 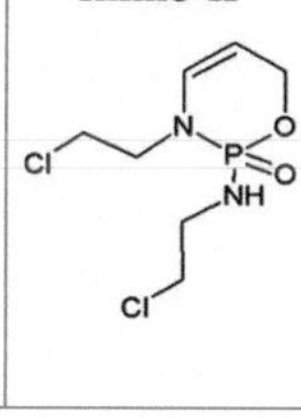	Česen et al. (2016)
	IF-TP7 IF-TP7 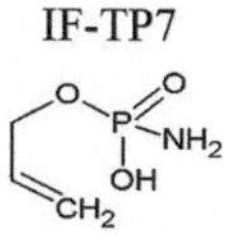			

(continued)

Table 12.11 (continued)

AOP	Identified TPs			Reference
UV/TIO$_2$ and UV/Pt-TiO$_2$	Inorganic species: NH^{4+} PO_4^{3-} Cl^-	IF-TP8 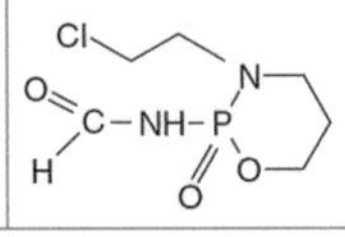	IF-TP9 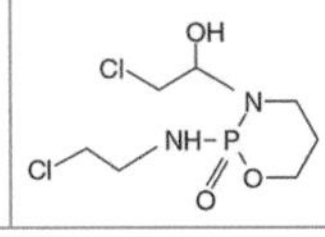	Ofiarska et al. (2016)
	2-dechloroethyl-IF 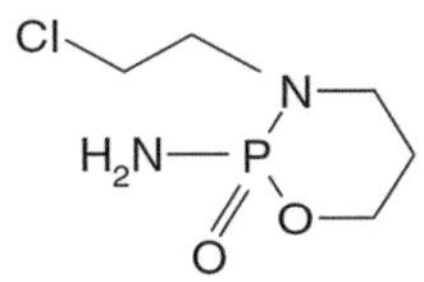			

namely, 2- and 3-dechloroethyl and imino derivatives of CP and IF formed during UV/H_2O_2 and UV/TiO_2 treatments (Tables 12.10 and 12.11). Certain treatments result in the same TPs, which is expected due to the similarity in the chemical structure CP and IF. These include, for example, a short chain TP (2-chloroethylamine), CP-TP4/IF-TP5 and CP-TP5/IF-TP7 (Table 12.10). Interestingly, Ofiarska et al. (2016) identified IF-TP9, which has the same molecular weight as CP-TP8, a TP identified by Zhang et al. (2017). Both TPs were identified as the hydroxylation products of parent compounds, where Ofiarska et al. (2016) left the exact position of hydroxyl group undetermined. Since no spectra are available for the comparison of both TPs, it is hard to conclude whether they share the same structural formula or not. As CP and IF typically occur together, the amount and potency of the formed species might be higher than one would assume based on the degradation of the individual compound. This should be investigated by studies addressing toxicity, where both compounds shall be treated simultaneously. Interestingly, Ofiarska et al. (2016) report the formation of NH^{4+}, PO_4^{3-}, and Cl^- formed from CP and IF when using UV/TiO_2 (Tables 12.10 and 12.11). As these inorganic species might have an adverse effects on aqueous biota, further studies addressing their formation during other treatments and an evaluation of the toxicity of UV/TiO_2-treated samples shall be studied in the future.

12.4 Conclusions

This chapter describes the analysis, occurrence, removal efficiency, and transformations of two cytostatic drug residues, CP and IF, in the aqueous environment. The most common method for the determination of CP and IF in aqueous samples is SPE with further LC-MS analysis. Their presence has been confirmed in

WWs on a global scale, while in SW, levels are typically below the LOD. Both compounds are recalcitrant to biodegradation and, for this reason, a number of studies have addressed their removal efficiency during abiotic treatments. So far, AOPs seem to be the most promising; however, their suitability for WW treatment is limited due to the high costs involved. Therefore, they require further optimization before they can be used in real world applications, for example, to treat highly contaminated hospital WWs. In addition, stable TPs have been confirmed during various abiotic treatments, which have structures similar to that of the parent compounds. These species might, besides CP and IF, also have adverse effects on aqueous biota. Therefore, environmental occurrence, fate, and effects of all CP and IF residues including identified TPs must be assessed in the future in order to evaluate the overall risks they pose to the environment.

References

Andreozzi R, Caprio V, Insola A, Marotta R (1999) Advanced oxidation processes (AOP) for water purification and recovery. Catal Today 53:51–59

Buerge IJ, Buser H-R, Poiger T, Müller MD (2006) Occurrence and fate of the cytostatic drugs cyclophosphamide and ifosfamide in wastewater and surface waters. Environ Sci Technol 40:7242–7250

Busetti F, Linge KL, Heitz A (2009) Analysis of pharmaceuticals in indirect potable reuse systems using solid-phase extraction and liquid chromatography–tandem mass spectrometry. J Chromatogr A 1216:5807–5818

Castiglioni S, Bagnati R, Calamari D, Fanelli R, Zuccato E (2005) A multiresidue analytical method using solid-phase extraction and high-pressure liquid chromatography tandem mass spectrometry to measure pharmaceuticals of different therapeutic classes in urban wastewaters. J Chromatogr A 1092:206–215

Česen M, Kosjek T, Laimou-Geraniou M, Kompare B, Širok B, Lambropolou D, Heath E (2015) Occurrence of cyclophosphamide and ifosfamide in aqueous environment and their removal by biological and abiotic wastewater treatment processes. Sci Total Environ 527–528:465–473

Česen M, Kosjek T, Busetti F, Kompare B, Heath E (2016) Human metabolites and transformation products of cyclophosphamide and ifosfamide: analysis, occurrence and formation during abiotic treatments. Environ Sci Pollut Res Int 23:11209–11223

Chen Z, Park G, Herckes P, Westerhoff P (2008) Physicochemical treatment of three chemotherapy drugs: irinotecan, tamoxifen, and cyclophosphamide. J Adv Oxid Technol 11:254–260

de Jongh CM, Kooij PJ, de Voogt P, ter Laak TL (2012) Screening and human health risk assessment of pharmaceuticals and their transformation products in Dutch surface waters and drinking water. Sci Total Environ 427–428:70–77

de Ridder DJ, McConville M, Verliefde ARD, Heijman LTJ, Verberk JQJC, Rietveld LC, vn Dijk JC (2009) Development of a predictive model to determine micropollutant removal using granular activated carbon. Drink Water Eng Sci 2:189–204

Delgado LF, Schetrite S, Gonzalez C, Albasi C (2010) Effect of cytostatic drugs on microbial behaviour in membrane bioreactor system. Bioresour Technol 101:527–536

Delgado LF, Faucet-Marquis V, Pfohl-Leszkowicz A, Dorandeu C, Marion B, Schetrite S, Albasi C (2011) Cytotoxicity micropollutant removal in a crossflow membrane bioreactor. Bioresour Technol 102:4395–4401

Eggen T, Vogelsang C, Eddy YZ (2015) Chapter 7: Occurrence and fate of pharmaceuticals and personal care products in wastewater. In: Comprehensive analytical chemistry. Elsevier, New York, pp 245–294

Fabiańska A, Ofiarska A, Fiszka-Borzyszkowska A, Stepnowski P, Siedlecka EM (2015) Electrodegradation of ifosfamide and cyclophosphamide at BDD electrode: decomposition pathway and its kinetics. Chem Eng J 276:274–282

Fernández LA, Hernández C, Bataller M, Véliz E, López A, Ledea O, Padrón S (2010) Cyclophosphamide degradation by advanced oxidation processes. Water Environ J 24:174–180

Ferrando-Climent L, Rodriguez-Mozaz S, Barceló D (2013) Development of a UPLC-MS/MS method for the determination of ten anticancer drugs in hospital and urban wastewaters, and its application for the screening of human metabolites assisted by information-dependent acquisition tool (IDA) in sewage samples. Anal Bioanal Chem 405:5937–5952

Ferrando-Climent L, Cruz-Morató C, Marco-Urrea E, Vicent T, Sarrà M, Rodriguez-Mozaz S, Barceló D (2015) Non conventional biological treatment based on Trametes versicolor for the elimination of recalcitrant anticancer drugs in hospital wastewater. Chemosphere 136:9–19

Ferre-Aracil J, Valcárcel Y, Negreira N, de Alda ML, Barceló D, Cardona SC, Navarro-Laboulais J (2016) Ozonation of hospital raw wastewaters for cytostatic compounds removal. Kinetic modelling and economic assessment of the process. Sci Total Environ 556:70–79

Franquet-Griell H, Medina A, Sans C, Lacorte S (2017a) Biological and photochemical degradation of cytostatic drugs under laboratory conditions. J Hazard Mater 323:319–328

Franquet-Griell H, Pueyo V, Silva J, Orera VM, Lacorte S (2017b) Development of a macroporous ceramic passive sampler for the monitoring of cytostatic drugs in water. Chemosphere 182:681–690

Garcia-Ac A, Segura PA, Viglino L, Fürtös A, Gagnon C, Prévost M, Sauvé S (2009) On-line solid-phase extraction of large-volume injections coupled to liquid chromatography-tandem mass spectrometry for the quantitation and confirmation of 14 selected trace organic contaminants in drinking and surface water. J Chromatogr A 1216:8518–8527

Garcia-Ac A, Broséus R, Vincent S, Barbeau B, Prévost M, Sauvé S (2010) Oxidation kinetics of cyclophosphamide and methotrexate by ozone in drinking water. Chemosphere 79:1056–1063

Glaze WH, Kang J-W, Chapin DH (1987) The chemistry of water treatment processes involving ozone, hydrogen peroxide and ultraviolet radiation. Ozone Sci Eng 9:335–352

Gómez-Canela C, Cortés-Francisco N, Oliva X, Pujol C, Ventura F, Lacorte S, Caixach J (2012) Occurrence of cyclophosphamide and epirubicin in wastewaters by direct injection analysis–liquid chromatography–high-resolution mass spectrometry. Environ Sci Pollut Res 19:3210–3218

von Gunten U, Janex-Habibi M-L, Ternes TA, Weber L (2006) Removal of PPCP during drinking water treatment. In: Ternes TA, Joss A (eds) Human pharmaceuticals, hormones and fragrances. IWA Publishing, London

Haddad T, Baginska E, Kümmerer K (2015) Transformation products of antibiotic and cytostatic drugs in the aquatic cycle that result from effluent treatment and abiotic/biotic reactions in the environment: an increasing challenge calling for higher emphasis on measures at the beginning of the pipe. Water Res 72:75–126

Huber MM, Korhonen S, Ternes TA, von Gunten U (2005) Oxidation of pharmaceuticals during water treatment with chlorine dioxide. Water Res 39:3607–3617

Hui-Hsiang Lin H, Yu-Chen Lin A (2013) Photocatalytic oxidation of 5-fluorouracil and Cyclophosphamide via UV/TiO_2 in an aqueous environment. Water Res 48:559–568

Janssens R, Mandal MK, Dubey KK, Luis P (2017) Slurry photocatalytic membrane reactor technology for removal of pharmaceutical compounds from wastewater: towards cytostatic drug elimination. Sci Total Environ 599–600:612–626

Jjemba PK (2008) The occurrence and fate of pharmaceuticals and personal care products in the environment. In: Pharma-ecology. Wiley, Hoboken, pp 251–293

Khetan SK (2007) Human pharmaceuticals in the aquatic environment: a challenge to green chemistry. Chem Rev 107:2319–2364

Kiffmeyer T, Götze H-J, Jursch M, Lüders U (1998) Trace enrichment, chromatographic separation and biodegradation of cytostatic compounds in surface water. Fresenius J Anal Chem 361:185–191

Kim I, Tanaka H (2009) Photodegradation characteristics of PPCPs in water with UV treatment. Environ Int 35:793–802

Kim I, Yamashita N, Tanaka H (2009a) Performance of UV and UV/H_2O_2 processes for the removal of pharmaceuticals detected in secondary effluent of a sewage treatment plant in Japan. J Hazard Mater 166:1134–1140

Kim I, Yamashita N, Tanaka H (2009b) Photodegradation of pharmaceuticals and personal care products during UV and UV/H_2O_2 treatments. Chemosphere 77:518–525

Klavarioti M, Mantzavinos D, Kassinos D (2009) Removal of residual pharmaceuticals from aqueous systems by advanced oxidation processes. Environ Int 35:402–417

Köhler C, Venditti S, Igos E, Klepiszewski K, Benetto E, Cornelissen A (2012) Elimination of pharmaceutical residues in biologically pre-treated hospital wastewater using advanced UV irradiation technology: a comparative assessment. J Hazard Mater 239–240:70–77

Kosjek T, Heath E (2011) Occurrence, fate and determination of cytostatic pharmaceuticals in the environment. TrAC Trends Anal Chem 30:1065–1087

Kosjek T, Heath E, Petrović M, Barceló D (2007) Mass spectrometry for identifying pharmaceutical biotransformation products in the environment. TrAC Trends Anal Chem 26:1076–1085

Kovalova L, Siegrist H, Singer H, Wittmer A, McArdell CS (2012) Hospital wastewater treatment by membrane bioreactor: performance and efficiency for organic micropollutant elimination. Environ Sci Technol 46:1536–1545

Kümmerer K (2001) Drugs in the environment: emission of drugs, diagnostic aids and disinfectants into wastewater by hospitals in relation to other sources – a review. Chemosphere 45:957–969

Kümmerer K, Steger-Hartmann T, Meyer M (1997) Biodegradability of the anti-tumour agent ifosfamide and its occurrence in hospital effluents and communal sewage. Water Res 31:2705–2710

Lai WW-P, Lin HH-H, Lin AY-C (2015) TiO_2 photocatalytic degradation and transformation of oxazaphosphorine drugs in an aqueous environment. J Hazard Mater 287:133–141

Legrini O, Oliveros E, Braun AM (1993) Photochemical processes for water treatment. Chem Rev 93:671–698

Li W, Nanaboina V, Chen F, Korshin GV (2016) Removal of polycyclic synthetic musks and antineoplastic drugs in ozonated wastewater: quantitation based on the data of differential spectroscopy. J Hazard Mater 304:242–250

Lin AY-C, Wang X-H, Lee W-N (2013) Phototransformation determines the fate of 5-fluorouracil and Cyclophosphamide in natural surface waters. Environ Sci Technol 47:4104–4112

Lin AY-C, Hsueh JH-F, Hong PKA (2015) Removal of antineoplastic drugs cyclophosphamide, ifosfamide, and 5-fluorouracil and a vasodilator drug pentoxifylline from wastewaters by ozonation. Environ Sci Pollut Res 22:508–515

Linden KG, Mohseni M (2014) Chapter 2.8: Advanced oxidation processes: applications in drinking water treatment. In: Ahuja S (ed) Comprehensive water quality and purification. Elsevier, Waltham, pp 148–172

Llewellyn N, Lloyd P, Jürgens MD, Johnson AC (2011) Determination of cyclophosphamide and ifosfamide in sewage effluent by stable isotope-dilution liquid chromatography–tandem mass spectrometry. J Chromatogr A 1218:8519–8528

Lutterbeck CA, Machado ÊL, Kümmerer K (2015) Photodegradation of the antineoplastic cyclophosphamide: a comparative study of the efficiencies of UV/H_2O_2, UV/Fe^{2+}/H_2O_2 and UV/TiO_2 processes. Chemosphere 120:538–546

Mahoney BP, Raghunand N, Baggett B, Gillies RJ (2003) Tumor acidity, ion trapping and chemotherapeutics: I. Acid pH affects the distribution of chemotherapeutic agents in vitro. Biochem Pharmacol 66:1207–1218

Martín J, Camacho-Muñoz D, Santos JL, Aparicio I, Alonso E (2011) Simultaneous determination of a selected group of cytostatic drugs in water using high-performance liquid chromatography–triple-quadrupole mass spectrometry. J Sep Sci 34:3166–3177

Mioduszewska K, Maszkowska J, Białk-Bielińska A, Krüger O, Kalbe U, Liberek B, Łukaszewicz P, Stepnowski P (2016) The leaching behavior of cyclophosphamide and ifosfamide from soil in the presence of co-contaminant—Mixture sorption approach. Sci Total Environ 542:915–922

Moldovan Z (2006) Occurrences of pharmaceutical and personal care products as micropollutants in rivers from Romania. Chemosphere 64:1808–1817

Momerency GVC, Slee K, Van Oosterom PH, De Bruijn AT, A E (1994) The determination of cyclophosphamide and its metabolites in blood plasma as stable trifluoroacetyl derivatives by electron capture chemical ionization gas chromatography/mass spectrometry. Biol Mass Spectrom 23:149–158

Mompelat S, Le Bot B, Thomas O (2009) Occurrence and fate of pharmaceutical products and by-products, from resource to drinking water. Environ Int 35:803–814

Negreira N, López de Alda M, Barceló D (2013) On-line solid phase extraction–liquid chromatography–tandem mass spectrometry for the determination of 17 cytostatics and metabolites in waste, surface and ground water samples. J Chromatogr A 1280:64–74

Negreira N, de Alda ML, Barceló D (2014) Cytostatic drugs and metabolites in municipal and hospital wastewaters in Spain: filtration, occurrence, and environmental risk. Sci Total Environ 497–498:68–77

Ofiarska A, Pieczyńska A, Fiszka Borzyszkowska A, Stepnowski P, Siedlecka EM (2016) Pt–TiO_2-assisted photocatalytic degradation of the cytostatic drugs ifosfamide and cyclophosphamide under artificial sunlight. Chem Eng J 285:417–427

Olalla A, Negreira N, Lopez de Alda M, Barcelo D, Valcarcel Y (2018) A case study to identify priority cytostatic contaminants in hospital effluents. Chemosphere 190:417–430

Oller I, Malato S, Sánchez-Pérez JA (2011) Combination of advanced oxidation processes and biological treatments for wastewater decontamination—A review. Sci Total Environ 409:4141–4166

Radjenović J, Petrović M, Ventura F, Barceló D (2008) Rejection of pharmaceuticals in nanofiltration and reverse osmosis membrane drinking water treatment. Water Res 42:3601–3610

Rakić V, Rac V, Krmar M, Otman O, Auroux A (2015) The adsorption of pharmaceutically active compounds from aqueous solutions onto activated carbons. J Hazard Mater 282:141–149

Ravikumar YVL, Kalyani S, Satyanarayana SV, Sridhar S (2014) Processing of pharmaceutical effluent condensate by nanofiltration and reverse osmosis membrane techniques. J Taiwan Inst Chem Eng 45:50–56

Russo C, Lavorgna M, Cesen M, Kosjek T, Heath E, Isidori M (2017) Evaluation of acute and chronic ecotoxicity of cyclophosphamide, ifosfamide, their metabolites/transformation products and UV treated samples. Environ Pollut 233:356–363

Saharan VK, Pinjari DV, Gogate PR, Pandit AB (2014) Chapter 3: Advanced oxidation technologies for wastewater treatment: an overview. In: Bhandari VVRM (ed) Industrial wastewater treatment, recycling and reuse. Butterworth-Heinemann, Oxford, pp 141–191

Seira J, Sablayrolles C, Montréjaud-Vignoles M, Albasi C, Joannis-Cassan C (2016) Elimination of an anticancer drug (cyclophosphamide) by a membrane bioreactor: comprehensive study of mechanisms. Biochem Eng J 114:155–163

Steger-Hartmann T, Kümmerer K, Schecker J (1996) Trace analysis of the antineoplastics ifosfamide and cyclophosphamide in sewage water by twostep solid-phase extraction and gas chromatography-mass spectrometry. J Chromatogr A 726:179–184

Steger-Hartmann T, Kümmerer K, Hartmann A (1997) Biological degradation of cyclophosphamide and its occurrence in sewage water. Ecotoxicol Environ Saf 36:174–179

Ternes TA (1998) Occurrence of drugs in German sewage treatment plants and rivers. Water Res 32:3245–3260

Thomas KV, C D, Schlabach M, Langford KH (2007) Source to sink tracking of selected human pharmaceuticals from two Oslo city hospitals and a wastewater treatment works. J Environ Monit 9:1410–1418

Valcárcel Y, González Alonso S, Rodríguez-Gil JL, Gil A, Catalá M (2011) Detection of pharmaceutically active compounds in the rivers and tap water of the Madrid Region (Spain) and potential ecotoxicological risk. Chemosphere 84:1336–1348

Venta MB, Castro CH, Garcia LA, Marzo AL, Lorenzo EV, Alvarez CA (2005) Effect of O_3/H_2O_2 molar concetration ratio at different pH values on cyclophosphamide degradation. J Water Supply Res Technol AQUA 54:403–410

von Gunten U, von Sonntag C (2012) Chemistry of ozone in water and wastewater treatment: from basic principles to applications. IWA Publishing, London

Wang L, Albasi C, Faucet-Marquis V, Pfohl-Leszkowicz A, Dorandeu C, Marion B, Causserand C (2009) Cyclophosphamide removal from water by nanofiltration and reverse osmosis membrane. Water Res 43:4115–4122

Wols BA, Hofman-Caris CHM, Harmsen DJH, Beerendonk EF (2013) Degradation of 40 selected pharmaceuticals by UV/H_2O_2. Water Res 47:5876–5888

Yin J, Shao B, Zhang J, Li K (2010a) A preliminary study on the occurrence of cytostatic drugs in hospital effluents in Beijing, China. Bull Environ Contam Toxicol 84:39–45

Yin J, Yang Y, Li K, Zhang J, Shao B (2010b) Analysis of anticancer drugs in sewage water by selective SPE and UPLC-ESI-MS-MS. J Chromatogr Sci 48:781–789

Zhang Y, Xiao Y, Zhang J, Chang VWC, Lim T-T (2017) Degradation of cyclophosphamide and 5-fluorouracil in water using UV and UV/H_2O_2: kinetics investigation, pathways and energetic analysis. J Environ Chem Eng 5:1133–1139

Zuccato E, Castiglioni S, Fanelli R (2005) Identification of the pharmaceuticals for human use contaminating the Italian aquatic environment. J Hazard Mater 122:205–209

Part III
Environment/Wastewater Treatment + Effects

Chapter 13
Fate and Effects of Cytostatic Pharmaceuticals in the Marine Environment

Maria João Bebianno and Tainá Garcia da Fonseca

Abstract Extensive efforts have been devoted to assessing the environmental fate, effects, and risks associated with the presence of pharmaceutical compounds in the marine environment. Higher standards of living combined with aging and the increase of the world population contribute to frequent and highly diverse needs for the use of pharmaceutical compounds. Therefore, inputs of these substances has begun to be detected in the marine environment. Their sources are hospital, industrial, and sewage effluents in which these compounds cannot be properly treated. Therefore, the substances reach the marine environment in their chemical form or metabolized, and their fate in the aquatic systems, after long-term exposure to organisms at environmentally relevant concentrations, should be assessed. One of the groups of these pharmaceutical compounds includes cytotoxic pharmaceuticals applied in chemotherapy. These compounds, although having mutagenic, genotoxic, and teratogenic properties, have received less attention in environmental risk assessment, despite the progressive enhancement of their use with the increase of cancer incidence in the human population. Cytotoxic drugs have variable chemical structures and are used to kill tumor cells or inhibit their proliferation by different modes of action (MoA). The aim of this chapter is to report data regarding acute and chronic responses of cytotoxic drugs on nontarget organisms. An integrative approach of molecular and cellular effects is reported as a result of single or mixture exposures to assess the ecotoxicological potential, synergistic, additive, and antagonist effects of these drugs in biological systems of nontarget species at realistic environmental concentrations. These data are integrated to contribute to the environmental risk assessment of these compounds in the marine environment. Furthermore, recommendations are made for suitable biological models to assess the ecotoxicological effects of these compounds in marine organisms.

Keywords Environmental risk assessment · Antineoplastics · Accumulation · Environmental levels

---

M. J. Bebianno (✉) · T. Garcia da Fonseca
CIMA, University of Algarve, Faro, Portugal
e-mail: mbebian@ualg.pt

E. Heath et al. (eds.), *Fate and Effects of Anticancer Drugs in the Environment*,
https://doi.org/10.1007/978-3-030-21048-9_13

13.1 Introduction

About 41% of the world population lives near the coast, where two thirds of the megacities (i.e., cities containing more than 10 million habitants) are located (UN 2016), and settlement is still growing rapidly (Burke et al. 2001; Martínez et al. 2007). Demographic projections indicate that by 2030 there will be 662 cities with at least 1 million residents, and 41 of these will be megacities, which corresponds to an increase of about 30% compared to 2016 (UN 2016) (Fig. 13.1).

Projections based on population growth indicate that human burden on the coast will increase to 3.1 billion people by 2025, and about 300 million people will settle in coastal megacities (Martínez et al. 2007; Von Glasow et al. 2013). Low-elevation coastal zones (LECZ), areas 10 m above sea level, represent about 2% of total land area where 13% of the global population inhabits (Barbier 2015). LECZ are expanding faster than any other area through immigration and demographic changes (Fig. 13.2).

This increase of population will broaden socioeconomic, cultural, and environmental impact underlying determinants fitted by risk factors, individually managed (e.g., inappropriate diet, tobacco, physical inactivity, harmful use of alcohol, infectious agents) or not (e.g., age, heredity) that are decisive in the distribution and occurrence of chronic noninfectious or noncommunicable diseases (NCDs) (Jemal et al. 2011;

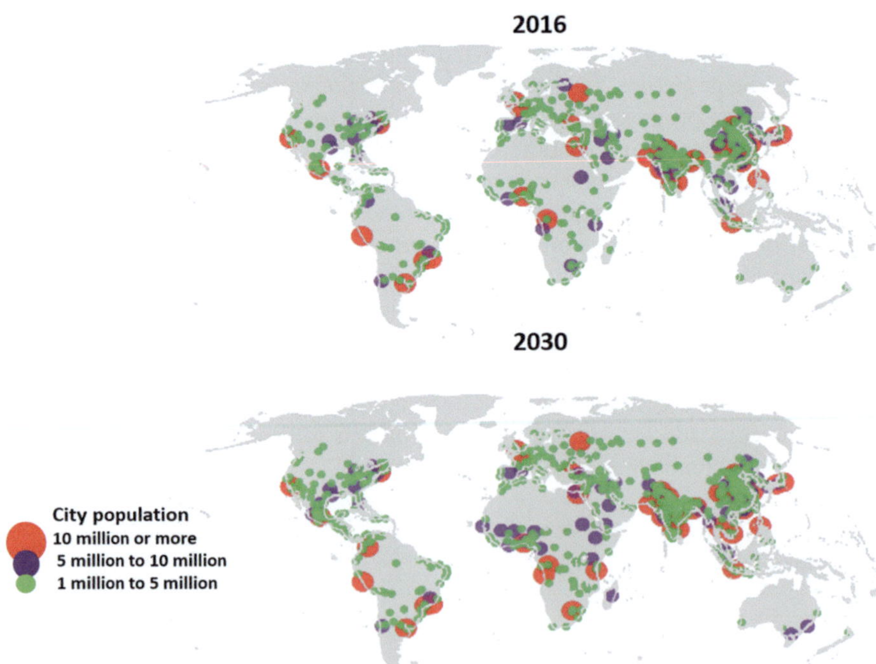

Fig. 13.1 Cities with 1 million or more inhabitants in 2016 and projections for 2030 (Adapted from UN 2016)

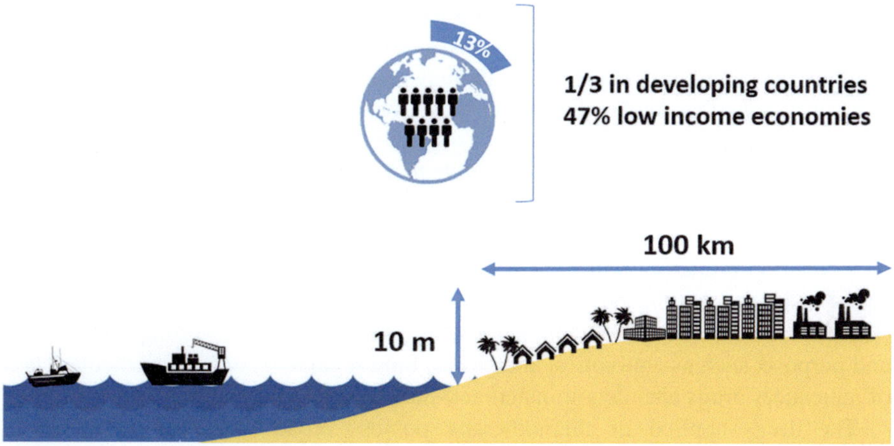

Fig. 13.2 Representation of urban settlement in low-elevation coastal zones (LECZ), where 13% of the world population lives

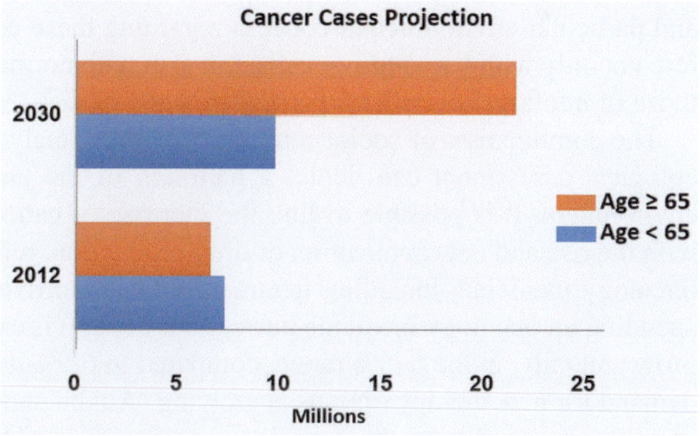

Fig. 13.3 Projection of cancer cases between 2012 and 2030 (millions of people). All types of cancer are considered, with the exception of non-melanoma skin cancer (Data extracted from GLOBOCAN; Ferlay et al. 2015)

WHO 2005). NCDs include cardiovascular and respiratory chronic illnesses, as well as neoplastic diseases (Ferrando-Climent et al. 2014), and accounted for 67.9% of total deaths worldwide in 2012. Neoplastic diseases are the most representative and concerning diseases, after cardiovascular illness, accounting for 21.7% of deaths and with worrying projections of increases by 2030 (WHO 2014) (Fig. 13.3).

With the improvement of living standards and the increase in diseases, the production and use of pharmaceutical compounds follow the same trend. Consequently, complex mixtures of parental molecules and metabolic products of pharmaceuticals are excreted and introduced into water bodies, predominantly via untreated, partially treated, or treated urban, hospital, and industrial effluents, with inadequate disposal of unused drugs, and from landfill leachates (Vieno et al. 2005; Claessens et al. 2013; Moreno-González et al. 2015; Heath et al. 2016; Isidori et al. 2016). Such continuous loads cause drugs to become persistent and ubiquitous in the aquatic environment (Kümmerer 2010) and, for that reason, are considered emerging contaminants of concern (Petrie et al. 2015).

In spite of significant efforts devoted to understanding the occurrence of pharmaceuticals in aquatic systems (e.g., lakes, rivers, marshlands, groundwater, and drinking water), their behavior and fate may change by virtue of abiotic and biotic transformations that may occur before they reach the marine environment (Weigel et al. 2002).

Pharmaceuticals are designed to provide therapeutic effects in biological systems by their key molecular structures, the active pharmaceutical ingredients (APIs), which are configured to persist through metabolic degradation processes to exert their function properly (Anderson et al. 2004; Kümmerer 2010). The wide range of drugs displayed by the pharmaceutical industry are grouped by different criteria (Daughton and Ternes 1999). Often, these drugs are classified according to their API and purpose, such as antibiotics, analgesics, lipid regulators, and the particular class of anticancer drugs also denominated anti-neoplastics (Kümmerer 2008).

The drugs applied in chemotherapy possess broad biological performance devoted to access and harm many cellular components such as DNA, RNA, proteins, membrane phospholipids, cytoskeletal microfilaments, and specific signaling pathways, ultimately causing irreversible harm effects, interruption of cell division, and cell death (Gonzalez et al. 2001; Emadi et al. 2009; Gorrini et al. 2013). The current and particular environmental concern regarding these compounds is that they interfere not only with target tumor cells, but also with normal proliferating cells, such as those of nontarget organisms present in water bodies (Mater et al. 2014).

The combination of socioeconomic structure, analytical technologies, and toxicological assessment can depict a hallmark of the anticancer drugs issue in the environment. It is possible to link the increase of cancer incidence in recent years with the rise and new application of drug production, reflected in global spending on oncology medicine, including treatment and supportive care (Fig. 13.4). Based on spending on oncology medicine throughout 2018, it is expected that this amount will grow annually, in the 6–8% range, compared to 6.5% during the past 5 years, as the demand for new therapy options is growing (Aitken and Kleinrock 2015). The input of such compounds to the aquatic environment worldwide could be neglected if sewage treatment plant technology could efficiently eliminate parent compounds and reactive by-products, because these chemicals may exert cytotoxic and cytostatic potential on every eukaryotic organism, even at trace levels (Johnson et al. 2008; Besse et al. 2012).

In this sense, the aim of this chapter is to report data regarding acute and chronic responses of cytotoxic drugs on marine nontarget organisms. An integrative approach of molecular and cellular effects is reported as a result of either single or mixture exposures to assess the ecotoxicological potential, synergistic, additive, and antagonist effects of these drugs in biological systems of nontarget species, at realistic environmental concentrations. These data are integrated to contribute to the environmental risk assessment of these compounds in the marine environment. Furthermore, recommendations are made about suitable biological models to assess the ecotoxicological effects of these compounds in marine organisms.

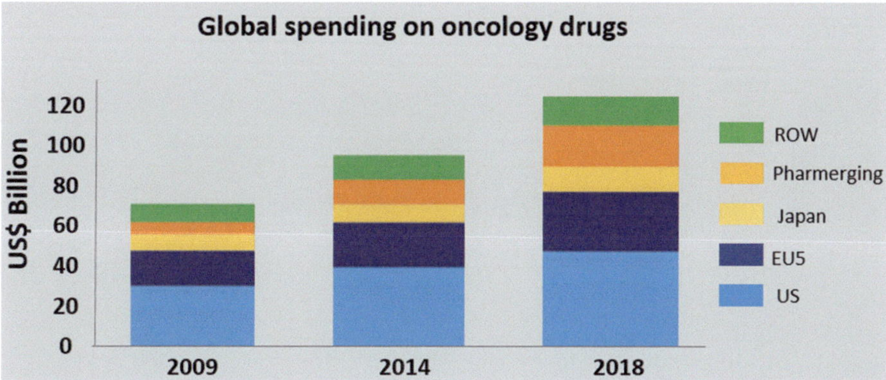

Fig. 13.4 Global forecast of spending on oncology pharmaceuticals: EU5 (Germany, Italy, France, Spain, UK); United States; Japan; pharmerging countries (Brazil, Mexico, Venezuela, Argentina, China, India, Indonesia, Vietnam, etc.); and ROW (rest of the world) (Adapted from Aitken and Kleinrock 2015)

13.2 Pharmaceutical Compounds in the Marine Environment

The occurrence of pharmaceuticals in the marine environment is relatively well documented, at levels ranging from nanograms per liter (ng l^{-1}) to micrograms per liter (μg l^{-1}) in coastal lagoons (Moreno-González et al. 2015), estuaries (Ashton et al. 2004), and marine waters (Buser et al. 1999; Weigel et al. 2002; Ashton et al. 2004; Togola and Budzinski 2008; Vidal-Dorsch et al. 2012; Zhang et al. 2012; Afonso-Olivares et al. 2013; Zhang et al. 2013a, b; Jiang et al. 2014; Pereira et al. 2016), and at nanograms per gram (ng g^{-1}) range in marine sediments (Silva et al. 2011; Long et al. 2013; Beretta et al. 2014; Lara-Martín et al. 2014; Moreno-González et al. 2015). The therapeutic groups most reported are antibiotics, anti-hypertensives, psychiatric drugs, nonsteroidal antiinflammatory drugs (NSAIDs), and analgesics, including household and hospital medicines, and those applied in livestock treatments and fish farms (Gonzalez-Rey and Bebianno 2012). Overall, pharmaceuticals consist of polar and highly ionizable molecules, with one or more dissociable functional groups, ranging from acid to alkaline. Electrolytes exist either as neutral or ionic molecules, depending on their dissociation constant (pK_a) and the pH of the water in which they are dissolved (Rendal et al. 2011). Therefore, part of the available information on physicochemical parameters is estimated by theoretical calculations based on their chemical structures (Zhang et al. 2013a, b). Neutral and non-ionized species indicate lower polarity and higher permeability into membranes and fatty acids compared to ionic forms (Rendal et al. 2011; Fu et al. 2009). Ibuprofen is anionic at pH 7 (above its pK_a), thus hydrophilic, and almost not sorbed onto sediments because of its negative charge, leading to an electrostatic repulsion (Oh et al. 2016).

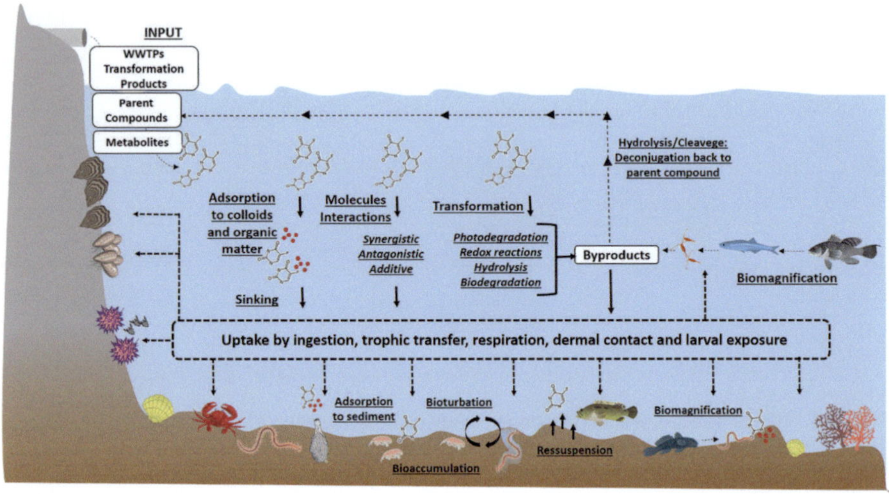

Fig. 13.5 Fate and behavior of pharmaceutical compounds, including anticancer drugs, in the marine environment

Once in the marine environment, pharmaceutical compounds go through different biotic and abiotic processes (Fig. 13.5) that are significantly different from limnic waters with different salinity, pH, temperature, turbidity, organic compounds, and microbial populations that may affect their chemical stability and persistence (Weigel et al. 2002; Kümmerer 2008; Lara-Martín et al. 2014). Because sorption mechanisms can be pH dependent and change from hydrophobic to ionic interaction, the sorption behavior of pharmaceuticals is often poorly predicted by the estimation of the octanol–water partition coefficient (K_{ow}) (Franco et al. 2009; Lara-Martín et al. 2014). In this sense, the application of lipophilicity corrected to pH (log D) is a more reliable strategy that enhances the prediction of environmental behavior of ionizable species (Schaffer et al. 2012). In contrast, sorption is strongly induced by hydrophobicity and neutral forms generated at pH below pK_a (Bui and Choi 2009). Moreover, drug distribution depends on currents and tidal events and on the proximity from point sources in such a way that levels in coastal waters may be orders of magnitude lower compared to the source (e.g., wastewater, river water) (Kim et al. 2017), reflecting the dilution that these compounds undergo in the marine environment (Vidal-Dorsch et al. 2012). As such, concentrations of pharmaceuticals are higher during ebb and low tide and reach minimum values during flood and high tide (Lara-Martín et al. 2014). Therefore, a clear negative relationship exists between pharmaceutical compound concentration and salinity, demonstrated by the decrease of the beta-blocker metoprolol upstream to downstream (from 16 ng l^{-1} to 8 ng l^{-1}) and the antiepileptic drug carbamazepine (10 ng l^{-1} to 6 ng l^{-1}, respectively) (Lara-Martín et al. 2014).

Although pharmaceuticals may be detected at low levels (ng l^{-1}), these concentrations may be significant to induce different consequences by virtue of transformation products (TPs) and combined effects that can occur independently of a similar or dissimilar mode of action (MoA) (Cleuvers 2003). TPs are formed in the environment from parent compounds and metabolites, through physicochemical (e.g., photodegradation, hydrolysis, redox reactions) and biological (e.g. biodegradation) pathways (Mompelat et al. 2009). Despite their potential importance, the formation and fate of TPs from pharmaceuticals have scarcely been investigated (Mompelat et al. 2009). Although a significant number of drugs are excreted and released as inactive conjugates, such as glucoronates and sulfates, deconjugation in the aquatic environment reverses metabolites to the biologically active parent compounds (Noppe et al. 2007; Behera et al. 2011). These species can also be as toxic or even more toxic than their parent compounds (Loffler et al. 2005; Li et al. 2014). Those effects can be a simple addition of a toxic effect (i.e., additive), less than the sum of the separate constituents (i.e., antagonism), or significantly higher than the sum of their individual toxic effects (i.e., synergism) (Láng and Kőhidai 2012; Daughton 2016). Conversely, it is important to establish whether toxicity effects on biota differ over an environmental pH change (Bostrom and Berglund 2015). Bioassays conducted with ionizable pharmaceutical compounds (2,4-dichlorophenol, 2,4,6-triclorophenol, pentachlorophenol, ibuprofen, fluoxetine) at different experimental environmental pHs (from 6 to 9) indicate that toxicity depends on the more toxic uncharged fraction (Nakamura et al. 2008; Xing et al. 2012), but when the ionic fraction increases, detrimental contributions may be shifted toward the ionic form (Bostrom and Berglund 2015).

The processes to which pharmaceuticals and respective products are subjected, after excretion until reaching marine waters, should not be disregarded because they are accountable for changing the kinetic and reactivity of drugs, particularly in respect to ionization, which is likely to contribute to unexpected behavior in aquatic compartments. Differing from many organic contaminants historically studied (e.g., persistent organic pollutants), the partitioning dynamics of the majority of pharmaceuticals are not only caused by hydrophobic interactions, but are also influenced by hydrogen bonding, cation exchange, cation bridging, and surface complexation, which hinder environmental modeling approaches of exposure, uptake (i.e., ingestion, dermal exposure, respiration exchange, larval stage hazard), trophic transfer, and bioaccumulation (Yamamoto et al. 2009; Du et al. 2014; Ribeiro et al. 2015).

Furthermore, there is a high possibility that hydrophobic drugs may be trapped in sediments in estuary or coastal conditions by high salinity. The settling of pharmaceuticals in sediments despite decrease from the water column represent another source of exposure to biota, mainly benthonic species (Oh et al. 2013) (Fig. 13.5). Thus, fine-grained sediments tend to accumulate large amounts of pharmaceutical compounds because of the high sorptive capacity.

Physical disturbances of sediments may occur during weather events, incoming tides, sediment transport, rainfall, and bioturbation by benthic organisms, in addition to dredging, fishing, and shipping. Those events remobilize contaminants and resuspend them back to the water column (Ferguson et al. 2013). More oxidative conditions to which contaminants are exposed in overlying water could result in desorption and transformation into more bioavailable or toxic compounds (Eggleton and Thomas 2004).

13.3 Cytotoxic Pharmaceutical Compounds

As the presence of pharmaceuticals in the environment increased during the past decades, new questions and issues began to emerge (Haddad et al. 2015), with recent attention given to drugs applied in anticancer therapies. Anticancer drugs are classified according to the anatomical therapeutic classification (ATC), where they are ranked as antineoplastic and immunomodulating agents, at Class L (Kovalova 2009; Booker et al. 2014). Further, four subgroups of anticancer drugs exist: L01 (antineoplastic agents); L02 (endocrine therapy); L03 (immunostimulants); and L04 (immunosuppressants), as seen in Table 13.1.

Antineoplastic agents are designed to interact directly or indirectly with DNA, by damaging its structure, or inhibiting, altering, and disrupting the mechanisms of its transcription, replication, and synthesis. Derivatives of nitrogen mustards were developed, including the DNA alkylators cyclophosphamide, chlorambucil, cisplatin, and melphalan, all of which are currently used in clinical therapeutics (Xie 2012; Cheung-Ong et al. 2013). Other examples of DNA-alkylating agents used in cancer treatment include nitrosoureas (e.g., carmustine, lomustine, semustine) and triazenes (e.g., dacarbazine, temozolomide) (Cheung-Ong et al. 2013). Most of these pharmaceuticals do not strictly act over tumor drivers, but also in mechanisms regarding all growing cells (Caley and Jones 2006; Kosjek and Heath 2011; Toolaram et al. 2014; Heath et al. 2016), usually leading to detrimental effects in patients from adverse toxicity to nontargeted tissues. Anti-metabolites represent a class of anticancer drugs that exert their activity by blocking nucleotide metabolism pathways and DNA replication, including the pyrimidine analogues 5-fluorouracil (5-FU), capecitabine, floxuridine, and gemcitabine, and the purine analogues 6-mercaptopurine, 8-azaguanine, fludarabine, and cladribine. In general, these compounds have great curative effects and prolong survival among patients, although not being effective for all types of cancer.

In contrast, there are compounds accountable for the disruption of biological processes through mechanisms of blocking cell replication factors, or indirectly by recruiting macrophages and monocytes cells (Besse et al. 2012; Caley and Jones 2012), thus, not directly involved with DNA. Endocrine therapy manipulates hormone-dependent receptors with tissue-selective interactions that lead to agonist or antagonist activities, represented by the group of selective estrogen receptor modulators (SERMs), such as tamoxifen, whose action is not elicited at DNA level (Paterni et al. 2014; An 2016). Attachment of these endocrine-disrupting compounds (EDCs) to respective binding sites acts competitively and subsequently inhibits the transcription of estrogen-responsive genes responsible for cell replication of estrogen receptor-positive cells, such as seen in breast and prostate cancer (Xie 2012; Pinto et al. 2014). Currently, the aforementioned groups of drugs are the main group used in cancer treatments. The evolution of drug discovery contributed to finding drug pathways downstream of checkpoint blockage, based on the insight that cancer is monitored by the immune system (Mahoney et al. 2015; Dhanak et al. 2017). A subset of cancer mutations generates sequences of protein during

Table 13.1 Classification of anticancer drugs

Classification	Subcategory		Compounds	MoA
L01: Antineoplastic agents	L01A: Alkylating agents	L01AA: Nitrogen mustard analogues	Cyclophosphamide, chlorambucil, ifosfamide	Alkylating agents react with nitrogen (N) and extracyclic oxygen (O) atoms of DNA bases to generate a variety of covalent adducts. Induce a range of cytotoxic and mutagenic adducts onto DNA
		L01AB: Alkyl sulfonates	Busulfan, treosulfan	
		L01AC: Ethylene imines	Thiotepa, triaziquone	
		L01 A Nitrosoureas	Carmustine, lomustine, nimustine	
		L01AG: Epoxides	Etoglucid	
		L01AG: Other alkylating agents	Mitobronitol, decarbazine, temozolomide	
	L01B: Anti-metabolites	L01BA: Folic acid analogues	Methotrexate	Structurally similar to endogenous nucleic acids, can be incorporated into the metabolic pathways instead of the endogenous purine and pyrimidines, thereby affecting the enzyme-dependent synthesis of DNA and cell reproduction
		L01BB: Purine analogues	Mercaptopurine, tioguanine, cladribine	
		L01BC: Pyrimidine analogues	Cytarabine, 5-fluorouracil, gemcitabine, capecitabine	
	L01C: Plant alkaloids and other natural products	L01CA: Vinca alkaloids and analogues	Vinblastine, vincristine, vinflunine	Interacts with microtubules or tubulins leading to inhibition of synthesis of proteins and nucleic acids, disruption of mitotic spindle, and eventually cell death
		L01CB: Podophyllotoxin derivatives	Etoposide, teniposide	
		L01CC: Colchicine derivatives	Demecolcine	
		L01CD: Taxanes	Paclitaxel, docetaxel	
		L01CX: Other plant alkaloids and natural products	Trabectedin	

(continued)

Table 13.1 (continued)

Classification	Subcategory		Compounds	MoA
	L01D: Cytotoxic antibiotics and related substances	L01DA: Actinomycines	Dactinomycin	Involve direct toxic action on cellular DNA, interfering with DNA replication and protein synthesis
		L01DB: Anthracyclines and related substances	Daunorubicin, doxorubicin, epirubicin	
		L01DC: Other cytotoxic antibiotics	Bleomycin, mitomycin	
	L01X: Other antineoplastic agents	L01XA: Platinum compounds	Cisplatin, carboplatin	e.g.: Cisplatin products highly electrophilic. Act toward nucleophilic sites in genomic and mitochondrial DNA, producing DNA inter- and intra-strand adducts that result in DNA distortion, inhibition of DNA replication, and interruption of cell division
		L01XB: Methylhydrazines	Procarbazine	
		L01XC: Monoclonal antibodies	Rituximab, trastuzumab	
		L01XD: Sensitizers used in photodynamics	Temoporfin	
L02: Endocrine therapy	L02A: Hormones and related agents	L02AA: Estrogens	Ethinylestradiol, diethylstilbestrol	Antitumor activity involves suppression of luteinizing hormone by inhibition of pituitary function
		L02AB: Progestogens	Megestrol, medroxyprogesterone	
		L02AE: Gonadotropin-releasing hormone analogues	Leuprorelin, buserelin	
	L02B: Hormone antagonists and related agents	L02BA: Anti-estrogens	Tamoxifen, toremifene	Diverse group of compounds bind to estrogen-α (ERα) and estrogen-β (ERβ) receptors and produce estrogen agonist effects in some tissues, but estrogen antagonist activity in others. Activity determined in part by formation of estrogen receptor–molecule complexes that vary in their ability to activate genes when bound to ERα or ERβ
		L02BB: Anti-androgens	Flutamide, nilutamide	
		L02BG: Aromatase inhibitors	Letrozole, anastrozole	
		L02BX: Other hormone antagonists and related agents	Abareliz, degarelix	

L03: Immunostimulants	L03A: Immunostimulants	L03AA: Colony-stimulating factors	Filgrastim, lenograstim	Immunomodulation is based on the stimulation of T-cell function with antibodies that block or activate regulatory receptors is sufficient to cause the regression of some tumors
		L03AB: Interferons	Interferon-α natural, interferon-β natural	
		L03 AC: Interleukins	Aldesleukin	
		L03AX: Other immunostimulants	Lentinan, roquinimex	
L04: Immunosuppressants	L04A: Immunosuppressants	L04AA: Selective immunosuppressants	Mycophenolic acid, sirolimus	
		L04AB: Tumor necrosis factor-alpha (TNF-α) inhibitors	Afelimomab, infliximab	
		L04AC: Interleukin inhibitors	Daclizumab, basiliximab	
		L04AD: Calcineurin inhibitors	Ciclosporin, tacrolimus	
		L04AX: Other immunosuppressants	Methotrexate, azathioprine	

mutagenesis called neoantigens that can be processed into peptide antigens and recognized as foreign by T cells. Targeting immune checkpoints such as programmed cell death protein 1 (PD1), programmed cell death 1 ligand 1 (PDL1), and cytotoxic T-lymphocyte antigen 4 (CTLA4) has achieved noteworthy benefit in multiple cancers by blocking immune inhibitory signals and enabling patients to produce an effective anti-tumor response (Mahoney et al. 2015; Guillerey et al. 2016; Dhanak et al. 2017).

Most chemotherapy regimes in clinical practice consist of a combination of several agents from different pharmaceutical groups, in an attempt to provide additive or synergistic effects to achieve maximal tumor cell death and avoid resistance (Caley and Jones 2006). In spite of their overall application in chemotherapy, there are strong differences in the chemical structure of the several anticancer groups that also have distinct MoA.

13.4 Sources and Cytotoxic Drugs in the Marine Environment

Cancer treatment is shifting from hospital treatment and medical facilities (where 80% of chemotherapies are administered in hospitals and medical facilities, mainly via intravenous or oral drug ingestion) to home treatment (i.e., as outpatients) (Johnson et al. 2008; Kosjek and Heath 2011; Besse et al. 2012). The generalization of home treatment and the availability of molecules in pharmacies are increasing to improve patient treatment comfort. Consequently, more diffuse parental and metabolized cytotoxic drugs are prone to succeed during home treatment compared to discharges from hospital wastewater effluents that are directly introduced into wastewater treatment plants (WWTPs) (Kümmerer 2008; Kümmerer 2010; Zhang et al. 2013a, b; Mater et al. 2014) (Fig. 13.6). Therefore, other inputs of cytotoxic drugs into the marine environment are through wastewater effluents of manufacturing processes, disposal of unused or expired drug products, and accidental spills during manufacturing or distribution (Díaz-Cruz et al. 2003).

WWTPs were not designed to treat complex mixtures of these compounds (Lenz et al. 2005; Rowney et al. 2009; Besse et al. 2012; Parrella et al. 2014b). Therefore, the extent to which WWTPs are successful in removing such mixtures depends on their technology and on the physicochemical properties of the effluent (Kümmerer 2008). Besides this scenario, sanitation in developing countries still relies on poorly managed septic tanks with direct discharges into the ground, and on surface waters that ultimately reach estuarine and marine ecosystems, most of the time untreated (Abessa et al. 2005; Pessatti et al. 2016). Currently, developing and underdeveloped regions are struggling with other more immediate problems such as infectious diseases, water supply, sanitation, waste disposal, war, and famine; thus, environmental problems of cytotoxic drugs may not be a major issue in the current scenario (Rahman et al. 2009). Installation and development of advanced wastewater

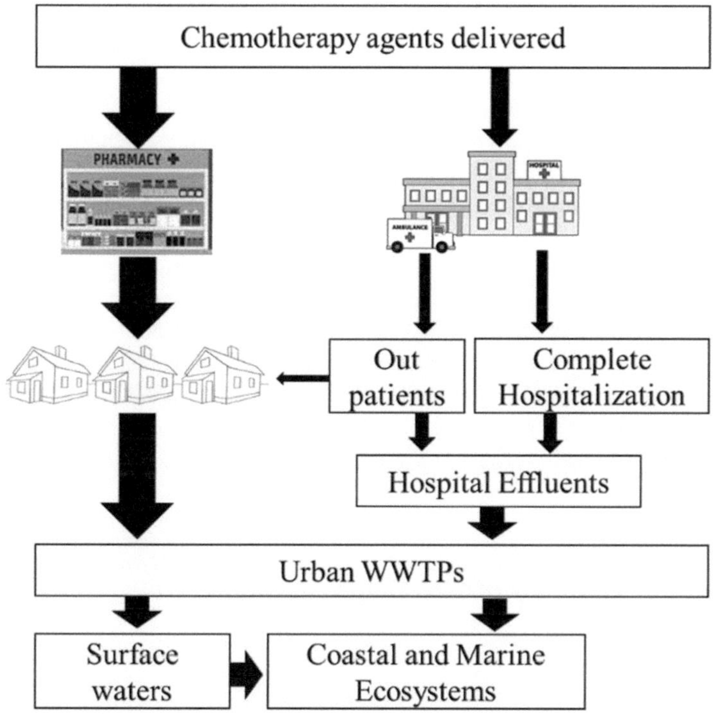

Fig. 13.6 Sources and pathways for anticancer drugs into coastal waters. (Adapted from Besse et al. 2012)

treatment technology require significant investment and skilled labor for operation, which is not considered a governmental priority in low-income regions (Rahman et al. 2009; Gaw et al. 2014). In those areas, because of landfill disposal of waste and insufficient removal of pharmaceuticals in WWTPs, chemicals may leach into groundwater aquifers and harm drinking water supplies or be introduced directly into marine systems. Additionally, those regions present higher trends of cancer risk factors with projections of new cases accounting for 60% (WHO 2014).

Therefore, most of the information available about the presence of cytotoxic molecules in aquatic systems focuses on the detection of these compounds at their source, such as hospital influents/effluents and sewage wastewater effluents (Steger-Hartmann et al. 1996, 1997; Ternes 1998; Castiglioni et al. 2005; Lenz et al. 2005; Buerge et al. 2006; Moldovan 2006; Mahnik et al. 2006, 2007; Kovalova 2009; Yin et al. 2010; Martín et al. 2011; Isidori et al. 2016) where they were detected at levels ranging from 6 to 7973 ng l^{-1} for methotrexate and ciprofloxacin, respectively. These compounds undergo transformation and degradation in the influent, through the WWTPs, and in the effluent, until they reach surface waters and the marine environment (Ferrando-Climent et al. 2014). The percentage of removal achieved at the end of a batch bioreactor WWTP for tamoxifen, ciprofloxacin, and etoposide is approximately 91%, 84%, and 100%, respectively. In contrast, ifosfamide, cyclophosphamide, vincristine, docetaxel, and paclitaxel are shown to be recalcitrant because of their inefficient elimination through the biological treatment in WWTPs. Moreover, Martín et al. (2012) identified removal rates of cytotoxic drugs in WWTPs to range from 0% (e.g., doxorubicin, gemcitabine, paclitaxel, cisplatin, cyclophosphamide, vinorelbine) to 100% (e.g., etoposide, fluorouracil, methrotrexate). The alkylating agent cyclophosphamide is considered resistant to conventional biological treatments and was detected in WWTP effluents in a range from 0.19 to 125 ng l^{-1} (Česen et al. 2015), also withstanding degradation processes over the aquatic pathway until concentrations detected in surface waters reach 0.15 to 17 ng l^{-1} (Buerge et al. 2006; Moldovan 2006; Busetti et al. 2009). Therefore, combinations of advanced and nonconventional technologies in WWTPs and water reclamation are considered alternatives to increase the percentage of removal of chemotherapy drugs: these include electrolysis, UV radiation with H_2O_2, ozonation, and membrane bioreactors (MBR) (Zhang et al. 2013a, b).

13.5 Presence of Cytotoxic Drugs in the Marine Environment

The first publication regarding the presence of anticancer drugs in the aquatic environment dates from the 1980s and depicted the distribution profile of several classes of pharmaceuticals in different aquatic compartments (e.g., sewage influent and effluent, surface waters and drinking water (Aherne et al. 1985; Richardson and Bowron 1985). These authors outlined that anticancer drugs are likely to yield risk, particularly to surface waters that receive discharges from WWTPs, or by the hazards identified in nurses after occupational exposure from pharmaceutical manipulation. Following the improvement of analytical technologies a set of cytotoxic pharmaceuticals were identified in environmental screening in the range of ng l^{-1} or less (Kummerer et al. 1997; Buerge et al. 2006; Zuccato et al. 2006; Kovalova et al. 2009; Toolaram et al. 2014). Data available to identify the fate and behavior of cytotoxics and to predict their pathways in aquatic compartments are based on their chemical structure, pK_a, bioconcentration factor (BCF), K_{ow}, organic carbon partition coefficient (K_{oc}), solubility, Henry's coefficient (KH), and vapor pressure (Kosjek and Heath 2011; Xie 2012; Toolaram et al. 2014). In addition, the search for metabolites and TPs, as well as mixtures of parental forms, recently has been evaluated (Fatta-Kassinos et al. 2011; Toolaram et al. 2014; Haddad et al. 2015; Zhang et al. 2017). Kovalova et al. (2009), Negreira et al. (2013, 2014a, 2014b, 2015), and Zhang et al. (2013a, b) contributed to the advances in analytical chemistry and processes that underline technologies in WWTPs able to remove anticancer drugs, and conform to molecule physicochemical properties, stability, and metabolism. By evaluating a set of parameters that may explain their mobility and partition into water, activated sludge, suspended solids, and sediments, it is possible to ensure that there are disparities in profiles of solubility and adsorption, even for compounds inside the same therapeutic group; thus, the evaluation of anticancer drug behavior must consider compounds individually. For instance, low levels of K_{ow} indicate chemicals to be permanent in solution and not prone to sorb onto a solid compartment or to bioaccumulate, as observed for cyclophosphamide and etoposide, with K_{ow} levels of 0.63 and 0.6, respectively. In this sense, these drugs are expected to pass unaltered through WWTPs into the receiving waters (Kosjek and Heath 2011). However, despite the low K_{ow} levels, cyclophosphamide was detected in sewage sludge (20 ng g^{-1}), indicating a capacity to bind to a solid organic matrix, so that according to coastal waters dynamics, aggregation and settling onto the bottom may become possible.

In UK estuaries, levels of tamoxifen range between 13 and 71 ng l^{-1}, suggesting that the physicochemical properties of this compound are such that it will persist for a long time in this ecosystem (Thomas and Hilton 2004). Also, cisplatin, which is relatively inert, once in aqueous solutions of low electrolyte concentration (as an intracellular medium), the chlorine ligands are gradually replaced by water in a stepwise process to form more reactive aqueous species (i.e., mono- and diaquacisplatin) (Curtis et al. 2010; Turner and Mascorda 2014). Similarly, electrostatic interactions in low-salinity environments provide a predominance of reactive complexes, whereas in seawater the ionic strength and presence of chlorine reduces their activity and induces adsorption onto estuarine particles (Curtis et al. 2010). This reaction decreases the levels in water and toxicity (Turner and Mascorda 2015), although bioavailability in deposit feeders may potentialize intake. Therefore, it is crucial to assess whether in marine waters these compounds become more or less reactive and what processes drive their bioavailability and toxicity.

13.6 Ecotoxicological Effects of Anticancer Drugs on Marine Organisms

Evidence of the unselective MoA of anticancer drugs triggered concern about their potential risks to nontarget aquatic organisms, even though they are prescribed in much lower quantities compared to other groups of pharmaceuticals. Occurrence in the environment is either below or at the current limits of analytical detection (below ng l^{-1}) (Johnson et al. 2013; Heath et al. 2016). Because the chemical MoA affects not only target cells, but also nontumor cells, any non human cells could be vulnerable to genetic impairment and DNA damage (Zounková et al. 2007; Parrella et al. 2014a), with potential effects even at trace levels (Bound and Voulvoulis 2004; Johnson et al. 2008; Kosjek and Heath 2011; Parrella et al. 2014b).

As mentioned previously, endocrine drugs applied in combination with cytotoxics in chemotherapy can act as EDCs in aquatic populations. Eventual homologies between hormone receptors of vertebrates and nontarget organisms that bind with respective molecules and induce antagonist or agonist mechanisms are reflected in alteration of metabolism, homeostatic control, or reproduction.

The application of a battery of selected biomarkers to access cytotoxic effects has the potential to offer additional and valuable information regarding impairments caused by anticancer drugs on nontarget marine organisms. This information should include the global impact of genotoxic insults caused by exposure to multiple chemotherapy agents (Suspiro and Prista 2011). Alterations in the set of scavenging systems [glutathione (GSH), superoxide dismutase (SOD), catalase (CAT)] and oxidative stress are the first line of action of those drugs, thus mandatory for assessing further processes that involve cell damage and death. Abnormalities affecting chromosome numbers or structure during mitosis, DNA strand breaks, and incomplete excision repair sites, together with cell viability, are representative approaches for studying mutagenicity, genotoxicity, and cytotoxicity caused by conventional anticancer drugs.

Although the application of some biomarkers may be nonspecific to reflect the influence of multiple environmental stress factors, the evaluation of biochemical and genotoxic alterations is rather consistent for monitoring anticancer drug purposes (Suspiro and Prista 2011). Clues for understanding how exposure and the effects of anticancer drugs are associated to respective cellular signaling pathways and interactions to specific receptors should be further evaluated through multiple gene transcriptional patterns (Sun et al. 2011). For full comprehension of toxicity and association of molecular effects of complex mixtures of cytotoxic compounds, large databases comprising expression profiles of individual pure chemicals are required. Transcriptomics and proteomics are complementary tools when combined and should be applied to reflect genomic changes and functional disturbance.

The generation of reactive oxygen species (ROS) in cancer cells, especially hydrogen peroxide (H_2O_2), is crucial to mediate particular signaling pathways in cell growth, proliferation, differentiation, protein synthesis, metabolism, and cell survival (Trachootham et al. 2009; Lord and Ashworth 2012; Gorrini et al. 2013). In contrast, such oxidative stress is counteracted by elevated antioxidant defense mechanisms from cancerous cells (Gorrini et al. 2013). Therefore, it is important to highlight that chemotherapeutic drugs are designed to exceed cellular ROS levels and overwhelm antioxidant defenses, aiming to induce irreparable damage and subsequently trigger tumor cell apoptosis (Liou and Storz 2010) (Fig. 13.7). Even cellular antioxidant mechanisms may be unable to prevent the interference of ROS provided by drugs on critical cellular mechanisms (Conklin 2004), combined with enhancement of lipid peroxidation products (LPO), inhibition of non enzymatic molecules (i.e., glutathione, flavonoids, vitamins A, C, E), and antioxidant enzymes (Gorrini et al. 2013). Taxanes (i.e., paclitaxel and docetaxel), vinka alkaloids (vincristine and vinblastine), and anti-metabolites (5-FU) induce ROS generation by the release of cytochrome *c* from mitochondria, leading to disruption of the electron transport chain and resulting in the production of superoxide radicals (Cairns et al. 2011). In addition, therapies that block glutathione (GSH) synthesis are also applied as a chemotherapy complement to modulate cancer cell sensitivity to drugs (Conklin 2004; Cairns et al. 2011), because GSH metabolism is actively

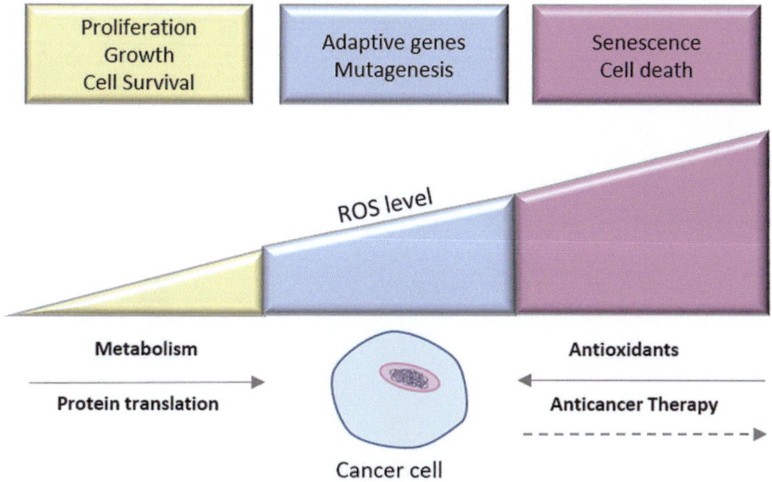

Fig. 13.7 Balance of induction and scavenging of reactive oxygen species (ROS) levels in cancer cells. Mutations and adaptations provide means for cancer cells to regulate ROS to moderate levels (*blue*), avoiding detrimental effects and cell death (*purple*) through strong antioxidant mechanisms, but also under ROS-mediated mutagenic events to promote tumorigenesis. The role of anticancer therapy is to overwhelm antioxidant levels and kill cancer cells via oxidative stress and apoptosis. (Adapted from Cairns et al. 2011)

involved in resistance mechanisms against anticancer compounds (i.e., anthracyclines, alkylating agents, and platinum-containing drugs).

Platinum-based compounds and alkylating agents are also known to increase ROS to extremely high levels in such a way that mitochondrial metabolism is impaired and may lead to caspase activation and cell death, DNA being a critical target for cytotoxicity (Liou and Storz 2010; Dasari and Bernard Tchounwou 2014). Reactive parental cytotoxic drugs or metabolites may covalently bind to DNA bases and cause alterations by single or cross-linked interactions (Lawley and Phillips 1996; Helleday et al. 2008), although cell ability to resist and repair is widely divergent over the exposed eukaryotic groups (Kondo et al. 2010). Lesions caused by direct contact of drugs with DNA, such as base alterations, cross-links, and single-strand breaks, may originate chromosome aberrations and further subsequent malignancies (Suspiro and Prista 2011). Although anticancer therapies widely aim to prevent cell replication through different cell cycle phases, the DNA damage following S-phase provides the ideal arrest of cell cycle because the replication of damaged DNA causes further cell deaths (Helleday et al. 2008).

Numerous studies were performed with marine species, using several classes of cytotoxic drugs, by exposing them individually, or as mixtures, prodrugs, metabolites, or TPs, via water or spiked into sediments (Aguirre-Martínez et al. 2013; Gonzalez-Rey and Bebianno 2014; Maranho et al. 2015; Aguirre-Martínez et al. 2016a, b; Pires et al. 2016; Trombini et al. 2016b). Information is still scarce regarding the ecotoxicological effects of anticancer agents and their metabolites in

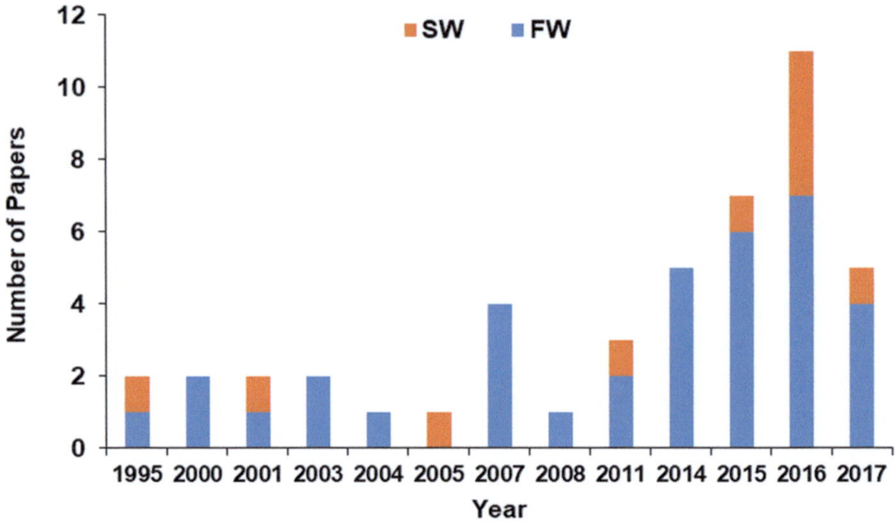

Fig. 13.8 Number of papers published from 1995 until December 2017 regarding ecotoxicological data of anticancer drugs using freshwater (*FW*) and seawater (*SW*) organisms

marine organisms, illustrating a critical discrepancy compared with freshwater biota (Fig. 13.8). Fortunately, despite the quantitative disparity, and the contrasts in the quality of the acquired data for marine and freshwater organisms and endpoints analyzed, marine experiments were aimed to assess biochemical effects to understand the conditions of oxidative stress that occur under chemotherapy and corroborate effects at relevant levels in the environment (Table 13.2).

In the marine eel *Anguilla anguilla* L. exposed to cisplatin (6.25–100 mg l^{-1}), an increasing number of sister chromatid exchanges (SCE) per metaphase occurs (Santos and Pacheco 1995), although concentrations are environmentally unrealistic. SCE results from symmetrical exchange of DNA replication products between sister chromatids at a given locus represent a sensitive test to detect genotoxic effects. The interruption of cell division caused by chemotherapy agents may also generate separation of nuclear material and dispersion in the cytoplasm in a small collection of micronuclei, as observed in freshwater fishes exposed to trace levels of 5-FU, bleomycin, and mytomicin (Grisolia and Cordeiro 2000; Kovacs et al. 2015).

Antioxidant system defenses [i.e., SOD, CAT, and glutathione peroxidase (GPX)] of the mussel *Mytilus galloprovincialis* exposed to 100 ng l^{-1} cisplatin for 14 days were shown to be a reliable scavenger mechanism that recovered LPO products to basal levels, in both gills and digestive glands. However, direct contact of reactive cisplatin products (monoaqua- and diaquacisplatin) to DNA indicated formation of intra- and interstrand breaks and DNA–protein cross-links (Trombini et al. 2016a). CisPt is designed to act over N7 centers of purine residues, especially guanine, mainly generating 1,2-intrastrand and 1,2-intrastrand adducts, besides the enhancement of oxidative stress and damage in mitochondrial proteins, which result in an overall failure of cellular function and apoptosis (Chu 1994; Dasari and

Table 13.2 Ecotoxicity effects of anticancer drugs in marine species

Drug	Species	Conc. range (ng l^{-1})[a]	Time (days)[b]	Endpoint	Effect	References
Cisplatin	*Nereis diversicolor*	0.1–100	14	Behavioral impairment, oxidative stress, oxidative damage, neurotoxicity	0.0001: AChE, SOD, CAT, GST *↓; GPx, MT, LPO *↑; burrowing behavior impaired	Fonseca et al. (2017)
	Mytilus galloprovincialis	100	14	Oxidative stress, oxidative damage, neurotoxicity, genotoxicity, oxidative damage	0.0001: (DG, G) AChE, LPO *↑; (G) SOD, T-GPx *↓; (DG) SOD, CAT, T-GPx, GST *↑; LPO *↓	Trombini et al. (2016b)
Cyclophosphamide	*Anguilla anguilla*	6.5–100[a]	72[b]	Sister chromatid exchange	12.5; 25; 50; 100: *↑	Santos and Pacheco (1995)
Methotrexate	*Ampelisca brevicornis*	1–10,000	10	Oxidative stress, oxidative damage, genotoxicity	0.0001 and 0.001: EROD, GST, GR, GPx, LPO, and DNA damage *↑	Moreira et al. (2016)
	Paracentrotus lividus	10–1,000,000,000	1[b]	Fertilization	$EC_{50} = 15{,}000$	Aguirre-Martínez et al. (2016a, b)
		10–1,000,000,000	2	Larval development	$EC_{50} = 1{,}500{,}000$	
	Isochrysis galbana	50–500,000,000	4	Growth inhibition	$EC_{50} = 84{,}000{,}000$	Aguirre-Martínez et al. (2016a, b)
	Ruditapes philippinarum	100–50,000	14	Oxidative stress, oxidative damage, neurotoxicity, genotoxicity, oxidative damage	100: DBF, DNA damage *↓; 1000: DBF, DNA damage *↓, GPx *↑; 10,000: DBF *↓, GPx *↑; 50,000: DBF, AChE *↓, GPx *↑	Aguirre-Martínez et al. (2016a, b)

(continued)

Table 13.2 (continued)

Drug	Species	Conc. range (ng l^{-1})[a]	Time (days)[b]	Endpoint	Effect	References
Tamoxifen	*Paracentrotus lividus*	10–1,000,000,000	1[b]	Fertilization	EC_{50} = 15,000,000	Aguirre-Martínez et al. (2016a, b)
		10–1,000,000,000	2	Larval development	EC_{50} = 1,000,000,000	
		10^{-8}–10^{-6} M	0.5[b]	Fertilization	10^{-7} M: $*\downarrow$	Pagano et al. (2001)
		10^{-8}–10^{-6} M	0.5[b]	Offspring quality	10^{-6} M: $*\downarrow$	
	Sphaerechinus granularis	10^{-8}–10^{-6} M	3	Larval development	10^{-6} M: Embryotoxicity $*\uparrow$	
	Strongylocentrotus purpuratus	–	4	Larval development	EC_{50} = 50	Roepke et al. (2005)
	Isochrysis galbana	1–500,000,000	4	Growth inhibition	EC_{50} = 35,000,000	Aguirre-Martínez et al. (2016a, b)
	Ruditapes philippinarum	100–50,000	14	Oxidative stress, oxidative damage, genotoxicity, oxidative damage	100: EROD, LPO $*\uparrow$; 1000: EROD, GR, LPO $*\uparrow$, GST $*\downarrow$; 10,000: EROD, LPO $*$; 50,000: EROD, GST, GR, LPO $*\uparrow$	Aguirre-Martínez et al. (2016a, b)
	Tautogolabrus adspersus	500,000–5,000,000[a]	17	Egg production	500,000: $*\downarrow$	Mills et al. (2015)
	Sparus aurata L.	100,000[a]	25	Gene expression	Vitellogenin $*\uparrow$	García-Hernández et al. (2016)
		100,000[a]	25	Gene expression	Immune response (ilb1, tnfa, tgfb1, mhc1a, tlr9) $*\uparrow$	
		100,000[a]	25	Sperm concentration	$*\uparrow$	
		100,000[a]	25	Sperm motility	$*\uparrow$	

[a]Drug exposure through intraperitoneal injection (Santos and Pacheco 1995) and oral administration (Mills et al. 2015), expressed in mg kg^{-1} body weight
[b]Time shown in hours

Bernard Tchounwou 2014). In contrast, no genotoxicity was observed in the benthic polychaete *Nereis diversicolor* exposed to the same CisPt concentration, as an outcome of the nucleophilic trapping capacity of metal-like proteins and GSH to the drug, widely recognized as a mechanism of drug resistance (Fonseca et al. 2017). In addition, inhibition of the antioxidant activity defense systems SOD, CAT, and GPX, and biotransformation of GST, do not give enough protection against ROS formation, thus altering the behavior of this species and contributing to neurotoxicity and oxidative damage. A 10-day sediment toxicity test, spiked with methotrexate (10, 100, and 1000 ng g^{-1}), showed significant relationships between amphipod mortality, biochemical disruption by induction of enzymes involved in oxidative stress (GR, GPx), in metabolism of phase I (EROD) and II (GST), membrane damage (LPO), and genotoxicity (DNA strand breaks) (Moreira et al. 2016).

To date, the chemotherapeutic drug with more available information is the selective estrogen-receptor modulator (SERM) tamoxifen (TAM). With these multiple activities, TAM is a SERM of the triphenylethylene chemical family that possess the ability to antagonize the proliferative action of estrogen through competitive binding to its respective receptor (Goldstein et al. 2000). TAM has become the most prescribed clinical drug in hormone therapy of primary and recurrent breast cancer around the world, irrespective of menopausal state. It has shown an extensive success on overall survival in clinical trials in women whose tumors are estrogen receptor positive and is effective in therapies of metastatic breast cancer. Its MoA is based on stable complex bindings with the receptors of 17 β-estradiol, causing sites for the respective molecule to bind to be unavailable. The drug has estrogen receptor antagonist properties, with no activation of the receptor to which it is attached, and finally counteracts the proliferative mechanism of estrogen. Nevertheless, it also has an agonist action in other tissues, with full activation of the binding receptor. TAM is metabolized in two more potent molecules, hydroxyl-dismethyltamoxifen and 4-hydroxytamoxifen (Besse et al. 2012), which are also disposed in the marine environment and are highly pertinent for ecotoxicity evaluation.

The clam *Ruditapes philippinarum* showed a decline in health status after a 14-day exposure to TAM (50 μg l^{-1}) by decrease in lysosomal membrane stability (LMS) followed by neurotoxicity, induction of antioxidant responses, and LPO (Aguirre-Martínez et al. 2016a, b). Environmental levels of TAM (10 ng l^{-1}) also caused teratogenic effects in larval development of the sea urchin *Paracentrotus lividus* after 48 h of exposure (Aguirre-Martínez et al. 2016a, b). Pagano et al. (2001) found that at therapeutic levels of the drug (10^{-6} M or 0.37 mg l^{-1}), in a dose-related manner, ROS formation and oxidation of DNA in *P. lividus* are likely to contribute to cytogenetic aberrations and cytolysis throughout embryogenesis. Responses were similar in the species *Strongylocentrotus purpuratus*, although binary incubations of TAM with other endocrine disruptor compounds [i.e., estradiol, ethynylestradiol (EE2), bisphenol A, octyphenol], at 0.02 ng ml^{-1}, significantly diminished the efficiency of their estrogenic activity and attenuated teratogenic potential on larval development, suggesting an antagonist and "protective" activity over estrogen receptors (ERs) (Roepke et al. 2005). Similarly, in the gilthead seabream *Sparus aurata* L. in which a diet containing a mixture of TAM (100 μg g^{-1} food) and EE2

(5 μg g^{-1} food), for 25 days, stable binding of TAM to ERs was indicated, thus retaining fewer available sites for EE2 docking (García-Hernández et al. 2016). Transcription of vitellogenin (VTG1, VTG2) and ERα were downregulated in a drug-dependent manner in females of freshwater *Oryzias latipes* exposed to a combination of TAM (30 μg l^{-1}) and EE2 (20 ng l^{-1}), in contrast to males where mRNA transcription was stimulated (Sun et al. 2011). Little is known about the interference of TAM with ligand-receptor binding of invertebrates. However, as extensively described for mammals, mechanisms of interaction of TAM in invertebrates were proposed to depend on organism gender, concentration, and tissue, and the amount of ER ligands to form stable complexes with the drug (Pagano et al. 2001; Chikae et al. 2004; Sun et al. 2011; García-Hernández et al. 2016), although this should be fully demonstrated.

These endpoints across different levels of biological organization reinforce the problem of anticancer agents as emergent contaminants of concern and the need to assess risks of priority molecules according to their hazardous impacts (Rowney et al. 2009; Ortiz de García et al. 2014; Lolic et al. 2015).

13.7 Environmental Risk Assessment of Cytotoxic Pharmaceuticals in the Marine Environment

An environmental risk assessment (ERA) is a set of tools to characterize the nature and magnitude of risk to human and ecological health from chemical contaminants and other stressors that may be present in the environment. Risk managers use this information to help to define protection measures (US EPA 1998). The ERA of active pharmaceutical ingredients (APIs) involves a two-tiered approach. Although anticancer drugs act by a distinctive MoA, they should also be integrated as part of the stepwise approach proposed by Environmental Risk Assessment of Medicinal Products for Human Use (EMEA), so that their commercialization may be authorized (EMEA 2006). However, EMEA guidelines only require environmental assessment for newly authorized pharmaceutical compounds (Besse et al. 2012). Therefore, several classes of anticancer agents that were introduced in the market before 2006 were not subjected to ERA procedures (EMEA 2006; Aguirre-Martínez et al. 2016a, b). Furthermore, ERA is only focused on pharmaceutical use and excludes storage, disposal, synthesis, and manufacture (Aguirre-Martínez et al. 2016a, b). ERA of pharmaceutical compounds covers sewage works and freshwater compartments, but the environmental risk in marine ecosystems is not addressed. The Marine Strategy Framework Directive (Directive 2008/56/2008) of the European Union aims to protect, by 2020, the coastal resources based upon which marine-related economic and social activities, according to the particular features of each European region, should identify and assess pharmaceuticals as a predominant pressure. The monitoring of prioritized pharmaceuticals and relevant metabolites in coastal aquatic resources may also be part of a global effort to achieve the sustainable development goals, proposed by the Agenda 2030 of the United Nations (Osborne et al. 2015), intersecting with other sustainable development goals of access to safe water and sanitation to marine life conservation.

In the first phase of the two-tiered ERA approach, the predicted environmental concentration (PEC) in the aquatic environment is calculated based on persistence, bioaccumulation, and toxicity (Koschorreck and de Knecht 2004). However, PEC only provides a rough insight of the overall situation at national or regional level (Zhang et al. 2013a, b) because it only accounts for the following parameters: annual drug consumption (mg year^{-1}), rates of metabolism and excretion from patients, dilution factors from effluents, predicted drug partitioning, and susceptibility to biotransformation/degradation, which can induce a slight overestimation and provide unreal and "worst-case" scenarios (Steger-Hartmann et al. 1996; Coetsier et al. 2009; Zhang et al. 2013a, b). Conversely, measured environmental concentrations (MECs) in water bodies (Buerge et al. 2006; Coetsier et al. 2009) represent a more realistic approach for performing a reliable ERA (Blasco and Delvalls 2008). Contaminant concentration surveys used to assess risk may also be achieved either by mathematical models that compile hydrological properties of catchments, chemical emissions profiles, and underlying assumptions used in fate and distribution calculations, or in monitoring data reflecting the real environmental complexity and hydrodynamic and chemical processes (i.e., MEC) (Versteeg et al. 2005). In general, PEC calculations assume that technologies for chemical removal and hospital infrastructure are similar among developed countries (Booker et al. 2014; Martín et al. 2014). The array of molecules likely to be present in the final effluent and to be released into receiving waters are considered for prioritization of potential risks before implementing a monitoring program (Johnson et al. 2008; Besse et al. 2012; Booker et al. 2014; Isidori et al. 2016; Santos et al. 2017). For both screening approaches, wide knowledge regarding pharmaceuticals inputs into the environment is required, including the form in which they are released, their removal and transformation, as well as their pathways and transport that ultimately determine their concentration in different compartments in the environment. Therefore, PECs of cytotoxic agents are important for decision making and for risk assessment, with acceptable compatibility and consistency. As all aquatic compartments are interconnected, an integrated analysis of overall matrices is necessary, subsidizing data gaps in the marine environment. The estimation of the behavior of each compound may also provide inferences about its bioavailability and routes of exposure to a range of organisms (Besse and Garric 2008; Johnson et al. 2008; Frédéric and Yves 2014) inhabiting the benthic and water column habitats.

If PEC or MEC is below 10 ng l^{-1}, according to EMEA, or below 1 μg l^{-1} according to FDA, the assessment stops because it is assumed that no environmental risk is expected (Koschorreck and de Knecht 2004). However, the present information about cytotoxic drugs in the marine environment suggests that these levels might be unrealistic for protecting the marine environment. Lists of priority anticancer drugs that were established based on PEC and MEC indicate that the following drugs may pose risk to surface waters: the alkylating agents ifosfamide and cyclophosphamide; the anti-metabolites capecitabine and methotrexate; the tyrosine kinase inhibitor imatinib; the anti-androgen bicalutamide; methotrexate; and the anti-estrogen tamoxifen (TAM) (Besse et al. 2012; Aguirre-Martínez et al. 2016a, b). In addition, the Canadian Environmental Protection Act (CCME 1999) identified TAM and its metabolites as priority for aquatic assessment because of their bioaccumulation, toxicity, and carcinogenicity for human and non human health.

If levels are equal or higher than those recommended by EMEA or FDA, phase II of the tiered approach is required, and should include the environmental fate and effects of cytotoxic compounds (Toolaram et al. 2014). Screening data of occurrences of active metabolites of anticancer drugs are compiled to ecotoxicity data, based on the lowest concentration of the compound accountable for adverse effects on wildlife (i.e., predicted no effect concentration, PNEC) (Schowanek et al. 2001; Länge and Dietrich 2002). Therefore, in this second phase, estimation of the hazard quotient that corresponds to the ratio between PEC and PNEC for various environmental compartments is needed. Although there is the possibility that MEC is not available to estimate PEC for several APIs, it is difficult to calculate PNEC for most of them because ecotoxicological data based on results of acute and chronic toxicity tests are still too scarce. Extrapolation of acute responses to predict chronic responses are performed, depicting an unrealistic paradigm when low concentrations and long-term exposures are likely to occur in the environment, thus limiting efficient and robust outcomes (EMEA 2006).

However, as environmental safety data for anticancer drugs are limited, and there are no safety thresholds for compounds with this MoA, which provide uncertainties and gaps in the ERA of some anticancer drugs, assessment with chemotherapy compounds is also incipient with only scarce ecotoxicological data available (Lenz et al. 2007; Besse et al. 2012). In addition, evaluations do not include transformation products and relevant active metabolites, often more toxic than parent compounds, and also responsible for the cytotoxic role. This integrated approach is important to subsidize management measures and focus on respective priorities efforts, proposing appropriate precautionary and safety measures in administration and disposal of products (EMEA 2006; Frédéric and Yves 2014). The European Medical Agency (EMEA 2006) proposes that PNEC should be estimated from chronic assays with organisms of each trophic level (fish, aquatic invertebrates, algae), and the lowest value should be used for risk characterization. Although subtle effects have been registered, together with acute endpoints, most of the studies have been performed with general toxicity responses, rather than a specific endpoint corresponding to the specific drug MoA (Bound and Voulvoulis 2004; Ferrando-Climent et al. 2014). If the PEC/PNEC ratio is less than 1, interpretation relies on an environmental concentration lower than levels resulting in adverse effects for wildlife. Alternatively, if the PEC/PNEC ratio is more than 1, levels present in environment exceed those likely to harm biota, wherefore subsequent approaches must be taken to analyze risk management options (Grung et al. 2008).

A need for risk assessment of genotoxic compounds is rare for (legal) residues, because in most cases genotoxicity is a criterion for exclusion of a legal approval (and thus use) of a chemical. In contrast, anticancer drugs may be genotoxic (carcinogen) and are not always completely unavoidable in consumer products, so that risk assessment approaches for this class of compounds had (and have) to be developed especially with marine species as a result of anticancer drug exposure. Genotoxicity and cytotoxicity may be triggered by long-term exposure to very low levels and have an inheritable and delayed-onset nature that may lead to major consequences at the population level (Llorente et al. 2012). A battery of genotoxicity bioassays is of particular importance for such assessments, but currently only mutagenicity tests performed with micronuclei in human lymphocytes, a bacterial reverse mutation test conducted with *Salmonella typhimurium* (Ames test), and *Escherichia coli* (SOS chromotest) are standardized (Giuliani et al. 1996; Yasunaga et al. 2006).The OSPAR commission reviews the genotoxicity tests with surface water and wastewater samples and lists the use of cell lines of fishes, algae, and protozoa to be evaluated by the Comet assay and sister chromatid exchange. So far only a few bioassays have been conducted with marine organisms exposed to anticancer agents, including genotoxic responses (Moreira et al. 2016; Trombini et al. 2016a; Fonseca et al. 2017), although studies assessing the impact of conventional pharmaceuticals have shown the reliable application of the following marine species: polychaete (Maranho et al. 2014), amphipod (Maranho et al. 2015), shrimp, bivalves (Aguirre-Martínez et al. 2016a, b; Mezzelani et al. 2016), and fish (Barreto et al. 2017) to evaluate DNA damage at low environmental levels.

Because they act in cell factors and on DNA structure and function, anticancer drugs possess mutagenic, cytotoxic, genotoxic, teratogenic, and carcinogenic properties that generally ensure the absence of safe doses in laws and environmental regulations, and as well no threshold values for lowest effect concentrations can be estimated (Kosjek and Heath 2011). In addition, the European Community bans the discharge of any chemicals and metabolites with carcinogenic or mutagenic potential into the wastewater system (EU Council Directive 2006). As the chemical MoA affects not only target cells, but also nontumor cells, any nonhuman cells could be vulnerable to genetic impairment and DNA damage (Zounková et al. 2007; Parrella et al. 2014a), with potential effects even at trace concentrations (Bound and Voulvoulis 2004; Johnson et al. 2008; Kosjek and Heath 2011; Parrella et al. 2014c). It is important to bear in mind that eukaryotic cells cover a broader range of genetic effects than prokaryotic cells, and interpretations are likely to be limited when the ERA is based only on bacteria. Levels of the tyrosine kinase inhibitor imatinib able to induce the SOS repair system were at least two orders of magnitude higher than those causing genotoxicity in daphniids, by Comet assay (Parrella et al. 2015). Fish cell lines (RTG-2 and ZFL) have been shown to be sufficiently or even more sensitive than mammalian cells in respect to genetic damage caused by imatinib (Novak et al. 2016), cyclophosphamide, ifosfamide, 5-FU, cisplatin, and their mixtures (Novak et al. 2017).

ERA on pharmaceuticals is concentrated in northern or central European countries as well as in the United States and Canada, where frameworks and guidelines for environmental assessment and monitoring of toxic compounds exist, although limited information is available in Mediterranean countries (Aguirre-Martínez et al. 2015). In spite of this, there is a marked absence of data and ERA for pharmaceuticals in developing regions, as notably Africa, South America, and small island nations in

Oceania, and in the marine environment in particular (Gaw et al. 2014). Cytotoxics are, therefore, far from being a priority group in these national environmental screening programs and for this reason there is a lack of information about the presence of cytotoxic drugs in these ecosystems (Besse and Garric 2008). Therefore, it is recommended that ERA be adapted and applied more in depth to cytotoxic drugs.

13.8 Conclusions

Evidence of escalation in cancer incidence over the world depicts a worrying situation in the health status of low-income and high-income nations, notwithstanding the environmental issues brought, especially in coastal zones. The increasing settlement of populations in coastal areas is associated with high volumes of anticancer drugs entering the marine environment. Progress regarding the potential risks of different therapeutic groups for anticancer purposes has filled gaps in comprehension of analytical detection of these compounds in aquatic compartments, physicochemical properties, and the eventual detrimental effects in nontarget biota after sequential processes occurred after human excretion until the chemicals reach water bodies. Unfortunately, although the focus on anticancer drugs in the environment is still expanding, detection of chemicals at trace levels in marine compartments (e.g. water, interstitial water, sediment, and biota) still demand improvement in analytical devices and diffusion of technologies. In addition, marine ecotoxicological assessment should consider that these compartments are subjected to oceanographic and hydrographic features that transport pharmaceuticals far from their point source and yield complex interactions in the trophic chain, including bioaccumulation, subtly harming key species of high ecological and economical relevance in coastal ecosystems.

Therefore, further approaches should integrate the cytotoxic effects related to the drug MoA, including mixtures and TPs, employing environmentally relevant concentrations by considering parameters that influence drug speciation and bioavailability to marine organisms. Efforts conducted to genotoxicity endpoints using species that live in the water column and in sediments, from tropical and temperate environments, support the interpretation of sensitivity on anticancer drugs in different maritime areas, subjected to different environmental protection frameworks, environmental priorities, and technologies for efficient effluent treatment containing anticancer drugs and hence contributing to improved ERA of anticancer drugs.

Acknowledgments M.J. Bebianno acknowledges the support from FCT through the grant UID/MAR/00350/2013 attributed to CIMA, University of Algarve. T.G. Fonseca is supported by the Brazilian National Council for Scientific and Technological Development (CNPq), through the Science Without Borders Program (202360/2014-8).

References

Abessa DMS, Carr RS, Rachid BRF, Sousa ECPM, Hortelani M, Sarkis JE (2005) Influence of a Brazilian sewage outfall on the toxicity and contamination of adjacent sediments. Mar Pollut Bull 50:875–885. https://doi.org/10.1016/j.marpolbul.2005.02.034

Afonso-Olivares C, Torres-Padrón E, Sosa-Ferrera Z, Santana-Rodríguez JJ (2013) Assessment of the presence of pharmaceutical compounds in seawater samples from coastal area of Gran Canaria Island (Spain). Antibiotics 2:274–287. https://doi.org/10.3390/antibiotics2020274

Aguirre-Martínez GV, Del Valls T, Martín-Díaz ML (2013) Identification of biomarkers responsive to chronic exposure to pharmaceuticals in target tissues of *Carcinus maenas*. Mar Environ Res 87-88:1–11. https://doi.org/10.1016/j.marenvres.2013.02.011

Aguirre-Martínez GV, Owuor MA, Garrido-Pérez C, Salamanca MJ, Del Valls TA, Martín-Díaz ML (2015) Are standard tests sensitive enough to evaluate effects of human pharmaceuticals in aquatic biota? Facing changes in research approaches when performing risk assessment of drugs. Chemosphere 120:75–85. https://doi.org/10.1016/j.chemosphere.2014.05.087

Aguirre-Martínez G, Okello C, Salamanca MJ, Garrido C, Del Valls TA, Martín-Díaz ML (2016a) Is the step-wise tiered approach for ERA of pharmaceuticals useful for the assessment of cancer therapeutic drugs present in marine environment? Environ Res 144:43–59

Aguirre-Martínez GV, Del Valls TA, Martín-Díaz ML (2016b) General stress, detoxification pathways, neurotoxicity and genotoxicity evaluated in *Ruditapes philippinarum* exposed to human pharmaceuticals. Ecotoxicol Environ Saf 124:18–31. https://doi.org/10.1016/j.ecoenv.2015.09.031

Aherne GW, English J, Marks V (1985) The role of immunoassay in the analysis of microcontaminants in water samples. Ecotoxicol Environ Saf 9:79–83

Aitken M, Kleinrock M (2015) Global medicines use in 2020 – outlook and implications. IMS Inst Healthc Informatics 47

An K (2016) Selective estrogen receptor modulators. Asian Spine J 10:787–791

Anderson PD, D'Aco VJ, Shanahan P, Chapra SC, Buzby ME, Cunningham VL, Duplessie BM, Hayes EP, Mastrocco FJ, Parke NJ, Rader JC, Samuelian JH, Schwab BW (2004) Screening analysis of human pharmaceutical compounds in U.S. surface waters. Environ Sci Technol 38:838–849

Ashton D, Hilton M, Thomas KV (2004) Investigating the environmental transport of human pharmaceuticals to streams in the United Kingdom. Sci Total Environ 333:167–184. https://doi.org/10.1016/j.scitotenv.2004.04.062

Barbier EB (2015) Climate change impacts on rural poverty in low-elevation coastal zones. Estuar Coast Shelf Sci 165:A1–A13. https://doi.org/10.1016/j.ecss.2015.05.035

Barreto A, Luis LG, Soares AMVM, Paíga P, Santos LHMLM, Hylland K, Loureiro S, Oliveira M (2017) Human pharmaceutical gemfibrozil induces genotoxicity in gilthead seabream *Sparus aurata*. Mutat Res 821:36–42. https://doi.org/10.1016/j.mrgentox.2017.05.011

Behera SK, Kim HW, Oh JE, Park HS (2011) Occurrence and removal of antibiotics, hormones and several other pharmaceuticals in wastewater treatment plants of the largest industrial city of Korea. Sci Total Environ 409:4351–4360. https://doi.org/10.1016/j.scitotenv.2011.07.015

Beretta M, Britto V, Tavares TM, da Silva SMT, Pletsch AL (2014) Occurrence of pharmaceutical and personal care products (PPCPs) in marine sediments in the Todos os Santos Bay and the north coast of Salvador, Bahia, Brazil. J Soils Sediments 14(7):1278–1286. https://doi.org/10.1007/s11368-014-0884-6

Besse J-P, Garric J (2008) Human pharmaceuticals in surface waters. Implementation of a prioritization methodology and application to the French situation. Toxicol Lett 176:104–123. https://doi.org/10.1016/j.toxlet.2007.10.012

Besse J-P, Latour J-F, Garric J (2012) Anticancer drugs in surface waters: what can we say about the occurrence and environmental significance of cytotoxic, cytostatic and endocrine therapy drugs? Environ Int 39:73–86. https://doi.org/10.1016/j.envint.2011.10.002

Blasco J, Delvalls A (2008) Impact of emergent contaminants in the environment. Environ Risk Assess 5:169–188

Booker V, Halsall C, Llewellyn NR, Johnson AC (2014) Prioritising anticancer drugs for environmental monitoring and risk assessment purposes. Sci Total Environ 473–474:159–170. https://doi.org/10.1016/j.scitotenv.2013.11.145

Bostrom ML, Berglund O (2015) Influence of pH-dependent aquatic toxicity of ionizable pharmaceuticals on risk assessments over environmental pH ranges. Water Res 72:154–161

Bound JP, Voulvoulis N (2004) Pharmaceuticals in the aquatic environment – a comparison of risk assessment strategies. Chemosphere 56:1143–1155. https://doi.org/10.1016/j.chemosphere.2004.05.010

Buerge IJ, Buser HR, Poiger T, Müller MD (2006) Occurrence and fate of the cytostatic drugs cyclophosphamide and ifosfamide in wastewater and surface waters. Environ Sci Technol 40:7242–7250. https://doi.org/10.1021/es0609405

Bui TX, Choi H (2009) Adsorptive removal of selected pharmaceuticals by mesoporous silica SBA-15. J Hazard Mater 168:602–608. https://doi.org/10.1016/j.jhazmat.2009.02.072

Burke L, Kura Y, Kassem K, Revenga C, Spalding M, McAllister D (2001) Pilot analysis of global ecosystems: coastal ecosystems, Pilot analysis of global ecosystems. World Resources Institute, Washington, DC. https://doi.org/10.1021/es0032881

Buser H, Mu MD, Wa C (1999) Occurrence and environmental behavior of the chiral pharmaceutical drug ibuprofen in surface waters and in wastewater. Environ Sci Technol 33:2529–2535

Busetti F, Linge KL, Heitz A (2009) Analysis of pharmaceuticals in indirect potable reuse systems using solid-phase extraction and liquid chromatography-tandem mass spectrometry. J Chromatogr A 1216:5807–5818. https://doi.org/10.1016/j.chroma.2009.06.001

Cairns R, Harris I, Mak T (2011) Regulation of cancer cell metabolism. Nat Rev Cancer 11:85–95. https://doi.org/10.1038/nrc2981

Caley A, Jones R (2006) Principles of cancer treatment by chemotherapy. Surgery (St. Louis) 24:66–69

Caley A, Jones R (2012) The principles of cancer treatment by chemotherapy. Surgery (St. Louis) 30:186–190. https://doi.org/10.1016/j.mpsur.2012.01.004

Canadian Council of Ministers of the Environment (CCME) (1999) Protocol for the derivation of Canadian sediment quality guidelines for the protection of aquatic life. CCME EPC-98E. Prepared by Environment Canada, Guidelines Division, Technical Secretariat of the CCME Task Group on Water Quality Guidelines, Ottawa. [Reprinted in Canadian environmental quality guidelines, Chapter 6, Canadian Council of Ministers of the Environment, Winnipeg

Castiglioni S, Bagnati R, Calamari D, Fanelli R, Zuccato E (2005) A multiresidue analytical method using solid-phase extraction and high-pressure liquid chromatography tandem mass spectrometry to measure pharmaceuticals of different therapeutic classes in urban wastewaters. J Chromatogr A 1092:206–215. https://doi.org/10.1016/j.chroma.2005.07.012

Česen M, Kosjek T, Laimou-Geraniou M, Kompare B, Širok B, Lambropolou D, Heath E (2015) Occurrence of cyclophosphamide and ifosfamide in aqueous environment and their removal by biological and abiotic wastewater treatment processes. Sci Total Environ 527–528:465–473. https://doi.org/10.1016/j.scitotenv.2015.04.109

Cheung-Ong K, Giaever G, Nislow C (2013) Perspective DNA-damaging agents in cancer chemotherapy: serendipity and chemical biology. Chem Biol 20:648–659. https://doi.org/10.1016/j.chembiol.2013.04.007

Chikae M, Ikeda R, Hasan Q, Morita Y, Tamiya E (2004) Effects of tamoxifen, 17-α-ethynylestradiol, flutamide, and methyltestosterone on plasma vitellogenin levels of male and female Japanese medaka (*Oryzias latipes*). Environ Toxicol Pharmacol 17:29–33. https://doi.org/10.1016/j.etap.2004.02.002

Chu G (1994) Cellular responses. J Biol Chem 269:787–790

Claessens M, Vanhaecke L, Wille K, Janssen CR (2013) Emerging contaminants in Belgian marine waters: single toxicant and mixture risks of pharmaceuticals. Mar Pollut Bull 71:41–50

Cleuvers M (2003) Aquatic ecotoxicity of pharmaceuticals including the assessment of combination effects. Toxicol Lett 142:185–194. https://doi.org/10.1016/S0378-4274(03)00068-7
Coetsier CM, Spinelli S, Lin L, Roig B, Touraud E (2009) Discharge of pharmaceutical products (PPs) through a conventional biological sewage treatment plant: MECs vs. PECs? Environ Int 35:787–792. https://doi.org/10.1016/j.envint.2009.01.008
Conklin KA (2004) Chemotherapy-associated oxidative stress: impact on chemotherapeutic effectiveness. Integr Cancer Ther 3:294–300. https://doi.org/10.1177/1534735404270335
Curtis L, Turner A, Vyas N, Sewell G (2010) Speciation and reactivity of cisplatin in river water and seawater. Environ Sci Technol 44:3345–3350. https://doi.org/10.1021/es903620z
Dasari S, Bernard Tchounwou P (2014) Cisplatin in cancer therapy: molecular mechanisms of action. Eur J Pharmacol 740:364–378. https://doi.org/10.1016/j.ejphar.2014.07.025
Daughton CG (2016) Pharmaceuticals and the Environment (PiE): evolution and impact of the published literature revealed by bibliometric analysis. Sci Total Environ 562:391–426. https://doi.org/10.1016/j.scitotenv.2016.03.109
Daughton CG, Ternes TA (1999) Pharmaceuticals and personal care products in the environment: agents of subtle change? Environ Health Perspect 107:907–938
Dhanak D, Edwards JP, Nguyen A, Tummino PJ (2017) Review: small-molecule targets in immuno-oncology. Cell Chem Biol 24:1148–1160. https://doi.org/10.1016/j.chembiol.2017.08.019
Díaz-Cruz MS, de Alda MJL, Barceló D (2003) Environmental behavior and analysis of veterinary and human drugs in soils, sediments and sludge. Trends Anal Chem 22:340–351. https://doi.org/10.1016/S0165-9936(03)00603-4
Du B, Haddad SP, Luek A, Scott WC, Saari GN, Kristofco LA, Connors KA, Rash C, Rasmussen JB, Chambliss CK, Brooks BW (2014) Bioaccumulation and trophic dilution of human pharmaceuticals across trophic positions of an effluent-dependent wadeable stream. Philos Trans R Soc Lond Ser B 369:1–10. https://doi.org/10.1098/rstb.2014.0058
Eggleton J, Thomas KV (2004) A review of factors affecting the release and bioavailability of contaminants during sediment disturbance events. Environ Int 30:973–980. https://doi.org/10.1016/j.envint.2004.03.001
Emadi A, Jones RJ, Brodsky RA (2009) Cyclophosphamide and cancer: golden anniversary. Nat Rev Clin Oncol 6:638–647. https://doi.org/10.1038/nrclinonc.2009.146
European Commission (EC) (2006) Directive 2006/11/EC of the European parliament and of the council of 15 February 2006 on pollution caused by certain dangerous substances discharged into the aquatic environment of the Community. Official Journal of the European Union L 64/52; 4.3.2006
European Medicines Agency (2006) Guideline on the environmental risk assessment of medicinal products for human use. EMEA/CHMP/SWP/4447/00
Fatta-Kassinos D, Vasquez MI, Kümmerer K (2011) Transformation products of pharmaceuticals in surface waters and wastewater formed during photolysis and advanced oxidation processes: degradation, elucidation of byproducts and assessment of their biological potency. Chemosphere 85:693–709. https://doi.org/10.1016/j.chemosphere.2011.06.082
Ferguson EM, Allinson M, Allinson G, Swearer SE, Hassell KL (2013) Fluctuations in natural and synthetic estrogen concentrations in a tidal estuary in South-Eastern Australia. Water Res 47:1604–1615
Ferlay J, Soerjomataram I, Dikshit R, Eser S, Mathers C, Rebelo M, Parkin DM, Forman D, Bray F (2015) Cancer incidence and mortality worldwide: sources, methods and major patterns in GLOBOCAN 2012. Int J Cancer 136:359–386. https://doi.org/10.1002/ijc.29210
Ferrando-Climent L, Rodriguez-Mozaz S, Barceló D (2014) Incidence of anticancer drugs in an aquatic urban system: from hospital effluents through urban wastewater to natural environment. Environ Pollut 193:216–223. https://doi.org/10.1016/j.envpol.2014.07.002
Fonseca TG, Morais MB, Rocha T, Abessa DMS, Aureliano M, Bebianno MJ (2017) Ecotoxicological assessment of the anticancer drug cisplatin in the polychaete *Nereis diversicolor*. Sci Total Environ 575:162–172. https://doi.org/10.1016/j.scitotenv.2016.09.185

Franco A, Fu W, Trapp S (2009) Influence of soil on the sorption of ionizable chemicals: modeling advances. Environ Toxicol Chem 28:458–464

Frédéric O, Yves P (2014) Pharmaceuticals in hospital wastewater: their ecotoxicity and contribution to the environmental hazard of the effluent. Chemosphere 115:31–39. https://doi.org/10.1016/j.chemosphere.2014.01.016

Fu W, Franco A, Trapp S (2009) Methods for estimating the bioconcentration factor of ionizable organic chemicals. Environ Toxicol Chem 28:1372–1379

García-Hernández MP, Rodenas MC, Cabas I, Garcíaa-Alcázar A, Chaves-Pozo E, García-Ayala A (2016) Tamoxifen disrupts the reproductive process in gilthead seabream males and modulates the effects promoted by 17α-ethynylestradiol. Comp Biochem Physiol Part C Toxicol Pharmacol 179:94–106. https://doi.org/10.1016/j.cbpc.2015.09.005

Gaw S, Thomas KV, Hutchinson TH (2014) Sources, impacts and trends of pharmaceuticals in the marine and coastal environment. Philos Trans R Soc Lond Ser B Biol Sci 369:20130572. https://doi.org/10.1098/rstb.2013.0572

Giuliani F, Koller T, Wurgler FE, Widmer RM (1996) Detection of genotoxic activity in native hospital waste water by the umuC test. Mutat Res 368:49–57

Goldstein SR, Siddhanti S, Ciaccia AV, Plouffe L (2000) A pharmacological review of selective oestrogen receptor modulators. Hum Reprod Update 6:212–224. https://doi.org/10.1093/humupd/6.3.212

Gonzalez VM, Fuertes MA, Alonso C, Perez JM (2001) Is cisplatin-induced cell death always produced by apoptosis? Mol Pharmacol 59:657–663

Gonzalez-Rey M, Bebianno MJ (2012) Does non-steroidal anti-inflammatory (NSAID) ibuprofen induce antioxidant stress and endocrine disruption in mussel *Mytilus galloprovincialis*? Environ Toxicol Pharmacol 33:361–371. https://doi.org/10.1016/j.etap.2011.12.017

Gonzalez-Rey M, Bebianno MJ (2014) Effects of non-steroidal anti-inflammatory drug (NSAID) diclofenac exposure in mussel *Mytilus galloprovincialis*. Aquat Toxicol 148:221–230. https://doi.org/10.1016/j.aquatox.2014.01.011

Gorrini C, Harris IS, Mak TW (2013) Modulation of oxidative stress as an anticancer strategy. Nat Rev Drug Discov 12:931–947. https://doi.org/10.1038/nrd4002

Grisolia CK, Cordeiro CMT (2000) Variability in micronucleus induction with different mutagens applied to several species of fish. Genet Mol Biol 23:235–239. https://doi.org/10.1590/S1415-47572000000100041

Grung M, Kallqvist T, Sakshaug S, Skurtveit S, Thomas KV (2008) Environmental assessment of Norwegian priority pharmaceuticals based on the EMEA guideline. Ecotoxicol Environ Saf 71:328–340. https://doi.org/10.1016/j.ecoenv.2007.10.015

Guillerey C, Huntington ND, Smyth MJ (2016) Targeting natural killer cells in cancer immunotherapy. Nature 17:1025–1036. https://doi.org/10.1038/ni.3518

Haddad T, Baginska E, Kummerer K (2015) Transformation products of antibiotic and cytostatic drugs in the aquatic cycle that result from effluent treatment and abiotic/biotic reactions in the environment: an increasing challenge calling for higher emphasis on measures at the beginning of the pipe. Water Res 72:75–126

Heath E, Filipi M, Kosjek T, Isidori M (2016) Fate and effects of the residues of anticancer drugs in the environment. Environ Sci Pollut Res 23:14687–14691. https://doi.org/10.1007/s11356-016-7069-3

Helleday T, Petermann E, Lundin C, Hodgson B, Sharma RA (2008) DNA repair pathways as targets for cancer therapy. Nat Rev Cancer 14:1291–1295. https://doi.org/10.1017/CBO9781107415324.004

Isidori M, Lavorgna M, Russo C, Kundi M, Zegura B, Novak M, Filipič M, Mišík M, Knasmueller S, de Alda ML, Barceló D, Zonja B, Cesen M, Scancar J, Kosjek T, Heath E (2016) Chemical and toxicological characterisation of anticancer drugs in hospital and municipal wastewaters from Slovenia and Spain. Environ Pollut 219:275–287. https://doi.org/10.1016/j.envpol.2016.10.039

Jemal A, Bray F, Center M, Ferlay J, Ward E, Forman D (2011) Global cancer statistics. CA Cancer J Clin 61:69–90. https://doi.org/10.3322/caac.20107.Available

Jiang JJ, Lee CL, Der Fang M (2014) Emerging organic contaminants in coastal waters: anthropogenic impact, environmental release and ecological risk. Mar Pollut Bull 85:391–399. https://doi.org/10.1016/j.marpolbul.2013.12.045

Johnson AC, Jürgens MD, Williams RJ, Kümmerer K, Kortenkamp A, Sumpter JP (2008) Do cytotoxic chemotherapy drugs discharged into rivers pose a risk to the environment and human health? An overview and UK case study. J Hydrol 348:167–175. https://doi.org/10.1016/j.jhydrol.2007.09.054

Johnson AC, Oldenkamp R, Dumont E, Sumpter JP (2013) Predicting concentrations of the cytostatic drugs cyclophosphamide, carboplatin, 5-fluorouracil, and capecitabine throughout the sewage effluents and surface waters of Europe. Environ Toxicol Chem 32:1954–1961. https://doi.org/10.1002/etc.2311

Kim H-U, Lee I-S, Oh J-E (2017) Human and veterinary pharmaceuticals in the marine environment including fish farms in Korea. Sci Total Environ 579:940–949

Kondo N, Takahashi A, Ono K, Ohnishi T (2010) DNA damage induced by alkylating agents and repair pathways. J Nucleic Acids 21:1–7. https://doi.org/10.4061/2010/543531

Koschorreck J, de Knecht J (2004) Environmental risk assessment of pharmaceuticals in the EU: a regulatory perspective. In: Kümmerer K (ed) Pharmaceuticals in the environment. Springer, Berlin

Kosjek T, Heath E (2011) Occurrence, fate and determination of cytostatic pharmaceuticals in the environment. Trends Anal Chem 30:1065–1087. https://doi.org/10.1016/j.trac.2011.04.007

Kovacs R, Csenki Z, Bakos K, Urbányi B, Horváth Á, Garaj-Vrhovac V, Gajski G, Geric M, Negreira N, Alda M, Barceló D, Heath E, Kosjek T, Zegura B, Novak M, Zajc I, Baebles S, Rotter A, Ramsak Z, Filipic M (2015) Assessment of toxicity and genotoxicity of low doses of 5-fluorouracil in zebrafish (*Danio rerio*): two-generation study. Water Res 77:201–212

Kovalova L (2009) Cytostatics in the aquatic environment. Doctorate Thesis. University Hospital Aachen, Germany

Kovalova L, McArdell CS, Hollender J (2009) Challenge of high polarity and low concentrations in analysis of cytostatics and metabolites in wastewater by hydrophilic interaction chromatography/tandem mass spectrometry. J Chromatogr A 1216:1100–1108. https://doi.org/10.1016/j.chroma.2008.12.028

Kümmerer K (2008) Pharmaceuticals in the Environment: sources, fate, effects and risks, Third edn. Springer, Berlin/Heidelberg

Kümmerer K (2010) Pharmaceuticals in the environment. Annu Rev Environ Resour 35:57–75. https://doi.org/10.1146/annurev-environ-052809-161223

Kummerer K, Steger-Hartmann T, Meyer M (1997) Biodegradability of the anti-tumour agent ifosfamide and its occurrence in hospital effluents and communal sewage. Water Res 31:2705–2710

Láng J, Kőhidai L (2012) Effects of the aquatic contaminant human pharmaceuticals and their mixtures on the proliferation and migratory responses of the bioindicator freshwater ciliate *Tetrahymena*. Chemosphere 89:592–601. https://doi.org/10.1016/j.chemosphere.2012.05.058

Länge R, Dietrich D (2002) Environmental risk assessment of pharmaceutical drug substances – conceptual considerations. Toxicol Lett 131:97–104

Lara-Martín PA, González-mazo E, Petrovic M, Barceló D, Brownawell BJ (2014) Occurrence, distribution and partitioning of nonionic surfactants and pharmaceuticals in the urbanized Long Island Sound Estuary (NY). Mar Pollut Bull 85:710–719

Lawley PD, Phillips DH (1996) DNA adducts from chemotherapeutic agents. Mutat Res 355:13–40. https://doi.org/10.1016/0027-5107(96)00020-6

Lenz K, Hann S, Koellensperger G, Stefanka Z, Stingeder G, Weissenbacher N, Mahnik SN, Fuerhacker M (2005) Presence of cancerostatic platinum compounds in hospital wastewater and possible elimination by adsorption to activated sludge. Sci Total Environ 345:141–152. https://doi.org/10.1016/j.scitotenv.2004.11.007

Lenz K, Mahnik SN, Weissenbacher N, Mader RM, Krenn P, Hann S, Koellensperger G, Uhl M, Knasmüller S, Ferk F, Bursch W, Fuerhacker M (2007) Monitoring, removal and risk assessment of cytostatic drugs in hospital wastewater. Water Sci Technol 56:141–149. https://doi.org/10.2166/wst.2007.828

Li Z, Maier MP, Radke M (2014) Screening for pharmaceutical transformation products formed in river sediment by combining ultrahigh performance liquid chromatography/high resolution mass spectrometry with a rapid data-processing method. Anal Chim Acta 810:61–70

Liou M-Y, Storz P (2010) Reactive oxygen species in cancer. Free Radic Res 44:1–31. https://doi.org/10.3109/10715761003667554

Llorente MT, Parra JM, Sánchez-Fortun S, Castaño A (2012) Cytotoxicity and genotoxicity of sewage treatment plant effluents in rainbow trout cells (RTG-2). Water Res 46:6351–6358. https://doi.org/10.1016/j.watres.2012.08.039

Loffler D, Rombke J, Meller M, Ternes T (2005) Environmental fate of pharmaceuticals in water/sediment systems. Environ Sci Technol 39:5209–5218

Lolic A, Paíga P, Santos LHMLM, Ramos S, Correia M, Delerue-Matos C (2015) Assessment of non-steroidal anti-inflammatory and analgesic pharmaceuticals in seawaters of North of Portugal: occurrence and environmental risk. Sci Total Environ 508:240–250. https://doi.org/10.1016/j.scitotenv.2014.11.097

Long ER, Dutch M, Weakland S, Chandramouli B, Benskin JP (2013) Quantification of pharmaceuticals, personal care products, and perfluoroalkyl substances in the marine sediments of Puget Sound, Washington, USA. Environ Toxicol Chem 32:1701–1710. https://doi.org/10.1002/etc.2281

Lord CJ, Ashworth A (2012) The DNA damage response and cancer therapy. Nature 481:287–294. https://doi.org/10.1038/nature10760

Mahnik SN, Rizovski B, Fuerhacker M, Mader RM (2006) Development of an analytical method for the determination of anthracyclines in hospital effluents. Chemosphere 65:1419–1425. https://doi.org/10.1016/j.chemosphere.2006.03.069

Mahnik SN, Lenz K, Weissenbacher N, Mader RM, Fuerhacker M (2007) Fate of 5-fluorouracil, doxorubicin, epirubicin, and daunorubicin in hospital wastewater and their elimination by activated sludge and treatment in a membrane-bio-reactor system. Chemosphere 66:30–37. https://doi.org/10.1016/j.chemosphere.2006.05.051

Mahoney KM, Rennert PD, Freeman GJ (2015) Combination cancer immunotherapy and new immunomodulatory targets. Nat Rev Drug Discov 14:561–584. https://doi.org/10.1038/nrd4591

Maranho LA, Baena-Nogueras RM, Lara-Martín PA, Delvalls TA (2014) Bioavailability, oxidative stress, neurotoxicity and genotoxicity of pharmaceuticals bound to marine sediments. The use of the polychaete *Hediste diversicolor* as bioindicator species. Environ Res 134:353–365

Maranho LA, Moreira LB, Baena-Nogueras RM, Lara-Martín PA, DelValls TA, Martín-Díaz ML (2015) A candidate short-term toxicity test using *Ampelisca brevicornis* to assess sublethal responses to pharmaceuticals bound to marine sediments. Arch Environ Contam Toxicol 68:237–258. https://doi.org/10.1007/s00244-014-0080-0

Martín J, Camacho-Muñoz D, Santos JL, Aparicio I, Alonso E (2011) Simultaneous determination of a selected group of cytostatic drugs in water using high-performance liquid chromatography-triple-quadrupole mass spectrometry. J Sep Sci 34:3166–3177. https://doi.org/10.1002/jssc.201100461

Martín J, Camacho-Muñoz MD, Santos JL, Aparicio I, Alonso E (2012) Distribution and temporal evolution of pharmaceutically active compounds alongside sewage sludge treatment. Risk assessment of sludge application onto soils. J Environ Manag 102:18–25. https://doi.org/10.1016/j.jenvman.2012.02.020

Martín J, Camacho-Muñoz D, Santos JL, Aparicio I, Alonso E (2014) Occurrence and ecotoxicological risk assessment of 14 cytostatic drugs in wastewater. Water Air Soil Pollut 225 (1896):1–10. https://doi.org/10.1007/s11270-014-1896-y

Martínez ML, Intralawan A, Vázquez G, Pérez-Maqueo O, Sutton P, Landgrave R (2007) The coasts of our world: ecological, economic and social importance. Ecol Econ 63:254–272. https://doi.org/10.1016/j.ecolecon.2006.10.022

Mater N, Geret F, Castillo L, Faucet-Marquis V, Albasi C, Pfohl-Leszkowicz A (2014) In vitro tests aiding ecological risk assessment of ciprofloxacin, tamoxifen and cyclophosphamide in range of concentrations released in hospital wastewater and surface water. Environ Int 63:191–200. https://doi.org/10.1016/j.envint.2013.11.011

Mezzelani M, Gorbi S, Da Ros Z, Fattorini D, Errico G, Milan M, Bargelloni L, Regoli F (2016) Ecotoxicological potential of non-steroidal anti-inflammatory drugs (NSAIDs) in marine organisms: bioavailability, biomarkers and natural occurrence in *Mytilus galloprovincialis*. Mar Environ Res 121:31–39. https://doi.org/10.1016/j.marenvres.2016.03.005

Mills LJ, Henderson WM, Jayaraman S, Gutjahr-Gobell RE, Zaroogian GE, Horowitz DB, Laws SC (2015) Approaches for predicting effects of unintended environmental exposure to an endocrine active pharmaceutical, Tamoxifen. Environ Toxicol 31:1834–1850. https://doi.org/10.1002/tox

Moldovan Z (2006) Occurrences of pharmaceutical and personal care products as micropollutants in rivers from Romania. Chemosphere 64:1808–1817. https://doi.org/10.1016/j.chemosphere.2006.02.003

Mompelat S, Le Bot B, Thomas O (2009) Occurrence and fate of pharmaceutical products and by-products, from resource to drinking water. Environ Int 35:803–814. https://doi.org/10.1016/j.envint.2008.10.008

Moreira LB, Maranho LA, Baena-Nogueras RM, Lara-Martín PA, Martín-Díaz ML (2016) Effects of novobiocin and methotrexate on the benthic amphipod *Ampelisca brevicornis* exposed to spiked sediments. Mar Environ Res 122:169–177. https://doi.org/10.1016/j.marenvres.2016.11.003

Moreno-González R, Rodriguez-Mozaz S, Gros M, Barceló D, León VM (2015) Seasonal distribution of pharmaceuticals in marine water and sediment from a Mediterranean coastal lagoon (SE Spain). Environ Res 138:326–344

Nakamura Y, Yamamoto H, Sekizawa J, Kondo T, Hirai N, Tatarazako N (2008) The effects of pH on fluoxetine in Japanese medaka (*Oryzias latipes*): acute toxicity in fish larvae and bioaccumulation in juvenile fish. Chemosphere 70:865–873. https://doi.org/10.1016/j.chemosphere.2007.06.089

Negreira N, Mastroianni N, Alda M, Barceló D (2013) Multianalyte determination of 24 cytostatics and metabolites by liquid chromatography-electrospray-tandem mass spectrometry and study of their stability and optimum storage conditions in aqueous solution. Talanta 116:290–299. https://doi.org/10.1016/j.talanta.2013.04.070

Negreira N, López de Alda M, Barceló D (2014a) Cytostatic drugs and metabolites in municipal and hospital wastewaters in Spain: filtration, occurrence, and environmental risk. Sci Total Environ 497–498:68–77

Negreira N, López de Alda M, Barceló D (2014b) Study of the stability of 26 cytostatic drugs and metabolites in wastewater under different conditions. Sci Total Environ 482–483:389–398

Negreira N, Regueiro J, López De Alda M, Barceló D (2015) Transformation of tamoxifen and its major metabolites during water chlorination: identification and in silico toxicity assessment of their disinfection byproducts. Water Res 85:199–207

Noppe H, Verslycke T, De Wulf E, Verheyden K, Monteyne E, Van Caeter P, Janssen CR, De Brabander HF (2007) Occurrence of estrogens in the Scheldt estuary: a 2-year survey. Ecotoxicol Environ Saf 66:1–8. https://doi.org/10.1016/j.ecoenv.2006.04.005

Novak M, Žegura B, Nunić J, Gajski G, Gerić M, Garaj-Vrhovac V, Filipič M (2016) Assessment of the genotoxicity of the tyrosine kinase inhibitor imatinib mesylate in cultured fish and human cells. Mutat Res 814:14–21. https://doi.org/10.1016/j.mrgentox.2016.12.002

Novak M, Žgura B, Modic B, Heath E, Filipic M (2017) Cytotoxicity and genotoxicity of anticancer drug residues and their mixtures in experimental model with zebra fish liver cells. Sci Total Environ 601–602:293–300

Oh S, Wang Q, Sik W, Song D (2013) Effect of salting out on the desorption-resistance of polycyclic aromatic hydrocarbons (PAHs) in coastal sediment. Chem Eng J 225:84–92

Oh S, Shin WS, Kim HT (2016) Effects of pH, dissolved organic matter, and salinity on ibuprofen sorption on sediment. Environ Sci Pollut Res 23:22882–22889. https://doi.org/10.1007/s11356-016-7503-6

Ortiz de García SA, Pinto Pinto G, García-Encina PA, Irusta-Mata R (2014) Ecotoxicity and environmental risk assessment of pharmaceuticals and personal care products in aquatic environments and wastewater treatment plants. Ecotoxicology 23:1517–1533. https://doi.org/10.1007/s10646-014-1293-8

Osborne D, Cutter A, Ullah F (2015) Universal sustainable development goals. Understanding the transformational challenge for developed countries. Stakehold Forum

Pagano G, De Biase A, Deeva IB, Degan P, Doronin YK, Iaccarino M, Oral R, Trieff NM, Warnau M, Korkina LG (2001) The role of oxidative stress in developmental and reproductive toxicity of tamoxifen. Life Sci 68:1735–1749. https://doi.org/10.1016/S0024-3205(01)00969-9

Parrella A, Kundi M, Lavorgna M, Criscuolo E, Russo C, Isidori M (2014a) Toxicity of exposure to binary mixtures of four anti-neoplastic drugs in *Daphnia magna* and *Ceriodaphnia dubia*. Aquat Toxicol 157:41–46

Parrella A, Lavorgna M, Criscuolo E, Russo C, Fiumano V, Isidori M (2014b) Acute and chronic toxicity of six anticancer drugs on rotifers and crustaceans. Chemosphere 115:59–66

Parrella A, Lavorgna M, Criscuolo E, Russo C, Isidori M (2014c) Estrogenic activity and cytotoxicity of six anticancer drugs detected in water systems. Sci Total Environ 486:216–222

Parrella A, Lavorgna M, Criscuolo E, Russo C, Isidori M (2015) Eco-genotoxicity of six anticancer drugs using comet assay in daphnids. J Hazard Mater 286:573–580

Paterni I, Granchi C, Katzenellenbogen J, Minutolo F (2014) Estrogen receptors alpha (ERα) and beta (ERβ): subtype-selective ligands and clinical potential. Natl Inst Health 15:1–45. https://doi.org/10.1016/j.steroids.2014.06.012.Estrogen

Pereira CDS, Maranho LA, Cortez FS, Pusceddu FH, Santos AR, Ribeiro DA, Cesar A, Guimarães LL (2016) Occurrence of pharmaceuticals and cocaine in a Brazilian coastal zone. Sci Total Environ 548–549:148–154. https://doi.org/10.1016/j.scitotenv.2016.01.051

Pessatti TB, Lüchmann KH, Flores-Nunes F, Mattos JJ, Sasaki ST, Taniguchi S, Bícego MC, Dias Bainy AC (2016) Upregulation of biotransformation genes in gills of oyster *Crassostrea brasiliana* exposed in situ to urban effluents, Florianópolis Bay, Southern Brazil. Ecotoxicol Environ Saf 131:172–180. https://doi.org/10.1016/j.ecoenv.2016.04.003

Petrie B, Barden R, Kasprzyk-hordern B (2015) A review on emerging contaminants in wastewaters and the environment: current knowledge, understudied areas and recommendations for future monitoring. Water Res 72:3–27

Pinto P, Estêvão M, Power D (2014) Effects of estrogens and estrogenic disrupting compounds on fish mineralized tissues. Mar Drugs 12:4474–4494. https://doi.org/10.3390/md12084474

Pires A, Almeida A, Correia J, Calisto V, Schneider RJ, Esteves VI, Soares AMVM, Figueira E, Freitas R (2016) Long-term exposure to caffeine and carbamazepine: impacts on the regenerative capacity of the polychaete *Diopatra neapolitana*. Chemosphere 146:565–573. https://doi.org/10.1016/j.chemosphere.2015.12.035

Rahman MF, Yanful EK, Jasim SY (2009) Occurrences of endocrine disrupting compounds and pharmaceuticals in the aquatic environment and their removal from drinking water: challenges in the context of the developing world. Desalination 248:578–585. https://doi.org/10.1016/j.desal.2008.05.105

Rendal C, Kusk KO, Trapp S (2011) Optimal choice of pH for toxicity and bioaccumulation studies of ionizing organic chemicals. Environ Toxicol Chem 30:2395–2406. https://doi.org/10.1002/etc.641

Ribeiro S, Torres T, Martins R, Santos MM (2015) Toxicity screening of diclofenac, propranolol, sertraline and simvastatin using *Danio rerio* and *Paracentrotus lividus* embryo bioassays. Ecotoxicol Environ Saf 114:67–74

Richardson ML, Bowron JM (1985) The fate of pharmaceutical chemicals in the aquatic environment. J Pharm Pharmacol 37:1–12. https://doi.org/10.1111/j.2042-7158.1985.tb04922.x

Roepke TA, Snyder MJ, Cherr GN (2005) Estradiol and endocrine disrupting compounds adversely affect development of sea urchin embryos at environmentally relevant concentrations. Aquat Toxicol 71:155–173. https://doi.org/10.1016/j.aquatox.2004.11.003

Rowney NC, Johnson AC, Williams RJ (2009) Cytotoxic drugs in drinking water: a prediction and risk assessment exercise for the Thames catchment in the United Kingdom. Environ Toxicol Chem 28:2733–2743. https://doi.org/10.1897/09-067.1

Santos MA, Pacheco M (1995) Mutagenicity of cyclophosphamide and kraft mill effluent and sediment on the eel *Anguilla anguilla* L. Sci Total Environ 171:127–130

Santos SF, Franquet-griell H, Lacorte S, Madeira LM (2017) Anticancer drugs in Portuguese surface waters: estimation of concentrations and identification of potentially priority drugs. Chemosphere 184:1250–1260

Schaffer M, Boxberger N, Börnick H, Licha T, Worch E (2012) Sorption influenced transport of ionizable pharmaceuticals onto a natural sandy aquifer sediment at different pH. Chemosphere 87:513–520. https://doi.org/10.1016/j.chemosphere.2011.12.053

Schowanek D, Fox K, Holt M, Schroeder FR, Koch V, Cassani G, Matthies M, Boeije G, Vanrolleghem P, Young A, Morris G, Gandolfi C, Feijtel TCJ (2001) GREAT-ER: a new tool for management and risk assessment of chemicals in river basins. Water Sci Technol 43:179–185

Silva BF, Jelic A, López-Serna R, Mozeto AA, Petrovic M, Barceló D (2011) Occurrence and distribution of pharmaceuticals in surface water, suspended solids and sediments of the Ebro River basin, Spain. Chemosphere 85:1331–1339. https://doi.org/10.1016/j.chemosphere.2011.07.051

Steger-Hartmann T, Kiimmerer K, Schecker J (1996) Trace analysis of the antineoplastics ifosfamide and cyclophosphamide in sewage water by two-step solid-phase extraction and gas chromatography-mass spectrometry. J Chromatogr A 726:179–184

Steger-Hartmann T, Kümmerer K, Hartmann A (1997) Biological degradation of cyclophosphamide and its occurrence in sewage water. Ecotoxicol Environ Saf 36:174–179. https://doi.org/10.1006/eesa.1996.1506

Sun L, Shao X, Hu X, Chi J, Jin Y, Ye W, Fu Z (2011) Transcriptional responses in Japanese medaka (*Oryzias latipes*) exposed to binary mixtures of an estrogen and anti-estrogens. Aquat Toxicol 105:629–639. https://doi.org/10.1016/j.aquatox.2011.08.024

Suspiro A, Prista J (2011) Biomarkers of occupational exposure do anticancer agents: a minireview. Toxicol Lett 207:42–52. https://doi.org/10.1016/j.toxlet.2011.08.022

Ternes TA (1998) Occurrence of drugs in German sewage treatment plants and rivers. Water Res 32:3245–3260. https://doi.org/10.1016/S0043-1354(98)00099-2

Thomas KV, Hilton MJ (2004) The occurrence of selected human pharmaceutical compounds in UK estuaries. Mar Pollut Bull 49:436–444. https://doi.org/10.1016/j.marpolbul.2004.02.028

Togola A, Budzinski H (2008) Multi-residue analysis of pharmaceutical compounds in aqueous samples. J Chromatogr A 1177:150–158. https://doi.org/10.1016/j.chroma.2007.10.105

Toolaram AP, Schneider M, Kummerer K (2014) Environmental risk assessment of anti-cancer drugs and their transformation products: a focus on their genotoxicity characterization. State of knowledge and shortcomings. Mutat Res 760:18–35

Trachootham D, Alexandre J, Huang P (2009) Targeting cancer cells by ROS-mediated mechanisms: a radical therapeutic approach? Nat Rev 8:579–591. https://doi.org/10.1038/nrd2803

Trombini C, Garcia da Fonseca T, Morais M, Lopes T, Blasco J, Bebianno MJ (2016a) Toxic effects of cisplatin cytostatic drug in mussel *Mytilus galloprovincialis*. Mar Environ 119:12–21

Trombini C, Hampel M, Blasco J (2016b) Evaluation of acute effects of four pharmaceuticals and their mixtures on the copepod *Tisbe battagliai*. Chemosphere 155:319–328

Turner A, Mascorda L (2014) Particle–water interactions of platinum-based anticancer drugs in river water and estuarine water. Chemosphere 119C:415–422. https://doi.org/10.1016/j.chemosphere.2014.06.074

Turner A, Mascorda L (2015) Particle–water interactions of platinum-based anticancer drugs in river water and estuarine water. Chemosphere 119:415–422

UN (2016) The world's cities in 2016. United Nations – Dep. Econ. Soc. Aff. Popul. Div. (2016). World's cities 2016 – data Bookl. (ST/ESA/SER.A/392)

US EPA - United States Environmental Protection Agency (1998) Guidelines for ecological risk assessment, EPA/630/R-95/002F. Washington, DC

Versteeg DJ, Alder AC, Cunningham VL, Kolpin DW, Murray-Smith R, Ternes T (2005) Environmental exposure modeling and monitoring of human pharmaceutical concentrations in the environment. In: Willams RT (ed) Human pharmaceuticals: assessing the impacts on aquatic ecosystems. SETAC Press, Pensacola

Vidal-Dorsch D, Bay S, Maruya KA, Snyder SA, Trenholm RA, Vanderford BJ (2012) Contaminants of emerging concern in municipal wastewater effluents and marine receiving water. Environ Toxicol Chem 31:2674–2682. https://doi.org/10.1002/etc.2004

Vieno NM, Tuhkanen T, Kronberg L (2005) Seasonal variation in the occurrence of pharmaceuticals in effluents from a sewage treatment plant and in the recipient water. Environ Sci Technol 39:8220–8226. https://doi.org/10.1021/es051124k

Von Glasow R, Jickells TD, Baklanov A, Carmichael GR, Church TM, Gallardo L, Hughes C, Kanakidou M, Liss PS, Mee L, Raine R, Ramachandran P, Ramesh R, Sundseth K, Tsunogai U, Uematsu M, Zhu T (2013) Megacities and large urban agglomerations in the coastal zone: interactions between atmosphere, land, and marine ecosystems. Ambio 42:13–28. https://doi.org/10.1007/s13280-012-0343-9

Weigel S, Kuhlmann J, Hühnerfuss H (2002) Drugs and personal care products as ubiquitous pollutants: occurrence and distribution of clofibric acid, caffeine and DEET in the North Sea. Sci Total Environ 295:131–141. https://doi.org/10.1016/S0048-9697(02)00064-5

WHO (World Health Organization) (2005) World Health Statistics 2005

WHO (World Health Organization) (2014) Global status report on noncommunicable diseases

Xie H (2012) Occurrence, ecotoxicology, and treatment of anticancer agents as water contaminants. J Environ Anal Toxicol. https://doi.org/10.4172/2161-0525.S2-002

Xing L, Liu H, Giesy J, Yu H (2012) pH-Dependent aquatic criteria for 2,4-dichlorophenol, 2,4,6-trichlorophenol and pentachlorophenol. Sci Total Environ 441:125–131

Yamamoto H, Nakamura Y, Moriguchi S, Nakamura Y, Honda Y, Tamura I, Hirata Y, Hayashi A, Sekizawa J (2009) Persistence and partitioning of eight selected pharmaceuticals in the aquatic environment: laboratory photolysis, biodegradation, and sorption experiments. Water Res 43:351–362. https://doi.org/10.1016/j.watres.2008.10.039

Yasunaga K, Kiyonari A, Nakagawa M, Yoshikawa K (2006) Investigation into the ability of the *Salmonella* umu test to detect DNA damage using antitumor drugs. Toxicol In Vitro 20:712–728. https://doi.org/10.1016/j.tiv.2005.10.007

Yin J, Yang Y, Li K, Zhang J, Shao B (2010) Analysis of anticancer drugs in sewage water by selective SPE and UPLC-ESI-MS-MS. J Chromatogr Sci 48:781–789

Zhang W, Isaacson CW, Rattanaudompol U, Powell TB, Bouchard D (2012) Fullerene nanoparticles exhibit greater retention in freshwater sediment than in model porous media. Water Res 46:2992–3004. https://doi.org/10.1016/j.watres.2012.02.049

Zhang H, Liu P, Feng Y, Yang F (2013a) Fate of antibiotics during wastewater treatment and antibiotic distribution in the effluent-receiving waters of the Yellow Sea, northern China. Mar Pollut Bull 73:282–290. https://doi.org/10.1016/j.marpolbul.2013.05.007

Zhang J, Chang V, Giannis A, Wang J (2013b) Removal of cytostatic drugs from aquatic environment: a review. Sci Total Environ 445–446:281–298

Zhang Y, Xiao Y, Zhang J, Chang VWC, Lim T-T (2017) Degradation of cyclophosphamide and 5-fluorouracil in water using UV and UV/H_2O_2: kinetics investigation, pathways and energetic analysis. J Environ Chem Eng 5:1133–1139. https://doi.org/10.1016/j.jece.2017.01.013

Zounková R, Odráska P, Dolezalová L, Hilscherová K, Marsálek B, Bláha L (2007) Ecotoxicity and genotoxicity assessment of cytostatic pharmaceuticals. Environ Toxicol Chem 26:2208–2214. https://doi.org/10.1897/07-137R.1

Zuccato E, Castiglioni S, Fanelli R, Reitano G, Bagnati R, Chiabrando C, Pomati F, Rossetti C, Calamari D (2006) Pharmaceuticals in the environment in Italy: causes, occurrence, effects and control. Environ Sci Pollut Res Int 13:15–21. https://doi.org/10.1065/espr2006.01.004

Chapter 14
5-Fluorouracil and Its Prodrug Capecitabine: Occurrence, Fate and Effects in the Environment

Milka Ljoncheva, Tina Kosjek, Marina Isidori, and Ester Heath

Abstract In this chapter, we examine the available literature on the cycling and effects of 5-flourouracil (5-FU) and capecitabine (CAP) residues in the aqueous environment. The aim is to understand better their environmental occurrence, fate and potential toxic effects. Physicochemical properties of 5-FU and CAP suggest that they are more likely to remain in aqueous environment than adsorbed to solid particles. Detectable levels have been reported in hospital effluents (< 122 μg/L) and in municipal wastewaters (< 280 ng/L), but rarely in surface waters (only 5-FU in one study: < 160 ng/L). Among different water treatments available, the most promising for removing 5-FU and CAP are the advanced oxidation processes (AOPs). So far, indirect photolysis has been most widely applied and is capable of almost completely removing both compounds (to < LOD) and in some cases resulting in complete mineralization. However, these treatments have been mostly tested in MilliQ or potable water and their suitability for complex matrices like wastewaters is questionable and biodegradation is still treatment of choice for these matrices. In other studies, a variety of transformation products has been identified adding to the overall environmental burden. Toxicity tests on single parent compounds have shown that they may have effects above the concentrations of environmental relevance. The studies of complex mixtures of parent compounds highlight that the actual ecological risk posed by mixtures of these compounds is difficult to evaluate. Overall, the main finding from this review is that a real need exists for further studies on the chemical and toxicological effects of environmental mixtures of cytotoxic compounds.

M. Ljoncheva
Department of Environmental Sciences, Jožef Stefan Institute, Ljubljana, Slovenia
e-mail: milka.ljoncheva@ijs.si

T. Kosjek · E. Heath (✉)
Department of Environmental Sciences, Jožef Stefan Institute, Ljubljana, Slovenia

Jožef Stefan International Postgraduate School, Ljubljana, Slovenia
e-mail: tina.kosjek@ijs.si; ester.heath@ijs.si

M. Isidori
Department of Environmental, Biological and Pharmaceutical Sciences and Technologies, University of Campania "L. Vanvitelli", Caserta, Italy
e-mail: marina.isidori@unicampania.it

E. Heath et al. (eds.), *Fate and Effects of Anticancer Drugs in the Environment*,
https://doi.org/10.1007/978-3-030-21048-9_14

Keywords 5-fluorouracil · Capecitabine · Occurrence · Instrumental analysis · Environmental fate · Ecotoxicity

14.1 Introduction

5-fluorouracil (5-FU) and its prodrug capecitabine (CAP) are among the most widely prescribed cytostatic agents worldwide (Kosjek et al. 2013; Kosjek and Heath 2011). The Food and Drug Administration Agency (FDA) approved their use for treating adjuvant and colon cancer, metastatic breast cancers, pancreatic endocrine cancers, and head and neck cancers (Faßbender and Braunbeck 2013). 5-FU is also applied topically for treating actinic sun keratosis and skin basal cell cancers. According to The Anatomical Therapeutic Chemical Classification System with Defined Daily Doses (ATC/DDD) (WHO 2018), both agents are classified as antimetabolites, pyrrolidine analogues. Their mechanism of action is based on inhibition of DNA synthesis, achieved through inhibition of *thymidylate synthetase*, an enzyme driving the biosynthesis of desoxythymidine monophosphate, thereby obstructing thymidine synthesis and ultimately blocking normal DNA replication. It also causes the synthesis of faulty DNA through incorporation of 5-FU in lieu of pyrimidine bases. Both mechanisms lead to cell apoptosis and subsequent slowing of tumor growth (Walko and Lindley 2005). The mean European consumption of 5-FU in 2012 was 0.0297 mg/cap/day and 0.0259 mg/cap/day for CAP (Johnson et al. 2013). While 5-FU is commonly administered intravenously, CAP is developed as a prodrug of 5-FU in order to improve patient tolerability and therefore quality of life. CAP undergoes tumor-specific conversion to 5-FU, and has fewer side effects, such as gastrointestinal toxicity, neutropenia and stomatitis due to cytotoxic nonselectivity (Kosjek and Heath 2011; Walko and Lindley 2005). This has led to a shift towards prescribing CAP over 5-FU (Mahnik et al. 2004).

5-FU is typically administered in hospitals intravenously. After administration, the drug is excreted in urine and feces as unchanged parent compound and metabolites. This means hospital effluents can be an important point source of cytostatic drug residues. Since 75% of outpatients leave hospitals within hours after receiving chemotherapy and an increasing number of outpatients receive oral chemotherapy at home (CAP), important amounts of administered drugs and metabolites are excreted into municipal sewage (Kosjek and Heath 2011). In both cases, wastewaters containing mixtures of parent compounds and metabolites, either from hospitals or households, arrive at wastewater treatment plants (WWTP). If not removed, they will end up in the aquatic environment, possibly even in drinking water supplies, posing a risk to human and aquatic organisms (Kosjek and Heath 2011).

Since 2011, the occurrence and effects of cytostatic agents in the environment have been extensively studied as part of two EU FP7 projects: CytoThreat and Pharmas. However, to date, no one has had the confidence to say whether cytostatic drugs have an impact on the environment even though the potential is there. This chapter gives an overview of the current scientific knowledge regarding the occurrence, cycling and fate of 5-FU and CAP in the environment, addressing also the analytical methods used and potential risk they pose to organisms.

14.2 Pharmacokinetical and Physico-Chemical Properties of 5-FU and CAP

5-FU is 5-fluoro-2,4-(1H,3H)-pyrimidindione, a fluorinated pyrimidine developed by Roche in the United States in 1957 and is still being used widely in systemic and local cancer treatment. CAP (5-deoxy-5-fluoro-N4-pentyloxycarbonyl cytidine) is a fluoropyrimidine carbamate prodrug of 5-FU (Fig. 14.1 and Table 14.1).

14.2.1 Pharmacokinetics

5-FU can be administered by intravenous (i.v.) bolus, by infusion, orally, or dermally due to variable gastrointestinal adsorption and its rapid degradation in daily doses of 750–3900 mg/day depending on the type of cancer (Drugbank 2018). Its plasma elimination half-life varies from 5–20 min after i.v. bolus (Weissbrodt et al. 2009), during which time rapid metabolism occurs. Between 60% and 90% of 5-FU is metabolized and excreted in a dose-dependent rate during 24 h as α-fluoro-β-alanine, FBAL (80%) (Fig. 14.1), as α-fluoro-β-ureidopropionic acid (10%) and as unchanged 5-FU (2–39%) (Johnson et al. 2013; Besse et al. 2012). Fecal excretion is negligible (Straub 2007).

CAP is administered orally in daily doses of 2000 mg/m^2 body surface, is readily absorbed from the gastrointestinal tract (~70–100%) and enzymatically converted to active 5-FU in tumor tissues, offering tumor-preferential activation of CAP (DrugBank 2018). Metabolism of CAP to 5-FU proceeds rapidly, resulting in a short CAP ($t_{1/2}$=38-45 min) and metabolite half-lives ($t_{1/2}$~46 min for 5-FU and $t_{1/2}$~3 h 15 min for FBAL) (Walko and Lindley 2005). Between 84–95% of CAP is excreted within 24 h mainly through the kidneys as FBAL (51–60% of the dose excreted), CAP (2.6–2.9%) and 5-FU (0.54%). Total fecal excretion of CAP is also negligible (2.6%) (Walko and Lindley 2005; DrugBank 2018; Straub 2007).

5-FU, when converted from CAP, seems to have a prolonged overall half-life and undergoes increased metabolism compared with 5-FU as the parent drug. This is due to tumor- and liver-specific conversion and that there is less free 5-FU in the bloodstream, that can be rapidly excreted. This is evidenced by a 10-fold decrease in 5-FU excreted after CAP administration (0.5% of CAP corresponds to 1.4% of converted 5-FU), compared to when 5-FU is administered directly (2–39%) (Straub 2007). Based on 5-FU and CAP pharmacokinetics, a mixture of parent compound and its metabolites, mostly FBAL and, to a lower extent, α-fluoro-β-ureidopropionic acid, are excreted from body and can potentially enter the environment.

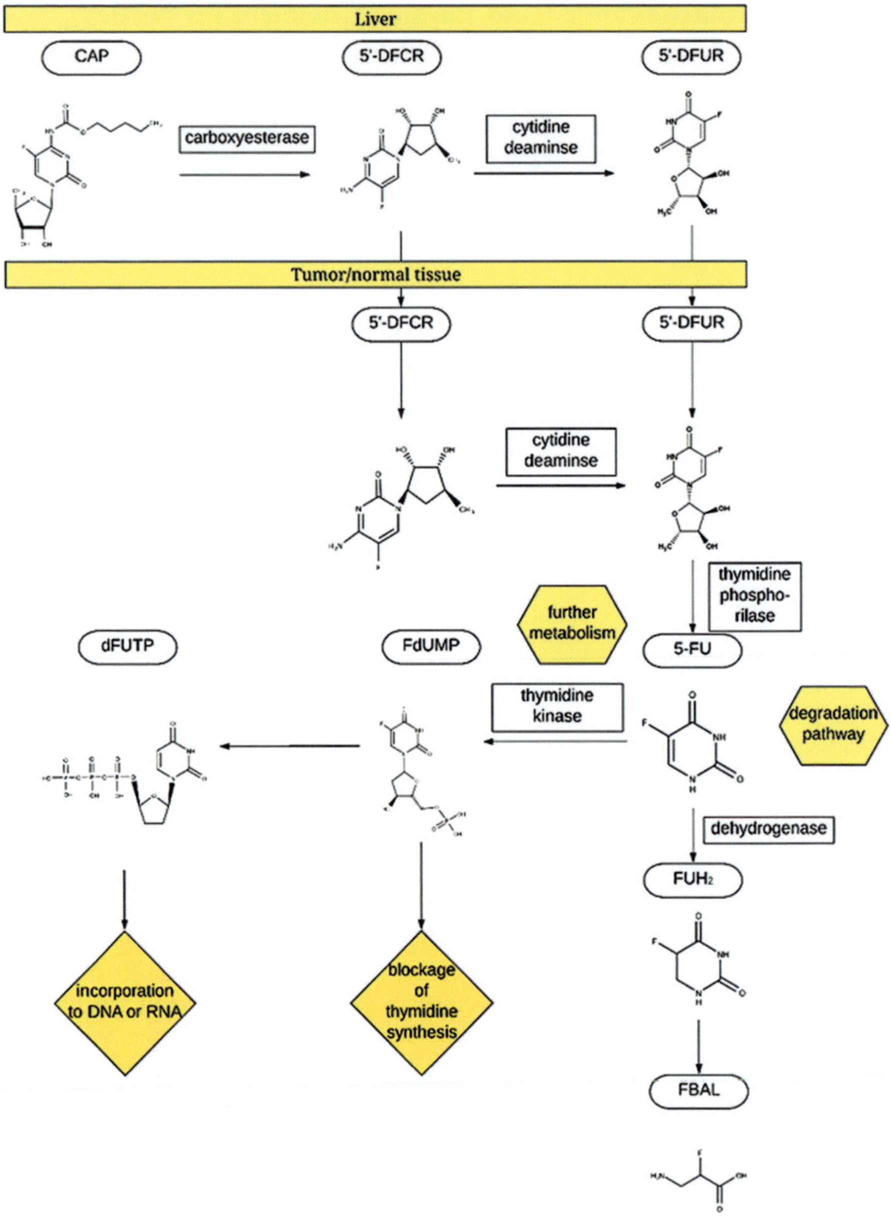

Fig. 14.1 Human metabolism of CAP and 5-FU. *5-FU* fluorouracil, *5'-DFCR* 5'-deoxy-5-fluorocytidine, *5'-DFUR* 5'-deoxy-5-fluorouridine, *CAP* capecitabine, *dFUTP* deoxyfluorouridine triphosphate, *FdUMP* fluorodeoxyuridine monophosphate, *FBAL* α-fluoro-β-alanine, FUH_2 dihydrofluorouracil

Table 14.1 Classification and chemical structure of 5-FU and CAP

Cytostatic agent	5-FU		CAP	
ATC code	L01BC02		L01BC06	
Group	Antimetabolite: pyrimidine analogue		Antimetabolite: pyrimidine analogue	
IUPAC name	5-fluoro-1H-pyrimidine-2,4-dione		Pentyl N-[1-[(2R,3R,4S,5R)-3,4-dihydroxy-5-methyloxolan-2-yl]-5-fluoro-2-oxopyrimidin-4-yl] carbamate	
CAS	51-21-8		154631-50-9	
Elemental formula/ monoisotopic mass	$C_4H_3FN_2O_2$	130.078 g/mol	$C_{15}H_{22}FN_3O_6$	359.354 g/mol
Chemical structure	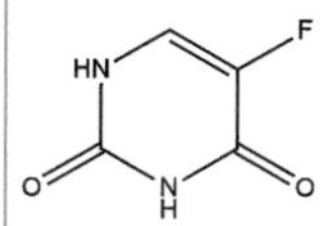		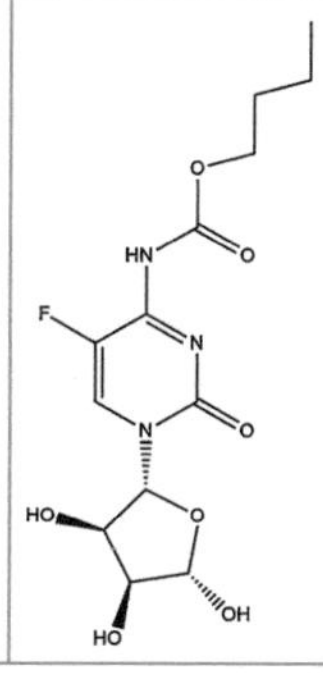	

14.2.2 *Physico-chemical Properties*

Physico-chemical properties are the most relevant factors for predicting the environmental distribution and fate of chemical substances (Kosjek and Heath 2011). Table 14.2 lists the most important parameters. The dissociation constant describes the degree of dissociation of a compound at a particular pH. Generally, as dissociation increases, so does the mobility of organic compounds in water. 5-FU ($pK_a1 = 7.93$, $pK_a2 = 13.0$) and CAP (pKa = 8.8) are weak acids and are not expected to be dissociated at an environmental pH (~7) (Toxnet 2018; Poyatos et al. 2009; Hoerger et al. 2009; Kovalova 2009; Roche 2007). Sorption process is also an important factor controlling cycling of organic compounds in the aqueous environment and is a function of the compound's chemical structure. It is defined by the octanol-water partition coefficient (log K_{ow}) and the soil organic carbon-water partitioning coefficient (K_{oc}). CAP's and 5-FU's log K_{ow} values suggest their high polarity and low adsorption to soil/sediment. In combination with their high water solubility (11.1 g/L for 5-FU and 0.8–26 g/L for CAP) (Toxnet 2018; Hoerger et al. 2009; Kovalova 2009; Roche 2007; Lin et al. 2014a; Zhang et al. 2013; Roche 2008; Gómez-Canela et al. 2014), 5-FU and CAP are more likely to be present in the aqueous environment than adsorbed on soil or sediments. In addition, low bioconcentration factor (BCF) values indicate a low potential for uptake (bioconcentration) of a chemical in aquatic organisms from surrounding

Table 14.2 Physico-chemical properties of 5-FU and CAP

Cytostatic agent	5-FU	CAP
pKa	$pK_a1 = 7.93$, $pK_a2 = 13.0$	8.8
$logK_{ow}$	−0.69–(−1.85)	−4.5–(−1.0)
K_{oc}	4.0–8.0	4.5–8.0
BCF	3.0–3.6	1.3–3.0
Solubility (g/L)	11.1	0.8–26
Henry's law constant (atm × m^3/ mole)	1.66×10^{-10}	2.9×10^{-19}
UV absorbance	256–266 nm	310 nm
Vapor pressure at 25 C (mm Hg)	2.68×10^{-6}	1.12×10^{-12}
Atmospheric OH rate (cm^3/ molecule × s)	5.8×10^{-12}	8.3×10^{-11}
References	Steger-Hartmann et al. (1996), Straub (2007), Poyatos et al. (2009), Besse et al. (2012), Lin and Lin (2014), Franquet-Griell et al. (2017), DrugBank (2018), TOXNET (2018)	Mahnik et al. (2007), Poyatos et al. (2009), Besse et al. (2012), Lin and Lin (2014), Barışçı et al. (2018), DrugBank (2018), TOXNET (2018)

environment. Other parameters, such as the Henry's law constant and vapor pressure have a minor effect on 5-FU and CAP's environmental distribution. Volatilisation of 5-FU or CAP is negligible.

5-FU and CAP absorb light at wavelengths 256–266 nm and 310 nm (Toxnet 2018), respectively, and are susceptible to direct photolysis by sunlight. The atmospheric OH rate coefficient indicates the potential for indirect photolysis by describing the rate of reaction between a compound and hydroxyl radicals (•OH) (Poyatos et al. 2009). A comparison of the OH rate coefficients of 5-FU and CAP with other cytostatic drugs, suggests a medium potential for indirect photolysis and other advanced oxidation processes (Kosjek and Heath 2011).

14.3 Analytical Methods

14.3.1 Sampling Techniques

For determining the true concentration profile of an organic contaminant in an environmental compartment, selecting and applying an appropriate sampling technique is crucial. In addition, population size (or more precisely number of patients receiving the drug), the excretion rate of non-metabolized drugs subjected to inter- and intra-individual variability and the amount of water entering the sewage system are important factors (Steger-Hartmann et al. 1996). Grab sampling, based on the collection of discrete samples at a specific point in time, is commonly applied in water sampling (Kosjek et al. 2013; Mahnik et al. 2007; Kovalova et al. 2009; Yu et al. 2012; Heath et al. 2015; Mendoza et al. 2016). The main reason is its simplicity of application, despite having significant drawbacks, i.e., the 'snapshot' does not represent the conditions either where levels of pollutants fluctuate or are not homogenous.

In the case of 5-FU and CAP, most studies use composite sampling, namely, time- or flow-proportional sampling (TPS or FPS). TPS (Kosjek et al. 2013; Gómez-Canela et al. 2014; Yu et al. 2006; Martín et al. 2011, 2014; Negreira et al. 2013a; Ferre-Aracil et al. 2016) consists of a collection of numerous individual discrete samples taken at regular intervals over a period of time, usually 24 h. FPS (Lin et al. 2014a; Gómez-Canela et al. 2014; Kovalova et al. 2009; Mahnik et al. 2004; Mullot et al. 2009; Tauxe-Wuersch et al. 2006; Negreira et al. 2014a) is based on sample collection at defined volume interval, i.e. aliquots are taken at fixed/uniform increments of volume metered past a flow measurement point. The final sample is prepared by combining a series of grab samples (aliquots) over the interval. Although frequently used, composite sampling techniques do not provide a true average of the concentration of the contaminants (Kosjek and Heath 2011). Rather, they represent only the average of snapshots in time and do not mimic the continuous discharge of contaminants.

An alternative technique is passive sampling, based on the continuous collection of analytes *in situ*, resulting in a time-weighted average (TWA) sorption and preconcentration of pollutants over time. Different types of passive sampling devices exist. Most common passive sampling devices are POCIS, a polar organic chemical integrative sampler, which allows for the *in situ* collection of a time-integrated average of hydrophilic organic contaminants and Chemcatcher - a highly versatile and cost-effective passive sampling device for monitoring a wide variety of pollutants in water (Kosjek and Heath 2011). So far, however, their use has neither been fully validated nor applied to the sampling of 5-FU and CAP.

Environmental samples must be stored correctly. In wastewater, 5-FU is stable at −20 °C and 4 °C for 2 months and at 25 °C for 9 days, whereas at 4 °C and 25 °C CAP is unstable (3 days). The acidification of samples is widely used to prevent the growth of bacteria that could biodegrade the compounds of interest. For instance, after acidification with HCl (pH = 2), 5-FU is stable at 25 °C for 9 days and at 4 °C and −20 °C for 3 months, while CAP is degraded at 4 °C (~30%) in 9 days (Negreira et al. 2014b).

14.3.2 Sample Preparation Methods

Environmental analysis requires the appropriate choice of sample extraction method. For aqueous samples this is usually solid-phase extraction (SPE). Cartridges with different SPE sorbents (reversed-phase or ion exchange), binding strength (degree of hydrophobicity), and binding capacities are available for selection, along with different solvents for cartridge conditioning, washing and elution.

Reversed-phase SPE (RP-SPE) are used for sample preparation prior to 5-FU and CAP analysis (Table 14.3). Despite RP-SPE sorbents not being the most appropriate choice for extraction of highly polar compounds, namely, 5-FU and CAP (Table 14.2), they are frequently used in multiresidual analyses, when samples contain many analytes with significant structural diversity (Gómez-Canela et al. 2014; Yu et al. 2012; Heath et al. 2015; Mendoza et al. 2016; Martín et al. 2011, 2014; Negreira et al. 2013a, b, 2014a). Most commonly applied RP sorbents are Oasis HLB (Gómez-Canela et al. 2014; Martín et al. 2011, 2014) and Isolute ENV^+ (Kosjek et al. 2013; Weissbrodt et al. 2009; Mahnik et al. 2004, 2007; Kovalova et al. 2009; Heath et al. 2015; Mullot et al. 2009; Tauxe-Wuersch et al. 2006). Isolute ENV^+ has higher affinity for polar compounds due to a retention mechanism based on specific interactions with the solvent and the analyte (Tauxe-Wuersch et al. 2006). Under optimized experimental conditions (sorbent mass, pH, etc.) (Kosjek et al. 2013), Isolute ENV^+ can be successfully applied for the analysis of 5-FU. Some studies investigating 5-FU have also used Strata X (Yu et al. 2006, 2012) for RP-SPE, but with no success in detection of the compound, possibly implying on low method sensitivity, which might be a result of improper selection of sorbent. In addition, when extracting 5-FU, the negative impact of matrix complexity on extraction efficiency has been noted (Kosjek et al. 2013).

When a sample contains ionic components that are very hydrophilic and RP-SPE retention can be problematic, it is suggested to use ion-pairing (IP-SPE). By adding an ion-pair reagent with an ionic end and a hydrophobic tail to the mobile phase, the hydrophobic tail of the reagent is retained by the stationary phase resulting in increasing retention of polar compounds. This principle has been applied when using an Isolute ENV+ sorbent to extract 5-FU from surface water by the addition of IP agent, Bu_4NCl. In total, an ~20% increase in extraction efficiency was achieved at pH > 10, however, IP extraction failed for more complex matrices like wastewaters (Kosjek et al. 2013).

Ion exchange SPE (IE-SPE) is a suitable alternative to RP-SPE for ionisable compounds, e.g. 5-FU at pH $\geq$ 10 (Lin et al. 2014a). Despite their potential, anion exchange sorbents (e.g. Oasis MAX) are inefficient except for low-volume simple matrix samples and even a slight increase in matrix complexity or ionic strength radically reduces extraction efficiency (Kosjek et al. 2013).

Alternatives for achieving satisfactory extraction of 5-FU include positioning a specific cartridge such as a Speedisk H_2O-philic SA-DVB after RP-SPE (Isolute ENV^+) (Kovalova et al. 2009) or applying dual layer SPE employing Superclean™ ENVI-Carb/NH_2 (Lin et al. 2014a). The former, when applied to a complex matrix – a hospital effluent, yielded LOD of 5 ng/L (Straub 2007). For the latter, the method failed to achieve the necessary sensitivity for extracting 5-FU from environmental samples (LOD = 250 ng) (Lin et al. 2014a).

Recent studies have also included online SPE for CAP extraction using a Symbiosis ™ Pico device with a cross-linked styrene-divinylbenzene polymer (PLRP) sorbent (Mendoza et al. 2016; Negreira et al. 2013a, 2014a; Ferre-Aracil et al. 2016). There are advantages to this technique such as automation of the extraction process, overall costs and the samples require less manipulation.

Usually, extracting disks are a viable alternative to conventional cartridges, as they allow enrichment of sample volumes as large as 12 L, resulting in enrichment

Table 14.3 Analytical methods for determination of 5-FU and CAP in different environmental compartments

Cytostatic agent	Matrix	Sample preparation: SPE Derivatization (where applied)	Analysis: Chromatographic separation	Analysis: Detection	LOQ (other limits, ionization) [matrix]	References
5-FU	WW from hospital collecting tanks	RP-SPE: Amberlyste® A-26	HPLC: RP C_{18} column	UV 267 nm	200 μg/L (LOD)	Kiffmeyer et al. (1998)
	Hospital effluent	RP-SPE: Isolute ENV⁺	CE	UV 265 nm	8.6 μg/L	Mahnik et al. (2004)
		Clean-up: centrifugation + filtration				Mahnik et al. (2007)
	WWTP influent and effluent	RP-SPE: Isolute ENV⁺	GC: RTX-5 column + deactivated fused silica pre-column	Q MS EI, CI (-) SIM	15 ng/L (CI (-))	Tauxe-Wuersch et al. (2006)
		Derivatization with PFBBr, catalyst K_2CO_3				
	Hospital effluent	Clean-up: SiOH cartridge			30 ng/L (EI)	
	Hospital effluent	RP-SPE: Isolute ENV⁺ + SPE Speedisc H_2O-philic SA-DVB	HPLC: HILIC column + ZIC-HILIC precolumn	QqQ MS/MS ESI(+)/MRM	5.0 ng/L	Kovalova et al. (2009)
	WWTP influent	RP-SPE: Isolute ENV⁺	GC: DB-5 MS column	IT MS/MS	1.6 ng/L [WW]	Kosjek et al. (2013)
	WWTP effluent	Derivatization with MTBSTFA		EI/external ionization source	0.54 ng/L [SW]	
	Hospital effluent					
	WWTP influent	SPE: Oasis HLB	HPLC: Eclipse XDB-C_{18} column	QqQ MS/MS	128 ng/L (MQL) [RW]	Martín et al. (2011)
	WWTP effluent			ESI (+)/(-)/MRM	70 ng/L (MQL) [WWE]	
	River water				114 ng/L (MQL) [WWI]	

(continued)

Table 14.3 (continued)

Cytostatic agent	Matrix	Sample preparation: SPE Derivatization (where applied)	Analysis: Chromatographic separation	Analysis: Detection	LOQ (other limits, ionization) [matrix]	References
	WWTP influent	SPE: Strata X	GC: DB-5MS column	Q MS EI (+) SIM	NR	Yu et al. (2006)
	WWTP effluent	Derivatization with PFBBr, catalyst K_2CO_3			NR	
	WWTP influent	SPE: Oasis HLB	HPLC: Eclipse XDB-C_{18} column	QqQ MS/MS	70.4 ng/L [WWE]	Martín et al. (2014)
	WWTP effluent			ESI (+)/MRM	128 ng/L [WWI]	
	WWTP influent	SPE: ENVI-Carb/NH_2	LC: SunFire C_{18} column	QqQ MS/MS	0.25 µg/L	Lin et al. (2014a)
	WWTP effluent			ESI (-)/MRM		
	River water					
	WWTP effluent	RP-SPE: Isolute ENV^+	HPLC: HILIC column + ZIC-HILIC precolumn	QqQ MS/MS ESI (+)/(-)/MRM	5 ng/L [HE]	Weissbrodt et al. (2009)
	WWTP influent					
	Hospital effluent					
	WWTP effluent	on-line SPE: HySphere Resin GP	LC: C_{18} column	QqQ MS/MS	100 ng/L	Heath et al. (2015)
	Hospital effluent			ESI (+)/MRM		
	River water	RP-SPE:Isolute ENV^+	GC: DB-5 MS column	IT, MS/MS	0.16 ng/L (LOD)	
		Derivatization with MTBSTFA		EI, external ionization source MRM		
	WWTP influent	SPE: Strata X	GC: DB-5 MS column	Q MS EI (+) SIM	17 ng/L (MDL)	Yu et al. (2012)
	WWTP effluent	Derivatization with PFBBr, catalyst K_2CO_3			51 ng/L (MRL)	

	Tap water	online SPE–LC device Symbiosis™ Pico, PLRP sorbent	LC: Purospher Star RP-18e column	QTRAP MS/MS ESI (+)/SRM	0.0025 µg/L		Mendoza et al. (2016)
	Hospital effluent	RP-SPE: Isolute ENV⁺ Derivatization with BSTFA +TMCS (99:1)	GC: VF-5MS column	QqQ MS/MS EI/SCAN and MRM	0.04 µg/L		Mullot et al. (2009)
CAP	Groundwater	online SPE–LC device Symbiosis™ Pico, PLRPS sorbent	LC: Purospher Star RP-18e column	QTRAP MS/MS ESI (+)/SRM	0.3 ng/L (LOD) [GW]	2.5 ng/L (Ldet) [GW]	Negreira et al. (2013a; 2014a)
	River water				0.5 ng/L (LOD) [SW]	3.5 ng/L (Ldet) [SW]	
	WWTP influent				0.5 ng/L (LOD) [WWE]	3.5 ng/L (Ldet) [WWE]	
	WWTP effluent				0.7 ng/L (LOD) WWI]	5.0 ng/L (Ldet) [WWI]	
	Hospital effluent	online SPE–LC device Symbiosis™ Pico, PLRPS sorbent	LC: Purospher Star RP-18e column	QTRAP MS/MS ESI (+)/SRM	3.5 ng/L (Ldet) [WWE] 5.0 ng/L (Ldet) [IWWI]		Ferre-Aracil et al. (2016)
	WWTP influent and effluent	SPE: Oasis HLB	LC: Luna C_{18} column	Orbitrap MS/MS ESI (+)/SCAN	15 ng/L (MDL)		Gómez-Canela et al. (2014)
	Hospital effluent						

GW ground water, HE hospital effluent, *Ldet* limit of determination, *MDL* method detection limit, *MQL* method quantification limit, *MRL* method reporting limit, *MRM* multiple reaction monitoring, *NR* not reported, *SRM* single reaction monitoring, *WWE* waste water effluent, *WWI* waste water influent factors (EF) of up to 75,000, compared to an EF of 250 for conventional cartridges (Kiffmeyer et al. 1998). To date, extracting disks have not been applied to the analysis of cytostatic compounds in natural waters, but for 5-FU, being so polar that extraction efficiency becomes much lower at higher volumes of extractions, disks do not seem a viable alternative.

14.3.3 Instrumental Analysis

14.3.3.1 GC-coupled Techniques

5-FU is frequently determined in environmental matrices using gas chromatography (GC) as separation technique coupled to mass spectrometry (MS) (Kosjek et al. 2013; Yu et al. 2006, 2012; Mullot et al. 2009; Tauxe-Wuersch et al. 2006). GC separation requires sufficient volatility and thermal stability of the analytes, which, for certain non-volatile compounds can be achieved by incorporating a derivatization step into the method. 5-FU can be analyzed by GC in its underivatized form, but with low method sensitivity and poor peak shape (Tauxe-Wuersch et al. 2006). Pentafluorobenzyl bromide (PFBBr) is the most commonly used derivatization agent for 5-FU. Completeness and repeatability of the derivatization reaction is achieved by adding different catalysts, e.g., K_2CO_3 (Yu et al. 2006, 2012; Tauxe-Wuersch et al. 2006). 5-FU undergoes syliation with syliating agents N-tert-butyldimethylsilyl-N-ethyltrifluoroacetamide (MTBSTFA) (Kosjek et al. 2013; Heath et al. 2015) and N, O-bistrifluoroacetamide/trimethylchlorosilane (BSTFA/TMCS) (Mullot et al. 2009), forming tert-butyl dimethyl silyl (TBDMS) and trimethylsilyl (TMS) derivatives, respectively. A major limitation of derivatization is the influence of the sample matrix complexity on the repeatability of the derivatization yield, which in turn affects the precision and accuracy of the analysis (Kosjek and Heath 2011).

The column typically used for the separation of 5-FU TMS- and PFB-ethers is a DB-5MS, which is ideal for GC-MS because of its low bleed characteristics (Kosjek et al. 2013; Yu et al. 2006, 2012; Heath et al. 2015). As an alternative, low polarity columns, such as a VF-5MS (Mullot et al. 2009) and RTX-5 column (Tauxe-Wuersch et al. 2006) can be used.

Different ionization techniques can be applied after GC separation, including electron-impact ionization (EI) (Kosjek et al. 2013; Heath et al. 2015; Yu et al. 2006, 2012; Mullot et al. 2009; Tauxe-Wuersch et al. 2006) and chemical ionization (CI) (Tauxe-Wuersch et al. 2006) in negative mode. EI results in extensive compound fragmentation suitable for identification purposes, along with robustness and sufficient sensitivity at low (ng/L) concentration ranges. CI in negative mode is more sensitive for detecting 5-FU due to high electronegativity of fluorine in its structure (Tauxe-Wuersch et al. 2006). A favourable derivatization agent in this case is PFBBr. To date, GC-coupled techniques have not been applied in the investigation of CAP. The likely reason is CAP being thermally labile at increased temperatures in GC injectors, where sugar cleaves off.

5-FU and CAP can be analyzed using GC fitted with a flame ionization detector (FID) and, because of the presence of a fluorine atom in their structure, an electron capture detector (ECD). Despite the better sensitivity, especially in case of ECD, these detectors do not offer the required specificity. In addition, with a FID, the sample is destroyed making it unsuitable for preparative analysis, while the ECD gives highly nonlinear responses. Nowadays, most GC based methods for 5-FU and CAP analysis employ low-resolution mass analyzers, such as the single quadrupole (Q) (Yu et al. 2006, 2012; Tauxe-Wuersch et al. 2006), ion trap (IT) (Kosjek et al. 2013; Heath et al. 2015) and triple quadrupole mass analysers (QqQ) (Mullot et al. 2009). The obtained limits of detection do not depend solely on the analyser, but also, especially in the case of polar compounds like 5-FU, on sample preparation. The lowest LOD / LOQ for 5-FU were 0.48 ng/L/1.6 ng/L in WW and 0.16 ng/L/ 0.54 ng/L in SW analysed by GC-IT-MS/MS (Kosjek et al. 2013). The IT is the only mass analyzer that can be coupled to GC and perform multiple stage (MS^n) analysis, together with the possibility of varying the reaction time in MRM mode. Regarding sensitivity, IT has higher sensitivity in product ion-scan mode, while the sensitivity of the QqQ's is higher in multiple reaction monitoring (MRM) mode.

14.3.3.2 LC-coupled Techniques

Liquid chromatography (LC) coupled techniques are mostly applied for 5-FU (Weissbrodt et al. 2009; Lin et al. 2014a; Kovalova et al. 2009; Heath et al. 2015; Martín et al. 2011, 2014; Kiffmeyer et al. 1998) and CAP (Gómez-Canela et al. 2014; Negreira et al. 2013a; Ferre-Aracil et al. 2016) analysis in aqueous matrices. In LC, separation of strongly polar analytes, such as 5-FU is less effective on C_{18} reverse-phase (RP) HPLC columns (Lin et al. 2014a; Heath et al. 2015; Mendoza et al. 2016; Martín et al. 2011, 2014; Kiffmeyer et al. 1998). For highly polar compounds, elaborated RP columns are used, namely, Purospher Star RP-18e for 5-FU and CAP (Mendoza et al. 2016; Negreira et al. 2013a, 2014a; Ferre-Aracil et al. 2016), Sunfire C_{18} for 5-FU (Lin et al. 2014a) and Luna C_{18} for CAP (Gómez-Canela et al. 2014). All three columns are suitable for the analysis of polar compounds, Sunfire C_{18} being preferable for separations at low pH (2–8), while the other two columns have a broad working pH range (1.5–10). Hydrophilic interaction liquid chromatography (HILIC) columns have also been widely used for hydrophilic compounds including 5-FU (Kovalova et al. 2009; Weissbrodt et al. 2009). HILIC uses semi-aqueous mobile phases, usually containing a high content of organic solvent (40–97%) in water and stationary phases comprised of bare silica or zwitterionic sulfobetaine groups bound to silica. The system retains polar compounds based on the hydrophilic partitioning of compounds into the water layer formed on the surface of the stationary phase or by weak electrostatic interactions with the zwitterions (Kosjek and Heath 2011).

LC coupled to MS offers a variety of different ionization techniques, i.e. atmospheric pressure ionization (API), atmospheric pressure photoionization (APPI), atmospheric pressure chemical ionization (APCI), matrix-assisted laser desorption / ionization (MALDI) and electrospray ionization (ESI). ESI remains the most commonly applied technique for 5-FU and CAP analysis (Weissbrodt et al. 2009; Lin et al. 2014a; Gómez-Canela et al. 2014; Kovalova et al. 2009; Heath et al. 2015; Mendoza et al. 2016; Martín et al. 2011, 2014; Negreira et al. 2013a, 2014a; Ferre-Aracil et al. 2016), offering good sensitivity (LOQ in low ng/L range) despite the expected poor fragmentation and high matrix influence (Martín et al. 2011). Poor fragmentation can be advantageous in the sense that the molecular ion (or more accurately a pseudo molecular ion) is always observed, however very little structural information can be gained from the simple mass spectrum obtained. This disadvantage can be overcome by coupling ESI with tandem mass spectrometry (ESI-MS/MS). Both ESI (+) and (−) modes have been used to analyze 5-FU, although, in contrast to most cytostatic agents, it is more readily ionized in the negative ion mode (Kosjek and Heath 2011; Negreira et al. 2013b).

For HPLC less expensive UV detection is also an option (Kiffmeyer et al. 1998). However, the higher amount of noise in the HPLC-UV system affects selectivity, accuracy and sensitivity and is the reason why it has been widely replaced by MS analyzers such as QqQ (Weissbrodt et al. 2009; Kovalova et al. 2009; Martín et al. 2011, 2014) and QTRAP (Mendoza et al. 2016; Negreira et al. 2013a, 2014a; Ferre-Aracil 2016). Despite the higher sensitivity and selectivity of QqQ and QTRAP over UV, some of published LC-QqQ (Weissbrodt et al. 2009; Kovalova et al. 2009; Martín et al. 2011, 2014) and LC-QTRAP (Mendoza et al. 2016; Negreira et al. 2014a) methods still fail to detect 5-FU in aqueous environmental samples, what is the consequence of demanding extraction and chromatography of this compound. LC can also be coupled to a high-resolution (HR) mass analyzer, such as a LTQ Orbitrap mass spectrometer (Gómez-Canela et al. 2014). However, HR MS instruments are more commonly applied to identify novel transformation products (see Sect. 14.6).

The main drawback of LC-coupled techniques is the influence of matrix compounds on the method specificity, which causes a significant degree of ion suppression or enhancement of the analytes of interest. Matrix effects are a major problem in the analysis of CAP in environmental samples such as WWTP influent and effluent compared to groundwater and river water (Negreira et al. 2013a). Even though comparable LOD/Q are achieved by both methods, e.g. LC-MS and GC-MS (Table 14.3), LC-MS is still the analytical instrument of choice for analyzing polar compounds including 5-FU and CAP in environmental samples.

14.3.3.3 Other Instrumental Techniques

Capillary Electrophoresis

Capillary electrophoresis has also been used to detect 5-FU in hospital WW samples (Table 14.3), and has LODs in the low μg/L range (Mahnik et al. 2004, 2007). The performance of a CE method strongly depends on the composition of the CE buffer. Since, 5-FU is a weak acid with a pK_a1 of 7.9 (Table 14.2), it requires buffer pH values between 9.0 and 11.0 (Mahnik et al. 2004) in order to dissociate and gain effective electrophoretic mobility. Despite its higher separation efficiency, its sensitivity is not comparable to either GC- and LC-coupled methods, and the concentration of many cytostatic agents is three orders of magnitude lower than method's LOQ (8.6 μg/L) (Mahnik et al. 2007). In addition, since there is little room to enhance further CE separation, improving detection may be an alternative. This could be achieved by replacing the stock UV or UV-Vis detector with a mass spectrometer (Kosjek and Heath 2011).

14.4 Occurrence in Waste Water and the Environment

Concentrations of cytostatic agents in hospital effluents depend upon the hospital's daily water consumption, number of patients, types of cytostatic agents used, dosing and excretion rates, and on the type and severity of specific neoplastic diseases. Hospital effluents are rarely treated prior to discharge into the sewage systems (Zhang et al. 2013). With the growing number of outpatients, household discharge is also becoming an important source of cytostatic drugs. In hospital effluents, 5-FU and its major urinary metabolites account for <30% of the total administered 5-FU (Kosjek and Heath 2011; Weissbrodt et al. 2009; Mahnik et al. 2007). This may seem as a paradox, since 5-FU and CAP have short half-lives and we would expect them to be excreted to a great extent during the patients' stay in the hospital. This deficit is, however, a likely result of 5-FU and CAP being excreted away from the hospital or, as in the case of CAP, a result of oral administration at home. Pharmaceutical production waste is also a potential source of these compounds, but data regarding discharges from pharmaceutical manufacturers and their environmental relevance is scarce (Zhang et al. 2013).

5-FU is commonly detected in hospital effluents (Kosjek et al. 2013; Weissbrodt et al. 2009; Mahnik et al. 2004, 2007; Kovalova et al. 2009; Heath et al. 2015; Lin et al. 2014a; Mullot et al. 2009) (Table 14.4). Concentrations as high as 122 μg/L have been detected in a hospital effluent from Austria (undiluted wastewater) (Mahnik et al. 2004), while lower levels: 8.3–440 ng/L, 90–4000 ng/L and 46–1500 ng/L and up to 27 ng/L, have been reported in Slovenian (Kosjek et al. 2013; Heath et al. 2015), French (Mullot et al. 2009), Taiwanese (Lin et al. 2014a) and Swiss hospital effluents, respectively (Weissbrodt et al. 2009). Several studies report levels of 5-FU as below LOD (Tauxe-Wuersch et al. 2006). Overall, the levels of 5-FU reported depend on not only the source characteristics, but also on the sensitivity of the analytical method used.

Considering that hospital effluent is about 10–100 times more concentrated than municipal WW (Weissbrodt et al. 2009; Ferre-Aracil et al. 2016), concentrations lower by one to two orders of magnitude are expected in these waters. As a result, many studies fail to detect 5-FU in WWTP influents (Kiffmeyer et al. 1998; Weissbrodt et al. 2009; Yu et al. 2006, 2012; Martín et al. 2011, 2014; Tauxe-Wuersch et al. 2006) and effluents (Kiffmeyer et al. 1998; Kosjek et al. 2013; Weissbrodt et al. 2009; Yu et al. 2006, 2012; Martín et al. 2011, 2014; Tauxe-Wuersch et al. 2006). Only two studies (Table 14.4) report levels of 5-FU above the LOD in WWTP influents: < 280 ng/L (Kosjek et al. 2013; Lin et al. 2014a) and

Table 14.4 Occurrence of 5-FU and CAP in aqueous environmental compartments

Cytostatic agent	Matrix	Sampling	Concentration		Country	References
5-FU	Hospital effluent	Grab sampling	3.9–112.8 μg/L		Austria	Mahnik et al. (2007)
	Hospital effluent	/	20–122 μg/L		Austria	Mahnik et al. (2004)
	WWTP influent	Composite: 24 h time-proportional	< LOD		Spain	Kiffmeyer et al. (1998)
	WWTP effluent		< LOD			
	Hospital effluent	Composite: 24 h flow-proportional	< LOQ-27 ng/L		Switzerland	Kovalova et al. (2009)
	Hospital effluent	Grab sampling	35–92 ng/L		Slovenia	Kosjek et al. (2013)
			< LOD			
	WWTP influent	Composite: 24 h time-proportional	14–47 ng/L			
	WWTP effluent	Composite: 24 h time-proportional	< LOD			
	Hospital effluent	Composite: 24 h flow-proportional	90–4000 ng/L		France	Mullot et al. (2009)
	Hospital effluent	Composite: 24 h flow-proportional	< LOD		Switzerland	Tauxe-Wuersch et al. (2006)
	WWTP influent		< LOD			
	WWTP effluent		< LOD			
	WWTP influent	Composite: 24 h time-proportional	< MDL		Spain	Martín et al. (2011)
	WWTP effluent		< MDL			
	River water		< MDL			
	WWTP influent	Grab sampling	< MDL		USA	Yu et al. (2012)
	WWTP effluent		< MDL			
	River water		< MDL			
	WWTP influent	Composite: 24 h time-proportional	< LOD		Spain	Martín et al. (2014)
	WWTP effluent		< LOD			
	Hospital effluent	Composite: 24 h flow-proportional	48–1500 ng/L	46–510 ng/ L	Taiwan	Lin et al. (2014a)
	River water		5–70 ng/L	35–160 ng/L		
	WWTP influent		280 ng/L	NR		
	WWTP effluent		80 ng/L			
	Hospital effluent	Composite: 18 h, 6 h and 24 h flow-proportional	< 5–27 ng/L		Switzerland	Weissbrodt et al. (2009)
	WWTP influent		< LOD			
	WWTP effluent	Composite: 24 h flow-proportional	< LOD			

(continued)

Table 14.4 (continued)

Cytostatic agent	Matrix	Sampling	Concentration		Country	References
	Spiked river water*	Grab sampling	150 ng/L	19 ng/L	Slovenia	Heath et al. (2015)
	Spiked hospital effluent**		560–610 ng/L	47–364 ng/L		
	Spiked WWTP effluent***		150 ng/L	38–41 ng/L		
	Hospital effluent		400–440 ng/L	8.3–15 ng/L		
	WWTP influent	Composite: 24 h time-proportional	< LOD		USA	Yu et al. (2006)
	WWTP effluent		< LOD			
	Tap water	Grab sampling	< LOD		Spain	Mendoza et al. (2016)
CAP	WWTP influent	Composite: 24 h time-proportional	< LOD-27 ng/L		Spain	Negreira et al. (2013a)
	Hospital effluent	Composite: 12 h time-proportional	< LOD-1139 ng/L		Spain	Ferre-Aracil et al. (2016)
	WWTP influent	Composite: 24 h flow-proportional	< LOQ-72.6 ng/L		Spain	Negreira et al. (2014a)
	WWTP effluent		< LOQ-36 ng/L			
	Hospital effluent		< LOQ			
	WWTP influent	Composite: 24 h time-proportional	< LOQ-49 ng/L		Spain	Gómez-Canela et al. (2014)
	WWTP effluent		< LOQ			
	Hospital effluent		< LOQ			

*River water spiked with 43 ng/L 5-FU; **Hospital WW spiked with 454 ng/L 5-FU; *** WWTP effluent spiked with 66 ng/L 5-FU; *LOD* limit of detection, *LOQ* limit of quantification, *MDL* method detection limit, *NR* not reported

effluents: 80 ng/L (Lin et al. 2014a). Reported levels of 5-FU in surface waters are from 5–160 ng/L (Lin et al. 2014a), but more frequently it is below the LOD (Yu et al. 2012; Martín et al. 2011, 2014). A recent study attempted to detect CAP in tap water with no success (Mendoza et al. 2016). So far, CAP has only been reported in Spanish hospital effluents (< LOD to 1139 ng/L) (Gómez-Canela et al. 2014; Ferre-Aracil et al. 2016; Negreira et al. 2014a), WWTP influents (< LOD to 72.6 ng/L) (Gómez-Canela et al. 2014; Negreira et al. 2013a, 2014a) and WWTP effluents (< LOQ to 36 ng/L) (Gómez-Canela et al. 2014; Negreira et al. 2014a).

PECs in hospital effluents are 0.008–95 μg/L and 0.012–0.117 μg/L for 5-FU and CAP, respectively (Straub 2007). PECs for WWTP effluents are much lower, i.e., 0.36–4.4 ng/L for 5-FU and 8.5–223 ng/L for CAP and even lower in European rivers: 0.01 ng/L for 5-FU and 0.21 ng/L for CAP (Johnson et al. 2013; Straub 2007; Gómez-Canela et al. 2014; Franquet-Griell et al. 2015). The measured concentrations of 5-FU and CAP in all three aqueous compartments are generally higher than their PECs, reflecting the variability and unpredictability in discharge patterns.

The most likely reasons why some studies fail to detect 5-FU and CAP in hospital effluents (Tauxe-Wuersch et al. 2006; Negreira et al. 2014a; Gómez-Canela et al. 2014), municipal WW (Kiffmeyer et al. 1998; Tauxe-Wuersch et al. 2006; Martín et al. 2011, 2014; Yu et al. 2006, 2012; Weissbrodt et al. 2009), surface waters (Martín et al. 2011; Yu et al. 2012) and tap water (CAP) (Mendoza et al. 2016) is low method sensitivity, that 5-FU and CAP at the time of sampling were not being administered, and/or they were eliminated completely during sewage treatment. A low administration dose, partial excretion rates in urine and natural attenuation (dilution, biodegradation and photodegradation) are other reasons for the low levels of 5-FU and CAP (Mendoza et al. 2016). Further investigations are needed regarding their presence in groundwater and drinking water, as well as in landfill leachates.

14.5 Environmental Fate of 5-FU and CAP

14.5.1 Biodegradation

The overall susceptibility of 5-FU and CAP to biodegradation correlates with their physico-chemical properties, e.g., hydrophobicity and BCF (Table 14.2), and with experimental conditions (biomass concentration, acclimation, and reaction time). Since both 5-FU and CAP are hydrophilic (log $K_{ow}<2$), they are more likely present in the aqueous environment than sorbed to sediments / soil or bioaccumulated and are thus more susceptible to biodegradation.

Biodegradation studies (Table 14.5) show that 5-FU is biodegraded in activated sludge (AS) (Kosjek et al. 2013; Roche 2007; Mahnik et al. 2007; Yu et al. 2006; Kiffmeyer et al. 1998; Onesios and Bouwer 2012; Lutterbeck et al. 2016) and imply that biodegradation is the main removal mechanism in batch AS experiments (Mahnik et al. 2007). Its removal during a non-standardized biodegradation process in AS inoculated bioreactor, a sewage inoculate biofilm column or a simulated sewage treatment plant (Kosjek et al. 2013; Mahnik et al. 2007; Yu et al. 2006; Onesios and Bouwer 2012), and in a series of standardized OECD tests (OECD 301D (Lutterbeck et al. 2016), 303A (Kiffmeyer et al. 1998; Roche 2007) and 308 (Roche 2007) is from 35% to 99% within 1–256 days. The process is dependent on both concentration of 5-FU and AS (Kosjek et al. 2013; Kiffmeyer et al. 1998). Its removal follows a concentration-dependent rate: the higher the initial concentration of 5-FU, the slower the degradation. For example, an AS concentration of 0.67 g/L resulted in a $t_{1/2} = 8$ h for 5-FU. Decreasing the AS concentration by 5 and 50-fold, the biodegradation slowed down even further to give a $t_{1/2} = 11$ h and 16 h for 0.14 and 0.014 g/L AS, respectively (Kosjek et al. 2013).

Some studies report different results showing that 5-FU in a closed bottle test (OECD 301D) remained undegraded after 28 days (Lutterbeck et al. 2015) and

Table 14.5 Biodegradation and adsorption of 5-FU and CAP

Cytostatic agent	Biodegradability	Concentration	Rate/Percentage (time)	References	Adsorption onto sludge, sediments	Concentration	Adsorption rate/ percentage	References
5-FU	No	–	OECD 302B: 2% (28 d)	Roche (2007)	No	–	–	Kümmerer and Al-Ahmad (1997), Mahnik et al. (2007)
			OECD 301D: 0% (40 d)					
		–	OECD 301D: negligible (28 d)	Lutterbeck et al. (2015)				
		854 mg/L	OECD 302B: 2% (28 d)	Kümmerer and Al-Ahmad (1997)				
	Yes	0.001 mg/L	< 60% (50 d) in AS	Yu et al. (2006)				
		0.01 mg/L						
		0.05 mg/L						
		–	OECD 301D: 35–44% (28 d)	Lutterbeck et al. (2016)				
		0.01 mg/L	≥ 95% (256 d) in sewage inoculate biofilm column	Onesios and Bouwer (2012)				
		–	OECD 303A: ≥ 25% (1 d), ≥ 97% (14 d)	Roche (2007)				
			OECD 308: $t_{1/2} \leq 2$ d					

(continued)

Table 14.5 (continued)

Cytostatic agent	Biodegradability	Concentration	Rate/Percentage (time)	References	Adsorption onto sludge, sediments	Concentration	Adsorption rate/ percentage	References
		0.001 mg/L	99.99% (40 h)	Kosjek et al. (2013)		0.1 mg/L	$K_d \geq 4786$ L/kg (5 g/L AS)	Fürhacker et al. (2006)
		1 mg/L	$t_{1/2} = 8$ h (0.67 g/L AS)					
		10 mg/L	$t_{1/2} = 11$ h (0.14 g/L AS)					
		20 mg/L	$t_{1/2}$=16 h (0.014 g/L AS)					
		100 mg/L						
		0.005 mg/L	> 90% (24 h) in AS	Mahnik et al. (2007)	Yes		< 10% in raw WW (12-15 g/L of mixed liquor suspended soils)	Mahnik et al. (2007)
		0.5 mg/L						
		5 mg/L	OECD 303 A: 100 ± 4% (2-10 d)	Kiffmeyer et al. (1998)			2-5% to activated sludge within 24 h	
		10 mg/L						
		20 mg/L (in mixture 3 mg/L)						
CAP	No	–	≤ 2% (6 h) in AS	Guo et al. (2015)	No	–	–	Hoerger et al. (2009)
	Yes	20 mg/L	100% (24 h) in AS	Franquet-Griell et al. (2017)		30 mg/L	inherent modified MITI-II test in AS	Studer (2005)

		30 mg/L	OECD 302C-modified: 27% mineralization (14 d), 41% (21 d)	Studer (2005)	Yes	–	$K_d = 272$ L/kg	Häner (2006)
		0.001 mg/L	> 99% (11 d) in simulated sewage treatment reactor	Kosjek et al. (2013)				
		1 mg/L						
		10 mg/L						
		20 mg/L						
		100 mg/L						
		–	OECD 302C: 29% (28 d) 44% (56 d) 55% (84 d)	Roche (2008)				
		–	OECD 302B: 15% (7d) 27% (14 d) 37% (21d) 58% (28 d)	ETT (2004)				
		–	OECD 302C-modified:	Häner (2006)				
			mineralization: 29% (28 d) 44% (56 d) 55% (84 d)					
			total removal: 68% (84 d)					

d day(s)

40 days (Roche 2007), although later studies report 35–44% biodegradation (Lutterbeck et al. 2016). The Zahn-Wellens test (OECD 302B) found negligible biodegradation (2%) after 28 days of inoculation (Kümmerer and Al-Ahmad 1997; Roche 2007). This inconsistency in the data is due to different experimental conditions. Also, almost all of the studies used levels of 5-FU between 0.001–854 mg/L (Table 14.5). In this range 5-FU is cytotoxic to AS microorganisms, potentially leading to a false negative results (Straub 2007). In addition, all tests showing rapid biodegradation were performed at higher concentrations of biomass (2.5 g/L dry mass (Roche 2007) and 5.4 g/L (Kosjek et al. 2013)), while the OECD 301D test is characterized by having a low bacterial density (500 CFU/mL), which may account for the discrepancies in the results. Therefore, the low biodegradability of 5-FU is likely associated with low bacteria density rather than toxicity to AS microorganisms (Martín et al. 2014). In addition, in hospital effluents, the presence of other cytostatic agents and antibiotics, that can have either antagonistic or synergistic effects on the cytotoxicity of 5-FU to AS microorganisms, can significantly affect biodegradation (Negreira et al. 2013a).

CAP is also biodegradable (Table 14.5), with 15–100% being removed (1–84 days) in standardized OECD tests (Roche 2008; ETT 2004; Studer 2005; Häner 2006) or in batch reactors with AS (Kosjek et al. 2013; Franquet-Griell et al. 2017). OECD 302C-like tests also report significant mineralization of CAP (58%, (ETT 2004) and 68%, Häner 2006) within 28 days and 84 days, respectively. Interestingly, one study reported CAP as being almost non-biodegradable in an AS inoculum (Guo et al. 2015). The experiment, however, only lasted for 6 h. Again, inconsistency in results from biodegradation experiments can be explained with the wide concentration range of CAP and biomass (variety of microorganisms and their concentration) used in different studies, e.g. CAP concentrations ranged from 0.001 to 100 mg/L (Table 14.5) Again, as concluded for 5-FU, higher concentrations of CAP may lead to false negative results (Kosjek et al. 2013).

Only one study compared 5-FU and CAP under similar experimental conditions and found that the rate of biodegradation of CAP was lower than 5-FU ($t_{1/2} = 77$ h), i.e., with >99% removal of CAP after 11 days compared with >99% of 5-FU within 40 h (Kosjek et al. 2013). This finding agrees with the majority of published studies. A plausible explanation is the presence of less biodegradable sugar molecule attached to pyrimidine ring (Kümmerer and Al-Ahmad 1997).

14.5.2 Adsorption and Mobility in Sludge, Sediment and Soils

Studies investigating the adsorption of 5-FU and CAP onto solids are sparse and contradictory, although the physico-chemical properties (Table 14.2) of both 5-FU and CAP suggest that adsorption to AS is unlikely to be an important sink during WW treatment. In contrast to the prediction, adsorption of 5-FU to WWTP biomass is reported as high as $K_d \geq 4786$ L/kg (Fürhacker et al. 2006) (Table 14.5). Alternatively, other studies find that adsorption of 5-FU is limited, e.g., < 10% in raw wastewater, 2–5% on AS (Mahnik et al. 2007), or there is no adsorption on AS during the Zahn-Wellens test (Kümmerer and Al-Ahmad 1997). In addition, there is no evidence of accumulation or the prolonged half-life of 5-FU in sediments (Straub 2007). CAP was shown to be either partially adsorbed ($K_d = 272$ L/kg AS) (Häner 2006) or not adsorbed to AS (Hoerger et al. 2009; Studer 2005).

14.5.3 Direct Photolysis, Indirect Photolysis and Advanced Treatment Processes

Sunlight-mediated photodegradation in environmental compartments occurs *via* direct and indirect pathways. Direct photolysis occurs when the compound adsorbs light directly, leading to chemical bond cleavage. In natural aqueous compartments, light can also interact with dissolved organic matter (DOM), chloride (Cl^-), bicarbonate (HCO_3^-), sulfates (SO_4^{2-}) and nitrates (NO_3^-) to produce transient reactive species, such as singlet oxygen ($^1O_2$), $^{\bullet}OH$, and triplet state natural organic matter (NOM) that react with contaminants in a process known as indirect photolysis. DOM can also absorb UV light and reduce UV transmittance, HCO_3^- reacts with $^{\bullet}OH$ and forms less reactive species, while Cl^- gives rise to the photodegradation of 5-FU (Zhang et al. 2017).

5-FU and CAP are susceptible to direct photolysis by sunlight (Table 14.2). At pH>7, the protonated form of 5-FU additionally absorbs light at 280–320 nm (Li et al. 2015). The differing experimental conditions described in literature make it difficult, if not impossible, to compare reported photodegradation rates, half-lives and removal percentages for 5-FU and CAP. For example, most studies attempt to simulate natural sunlight (SNS), i.e., 200–600 nm, while others concentrate on using UV light (UV C: 100–280 nm, UV B: 280–315 nm, UV A: 315–400 nm), what makes results difficult to compare due to different intensity of these light sources. Additionally, times of irradiation (12 min-14 h) and cytostatic concentrations (5-FU: 0.00027-19.13 mg/L and CAP: 0.05-1000 mg/L, Table 14.6) vary greatly between the experiments, confirming the complexness of the studies.

For 5-FU, the reported and calculated (where possible) values for photolytic rate constant are from 0.00016 min^{-1} to 0.16 min^{-1}, half-lives from 0.071 h–71 h and photolytic removal efficiencies of 24.32–100% (Kosjek et al. 2013; Lutterbeck et al. 2015, 2016; Zhang et al. 2017; Lin et al. 2013; Gomez-Canela et al. 2017; Koltsakidou et al. 2017; Governo et al. 2017; Miolo et al. 2011). Such wide ranges are a result of different experimental conditions, but indicate that with optimization, high removal rates and even mineralization of 5-FU could be achieved. Only three papers address direct photolysis of CAP (Kosjek et al. 2013; Franquet-Griell et al. 2017; Guo et al. 2015). They show that CAP degrades faster, when irradiated with UV light (half-lives: 0.023 h–0.76 h) (Kosjek et al. 2013; Franquet-Griell et al. 2017; Guo et al. 2015), than with simulated sunlight (4.81 h) (Franquet-Griell et al. 2017), and is either partially (Kosjek et al. 2013; Franquet-Griell et al. 2017) or completely

Table 14.6 Direct photolysis and advanced treatment processes susceptibility of 5-FU and CAP

Cytost. agent	Direct photolysis	Concentration	Photolysis rate	References	Advanced treatment processes	Concentration	Removal rate	References
5-FU	Yes	10 mg/L	**SNS** 2% (45 min)	Koltsakidou et al. (2017)	Yes	0.00027 mg/L	**O_3** 100% (45 min) k = 0.049 min^{-1} $t_{1/2}$ = 0.234 h	Rey et al. (1999)
		20 mg/L	**UV** 100% (256 min) k = 0.034 min^{-1} $t_{1/2}$ = 0.34 h	Lutterbeck et al. (2015)		5 mg/L	**O_3** 100% (< 2 min) k = 0.499 min^{-1} $t_{½}$ = 0.023 h	Lin et al. (2014b)
		0.13 mg/L	**UV** 24.32% k = 0.109 min^{-1} $t_{1/2}$ = 0.106 h	Zhang et al. (2017)		20 mg/L	**$UV/Fe^{2+}/H_2O_2$** 100% (< 2 min)	Lutterbeck et al. (2015)
							UV/H_2O_2 > 99% (8 min) k = 0.41 min^{-1} $t_{½}$ = 0.028 h	
		10 mg/L	**UV** 99.9% (50 min) k = 0.16 min^{-1} $t_{1/2}$ = 0.071 h	Gómez-Canela et al. (2017)			**UV/TiO_2** > 99% (16 min) k = 0.13 min^{-1} $t_{1/2}$ = 0.089 h	
						49.43 mg/L	**$UV/Fe^{2+}/H_2O_2$** ~80% (15 min) 100% (2 h)	Governo et al. (2017)
		20 mg/L	**SNS** 32% (256 min) k = 0.0015 min^{-1} $t_{1/2}$ = 7.7 h	Lutterbeck et al. (2016)			**UV/H_2O_2** 100% (30 min)	
							$UV/Fe^{2+}/H_2O_2$	
			UV 100% (32 min) k = 0.055 min^{-1} $t_{1/2}$ = 0.21 h				**Fenton oxidation $H_2O_2/Fe2^+$** 100% (1 h)	

		49.43 mg/L	**UV** > 99% (15 min)	Governo et al. (2017)		0.2 mg/L	**UV/ZnO** 87.9% (8 h) k = 0.0041 min^{-1} $t_{1/2}$ = 2.817 h	Lin and Lin (2014)
							UV/TiO_2 > 99.9% (120 min) k = 0.0365 min^{-1} $t_{1/2}$ = 0.316 h	
		1.3 mg/L	**UVB** 5, 15 and 30 J/cm^2 30 ± 2.5%, 65 ± 3.1 % and > 95%	Miolo et al. (2011)		13.01 mg/L	**Biological treatment + Electro Fenton** 100% (2 h)	Ganzenko et al. (2017)
		1 mg/L	**UV** > 99% (190 min) k = 0.045 min^{-1} $t_{1/2}$ = 0.257 h	Kosjek et al. (2013)		0.13 mg/L	**UV/H_2O_2** 94% (3 min) k = 0.919 min^{-1} $t_{1/2}$ = 0.0126 h	Zhang et al. (2017)
		0.0049 mg/L 0.049 mg/L 0.49 mg/L 10.02 mg/L	**SNS,** river water k = 0.0011–0.0013 min^{-1} $t_{1/2}$ = 9.8 ± 0.7 h	Lin et al. (2013)		1 mg/L	**UV/H_2O_2** 99.6% (10 min) k = 0.0369 min $t_{1/2}$ = 0.313.h	Kosjek et al. (2013)
			SNS, deionized water k = 0 00016–0.00021 min^{-1} $t_{1/2}$ = 63 ± 8 h			10 mg/L	**SNS/co-doped N/S-TiO_2** 100% (360 min)	Koltsakidou et al. (2017)
						0.2 mg/L	chlorination/bromination 100% (30 min) 1 mg/L Cl^- + 0.008 mg/L Br^-	Li et al. (2015)

(continued)

Table 14.6 (continued)

Cytost. agent	Direct photolysis	Concentration	Photolysis rate	References	Advanced treatment processes	Concentration	Removal rate	References
CAP	Yes	1 mg/L	**UV** remains present	Kosjek et al. (2013)	Yes	1000 mg/L (single solutions)	**O_3** 100%	Ferre-Aracil et al. (2016)
			(14 h)			25 mg/L total concentration of 17 cytostatics in DMSO	**O_3/H_2O_2** 100% (10 min)	
			$k = 0.015\ min^{-1}$					
			$t_{1/2} = 0.76$ h					
		0.05 mg/L	**SNS** 40% (240 min) $k = 0.0024\ min^{-1}$ $t_{1/2} = 4.81$ h	Franquet-Griell et al. (2017)		1 mg/L	**UV/H_2O_2** 97% (10 min) $k = 0.0379\ min^{-1}$ $t_{1/2} = 0.31$ h	Kosjek et al. (2013)
			UVC ~90% (90 min) $k = 0.0532\ min^{-1}$ $t_{½} = 0.22$ h			0.5 mg/L	**electrochemical oxidation with $Ti/IrO_2 – RuO_2$ electrode** 15% (5 min) 85% (30 min)	Barışçı et al. (2018)
		20 mg/L	**UV** 99.9% (12 min) $k = 0.5012\ min^{-1}$ $t_{½} = 0.023$ h	Guo et al. (2015)		0.05 mg/L	**UV/H_2O_2** 81% (1 min) 100% (2 min)	Franquet-Griell et al. (2017)
			biodegradation +green algae +UV 6.73%					

(Guo et al. 2015) removed. Again, intensity of the light sources and times of irradiation differ greatly and make the results difficult to compare.

Advanced oxidation processes (AOPs) are treatment strategies designed to remove recalcitrant organic contaminants. They include photochemical degradation (UV/O_3, UV/H_2O_2), photocatalysis (TiO_2/UV, $TiO_2/UV/H_2O_2$, ZnO/UV, $UV/Fe^{2+}/H_2O_2$, photo-Fenton), chemical oxidation (O_3, O_3/H_2O_2, H_2O_2/Fe^{2+}), electrochemical oxidation and electro-Fenton. All AOPs produce $^{\bullet}OH$ radicals - reactive species with low selectivity and capable of removing many of the most recalcitrant organic contaminants. However, even partial decomposition of non-biodegradable organic pollutants can lead to biodegradable intermediates (Poyatos et al. 2009).

The most studied AOP regarding the removal of 5-FU is UV/H_2O_2. Indirect photolysis rate constants are from 0.0369 min^{-1} to 0.919 min^{-1}. The half-lives are from 0.0126 h to 0.313 h and removal rates are between 94 and 100% (Table 14.6). Mineralization is from 26–100% (Kosjek et al. 2013; Lutterbeck et al. 2015; Zhang et al. 2017; Governo et al. 2017). In comparison, direct UV irradiation (200–400 nm) results in comparable removal rates (~93%) but in low mineralization (~5–40%) during UV treatment (Lutterbeck et al. 2015, 2016; Lin et al. 2013; Koltsakidou et al. 2017; Governo et al. 2017). Ozonation (O_3) is also efficient at removing 5-FU from distilled water within in 2–45 min (Lin et al. 2014b; Rey et al. 1999) and in real wastewater samples at natural pH (~7) (Lin et al. 2014b). This is due to the presence of electron-rich sites on the 5-FU molecule that can directly react with O_3, or, alternatively by producing $^{\bullet}OH$ radicals. Other efficient AOPs for removing 5-FU are Fenton oxidation (Governo et al. 2017), photo-Fenton reaction (Governo et al. 2017), photocatalytic degradation with TiO_2 (Lutterbeck et al. 2015; Lin and Lin 2014), ZnO (Lin and Lin 2014), co-doped-N/S-TiO_2 with SNS (Koltsakidou et al. 2017) and electro-Fenton reaction (Ganzenko et al. 2017). These are all capable of completely removing 5-FU within 6–360 min. Regarding mineralization, photo-Fenton and electro-Fenton are the two most efficient AOPs. Studies reported 5-FU to be 75% mineralized after 256 min of photo-Fenton treatment (Governo et al. 2017) and 94% mineralized with electro-Fenton oxidation within 3 h (Ganzenko et al. 2017). Unfortunately, there are no studies investigating the electrochemical oxidation of 5-FU. 5-FU also interacts readily with free chlorine and bromine and is completely degraded within 30 min, while in absence of background bromide, 50% of 5-FU is removed within 1 h at pH~7.3 and the reaction follows second-order kinetics (Li et al. 2015). Bromination of 5-FU occurs at rate 2–3 orders of magnitude faster (Li et al. 2015). From the available published studies investigating the removal of CAP, UV/H_2O_2 (>97% in max. 10 min) (Kosjek et al. 2013; Franquet-Griell et al. 2017) and O_3/H_2O_2 (100% in 10 min) (Ferre-Aracil et al. 2016) were reportedly more successful at removing CAP than electrochemical oxidation with Ti/IrO_2-RuO_2 (85% within 30 min) (Barışçı et al. 2018).

Based on mainly laboratory scale studies, it is evident that optimized advanced treatments can achieve high removal rates of both, 5-FU and CAP. Complete mineralization, however, is rarely achieved and various transformation products of 5-FU (Kosjek et al. 2013; Lutterbeck et al. 2015, 2016; Zhang et al. 2017; Lin et al. 2013; Gomez-Canela et al. 2017; Koltsakidou et al. 2017; Miolo et al. 2011; Lin and

Lin 2014; Franquet-Grill et al. 2017) and CAP (Kosjek et al. 2013; Barışçı et al. 2018) have been reported (see Sect. 14.6). The removal of 5-FU and CAP by direct and indirect photolysis and other AOPs at environmentally relevant concentrations still require investigation since current findings are based on unrealistically high concentrations (1–20 mg/L). Overall, data suggests that AOPs are very promising and more efficient at removing 5-FU and CAP than direct photolysis under UV light and SNS, but have mostly been applied in simple matrices like MilliQ and potable water and have neither been applied nor are suitable for complex matrices like wastewaters.

14.6 Transformation Products of 5-FU and CAP

In 2011, a review paper on the occurrence, fate and determination of cytostatic drugs in the environment found only three studies identifying three human metabolites and one TP for all cytostatic drugs (Kosjek and Heath 2011). Of these, the authors confirmed only one TP in a real sample, namely, a hospital effluent. Since then, the number of studies has increased significantly and there have now been 11 studies investigating the formation of 5-FU and/or CAP TPs during water treatment.

Many of the laboratory-scale advanced treatments investigated either direct (UV only) (Kosjek et al. 2013; Lin et al. 2013; Lutterbeck et al. 2016; Gomez-Canela et al. 2017; Miolo et al. 2011) or indirect (addition of H_2O_2, TiO_2, or Fe^{2+}/H_2O_2) photolysis (Kosjek et al. 2013; Lutterbeck et al. 2015; Zhang et al. 2017; Koltsakidou et al. 2017; Lin and Lin 2014; Barışçı et al. 2018) (Table 14.7) and confirmed the presence of TPs as a result of either no or incomplete mineralization (Negreira et al. 2013b; Lutterbeck et al. 2015; Lin et al. 2013; Governo et al. 2017; Lin and Lin 2014) and/or changes in acute toxicity of treatment mixture (Lutterbeck et al. 2015, 2016; Guo et al. 2015; Miolo et al. 2011; Barışçı et al. 2018).

The most commonly identified phototransformation reactions of 5-FU are the addition of water, defluorination, hydroxylation and the formation of diastereomeric photoproducts. The first step in the transformation of 5-FU is hydroxylation, with addition of a single water molecule on the C_5-C_6 bond. This single hydroxylation step forms the most commonly detected TP during water treatment with a characteristic MS fragment with a m/z of 148, representing the two isomers, 5-fluoro-6-hydroxy-5,6-dihydrouracil and 5-fluoro-5-hydroxy-5,6-dihydrouracil (Kosjek et al. 2013; Lutterbeck et al. 2015, 2016; Koltsakidou et al. 2017) or with m/z 147 (Kosjek et al. 2013; Lutterbeck et al. 2015; Zhang et al. 2017). Further dihydroxylation occurs at the C_5-C_6 bond, forming cis/trans-5,6-dihydroxyuracil (m/z 143) (Kosjek et al. 2013; Gomez-Canela et al. 2017; Koltsakidou et al. 2017; Miolo et al. 2011), or alternatively, on the C_6-N_1 bond, forming TP-165 (Lutterbeck et al. 2015) (Table 14.7). This reaction may be followed by the opening of the pyrimidine ring

Table 14.7 Transformation products of 5-FU and CAP

Cytostatic. agent	Transformation product	Instrument	Identification parameters	Molecular formula	Degradation process	References
5-FU	TP 253 (5-hydroxyuracil dimer)	UPLC-QqTOF-MS/MS	$[M-H]^- = 253$ (127, 84)	$C_4H_4N_2O_4$	UV	Kosjek et al. (2013)
	TP 165	LC-IT-MS/MS	$[M+H]^+ = 165$ (144, 130, 121, 107, 98, 79)	–	$UV/Fe^{2+}/H_2O_2$	Lutterbeck et al. (2015)
	TP 168	LC-IT-MS/MS	$[M+H]^+ = 168$ (151, 138, 123, 109, 95, 83, 71)	–	UV/H_2O_2	Lutterbeck et al. (2015)
	TP 152	LC-IT-MS/MS	$[M+H]^+ = 152$ (132, 108, 95, 80, 57)	–	UV/H_2O_2	Lutterbeck et al. (2015)
					UV/TiO_2	
	TP 149	LC-IT-MS/MS	$[M+H]^+ = 149$ (131, 119, 114, 106, 95, 88, 63)	–	UV	Lutterbeck et al. (2016)
	TP-147 (5-fluoro-6-hydroxy-5,6-dihydrouracil)	UPLC-QqToF-MS/MS	$[M-H]^- = 147$ (194)	$C_4H_4N_2O_3F$	UV/H_2O_2	Kosjek et al. (2013)
		LC-IT-MS/MS	$[M+H]^+ = 147$ (130, 104, 92)		UV/TiO_2	Lutterbeck et al. (2015)
		LC-IT-MS/MS	$[M+H]^+ = 148$ (130, 119, 114, 106, 95, 88, 63)		UV	Lutterbeck et al. (2016)
		UHPLC-TOF-MS	$[M+H]^+ = 147$		SNS/TiO_2	Koltsakidou et al. (2017)
		LC-QqQ-MS/MS	$[M-H]^- = 147$		UV/H_2O_2	Zhang et al. (2017)
	ISO-TP-147 (5-fluoro-6-hydroxy-5,6-dihydrouracil)	UPLC-QqToF-MS/MS	$[M-H]^- = 147$ (104)	$C_4H_4N_2O_3F$	UV/H_2O_2	Kosjek et al. (2013)

(continued)

Table 14.7 (continued)

Cytostatic. agent	Transformation product	Instrument	Identification parameters	Molecular formula	Degradation process	References
	TP-143 (*cis/trans*-5,6-dihydroxyuracil)	UPLC-QqTOF-MS/MS	$[M-H]^- = 143$ (104)	$C_4H_4N_2O_4$	UV	Kosjek et al. (2013)
		LC-TOF-MS	$[M-H]^- = 143$			
		LC-LTQ-Orbitrap	$[M-H]^- = 143$		UV	Gómez-Canela et al. (2017)
		UHPLC-TOF-MS/MS			SNS/TiO_2	Koltsakidou et al. (2017)
		LC-TOF-MS			UV	Miolo et al. (2011)
	ISO-TP-143 (*cis/trans*-5,6-dihydroxyuracil)	UPLC-QqTOF-MS/MS	$[M-H]^- = 143$ (104)	$C_4H_4N_2O_4$	UV	Kosjek et al. (2013)
		LC-TOF-MS	$[M-H]^- = 143$		UV	Miolo et al. (2011)
	TP-127 (5-hydroxyuracil)	UPLC-QqTOF-MS/MS	$[M-H]^- = 127$ (84)	$C_4H_4N_2O_3$	UV	Lutterbeck et al. (2016)
		LC-TOF-MS	$[M-H]^- = 127$		UV	Kosjek et al. (2013)
		UHPLC-TOF-MS	$[M+H]^+ = 127$ (73)		SNS/TiO_2	Koltsakidou et al. (2017)
	TP-116	LC-IT-MS/MS	$[M+H]^+ = 116$ (98)	–	UV/H_2O_2 $UV/Fe^{2+}/H_2O_2$	Lutterbeck et al. (2015)
	TP-140	LC-TOF-MS	$[M-H]^- = 140$	$C_4HN_2O_4$	SNS/TiO_2 $SNS/TiO_2/H_2O_2$	Koltsakidou et al. (2017)
	TP-145	LC-TOF-MS	$[M+H]^+ = 145$	$C_4H_5N_2O_4$	SNS/TiO_2	Koltsakidou et al. (2017)
	TP-163	LC-TOF-MS	$[M+H]^+ = 163$	$C_4H_7N_2O_5$	SNS/TiO_2	Koltsakidou et al. (2017)

	TP-159	LC-TOF-MS	$[M - H]^- = 159$	$C_4H_3N_2O_5$	SNS/TiO_2 visible light/TiO_2	Koltsakidou et al. (2017)
	TP-105	LC-QqQ-MS/MS	$[M - H]^- = 104$	$C_3H_4FNO_2$	SNS	Lin et al. (2013)
		LC-IT-MS/MS	$[M - H]^- = 104$ (84)		UV	Lutterbeck et al. (2016)
		LC-IT-MS/MS	$[M + H]^+ = 104$ (74)		UV/TiO_2	Lutterbeck et al. (2015)
		HPLC-QqQ-MS/MS	$[M - H]^- = 105$		UV/TiO_2	Lin and Lin (2014)
		LC-MS/MS	$[M - H]^- = 104$		UV/H_2O_2	Zhang et al. (2017)
CAP	ISO-CAP	UPLC-QqTOF-MS/MS	$[M + H]^+ = 360$ (244, 174, 270)	$C_{15}H_{22}N_3O_6F$	UV	Kosjek et al. (2013)
	TP-244	UPLC-QqTOF-MS/MS	$[M + H]^+ = 244$ (174, 130, 154)	$C_{10}H_{14}N_3O_3F$	UV	Kosjek et al. (2013)
		LC-TOF-MS/MS	$[M + H]^+ = 244$		Electrochemical oxidation	Barışçı et al. (2018)
	TP-226	UPLC-QqTOF-MS/MS	$[M + H]^+ = 226$ (156, 112)	$C_{10}H_{15}N_3O_3$	UV	Kosjek et al. (2013)
	TP-242	UPLC-QqTOF-MS/MS	$[M + H]^+ = 242$ (172, 128)	$C_{115}N_3O_4$	UV	Kosjek et al. (2013)
	TP-358	UPLC-QqTOF-MS/MS	$[M + H]^+ = 358$ (242, 172, 128)	$C_{15}H_{23}N_3O_7$	UV	Kosjek et al. (2013)

(continued)

Table 14.7 (continued)

Cytostatic. agent	Transformation product	Instrument	Identification parameters	Molecular formula	Degradation process	References
	TP-378/L	UPLC-QqTOF-MS/MS	$[M + H]^+ = 378$ (262, 219, 192, 149, 134, 217, 129)	$C_{15}H_{24}N_3O_7F$	UV	Kosjek et al. (2013)
	TP-378/M	UPLC-QqTOF-MS/MS	$[M + H]^+ = 378$ (219, 192, 149, 129)	$C_{15}H_{24}N_30_7F$	UV	Kosjek et al. (2013)
	TP-378/R	UPLC-QqTOF-MS/MS	$[M + H]^+ = 378$ (262, 219, 192, 149, 129)	$C_{15}H_{24}N_3O_7F$	UV	Kosjek et al. (2013)
	TP-378/A	UPLC-QqTOF-MS/MS	$[M + H]^+ = 378$	$C_{15}H_{24}N_3O_7F$	UV	Kosjek et al. (2013)
	TP-260	UPLC-QqTOF-MS/MS	$[M + H]^+ = 238$ (220, 150, 106)	$C_9H_{16}NO_5F$	UV	Kosjek et al. (2013)
	5-FU	UPLC-QqTOF-MS/MS	$[M + H]^+ = 129$	$C_4H_3N_2O_2F$	UV	Kosjek et al. (2013)
		LC-TOF-MS/MS	$[M + H]^+ = 130$		Electrochemical oxidation	Barışçı et al. (2018)
	TP-155	LC-TOF-MS/MS	$[M + H]^+ = 155$	–	Electrochemical oxidation	Barışçı et al. (2018)
	TP-174	LC-TOF-MS/MS	$[M + H]^+ = 174$	–	Electrochemical oxidation	Barışçı et al. (2018)
	TP-129 (5-FU)	LC-TOF-MS/MS	$[M + H]^+ = 129$	–	Electrochemical oxidation	Barışçı et al. (2018)
	TP-115	LC-TOF-MS/MS	$[M + H]^+ = 115$	–	Electrochemical oxidation	Barışçı et al. (2018)

QqTOF quadrupole time-of-flight, *TOF* time-of-flight, *UPLC* ultra performance liquid chromatography and the formation of dicarboxylic acid TPs with a m/z of 152 and 168 (Lutterbeck et al. 2015), as well as with a m/z of 163 (Koltsakidou et al. 2017).

Another transformation pathway begins with defluorination, forming 5-hydroxyuracil (m/z 127) (Kosjek et al. 2013; Lutterbeck et al. 2016; Gomez-Canela et al. 2017), followed by a C_5-hydroxylation (Kosjek et al. 2013), subsequent deamination (Lutterbeck et al. 2015), reduction of ketone group (TP-131) or its loss (TP-115) (Gomez-Canela et al. 2017), and the formation of a carboxylic acid TP (m/z 104/5) in the final step during SNS, UV, UV/H_2O_2 and UV/TiO_2 treatment (Lutterbeck et al. 2015, 2016; Zhang et al. 2017; Lin et al. 2013; Lin and Lin 2014). This is the most commonly identified TP of 5-FU. Also, a dimer of 5-hydroxyuracil is formed, with a m/z 253 (Kosjek et al. 2013). Table 14.7 lists the reported TPs of 5-FU during photolysis.

Of the two studies investigating the TPs of CAP, one involves indirect photolysis (direct: UV, indirect: with UV/H_2O_2) (Kosjek et al. 2013) and one electrochemical oxidation (Barışçı et al. 2018). The different treatments produced different TPs. During UV/H_2O_2 photolysis, a variety of TPs have been identified as a result of ribofuranose cleavage (TP-244), subsequent dissociation of the pentyl chain (TP-174) and the dissociation of CO_2, forming 5-FU as secondary TP of CAP. Alternatively, after cleavage of the ribofuranose ring, defluorination (TP-226) and the subsequent hydroxylation (TP-242) or dissociation of the pentyl chain (TP-156) can occur (Kosjek et al. 2013). Photoaddition of water on the 5-fluoropyrimidin-2-(1H)-one segment in CAP molecule gives three TPs with a m/z 378 (Kosjek et al. 2013). Diastereoisomer of CAP (ISO-CAP) is also identified (Kosjek et al. 2013).

Alternatively, electrochemical oxidation of CAP led to identification of produced 5-deoxy-fluorocytidine and 5-FU (Barışçı et al. 2018). Finally, none of the TPs identified during 5-FU transformation are observed during CAP transformation. During final transformation, both 5-FU and CAP are expected to be transformed into tartronic acid and mesoxalic acid before complete mineralization (Koltsakidou et al. 2017).

Since in most cases mineralization of the parent drugs is incomplete, identifying stable TPs is important because they could contribute to overall toxicity. In addition, these compounds could have clinical significance (Kosjek et al. 2013). Unfortunately, the majority of TPs have only been tentatively identified and more work is needed to identify their structure including their isolation, concentration and identification using MS and complementary analytical methods such as nuclear magnetic resonance spectroscopy (NMR) as well as the synthesis of authentic standards.

Even though many 5-FU and CAP TPs have been recently proposed, their occurrence in waste and environmental waters is yet to be determined. There is also limited knowledge regarding their acute toxicity, since few studies test the toxicity of treatment effluents (photolysis). Generally, TP mixtures of 5-FU have lower (Lutterbeck et al. 2015, 2016; Barışçı et al. 2018) or unchanged (Lutterbeck et al. 2015, 2016) toxicity when compared to 5-FU. TPs of CAP formed during UV irradiation are reported to be more toxic than CAP (Guo et al. 2015), while TPs formed during electrochemical oxidation are significantly less toxic (Barışçı et al.

2018). To our knowledge, no attempts have been made to determine single TP toxicity, which is mostly due to the lack of authentic standards.

14.7 Ecotoxicity

Based on 5-FU and CAP mechanism of action, these compounds may exert acute and/or chronic toxic effects in non-target organisms. Drug development and mammalian toxicity data confirm that both 5-FU and CAP are carcinogenic and mutagenic (Gómez-Canela et al. 2014). CAP is also suspected to be teratogenic in mammals, and 5-FU has been shown to be embryotoxic and teratogenic in mammals (Roche 2007, 2008). Unspecific effects of 5-FU have also been reported. For example, it can affect the surface properties of algae cells disrupting cell aggregation (Elersek et al. 2016). Such an effect could have consequences for aquatic ecosystems.

14.7.1 Acute Toxicity

Fish Toxicity
5-FU and CAP are barely toxic for rainbow trout (*Oncorhynchus mykiss*), with NOEC > 800 mg/L during 48 h- and 96 h-exposure (Roche 2007, 2008).

Invertebrate/daphnid Toxicity
5-FU is reported as harmful for the planktonic crustaceans (*Daphnia magna*), causing growth impairment and/or immobilization (Poyatos et al. 2009; Zounkova et al. 2007, 2010; Parrella et al. 2014a), and EC_{50} values between 15 and 36 mg/L during 48 h-exposure. Its main urinary metabolite, FBAL is nontoxic to crustaceans (Zounkova et al. 2010). However, 5-FU has higher toxicity to *D.magna* than cytarabine and gemcitabine (Zounkova et al. 2010), cyclophosphamide (CP) (Zounkova et al. 2007), CAP (Parrella et al. 2014a), cisplatin (CDDP), doxorubicin (DOX), etoposide (ET) (Zounkova et al. 2007), tamoxifen (TAM) (Białk-Bielińska et al. 2017) and the main urinary metabolite FBAL (Zounkova et al. 2010). On the contrary, 5-FU does not show any acute toxicity below 100 mg/L towards *D. magna* (Białk-Bielińska et al. 2017). 5-FU is very toxic to another freshwater crustacean, *Thamnocephalus platyurus* (EC_{50} = 0.26–0.29 mg/L) (Parrella et al. 2014a).

Toxicity studies of CAP report an acute effect at hundreds of mg/L in *Ceriodaphnia dubia* and *Daphnia magna* (Mahnik et al. 2007; Lutterbeck et al. 2016), with a NOEC = 500 mg/L (Roche 2008). In *Thamnocephalus platyurus* the EC_{50} 95% confidence range was from 174.7 to 223 mg/L (Parrella et al. 2014a), but no effects at levels below 500 mg/L were observed for the rotifier *Brachionus calycifluorus* (Parrella et al. 2014a).

Algal Toxicity

5-FU is very toxic to algae and inhibits the growth of green microalgae *Pseudokirchneriella subcapitata*. After 72 h, EC_{50} values are from 0.075 to 0.13 mg/L (Roche 2007; Brezovšek et al. 2014; Ud-Daula et al. 2012) and after 96 h 0.11 mg/L (Zounkova et al. 2007). 5-FU is less toxic to the green algae *Scenodesmus vacuolatus* (NOEC = 0.08 mg/L, 72 h) (Ud-Daula et al. 2012) and to *Synechococcus leopoliensis* (NOEC = 0.12 mg/L, 72 h) (Brezovšek et al. 2014).

5-FU is more toxic to *Desmodesmus subspicatus* (EC_{50} of 48 mg/L, 72 h) than cytarabine (Zounkova et al. 2007) and CDDP, CP, DOX and ET to *P. subcapitata* (Zounkova et al. 2010). 5-FU is also more toxic than the main urinary metabolite FBAL to *D. subcapicatus* (Zounkova et al. 2007).

CAP was reported as not toxic to microalgae (NOEC = 14 mg/L, 72 h) (Roche 2008).

Microorganism Toxicity

5-FU (Zounkova et al. 2007, 2010) is more toxic to bacteria *Pseudomonas putida* (EC_{50}: 0.027–0.044 mg/L) than cytarabine, gemcitabine (Zounkova et al. 2007), CDDP, CP, DOX and ET (Zounkova et al. 2010). This contradicts previous studies that report 5-FU having no effect on the growth of *Pseudomonas putida* in the range 1–256 mg/L (Kümmerer and Al-Ahmad 1997).

As expected, 5-FU is highly toxic to rapidly dividing photosynthesizing bacteria, such as cyanobacteriae *Anabaena flosaquae* (growth rate: EC_{50} = 24 μg/L, 72 h) (Straub 2007) and *Vibrio fischeri* (EC_{50} = 122 μg/L, 24 h) (Backhaus and Grimme 1999), although recent studies report its lack of acute toxicity (Białk-Bielińska et al. 2017; Załęska-Radziwiłł et al. 2014). Microbial assays used for toxic risk assessment, reveal that 5-FU is also toxic to many human pathogenic and non-pathogenic bacteria. It is harmful to *Brevundimonas diminuta* and toxic to *Comamonas testosteroni* and *Staphylococcus warneri*, very toxic to *Pseudomonas aurantiaca, Citrobacter freundii, Pichia anomala, Serratia rubidaea, Delftia acidovorans, Microbacterium spp., Enterococcus casseliflavus, Pseudomonas fluorescens* and Kurthia gibsonii, which agrees with the EU's criteria of harmfulness to aquatic biocenoses (Załęska-Radziwiłł et al. 2014).

Studies find that CAP does not inhibit bacterial respiration of aerobic bacteria (Roche 2008; ETT 2004).

Toxicity to Other Organisms

5-FU has been shown to impair the reproductive and embryonic development of the nematode *Caenorhabditis elegans*, with induced cell-cycle arrest and apoptosis of germline cells, reduction in the number of mitotic nuclei, a dose-dependent decrease in body size, germ cell death, vulval defects and embryonic and larval development defects at levels of ~0.6 mg/L (Kumar et al. 2010). Metaestodes of tapeworm (*Echinococcus granulosus*) exhibit tumor-like properties, which is evidenced by their seemingly unlimited growth, proliferation potential and their ability to modulate their immune response and to form metastases. 5-FU reduces cell viability, number and size of tapeworm cells in a dose- and time-dependent manner (Pensel et al. 2014).

14.7.2 Chronic Toxicity

Chronic toxic effects of 5-FU and CAP on the reproduction and growth of aquatic organisms are more limited, but more consistent. Generally, studies on invertebrates (*Daphnia magna, Ceriodaphnia dubia* and *Brachionus calyciflorus*) and fish (*Danio rerio*) report significant effects on reproduction and embryonal development, respectively, at much lower concentrations than reported in acute toxicity studies.

Chronic exposure to 5-FU caused reproduction inhibition in *D. magna* at $EC_{50} = 0.1$ mg/L (Negreira et al. 2013b) and 0.026 mg/L (Parrella et al. 2014a), after 21 days, several orders of magnitude lower than EC_{50} values observed in acute toxicity studies. 5-FU is also very toxic to the crustacean *Ceriodaphnia dubia* and to the rotifer *Brachinous calyciflorus*, with EC_{50} values of 0.003 mg/L (7 days) and 0.322 mg/L (48 h), respectively (Parrella et al. 2014a).

5-FU is nontoxic to zebrafish (*Danio rerio*), with lowest effective concentration for overall subchronic effects of 100 mg/L (Straub 2007).

Only one study reported the chronic toxicity of CAP as harmful to *D. magna* ($EC_{50} = 20.5$ mg/L), and *B. calyciflorus* ($EC_{50} = 15.4$ mg/L) and toxic to *C. dubia* ($EC_{50} = 2.4$ mg/L), 2–3 orders of magnitude higher than 5-FU, CDDP, DOX, ET and IM (imatinib mesylate) EC_{50}s (Parrella et al. 2014a).

The EU Commission Directive 93/67/EEC (1993) classifies chemical substances according to their EC_{50} values against aquatic organisms into very toxic ($EC_{50} < 1$ mg/L), toxic ($1 \leq EC_{50} \leq 10$ mg/L) and harmful ($10 \leq EC_{50} \leq 100$ mg/L) and substances with an $EC_{50} > 100$ mg/L are classified as non-toxic. Following these criteria, acute toxicity data are considered when chronic data are not available. Thus, 5-FU is classified as:

- Very toxic to *P. aurantiaca, C. freundii, P. anomala, S. rubidaea, D. acidovorans, Microbacterium spp., E. casseliflavus, P. fluorescens* and *K. gibsonii* (Załęska-Radziwiłł et al. 2014), *A. flos-aquae* (Straub 2007), *P. putida* (Zounkova et al. 2010), *D. subscapicatus* (Zounkova et al. 2007), *P. subcapitata* (Roche 2007; Ud-Daula et al. 2012), *B. calyciflorus* (Parrella et al. 2014a), *D.magna* (Straub 2007; Kümmerer and Al-Ahmad 1997; Parrella et al. 2014a) and *C. dubia* (Parrella et al. 2014a);
- Toxic to *C. testosteroni* and *S. warneri* (Załęska-Radziwiłł et al. 2014), *S. leopoliensis* (Brezovšek et al. 2014);
- Harmful to *B.diminuta* (Białk-Bielińska et al. 2017; Załęska-Radziwiłł et al. 2014), and *S.vacuolatus* (Ud-Daula et al. 2012);
- Non-toxic to: *O. mykiss* (Roche 2007) and *V. fischeri* (Załęska-Radziwiłł et al. 2014)

CAP is generally less toxic than 5-FU, since it is classified as toxic to *C. dubia* (Parrella et al. 2014a), harmful to *P. subcapitata* (Mahnik et al. 2007), *B. calyciflorus* (Parrella et al. 2014a) and *D. magna* (Parrella et al. 2014a), non-toxic to *O. mykiss* (Mahnik et al. 2007) and *T. platyurus* (Parrella et al. 2014a).

Based on the available data here shown, 5-FU and CAP are toxic to decomposers (bacteriae), producers (green algae), primary and secondary consumers (crustaceans and fish). Acute (Straub 2007; Roche 2007, 2008; Elersek et al. 2016; Zounkova et al. 2007, 2010; Białk-Bielińska et al. 2017; Ud-Daula et al. 2012; Pensel et al. 2014; Novak et al. 2017; Parrella et al. 2015; Gačić et al. 2014) and chronic toxicity data (Parrella et al. 2014a; Kumar et al. 2010; Kovács et al. 2015) have been compared to 5-FU and CAP PECs (Johnson et al. 2013; Gomez-Canela et al. 2014; Franquet-Griell et al. 2015) depicting a potential risk to aquatic organisms. Nevertheless, other studies state that these compounds do not pose any environmental risk (Johnson et al. 2013; Straub 2007; Brezovšek et al. 2014).

14.7.3 Toxicity of 5-FU in Mixture with Other Cytostatic Agents

In the aquatic environment, cytostatic drugs are expected to occur in mixtures which can exert antagonistic, additive or synergistic toxic effects in concentrations a few orders of magnitude lower than the effective concentrations of the individual cytostatic agents (Elersek et al. 2016; Ud-Daula et al. 2012; Parrella et al. 2014b; Kovács et al. 2015). For example, exposure of *Daphnia magna* to a combination of 0.27 μg/L of CDDP and 5.0 μg/L of 5-FU produced the same response as observed at single exposures of 0.5 μg/L of CDDP and 12.2 μg/L of 5-FU (Parrella et al. 2014b). Even though the concentration of the individual compounds might be low, their effects in mixtures might be of ecotoxicological significance (Ud-Daula et al. 2012). This is relevant because many cancer treatments use multiple combinations of cytostatic agents to achieve an increased therapeutic effect. The possibility of mixtures of cytostatic drugs occurring in the environment will, therefore, potentially increase the threat posed to non-target organisms (Besse et al. 2012).

A triple mixture of 5-FU + IM + ET in different concentrations (Elersek et al. 2016) caused comparable levels of growth inhibition in green algae and cyanobacteria as a mixture of 5-FU + IM (Elersek et al. 2016; Brezovšek et al. 2014). This is a likely result of ET's ability to suppress the toxic activity of 5-FU (Franquet-Griell et al. 2015). This antagonism is confirmed for binary mixtures of 5-FU and ET on the crustacean *Ceriodaphnia dubia* (Parrella et al. 2014b).

The mixture of CP, IF, CDDP and 5-FU in environmentally relevant concentrations (all: 0.09–120 μg/L, 5-FU: 0.09–0.9 μg/L) showed no cytotoxicity in a zebrafish liver cell line after 72 h (Novak et al. 2017).

14.8 Genotoxicity

Cytostatic drugs, interfering directly or indirectly with DNA, may affect non-target organisms. In particular, 5-FU is genotoxic to early-stage zebrafish (*Danio rerio*) in concentrations above 32 mg/L during a 35-day early life-stage exposure (Straub 2007). Furthermore, exposure in a two-generation zebrafish assay to 5-FU at relevant environmental concentrations (0.01–100 μg/L) DNA strand breaks, micronuclei, whole transcriptome changes and histopathological changes were observed. A dose-dependent increase in the number of differentially expressed DNA-damage responsive genes and oncogenes was also observed, leading to lipidosis, regressive liver degeneration and oncogenesis (Kovács et al. 2015). Strong DNA damage in liver and blood cells were confirmed with liver cell line studies (Novak et al. 2017).

Lower down the aquatic trophic chain, genotoxic damage caused by 5-FU and CAP were observed in mussels and crustaceans. 5-FU induced DNA strain breaks at concentrations $\geq$ 52 μg/L after 72 h treatment *in vivo* in freshwater mussels *Unio pictorum* and *Unio tumidis* (Gačić et al. 2014). Regarding crustaceans, 5-FU caused DNA strain breaks in *Ceriodaphna dubia*, after 24-h exposure with the lowest observed adverse effect concentration (LOAEC) in the order of tens of ng/L lower than that of CDDP, ET, IM and CAP on the same organism (Parrella et al. 2015). A LOAEC of hundreds of ng/L was observed for 5-FU in *D. magna*. In the yeast *S. cerevisiae*, 5-FU caused genotoxic effects at 0.02 mg/L (Zounkova et al. 2010).

5-FU and CAP determined DNA damages in microorganisms at high concentrations (mg/L). CAP showed genotoxicity in *Escherichia coli* at concentrations $\geq$ 75 mg/L (Parrella et al. 2015). 5-FU was genotoxic towards *Escherichia coli* starting from 1.4 mg/L (Martín et al. 2011) and 2.5 mg/L (Parrella et al. 2015), whereas its metabolite FBAL was genotoxic to *Salmonella cholarasius subsp. Chol* at 667 mg/L (Zounkova et al. 2010).

CAP exerted a mutagenic activity in *Salmonella typhimurium* from 300 mg/L, while 5-FU was mutagen at concentration $\geq$ 10 mg/L after 72 h exposure (Parrella et al. 2015).

Unlike the prodrug CAP, that showed the lowest genotoxic activity when compared to other cytostatic drugs such as CDDP, IM and ET, its active metabolite 5-FU was the most active compound. The enzymatic metabolization of CAP in 5-FU in non-target organisms is still unknown, so that the direct exposure of such organisms to 5-FU could rapidly induce DNA strand breaks and other DNA lesions, causing its stronger genotoxicity.

Genotoxicity of 5-FU in Mixture with Other Cytostatic Agents

Genotoxic effects of mixtures of cytostatic agents are generally compound-, dose- and organism-dependent, and many contradicting results are found in the literature. An example is the 5-FU and IM mixture, which has synergistic genotoxic effects on green algae (*P. subcapitata*) and cyanobacterium (*S. leopoliensis*) (Brezovšek et al. 2014), but expresses antagonistic genotoxicity to the crustacean *D. magna* (Kundi et al. 2015). Synergistic genotoxic effects are confirmed for binary mixtures of 5-FU and CDDP on green algae and cyanobacteria (Brezovšek et al. 2014), while antagonistic effects were observed in *C. dubia* and additivity was found in *D. magna* (Kundi et al. 2015). Indeed, when *D. magna* was exposed to a combination of 0.07 μg/L CDDP and 10 μg/L 5-FU, the same response was observed as for single exposures of 1.1 μg/L CDDP or 106 μg/L 5-FU (Kundi et al. 2015). The synergism arises from the interaction of 5-FU with CDDP *via* 5-FU mediated suppression of the repair of CDDP-induced DNA adducts and crosslinks, which in turn increases the cytotoxicity of CDDP.

Antagonism is confirmed for binary mixtures of 5-FU and ET in *C. dubia* (Kundi et al. 2015).

Since residues of anticancer drugs are released into the environment as complex mixtures of parent compounds and their metabolites, some studies have focused on investigating genotoxic effects of multiple combinations of different anticancer drugs. For example, a mixture of CP, IF, CDDP and 5-FU at environmentally relevant concentrations (all: 0.09–120 μg/L, 5-FU: 0.09–0.9 μg/L) revealed no genomic instability after 72 h to zebrafish liver cell lines, but induced a significant increase in the formation of DNA strand breaks at concentrations several orders of magnitude lower from the ones effective when tested as individual compounds (Novak et al. 2017).

14.9 Conclusion

5-FU and CAP are polar compounds, which, according to their physico-chemical properties, are expected to be present in aquatic environment and are unlikely to bioconcentrate or sorb onto organic matter. Sensitivity of the analytical methods for determining 5-FU and CAP in aqueous samples vary depending on the specific method (sample preparation, instrumentation) and complexity of the investigated matrix. Sensitivity ranges from μg/L (HPLC-UV) to low ng/L range (GC-MS and LC-MS) and performance of analytical methods is further improved in terms of sensitivity, selectivity and structural information, when high resolution mass analyzers are applied. Despite relatively limited data on their occurrence in the aqueous environment, point sources include hospital effluents (μg/L) and municipal wastewaters (ng/L). With one exception, where 5-FU was determined in a Taiwanese surface water, 5-FU and CAP are yet to be determined in surface waters. This is a result of low levels of these compounds in WW effluents and high dilution factors resulting in concentrations of cytostatic drug residues in surface waters typically below the limits of detection of existing methods.

Between the most commonly applied wastewater treatments for removing 5-FU and CAP, UV based treatments (direct and indirect photolysis) and ozonation have been the most studied and give the highest removal and mineralization rates. Again, these treatments are neither studied nor suitable for complex matrices like wastewaters. Recent studies have also reported the formation of stable transformation product of 5-FU and CAP indicating defluorination, mono/poly-hydroxylation, pyrrolidine ring opening and sugar cleavage, pentyl chain dissociation and CO_2 removal as main transformation mechanism of 5-FU ad CAP, respectively. The environmental relevance of these transformation products is yet to be confirmed.

Both predicted and measured concentrations of 5-FU and CAP in the aquatic environment are lower than those expected to cause adverse effects in aquatic organisms, where CAP is less toxic than its metabolite 5-FU. However, we should be aware that environmental samples contain mixtures of pharmaceuticals residues (including cytostatic drugs), their metabolites and transformation products, what can result in synergistic and additive effects resulting in lower effect concentrations than those determined for single compound.

Although few studies suggest that there is no exposure level of concern in the aquatic environment for CAP and 5-FU and that they do not pose any risk, on the other hand, many available data show that these drugs are toxic to different trophic levels of ecosystems. Comparing 5-FU and CAP concentrations used in acute and chronic toxicity studies to their respective PECs and experimentally obtained results, it appears that 5-FU and CAP exert a potential risk to aquatic organisms. However, it is premature to conclude that environmental concentrations of 5-FU and CAP are safe, without further long-term exposure studies on non-target organisms. Moreover, single compound toxicity data are not sufficient for predicting the toxicity of anticancer drug mixtures frequently present in the environment.

References

Backhaus T, Grimme L (1999) The toxicity of antibiotic agents to the luminescent bacterium Vibrio fischeri. Chemosphere 38:3291–3301. https://doi.org/10.1016/s0045-6535(98)00560-8

Barışçı S, Turkay O, Ulusoy E, Şeker MG, Yüksel E, Dimoglo A (2018) Electro-oxidation of cytostatic drugs: experimental and theoretical identification of by-products and evalua-tion of ecotoxicological effects. Chem Eng J 334:1820–1827. https://doi.org/10.1016/j.cej.2017.11.105

Białk-Bielińska A, Mulkiewicz E, Stokowski M, Stolte S, Stepnowski P (2017) Acute aquatic toxicity assessment of six anti-cancer drugs and one metabolite using biotest battery – biological effects and stability under test conditions. Chemosphere 189:689–698. https://doi.org/10.1016/j.chemosphere.2017.08.174

Besse J-P, Latour J-F, Garric J (2012) Anticancer drugs in surface waters. Environ Int 39:73–86. https://doi.org/10.1016/j.envint.2011.10.002

Brezovšek P, Eleršek T, Filipič M (2014) Toxicities of four anti-neoplastic drugs and their binary mixtures tested on the green alga Pseudokirchneriella subcapitata and the cyanobacterium Synechococcus leopoliensis. Water Res 52:168–177. https://doi.org/10.1016/j.watres.2014.01.007

DrugBank (2018) In: DrugBank. https://www.drugbank.ca/. Accessed 24 Jan 2018

Elersek T, Milavec S, Korošec M, Brezovsek P, Negreira N, Zonja B, Alda MLD, Barceló D, Heath E, Ščančar J, Filipič M (2016) Toxicity of the mixture of selected antineoplastic drugs against aquatic primary producers. Environ Sci Pollut Res 23:14780–14790. https://doi.org/10.1007/s11356-015-6005-2

ETT (2004) Zahn-Wellens/EMPA Test OECD method 302B, Capecitabine (Ro 091978), Manufacturer Roche Carolina ETT Environmental, 24770A-C

European Commission (1993) Commission Directive 93/67/EEC of 20 July 1993 laying down the principles for assessment of risks to man and the environment of subtances notified in accordance with Council Directive 67/548/EEC, Off J Eur Communities: Legis 227:9–18. http://data.europa.eu/eli/dir/1993/67/oj. Accessed 24 Jan 2018

Faßbender C, Braunbeck T (2013) Reproductive and genotoxic effects in zebrafish after chronic exposure to methyl methanesulfonate in a multigeneration study. Ecotoxicology 22:825–837. https://doi.org/10.1007/s10646-013-1057-x

Ferre-Aracil J, Valcárcel Y, Negreira N, Alda MLD, Barceló D, Cardona S, Navarro-Laboulais J (2016) Ozonation of hospital raw wastewaters for cytostatic compounds removal. Kinetic modelling and economic assessment of the process. Sci Total Environ 556:70–79. https://doi.org/10.1016/j.scitotenv.2016.02.20

Franquet-Griell H, Gómez-Canela C, Ventura F, Lacorte S (2015) Predicting concentrations of cytostatic drugs in sewage effluents and surface waters of Catalonia (NE Spain). Environ Res 138:161–172. https://doi.org/10.1016/j.envres.2015.02.015

Franquet-Griell H, Medina A, Sans C, Lacorte S (2017) Biological and photochemical degradation of cytostatic drugs under laboratory conditions. J Hazard Mater 323:319–328. https://doi.org/10.1016/j.jhazmat.2016.06.057

Fürhacker M, Lenz K, Mahnik S, Weissenbacher N, Mader R, Knasmüller S, Ferk F, Uhl M, Bursch W, Kollensperger G, Hann S (2006) Chemische Analyse, Risikobewertung und Entfernung von ausgewählten Zytostatika ausAbwasserströmen aus Krankenhäusern; Teil II: Risikoabschätzung undRisikomanagement, Mai 2006. Vienna (AT): Lebensminist erium; [cited 2008 June]. Available from: http://gpool.lfrz.at/gpoolexport/media/file/Chemische_Analyse_-_Risikobewertung_-_Zytostatika_Endbericht.pd. Accessed 12 Feb 2018

Gačić Z, Kolarević S, Sunjog K, Kračun-Kolarević M, Paunović M, Knežević-Vukčević J, Vuković-Gačić B (2014) The impact of in vivo and in vitro exposure to base analogue 5-FU on the level of DNA damage in haemocytes of freshwater mussels Unio pictorum and Unio tumidus. Environ Pollut 191:145–150. https://doi.org/10.1016/j.envpol.2014.04.024

Ganzenko O, Oturan N, Sirés I, Huguenot D, Hullebusch EDV, Esposito G, Oturan MA (2017) Fast and complete removal of the 5-fluorouracil drug from water by electro-Fenton oxidation. Environ Chem Lett. https://doi.org/10.1007/s10311-017-0659-6

Gómez-Canela C, Ventura F, Caixach J, Lacorte S (2014) Occurrence of cytostatic compounds in hospital effluents and wastewaters, determined by liquid chromatography coupled to high-resolution mass spectrometry. Anal Bioanal Chem 406:3801–3814. https://doi.org/10.1007/s00216-014-7805-9

Gómez-Canela C, Bolivar-Subirats G, Tauler R, Lacorte S (2017) Powerful combination of analytical and chemometric methods for the photodegradation of 5-Fluorouracil. J Pharm Biomed Anal 137:33–41. https://doi.org/10.1016/j.jpba.2017.01.019

Governo M, Santos MSF, Alves A, Madeira LM (2017) Degradation of the cytostatic 5-Fluorouracil in water by Fenton and photo-assisted oxidation processes. Environ Sci Pollut Res 24:844–854. https://doi.org/10.1007/s11356-016-7827-2

Guo R, Zheng F, Chen J (2015) Evaluation of the aquatic toxic effect varied during the degradation of capecitabine under the environmental abiotic and biotic processes. RSC Adv 5:76772–76778. https://doi.org/10.1039/c5ra17315a

Häner A (2006) Capecitabin, inherent biodegradability, evaluation of the aerobic biodegradability in an aqueous medium: Modified MITI (II) test. BMG 1173/05, on behalf of F.Hoffmann-La Roche Ltd (unpublished)

Heath E, Česen M, Negreira N, Alda MLD, Ferrando-Climent L, Blahova L, Nguyen TV, Adahchour M, Ruebel A, Llewellyn N, Ščančar J, Novaković S, Mislej V, Stražar M, Barceló D, Kosjek T (2015) First inter-laboratory comparison exercise for the determination of anticancer drugs in aqueous samples. Environ Sci Pollut Res 23:14692–14704. https://doi.org/10.1007/s11356-015-4982-9

Hoerger CC, Dörr B, Schlienger C, Straub JO (2009) Environmental risk assessment for the Galenical formulation of solid medicinal products at Roche Basle, Switzerland. Integr Environ Assess Manag 5:331. https://doi.org/10.1897/ieam_2008-049.1

Johnson AC, Oldenkamp R, Dumont E, Sumpter JP (2013) Predicting concentrations of the cytostatic drugs cyclophosphamide, carboplatin, 5-fluorouracil, and capecitabine throughout the sewage effluents and surface waters of europe. Environmental Toxicology and Chemistry 32:1954–1961. https://doi.org/10.1002/etc.2311

Kiffmeyer T, Götze H-J, Jursch M, Lüders U (1998) Trace enrichment, chromatographic separation and biodegradation of cytostatic compounds in surface water. Fresen J Anal Chem 361:185–191. https://doi.org/10.1007/s002160050859

Koltsakidou A, Antonopoulou M, Evgenidou E, Konstantinou I, Giannakas A, Papadaki M, Bikiaris D, Lambropoulou D (2017) Photocatalytical removal of fluorouracil using TiO2-P25 and N/S doped TiO2 catalysts: a kinetic and mechanistic study. Sci Total Environ 578:257–267. https://doi.org/10.1016/j.scitotenv.2016.08.208

Kosjek T, Heath E (2011) Occurrence, fate and determination of cytostatic pharmaceuticals in the environment. TrAC Trends Anal Chem 30:1065–1087. https://doi.org/10.1016/j.trac.2011.04.007

Kosjek T, Perko S, Žigon D, Heath E (2013) Fluorouracil in the environment: analysis, occurrence, degradation and transformation. J Chromatogr A 1290:62–72. https://doi.org/10.1016/j.chroma.2013.03.046

Kovács RCB, Csenki Z, Bakos K, Urbányi B, Horváth Á, Garaj-Vrhovac V, Gajski G, Gerić M, Negreira N, Alda MLD, Barceló D, Heath E, Kosjek T, Žegura B, Novak M, Zajc I, Baebler Š, Rotter A, Ramšak Ž, Filipič M (2015) Assessment of toxicity and genotoxicity of low doses of 5-fluorouracil in zebrafish (Danio rerio) two-generation study. Water Res 77:201–212. https://doi.org/10.1016/j.watres.2015.03.025

Kovalova L (2009) Cytostatics in the aquatic environment: analysis, occurrence, and possibilities for removal (PhD dissertation). Institute of Hygiene and Environmental Medicine at the University Hospital Aachen, Germany

Kovalova L, Mcardell CS, Hollender J (2009) Challenge of high polarity and low concentrations in analysis of cytostatics and metabolites in wastewater by hydrophilic interaction chromatography/tandem mass spectrometry. J Chromatogr A 1216:1100–1108. https://doi.org/10.1016/j.chroma.2008.12.028

Kumar S, Aninat C, Michaux G, Morel F (2010) Anticancer drug 5-fluorouracil induces reproductive and developmental defects in Caenorhabditis elegans. Reprod Toxicol 29:415–420. https://doi.org/10.1016/j.reprotox.2010.02.006

Kümmerer K, Al-Ahmad A (1997) Biodegradability of the anti-tumour agents 5-Fluorouracil, Cytarabine, and Gemcitabine: impact of the chemical structure and synergistic toxicity with hospital effluent. Acta Hydrochim Hydrobiol 25:166–172. https://doi.org/10.1002/aheh.19970250402

Kundi M, Parrella A, Lavorgna M, Criscuolo E, Russo C, Isidori M (2015) Prediction and as sessment of ecogenotoxicity of antineoplastic drugs in binary mixtures. Environ Sci Pollut Res 23:14771–14779. https://doi.org/10.1007/s11356-015-4884-x

Li W, Tanumihardja J, Masuyama T, Korshin G (2015) Examination of the kinetics of degradation of the antineoplastic drug 5-fluorouracil by chlorine and bromine. J Hazard Mater 282:125–132. https://doi.org/10.1016/j.jhazmat.2014.05.090

Lin HH-H, Lin AY-C (2014) Photocatalytic oxidation of 5-fluorouracil and cyclophosphamide via UV/TiO2 in an aqueous environment. Water Res 48:559–568. https://doi.org/10.1016/j.watres.2013.10.011

Lin AY-C, Wang X-H, Lee W-N (2013) Phototransformation determines the fate of 5-Fluorouracil and cyclophosphamide in natural surface waters. Environ Sci Technol 47:4104–4112. https://doi.org/10.1021/es304976q

Lin AY-C, Lin Y-C, Lee W-N (2014a) Prevalence and sunlight photolysis of controlled and chemotherapeutic drugs in aqueous environments. Environ Pollut 187:170–181. https://doi.org/10.1016/j.envpol.2014.01.005

Lin AY-C, Hsueh JH-F, Hong PKA (2014b) Removal of antineoplastic drugs cyclophosphamide, ifosfamide, and 5-fluorouracil and a vasodilator drug pentoxifylline from wastewaters by ozonation. Environ Sci Pollut Res 22:508–515. https://doi.org/10.1007/s11356-014-3288-7

Lutterbeck CA, Wilde ML, Baginska E, Leder C, Machado ÊL, Kümmerer K (2015) Degradation of 5-FU by means of advanced (photo)oxidation processes: UV/H2O2, UV/Fe2 /H2O2 and UV/TiO2 — comparison of transformation products, ready biodegradability and toxicity. Sci Total Environ 527–528:232–245. https://doi.org/10.1016/j.scitotenv.2015.04.111

Lutterbeck CA, Wilde ML, Baginska E, Leder C, Machado ÊL, Kümmerer K (2016) Degradation of cyclophosphamide and 5-fluorouracil by UV and simulated sunlight treatments: assessment of the enhancement of the biodegradability and toxicity. Environ Pollut 208:467–476. https://doi.org/10.1016/j.envpol.2015.10.016

Mahnik S, Rizovski B, Fuerhacker M, Mader R (2004) Determination of 5-fluorouracil in hospital effluents. Anal Bioanal Chem. https://doi.org/10.1007/s00216-004-2727-6

Mahnik S, Lenz K, Weissenbacher N, Mader R, Fuerhacker M (2007) Fate of 5-fluorouracil, doxorubicin, epirubicin, and daunorubicin in hospital wastewater and their elimination by activated sludge and treatment in a membrane-bio-reactor system. Chemosphere 66:30–37. https://doi.org/10.1016/j.chemosphere.2006.05.051

Martín J, Camacho-Muñoz D, Santos JL, Aparicio I, Alonso E (2011) Simultaneous determination of a selected group of cytostatic drugs in water using high-performance liquid chromatography-triple-quadrupole mass spectrometry. J Sep Sci 34:3166–3177. https://doi.org/10.1002/jssc.201100461

Martín J, Camacho-Muñoz D, Santos JL, Aparicio I, Alonso E (2014) Occurrence and ecotoxicological risk assessment of 14 cytostatic drugs in wastewater. Water Air Soil Pollut. https://doi.org/10.1007/s11270-014-1896-y

Mendoza A, Zonja B, Mastroianni N, Negreira N, Alda MLD, Pérez S, Barceló D, Gil A, Valcárcel Y (2016) Drugs of abuse, cytostatic drugs and iodinated contrast media in tap water from the Madrid region (central Spain): a case study to analyse their occurrence and human health risk characterization. Environ Int 86:107–118. https://doi.org/10.1016/j.envint.2015.11.001

Miolo G, Marzano C, Gandin V, Palozzo AC, Dalzoppo D, Salvador A, Caffieri S (2011) Photoreactivity of 5-Fluorouracil under UVB Light: photolysis and cytotoxicity studies. Chem Res Toxicol 24:1319–1326. https://doi.org/10.1021/tx200212z

Mullot J-U, Karolak S, Fontova A, Huart B, Levi Y (2009) Development and validation of a sensitive and selective method using GC/MS-MS for quantification of 5-fluorouracil in hospital wastewater. Anal Bioanal Chem 394:2203–2212. https://doi.org/10.1007/s00216-009-2902-x

Negreira N, Alda MLD, Barceló D (2013a) On-line solid phase extraction–liquid chromatography–tandem mass spectrometry for the determination of 17 cytostatics and metabolites in waste, surface and ground water samples. J Chromatogr A 1280:64–74. https://doi.org/10.1016/j.chroma.2013.01.031

Negreira N, Mastroianni N, Alda MLD, Barceló D (2013b) Multianalyte determination of 24 cytostatics and metabolites by liquid chromatography–electrospray–tandem mass spectrometry and study of their stability and optimum storage conditions in aqueous solution. Talanta 116:290–299. https://doi.org/10.1016/j.talanta.2013.04.070

Negreira N, Alda MLD, Barceló D (2014a) Cytostatic drugs and metabolites in municipal and hospital wastewaters in Spain: filtration, occurrence, and environmental risk. Sci Total Environ 497-498:68–77. https://doi.org/10.1016/j.scitotenv.2014.07.101

Negreira N, Alda MLD, Barceló D (2014b) Study of the stability of 26 cytostatic drugs and metabolites in wastewater under different conditions. Sci Total Environ 482-483:389–398. https://doi.org/10.1016/j.scitotenv.2014.02.131

Novak MCBE, Žegura B, Modic B, Heath E, Filipič M (2017) Cytotoxicity and genotoxicity of anticancer drug residues and their mixtures in experimental model with zebrafish liver cells. Sci Total Environ 601-602:293–300. https://doi.org/10.1016/j.scitotenv.2017.05.115

Onesios KM, Bouwer EJ (2012) Biological removal of pharmaceuticals and personal care products during laboratory soil aquifer treatment simulation with different primary substrate concentrations. Water Res 46:2365–2375. https://doi.org/10.1016/j.watres.2012.02.001

Parrella A, Lavorgna M, Criscuolo E, Russo C, Fiumano V, Isidori M (2014a) Acute and chronic toxicity of six anticancer drugs on rotifers and crustaceans. Chemosphere 115:59–66. https://doi.org/10.1016/j.chemosphere.2014.01.013

Parrella A, Kundi M, Lavorgna M, Criscuolo E, Russo C, Isidori M (2014b) Toxicity of exposure to binary mixtures of four anti-neoplastic drugs in Daphnia magna and Ceriodaphnia dubia. Aquat Toxicol 157:41–46. https://doi.org/10.1016/j.aquatox.2014.09.012

Parrella A, Lavorgna M, Criscuolo E, Russo C, Isidori M (2015) Eco-genotoxicity of six anticancer drugs using comet assay in daphnids. J Hazard Mater 286:573–580. https://doi.org/10.1016/j.jhazmat.2015.01.012

Pensel P, Albani C, Gamboa GU, Benoit J, Elissondo M (2014) In vitro effect of 5-fluorouracil and paclitaxel on Echinococcus granulosus larvae and cells. Acta Trop 140:1–9. https://doi.org/10.1016/j.actatropica.2014.07.013

Poyatos JM, Muñio MM, Almecija MC, Torres JC, Hontoria E, Osorio F (2009) Advanced oxidation processes for wastewater treatment: state of the art. Water Air Soil Pollut 205:187–204. https://doi.org/10.1007/s11270-009-0065-1

Rey RPCA, Padrón AS, León LG, Pozo MM, Baluja C (1999) Ozonation of cytostatics in water medium. Nitrogen bases. Ozone Sci Eng 21:69–77. https://doi.org/10.1080/01919519908547260

Roche (2007) Safety data sheet for Fluorouracil. F. Basle (CH): Hoffmann-La Roche. Available from: http://www.roche.com/corporate_responsibility/environment/safety_data_sheets_msds/safety_data_sheets-row.htm. Accessed 12 Feb 2018

Roche (2008) Safety data sheet for Capecitabine. F. Basle (CH): F. Hoffmann-La Roche. Available from: http://www.roche.com/corporate_responsibility/environment/safety_data_sheets_msds/safety_data_sheets-row.htm. Accessed 12 Feb 2018

Steger-Hartmann T, Kümmerer K, Schecker J (1996) Trace analysis of the antineoplastics ifosfamide and cyclophosphamide in sewage water by twostep solid-phase extraction and gas chromatography-mass spectrometry. J Chromatogr A 726:179–184. https://doi.org/10.1016/0021-9673(95)01063-7

Straub JCBCO (2007) Combined environmental risk assessment for 5-fluorouracil and capecitabine in Europe. Integr Environ Assess Manag preprint:1. https://doi.org/10.1897/ieam_2009-073.1

Studer M (2005) Ecotoxicity assessment capecitabine/xeloda, E-05/04. Roche Basle, Environmental Laboratories, 2005.

TOXNET (2018) In: U.S. National Library of Medicine. https://toxnet.nlm.nih.gov/. Accessed 12 Feb 2018

Tauxe-Wuersch A, Alencastro LFD, Grandjean D, Tarradellas J (2006) Trace determination of tamoxifen and 5-fluorouracil in hospital and urban wastewaters. Int J Environ Anal Chem 86:473–485. https://doi.org/10.1080/03067310500291502

Ud-Daula A, Pfister G, Schramm K-W (2012) 5-Fluorouracil accumulation in green microalgae and its biogenetic transfer into ciliate protozoan. Bull Environ Contam Toxicol 88:548–554. https://doi.org/10.1007/s00128-012-0536-8

Walko CM, Lindley C (2005) Capecitabine: a review. Clin Ther 27:23–44. https://doi.org/10.1016/j.clinthera.2005.01.005

Weissbrodt D, Kovalova L, Ort C, Pazhepurackel V, Moser R, Hollender J, Siegrist H, Mcardell CS (2009) Mass flows of X-ray contrast media and cytostatics in hospital wastewater. Environ Sci Technol 43:4810–4817. https://doi.org/10.1021/es8036725

WHO (2018) The Anatomical Therapeutic Chemical Classification System with Defined Daily Doses (ATC/DDD). http://www.who.int/classifications/atcddd/en/. Accessed 18 Jan 2018

Yu JT, Bouwer EJ, Coelhan M (2006) Occurrence and biodegradability studies of selected pharmaceuticals and personal care products in sewage effluent. Agric Water Manag 86:72–80. https://doi.org/10.1016/j.agwat.2006.06.015

Yu JT, Bisceglia KJ, Bouwer EJ, Roberts AL, Coelhan M (2012) Determination of pharmaceuticals and antiseptics in water by solid-phase extraction and gas chromatography/mass spectrometry:

analysis via pentafluorobenzylation and stable isotope dilution. Anal Bioanal Chem 403:583–591. https://doi.org/10.1007/s00216-012-5846-5

Załęska-Radziwiłł M, Affek K, Rybak J (2014) Ecotoxicity of chosen pharmaceuticals in relation to microorganisms—risk assessment. Desalin Water Treat 52:3908–3917. https://doi.org/10.1080/19443994.2014.887503

Zhang J, Chang VW, Giannis A, Wang J-Y (2013) Removal of cytostatic drugs from aquatic environment: a review. Sci Total Environ 445–446:281–298. https://doi.org/10.1016/j.scitotenv.2012.12.061

Zhang Y, Xiao Y, Zhang J, Chang VW, Lim T-T (2017) Degradation of cyclophosphamide and 5-fluorouracil in water using UV and UV/H2O2: Kinetics investigation, pathways and energetic analysis. J Environ Chem Eng 5:1133–1139. https://doi.org/10.1016/j.jece.2017.01.013

Zounkova R, Odraska P, Dolezalova L, Hilscherova K, Marsalek B, Blaha L (2007) Ecotoxicity and genotoxicity assessment of cytostatic pharmaceuticals. Environ Toxicol Chem preprint:1. https://doi.org/10.1897/07-137

Zounkova R, Kovalova L, Blaha L, Dott W (2010) Ecotoxicity and genotoxicity assessment of cytotoxic antineoplastic drugs and their metabolites. Chemosphere 81:253–260. https://doi.org/10.1016/j.chemosphere.2010.06.029

Part IV
Toxicity (Effects)

Chapter 15
Toxicity of Anticancer Drug Residues in Organisms of the Freshwater Aquatic Chain

Chiara Russo, Margherita Lavorgna, Concetta Piscitelli, and Marina Isidori

Abstract Antineoplastic drug residues released in the aquatic system represent a potential risk to exposed non-target organisms. Antineoplastic drugs are present in aquatic environments at lower concentration levels than in other therapeutic classes. In particular, antineoplastic drugs are known for their continuous release and subsequent exposure throughout the life span of aquatic organisms and will yield long-term toxicity rather than pose an immediate threat to the environment and human health. Furthermore, these drugs interfere directly or indirectly with DNA and their eco-toxicological effects, such as reproductive inhibition, may modify genetic material. In light of these observations, this chapter aims to investigate the acute and chronic effects of the most commonly used anticancer drugs in organisms of the freshwater trophic chain. The drugs investigated belong to the subgroup L01 of the Anatomical Therapeutic Chemical Classification System and have been chosen considering their occurrence, chemical structures, and modes of action.

Keywords Antineoplastic drugs · Aquatic organisms · Acute toxicity · Chronic toxicity · PEC

15.1 Introduction

It is known that pharmaceutical residues may induce a biological response in non-target organisms that are environmentally exposed. Prioritization methodologies are fundamental to assess potential toxicity of pharmaceuticals.

Both the European Medicines Agency (EMA) and the American Food and Drug Administration (FDA) have established guidelines on Environmental Risk Assessment (ERA) of human drugs, setting a Predicted Environmental Concentration (PEC) as a trigger value for environmental risk assessment. The trigger values set by EMA and

C. Russo · M. Lavorgna · C. Piscitelli · M. Isidori (✉)
Department of Environmental, Biological and Pharmaceutical Sciences and Technologies, University of Campania "L. Vanvitelli", Caserta, Italy
e-mail: marina.isidori@unicampania.it

E. Heath et al. (eds.), *Fate and Effects of Anticancer Drugs in the Environment*,
https://doi.org/10.1007/978-3-030-21048-9_15

FDA differ in two orders of magnitude: 0.01 μg/L and 1 μg/L, respectively. Antineoplastic drug residues occur in the aquatic environment at lower concentrations than in other pharmaceuticals, and the EMA trigger value of 0.01 μg/L is rarely reached. Nevertheless, antineoplastic drugs that directly or indirectly interfere with DNA pose a significant hazard to the environment with sublethal chronic effects on non-target organisms and human health through intake of drinking water. We, thus, need to assess the relevance of trigger values for compounds that interfere directly or indirectly with DNA affecting both cancerous and healthy cells. The consumption of anticancer drugs will presumably increase considering an aging population, a potential increase in cancer incidence (US National Institute of Health, www.cancer.gov), early diagnosis, treatment duration in relation to survival time, and higher dosages to reduce side effects (Suhail et al. 2012; Kümmerer et al. 2016). Antineoplastic residues are therefore likely to occur at higher concentrations in the aquatic systems over the next few years, and despite advances in analytical chemistry, which may detect such compounds at increasingly low concentrations (ng/L), the assessment of the potential risk to the aquatic environment of mutagenic, genotoxic, and carcinogenic compounds is still required.

The various prioritization methodologies include risk-based approaches to identify drugs with unintended effects on non-target organisms (Burns et al. 2017). One of the most common procedures is based on the estimation of the PEC or the quantification of the Measured Environmental Concentration (MEC) compared to the Predicted No-Effect Concentrations (PNEC). PEC value is based on the consumption of pharmaceuticals, human metabolism, excretion fraction, removal in wastewater treatment plants (WWTP), and dilution in the aquatic systems, while MEC is the actual concentration detected in the environment obtained by chemical analysis. PNEC is the concentration expected to be free of adverse effects and considers acute and/or chronic toxicity data in non-target organisms from different trophic levels.

Studies on the occurrence of antineoplastic drugs in aquatic systems have increased considerably, with 121 papers published in 2002 and 303 in 2016. However, few studies have investigated the acute and/or chronic effects of antineoplastic drugs in aquatic organisms with a total of 15 publications from 2007 to 2017 (www.scopus.com).

These studies have been performed on several organisms belonging to different trophic levels of the freshwater aquatic chain, i.e., decomposers (bacteria), producers (algae and aquatic plants), and consumers (rotifers, crustaceans, fish), in accordance with the international standardization guidelines (ASTM, ISO, OECD) (Table 15.1).

Acute toxicity endpoints are expressed as median lethal or effect (i.e., immobilization, luminescence inhibition, inhibition of growth) concentrations: L(E)C50, while chronic toxic effects such as population growth inhibition or reproduction inhibition at different times of exposure are expressed as EC50. In long-term toxicity testing, the Lowest Observed Effect Concentration (LOEC) and the No Observed Effect Concentration (NOEC) have been measured for specific adverse effects.

According to the Anatomical Therapeutic Chemical Classification System (ATC, www.whocc.no), the active substances are divided into different groups based on the

Table 15.1 International Standardization guidelines used for the evaluation of acute and chronic toxicity of several organisms from different levels of the aquatic food chain

Toxicity	Species	Organism	Guideline
Acute	Bacteria	*Vibrio fischeri*	DIN 38412-L34 (1991)
		Pseudomonas putida	ISO 10712 (1995)
	Rotifers	*Brachionus calyciflorus*	ASTM E1440-91 (2004)
	Crustaceans	*Ceriodaphnia dubia*	US EPA-600-4-90 (1993)
		Daphnia magna	OECD 202 (2004); ISO 6341 (1996)
		Thamnocephalus platyurus	ISO 14380 (2011)
	Fish	*Danio rerio*	OECD 203 (1992)
Chronic	Bacteria	*Vibrio fischeri*	DIN 38412-L37 (1991)
	Cyanobacteria	*Synechococcus leopoliensis*	OECD 201 (2011)
	Algae	*Pseudokirchneriella subcapitata*	OECD 201 (2011)
	Rotifers	*Brachionus calyciflorus*	ISO 20666 (2008)
	Crustaceans	*Ceriodaphnia dubia*	ISO 20665 (2008)
		Daphnia magna	OECD 211 (2008); ISO 10706 (2000)
	Plants	*Lemna minor*	OECD 221 (2006)

target organ and their therapeutic, pharmacological, and chemical properties. Antineoplastic drugs, L01, are a subgroup of L (antineoplastic and immunomodulating agents) and are subsequently divided into five subgroups as reported in Scheme 15.1.

Some antineoplastic drugs belonging to each subgroup of L01 are investigated in this chapter. The subgroup L01A is represented by alkylating agents, which are able to attach directly to an alkyl group (CnH_2n_{+1}) of the guanine base forming DNA adducts causing cell growth inhibition or apoptosis stimulation. The antimetabolites (L01B) are chemicals that alter the DNA replication process by incorporating chemically altered nucleotides in genetic material or by depleting the supply of deoxy-nucleotides. The subgroup of plant alkaloids and other natural products (L01C) include drugs that interfere with the DNA replication as well as cell division, inhibiting the topoisomerase enzymes, blocking the formation of the spindle in the mitosis or inhibiting its removal.

Cytotoxic antibiotics and related substances (L01D) may bind tightly and intercalate into the double-stranded DNA and may interfere with RNA synthesis. In the group of other antineoplastic agents (L01X), there are compounds with diverse chemical structures, diverse specific activities, and different mechanisms of action, i.e., protein kinase inhibitors targeting specific protein kinases and monoclonal antibodies acting against cancer-specific antigens.

An extensive literature analysis was conducted herein to examine the ecotoxicity of the antineoplastic drugs, reported in Table 15.2, selected according to their chemical structures, modes of action, increasing consumption levels, and their environmental measured concentration in waterbodies.

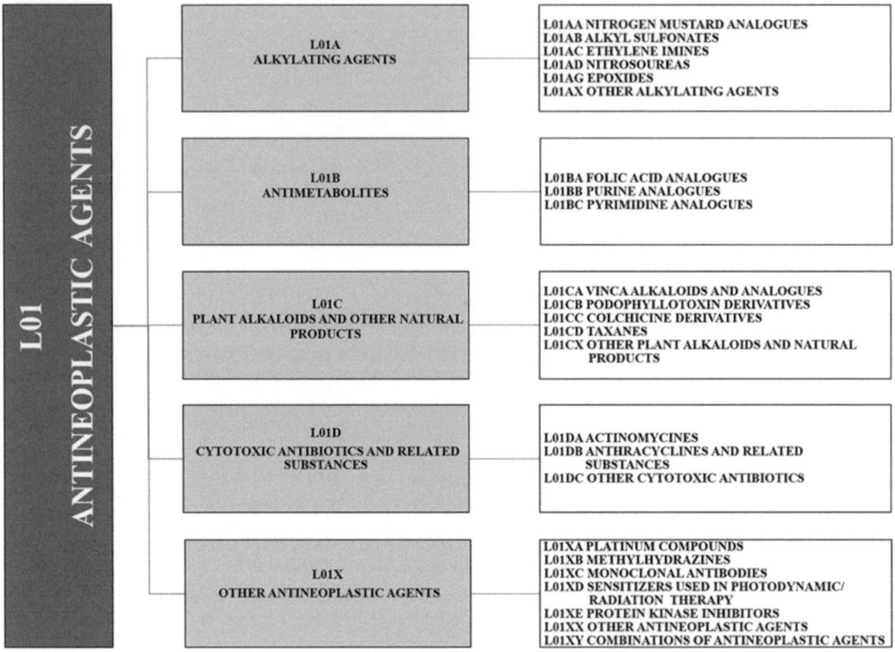

Scheme 15.1 Classification of antineoplastic drugs

15.2 Acute and Chronic Toxicity of Selected Antineoplastic Drugs

15.2.1 L01A: Alkylating Agents

15.2.1.1 Subgroup L01AA: *Cyclophosphamide and Ifosfamide*

Cyclophosphamide (CP, L01AA01) and ifosfamide (IFO, L01AA06) belong to nitrogen mustard analogues (L01AA), which are chemotherapeutics used to treat various forms of cancer: lymphoma, leukemia, and several solid tumors such as lung and bronchus, breast, and ovarian cancers (Xie 2012). CP and IFO are widely used antineoplastic drugs in cancer therapy. They are polar compounds, highly soluble in water, with a low octanol/water partition coefficient (log Kow of 0.6 and 0.9 for CP and IFO, respectively, Table 15.2) and low adsorption potential to sewage sludge or sediments (Kosjek and Heath 2011; Xie 2012). Based on their mode of action, CP and IFO are pro-drugs, which, following administration, are metabolized and excreted together with their metabolites and released into the sewage systems, subsequently entering the wastewater treatment plants. They are only partially removed and eventually flow into the water bodies (Delgado et al. 2011; Köhler et al. 2012; Kovalova et al. 2012; Lutterbeck et al. 2015a). Indeed,

Table 15.2 Anticancer drugs classified by their name, chemical structures, class, selected physico-chemical parameters, mode of action and cancer treatments

ATC	Name	Abbr.	Chemical Structure	Selected physico-chemical properties	Mode of action	Cancer treatment
L01AA01	Cyclophosphamide	CP	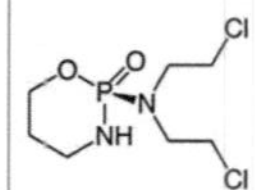	M.W. = 261.09 $Logk_{OW} = 0.6$	Interaction with DNA, causing cross-links and strand breaks with consequent cell death	Lymphoma, leukemia and several solid tumors
L01AA06	Ifosfamide	IFO	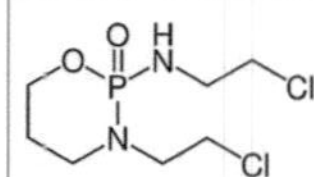	M.W. = 261.09 $LogK_{OW} = 0.9$	Interaction with DNA, causing cross-links and strand breaks with consequent cell death	Lymphoma, leukemia and several solid tumors
L01BA01	Methotrexate	MTX	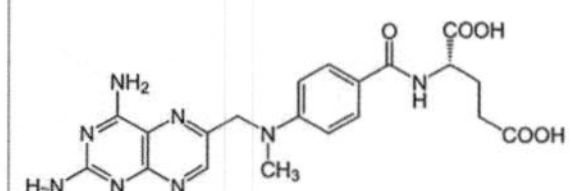	M.W. = 454.44 $LogK_{OW} = -1.8$	Inhibition of dihydrofolate reductase, bloking purine and pyrimidine bases pathways and the consecutive DNA synthesis	Breast, skin, head and lung cancers, rheumatoid arthritis treatment
L01BC02	5- Fluorouracil	5-FU	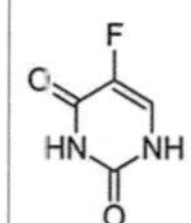	M.W. = 130.08 $LogK_{OW} = -1.0$	Inhibition of the DNA polymerase and the DNA synthesis in the S phase, inducing cell cycle arrest and apoptosis	Breast, colorectal, oesophageal, stomach, pancreatic and skin cancers
L01BC06	Capecitabine	CAP	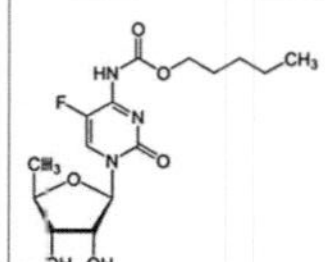	M.W. = 359.35 $LogK_{OW} = 0.6$	Inhibition of the DNA polymerase and the DNA synthesis in the S phase, inducing cell cycle arrest and apoptosis	Colorectal, breast, gastric, oesophageal cancers

(continued)

Table 15.2 (continued)

ATC	Name	Abbr.	Chemical Structure	Selected physico-chemical properties	Mode of action	Cancer treatment
L01CB01	Etoposide	ET	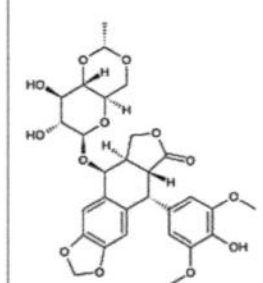	M.W. = 588.56 $LogK_{OW} = 0.6$	Topoisomerase II inhibitor: induction of DNA damage by blocking re-ligation of DNA double-strand breaks	Kaposi's sarcoma, Ewing's sarcoma; lung, testicular and lymphoma cancers
L01DBO1	Doxorubicin	DOX	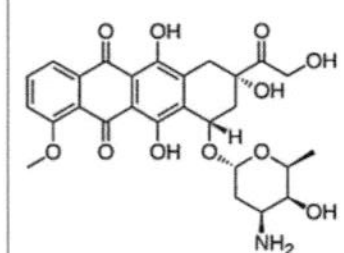	M.W. = 543.52 $LogK_{OW} = 1.3$	Topoisomerase II inhibitor: induction of DNA damage, intercalating between base pairs in the DNA double helix	Breast, bladder cancers, Kaposi's sarcoma, lymphoma and acute lymphocytic leukemia
L01XA01	Cisplatin	CDDP	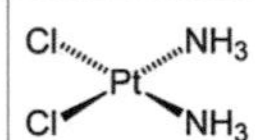	M.W. = 300.01 $LogK_{OW} = -2.2$	Platinum complexes: inhibition of DNA replication, binding *in vivo* nucleophilic groups of DNA and inducing DNA protein cross-links	Lymphoma, lung and ovarian, cancers
L01XE01	Imatinib	IM	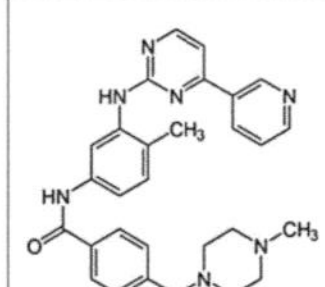	M.W. = 493.60 $LogK_{OW} = 2.89$	Protein kinase inhbition: interaction with oral tyrosine kinase, involved in the regulation of many biological processes (cell growth, migration, survival)	Chronic myeloid leukemia, gastrointestinal stromal tumors and anaplastic thyroid carcinoma

CP and IFO have been detected in different compartments including hospital wastewaters, wastewater treatment plant influents, effluents, and surface water at concentrations ranging from ng/L to μg/L (Brooker et al. 2014; Negreira et al. 2014; Česen et al. 2016a).

As regards CP and IFO, acute toxicity assays were performed using the bioluminescent bacterium *Vibrio fischeri* with no luminescence inhibition after 30 min for each compound (Lutterbeck et al. 2014, 2015a; Białk-Bielińska et al. 2017). Furthermore, the *Pseudomonas putida* growth inhibition test was performed on CP by Zounková et al. 2007, showing no change in the absorption (590 nm) of bacterial culture after 16-h exposure. The acute toxic effect of these two alkylating agents was also tested in primary consumers: the rotifer *Brachionus calyciflorus*, the anostracan crustacean *Thamnocephalus platyurus*, and the cladoceran crustaceans *Daphnia magna* and *Ceriodaphnia dubia*, with L(E)C50 values at concentrations ranging from 196.4 to 1924 mg/L, several orders of magnitude higher than concentrations found in the aquatic environment (Česen et al. 2015; Isidori et al. 2016a; Białk-Bielińska et al. 2017; Russo et al. 2018). Furthermore, IFO showed an increasing acute toxic effect along the aquatic trophic chain (*B. calyciflorus* < *T. platyurus* < *C. dubia*) (Russo et al. 2018).

Regarding chronic toxicity, Russo et al. (2018) found that CP and IFO, at the highest concentration tested (200 mg/L), showed no significant growth inhibition compared to the negative control (after 72 h of incubation), when tested on the green alga *Pseudokirchneriella subcapitata* (also known as *Raphidocelis subcapitata* and *Selenastrum capricornutum*). These results are in line with data presented in other studies (Zounková et al. 2007; Grung et al. 2008; Česen et al. 2016b; Białk-Bielińska et al. 2017). No toxic effect up to 320 mg/L was observed in the cyanobacterium *Synechococcus leopoliensis* (Česen et al. 2016b). The absence of toxicity found for reducers and primary producers may be due to the fact that these drugs are pro-drugs requiring metabolic activation by liver enzymes. Literature also reports chronic studies on primary consumers such as *B. calyciflorus* and *C. dubia* considering offspring reduction over 48-h and 7-day exposure, respectively, as the test endpoint. The chronic exposure results indicate that IFO was more toxic than CP with a median offspring reduction at 15.84 mg/L in *C. dubia* (vs. CP EC50 = 58.03 mg/L) after 7-day exposure and at 76.05 mg/L in *B. calyciflorus* (vs. CP EC50 = 89.84 mg/L) after 48-h exposure. The higher toxicity observed for crustaceans compared to primary producers could be explained by the presence of the cytochrome P450 gene superfamily in daphnids, as well as in other organisms of the phylogenetic tree such as insects and nematodes that play a fundamental role in the metabolism of and tolerance to anthropogenic chemicals (Baldwin et al. 2009). Białk-Bielińska et al. (2017) conducted a study on the toxicity of the CP and IFO in the aquatic plant *Lemna minor* and did not observe any toxic effects up to 100 mg/L. The data of the described reports are listed in Table 15.3.

Table 15.3 Available ecotoxicological data (L(E)C50 in mg/L) for the selected anticancer drugs towards selected organisms. Values in the parentheses indicate 95% confidence interval

Compound	Organism tested in acute or chronic toxicity			Toxicity endpoint	Toxicity of compound (L(E)C50 in mg/L)	References
CP	Acute toxicty	Bacteria	*V. fischeri*	Luminescence inhibition 30 min	> 100	Białk-Bielińska et al. (2017)
			P. putida	Growth inhibition 16 h	> 1000	Zounková et al. (2007)
			B. calyciflorus	Mortality 24 h	1924 (1210–3036)	Russo et al. (2018)
		Rotifers	*C. dubia*	Mortality 24 h	986.6 (765.3–1272)	Russo et al. (2018)
		Crustaceans	*D. magna*	Immobilization 48 h	> 1000	Zounková et al. (2007)
					>100	Białk-Bielińska et al. (2017)
			T. platyurus	Mortality 24 h	1396 (1304–1494)	Russo et al. (2018)
	Chronic toxicty	Bacteria	*V. fischeri*	Luminescence inhibition 24 h	>120	Lutterbeck et al. (2015a)
		Cyanobacteria	*S. leopoliensis*	Growth inhibition 72 h	>320	Česen et al. (2016b)
		Algae	*P. subcapitata*	Growth inhibition 72 h	930 (700–1100)	Zounková et al. (2007)
					>320	Česen et al. (2016a, b)
					>100	Białk-Bielińska et al. (2017)
					>200	Russo et al. (2018)
		Rotifers	*B.calyciflorus*	Reproduction inhibition 48 h	89.84 (67.62–119.4)	Russo et al. (2018)
		Crustaceans	*C. dubia*	Reproduction inhibition 7 day	58.03 (37.43–89.98)	Russo et al. (2018)
			D. magna	Reproduction inhibition 21 day	>100	Grung et al. (2008)
		Plants	*L. minor*	Growth inhibition 7 day	>100	Białk-Bielińska et al. (2017)

IFO	Acute toxicty	Bacteria	*V. fischeri*	Luminescence inhibition 30 min	>100	Białk-Bielińska et al. (2017)
		Rotifers	*B. calyciflorus*	Mortality 24 h	996.3 (767.3–1294)	Russo et al. (2018)
		Crustaceans	*C. dubia*	Mortality 24 h	196.4 (149.1–258.7)	Russo et al. (2018)
			D. magna	Immobilization 48 h	>100	Białk-Bielińska et al. (2017)
			T. platyurus	Mortality 24 h	771.5 (627.9–947.8)	Russo et al. (2018)
	Chronic toxicty	Cyanobacteria	*S. leopoliensis*	Growth inhibition 72 h	>320	Česen et al. (2016b)
		Algae	*P. subcapitata*	Growth inhibition 72 h	>100	Grung et al. (2008)
					>320	Česen et al. (2016b)
					>100	Białk-Bielińska et al. (2017)
					>200	Russo et al. (2018)
		Rotifers	*B.calyciflorus*	Reproduction inhibition 48 h	76.05 (48.04–120.4)	Russo et al. (2018)
		Crustaceans	*C. dubia*	Reproduction inhibition 7 day	15.84 (13.03–19.27)	Russo et al. (2018)
		Plants	*L. minor*	Growth inhibition 7 day	>100	Białk-Bielińska et al. (2017)
MTX	Acute toxicty	Bacteria	*V. fischeri*	Luminescence inhibition 30 min	>100	Białk-Bielińska et al. (2017)
		Crustaceans	*D. magna*	Immobilization 48 h	>100	Białk-Bielińska et al. (2017)
	Chronic toxicty	Algae	*P. supcapitata*	Growth inhibition 72 h	9.51 (9.44–9.58)	Białk-Bielińska et al. (2017)
		Plants	*L. minor*	Growth inhibition 7 day	0.08 (0.07–0.09)	Białk-Bielińska et al. (2017)
5-FU	Acute toxicty	Bacteria	*V. fisheri*	Luminescence inhibition 30 min	>100	Białk-Bielińska et al. (2017)

(continued)

Table 15.3 (continued)

Compound	Organism tested in acute or chronic toxicity			Toxicity endpoint	Toxicity of compound (L(E)C50 in mg/L)	References
			P. putida	Growth inhibition 16 h	0.044 (0.025–0.077)	Zounková et al. (2010)
		Rotifers	*B. calyciflorus*	Mortality 24 h	>200	Parrella et al. (2014)
		Crustaceans	*C. dubia*	Mortality 24 h	501 (351–854)	Parrella et al. (2014)
			D. magna	Immobilization 48 h	36 (12–70)	Zounková et al. 2007
					15 (5.2–45)	Zounková et al. (2010)
					20.84 (18.07–24.04)	Parrella et al. (2014)
					>100	Białk-Bielińska et al. (2017)
			T. platyurus	Mortality 24 h	0.28 (0.26–0.29)	Parrella et al. (2014)
		Fish	*D. rerio*	Mortality 96 h	>100	Kovács et al. (2016)
	Chronic toxicty	Cyanobacteria	*S. leopoliensis*	Growth inhibition 72 h	1.20 (0.72–2.02)	Brezovšek et al. (2014)
		Algae	*D. subspicatus*	Growth inhibition 72 h	48 (44–51)	Zounková et al. (2010)
			P. subcapitata	Growth inhibition 72 h	0.13 (0.07–0.22)	Brezovšek et al. (2014)
					0.075 (0.067–0.084)	Białk-Bielińska et al. (2017)
		Rotifers	*B. calyciflorus*	Reproduction inhibition 48 h	0.322 (0.285–0.364)	Parrella et al. (2014)
		Crustaceans	*C. dubia*	Reproduction inhibition 7 day	3.35×10^{-3} (2.20×10^{-3}–5.09×10^{-3})	Parrella et al. (2014)
			D. magna	Reproduction inhibition 21 day	0.1	Zounková et al. (2010)
					26.4×10^{-3} (20.6×10^{-3}–33.9×10^{-3})	Parrella et al. (2014)
		Plants	*L. minor*	Growth inhibition 7 day	0.91 (0.70 –1.11)	Załeska-Radziwiłł et al. (2011)
					2.45 (2.19–2.76)	Białk-Bielińska et al. (2017)

CAP	Acute toxicty	Rotifers	*B. calyciflorus*	Mortality 24 h	>500	Parrella et al. (2014)
		Crustaceans	*C. dubia*	Mortality 24 h	1.23×10^3 (0.9×10^3–1.6×10^3)	Parrella et al. (2014)
			D. magna	Immobilization 48 h	>850	Straub (2010)
					224 (118–404)	Parrella et al. (2014)
			T. platyurus	Mortality 24 h	197.7 (174.7–223.7)	Parrella et al. (2014)
	Chronic toxicty	Rotifers	*B. calyciflorus*	Reproduction inhibition 48 h	15.4 (11.3–21.1)	Parrella et al. (2014)
		Crustaceans	*C. dubia*	Reproduction inhibition 7 day	2.4 (2.0–2.8)	Parrella et al. (2014)
			D. magna	Reproduction inhibition 21 day	20.5 (15.5–27.2)	Parrella et al. (2014)
ET	Acute toxicty	Bacteria	*P. putida*	Growth inhibition 16 h	630 (580–830)	Zounková et al. (2007)
		Rotifers	*B. calyciflorus*	Mortality 24 h	>120	Parrella et al. (2014)
		Crustaceans	*C. dubia*	Mortality 24 h	16% at 120	Parrella et al. (2014)
			D. magna	Immobilization 48 h	25% at 120	Parrella et al. (2014)
					30(16–40)	Zounková et al. (2007)
			T. platyurus	Mortality 24 h	74.85 (56.36–99.40)	Parrella et al. (2014)
		Fish	*D. rerio*	Mortality 96 h	>100	Kovács et al. (2016)
	Chronic toxicty	Cyanobacteria	*S. leopoliensis*	Growth inhibition 72 h	n.d.	Brezovšek et al. (2014)
		Algae	*P. subcapitata*	Growth inhibition 72 h	250 (120–460)	Zounková et al. (2007)
		Rotifers	*B. calyciflorus*	Reproduction inhibition 48 h	3.7 (2.7–5.3)	Parrella et al. (2014)
		Crustaceans	*C. dubia*	Reproduction inhibition 7 day	0.204 (0.152–0.256)	Parrella et al. (2014)
			D. magna	Reproduction inhibition 21 day	0.239 (0.181–0.299)	Parrella et al. (2014)
DOX	Acute toxicty	Bacteria	*P. putida*	Growth inhibition 16 h	>1000	Zounková et al. (2007)
		Rotifers	*B. calyciflorus*	Mortality 24 h	12.69 (10.25–16.57)	Parrella et al. (2014)
		Crustaceans	*C. dubia*	Mortality 24 h	5.18 (4.44–6.04)	Parrella et al. (2014)

(continued)

Table 15.3 (continued)

Compound	Organism tested in acute or chronic toxicity			Toxicity endpoint	Toxicity of compound (L(E)C50 in mg/L)	References
			D. magna	Immobilization 48 h	2.0 (0.52–4.8)	Zounková et al. (2007)
					2.14 (1.55–2.46)	Parrella et al. (2014)
			T. platyurus	Mortality 24 h	0.31 (0.12–0.83)	Parrella et al. (2014)
		Fish	*D. rerio*	Mortality 72 h	16.96	Han et al. (2015)
	Chronic toxicty	Algae	*P. subcapitata*	Growth inhibition 72 h	13 (12–17)	Zounková et al. (2007)
		Rotifers	*B. calyciflorus*	Reproduction inhibition 48 h	7.7 (5.9–9.9)	Parrella et al. (2014)
		Crustaceans	*C. dubia*	Reproduction inhibition 7 day	n.d.	Parrella et al. (2014)
			D. magna	Reproduction inhibition 21 day	n.d.	Parrella et al. (2014)
CDDP	Acute toxicty	Bacteria	*P. putida*	Growth inhibition 16 h	1.2 (1.0–1.4)	Zounková et al. (2007)
		Rotifers	*B. calyciflorus*	Mortality 24 h	6.52 (4.31–9.86)	Parrella et al. (2014)
		Crustaceans	*C. dubia*	Mortality 24 h	2.50 (2.13–2.97)	Parrella et al. (2014)
			D. magna	Immobilization 48 h	0.94 (0.90–0.97)	Parrella et al. (2014)
			T. platyurus	Mortality 24 h	8.44 (7.18–9.91)	Parrella et al. (2014)
		Fish	*D. rerio*	Mortality 96 h	64.5 (59.1–69.2)	Kovács et al. (2016)
	Chronic toxicty	Cyanobacteria	*S. leopoliensis*	Growth inhibition 72 h	0.67 (0.336–1.339)	Brezovšek et al. (2014)
		Algae	*P. subcapitata*	Growth inhibition 72 h	2.3 (1.7–2.9)	Zounková et al. (2007)
					1.52 (1.26–1.83)	Brezovšek et al. (2014)
		Rotifers	*B. calyciflorus*	Reproduction inhibition 48 h	0.440 (0.283–0.728)	Parrella et al. (2014)
		Crustaceans	*C. dubia*	Reproduction inhibition 7 day	16.8×10^{-3} (12.5×10^{-3}–22.6×10^{-3})	Parrella et al. (2014)
			D. magna	Reproduction inhibition 21 day	1.63×10^{-3} (1.23×10^{-3}–2.18×10^{-3})	Parrella et al. (2014)

IM	Acute toxicty	Bacteria	*P. putida*	Growth inhibition 30 min	23.06 (22.00–24.16)	Białk-Bielińska et al. (2017)
		Rotifers	*B. calyciflorus*	Mortality 24 h	3.82 (3.63–4.04)	Parrella et al. (2014)
		Crustaceans	*C. dubia*	Mortality 24 h	31.92 (27.61–36.98)	Parrella et al. (2014)
			D. magna	Immobilization 48 h	11.97 (9.37–15.45)	Parrella et al. (2014)
					72.43 (66.54–n. a.)	Białk-Bielińska et al. (2017)
			T. platyurus	Mortality 24 h	43.27 (31.39–59.65)	Parrella et al. (2014)
		Fish	*D. rerio*	Mortality 96 h	70.8 (63.7–81.1)	Kovács et al. (2016)
	Chronic toxicty	Cyanobacteria	*S. leopoliensis*	Growth inhibition 72 h	5.36 (4.29–6.70)	Brezovšek et al. (2014)
		Algae	*P. subcapitata*	Growth inhibition 72 h	2.29 (1.88–2.77)	Brezovšek et al. (2014)
					5.08 (4.94–5.24)	Białk-Bielińska et al. (2017)
		Rotifers	*B. calyciflorus*	Reproduction inhibition 48 h	0.740 (0.550–0.980)	Parrella et al. (2014)
		Crustaceans	*C. dubia*	Reproduction inhibition 7 day	0.115 (0.063–0.209)	Parrella et al. (2014)
		Plants	*D. magna*	Reproduction inhibition 21 day	0.308 (0.147–0.872)	Parrella et al. (2014)
			L. minor	Reproduction inhibition 7 day	61.05 (58.81–63.37)	Białk-Bielińska et al. (2017)

15.2.2 L01B: Antimetabolites

15.2.2.1 Subgroup L01BA: *Methotrexate*

Methotrexate (MTX, L01BA01) is an antimetabolite belonging to the class of folic acid analogues widely used as a chemotherapeutic agent in the treatment of bronchial, breast, and ovarian cancers, lymphomas, and leukemia. It acts by inhibiting the dihydrofolate reductase (DHFR), and blocking purine and pyrimidine bases pathways, crucial for DNA synthesis. Methotrexate is commonly used to treat autoimmune diseases. It is a polar compound (log Kow = −1.8, Table 15.2), it is excreted in the urine by 80–100%, and it is characterized by a high biodegradability and a direct degradation by photolysis with few transformation products (Xie 2012; Lutterbeck et al. 2015b). Due to its physical and chemical properties, MTX is unlikely to be adsorbed by wastewater sludge or sediments and may be detected in the water cycle (Lutterbeck et al. 2015b). In fact, this antimetabolite has been observed in different environmental compartments including hospital and wastewater treatment plant effluents and surface waters at concentrations ranging from tens of ng/L to μg/L (Isidori et al. 2016a). Considering its aquatic concentrations, MTX is unlikely to pose a risk of acute toxicity on non-target organisms. In fact, testing the parent compound up to 100 mg/L, no median short-term effects such as luminescence inhibition during 30 min and immobilization during 48-h exposure were observed in *V. fischeri* and *D. magna*, respectively (Białk-Bielińska et al. 2017). On the contrary, the same authors found ~~a~~ long-term toxicity in *P. subcapitata* and *L. minor* at concentrations in the order of units of mg/L and tens of μg/L, respectively, as shown in Table 15.3. These results on aquatic plants may cause environmental concern considering that MTX may be found at concentrations only one order of magnitude lower than those causing growth inhibition after 7-day exposure.

15.2.2.2 Subgroup L01 BC: *5-Fluorouracil and Capecitabine*

5-Fluorouracil (5-FU, L01 BC02) and its prodrug capecitabine (CAP, L01 BC06) are two pyrimidine analogues with a potential to inhibit the DNA polymerase and the DNA synthesis in the S phase, inducing cell cycle arrest and apoptosis (Straub 2010). CAP is used in the treatment of metastatic breast and colorectal cancers, and after oral administration, it undergoes rapid enzymatic metabolization in the active 5-FU by thymidine phosphorylase mainly in tumor cells. Differently from CAP, 5-FU is administered intravenously and is used to treat breast, colorectal, esophageal, stomach, pancreatic, and skin cancers. Of the administered CAP, 95% is recovered in the urine (only 3% as an unchanged drug), while 5-FU is excreted in the urine by 85–90% as metabolites, and the remaining 10–15% is unchanged. In aqueous solution, both polar compounds show high solubility with low log Kow values equal to −1 and 0.6, for 5-FU and CAP, respectively (Table 15.2), presenting low adsorption to suspended solids and to sewage sludge and adequate biodegradability (Straub 2010; Xie 2012). These antimetabolites are found in hospital effluents and wastewater treatment plants at concentrations in the order of units/tens of ng/L for 5-FU and units/tens of μg for CAP (Straub 2010; Isidori et al. 2016a; Olalla et al. 2018).

5-FU was particularly toxic to the microorganism *P. putida* with an acute EC50 value equal to 0.044 mg/L (Zounková et al. 2010), differently from *V. fischeri* that did not show luminescence inhibition after 30-min exposure at concentrations up to 100 mg/L (Białk-Bielińska et al. 2017). In acute toxicity testing toward consumers, 5-FU was particularly toxic in *T. platyurus* with LC50 equal to 0.28 mg/L, while in *D. magna*, the EC50 was found at a concentration of two orders of magnitude higher (20.84 mg/L). CAP presented fewer toxic effects for all consumers tested with L(E)C50 values two or three orders of magnitude higher than 5-FU (Straub 2010; Parrella et al. 2014) in line with the results reported by Zounková et al. (2007, 2010).

As regards chronic toxicity, expressed as reproduction inhibition, very low EC50 values, in the order of μg/L and mg/L for 5-FU and CAP, respectively, were observed in both rotifers and crustaceans.

In addition, 5-FU was able to inhibit the growth of green alga (Zounková et al. 2010; Brezovšek et al. 2014; Białk-Bielińska et al. 2017), *L. minor* (Załeska-Radziwiłł et al. 2011; Białk-Bielińska et al. 2017), and cyanobacteria (Brezovšek et al. 2014) although at concentrations higher than those found for rotifers and crustaceans. Differently from the pro-drug CAP, which is of less environmental concern due to its toxic effects at very high concentrations (mg/L), 5-FU is more hazardous since, as reported by Xie (2012), it is converted into different metabolites forming stable complexes that interfere with DNA synthesis.

15.2.3 L01C: Plant Alkaloids and Other Natural Products

15.2.3.1 Subgroup L01CB: *Etoposide*

Etoposide (ET, L01CB01) is a semisynthetic derivative of podophyllotoxin, a chemotherapeutic agent prescribed in clinical therapy to treat a wide spectrum of human cancers, especially small cell lung cancer and testicular cancer. ET is a topoisomerase II inhibitor and, acting in the late S and early G2 phases of the cell cycle, it binds the DNA forming a covalent cleavage complex, preventing DNA from relegation, causing DNA damage due to the generation of permanent DNA strand breaks and chromosomal translocations, which lead to cell apoptosis (Baldwin and Osheroff 2005; Xie 2012). After administration, 5–22% of ET is excreted in the urine in unchanged form (Zhang et al. 2013). This drug is neither a readily water-soluble compound nor a highly biodegradable compound; in fact, under dark conditions, ET is degraded for approximately 40% in 7 days and subjected to slow hydrolysis in water (Kosjek et al. 2016), and its log Kow is equal to 0.6 (Table 15.2). It is present in hospital effluents from 3.4 to 714 ng/L (Catastini et al. 2008; Yin et al. 2010; Ferrando-Climent et al. 2014). Acute toxicity was observed in crustaceans

(*D. magna*, *C. dubia*, *T. platyurus*), rotifers (*B. calyciflorus*), and fish (*D. rerio*) and was found at concentrations far higher (dozens-hundreds of mg/L) than those of environmental concern (Zounková et al. 2007; Parrella et al. 2014; Kovács et al. 2016). Similarly, the inhibition of growth of the green alga *P. subcapitata* and of different species of bacteria such as the cyanobacterium *S. leopoliensis* and the gram-negative bacterium *P. putida* was found at concentrations in the order of dozens-hundreds of mg/L (Zounková et al. 2007; Brezovšek et al. 2014). In addition, according to Parrella et al. (2014) and Kovács et al. (2016), chronic toxicity [evaluated as offspring reduction for crustaceans and rotifers or as embryo toxicity for fish (OECD 2006)] was observed in *D. magna* (EC50 = 0.239 mg/L), *C. dubia* (EC50 = 0.204 mg/L), *B. calyciflorus* (EC50 = 3.7 mg/L), and *D. rerio* (LC50 > 300 mg/L). The highest chronic toxic effects did not increase as the aquatic trophic chain level increased, as expected, thus indicating the highest susceptibility of crustaceans to ET.

15.2.4 L01D: Cytotoxic Antibiotic and Related Substances

15.2.4.1 Subgroup L01DB: *Doxorubicin*

Doxorubicin (DOX, L01DB01) is an anthracycline that is a cytotoxic antibiotic mainly used in chemotherapy against several solid cancers. This drug consists in a planar moiety intercalating between two base pairs and in a daunosamine moiety forming a complex with the immediately adjacent base pairs to the intercalation site. This particular structure allows DOX to interact with DNA, blocking DNA replication and preventing DNA relegation inhibiting the progression of the topoisomerase II (Xie 2012). DOX is excreted 5–12% in urine and 40–50% in feces. DOX water solubility is relatively low; the log Kow is equal to 1.27 (Table 15.2) and is easily adsorbed to sewage sludge. This drug was measured in Austrian hospital wastewaters in concentrations ranging from 0.26 μg/L to 1.35 μg/L (Lenz et al. 2007; Mahnik et al. 2007), while in both Spanish and Slovenian municipal and hospital wastewaters, DOX measured concentrations were below detection limit, LOD = 0.7–0.8 ng/L (Isidori et al. 2016a). Acute toxicity was observed in *P. putida* at high concentrations (EC50 > 1000 mg/L) and in crustaceans at concentrations ranging from 0.31 to 5.18 mg/L (Zounková et al. 2007; Parrella et al. 2014). Rotifers were less susceptible to DOX with EC50 found at few dozens of mg/L (Parrella et al. 2014).

The growth inhibition was observed in *P. subcapitata* at dozens of mg/L (Zounková et al. 2007). The same order of magnitude in toxicity was found by Han and coauthors (Han et al. 2015) in *Danio rerio* embryos (LC50 = 16.96 mg/L). In addition, the chronic toxicity in the rotifer *B. calyciflorus* was observed by Parrella et al. (2014) at units of mg/L (EC50 = 7.7 mg/L). DOX, as well as other anthracyclines, shows ecotoxicity at relatively high concentrations in the spectrum of the species tested. Moreover, anthracyclines may be effectively removed from wastewater treatment plants and play a less significant role in the hazard of antineoplastic drugs.

15.2.5 L01X: Other Antineoplastic Agents

15.2.5.1 Subgroup L01XA: *Cisplatin or Cis-diamminedichloroplatinum*

Cisplatin (CDDP, L01XA01) is a square planar Pt (II) complex used in chemotherapy to treat ovarian, lung, head, and neck cancers and in particular testicular cancer. After administration, one of the two chloride atoms is displaced by water to give the aquo-complex cis-[PtCl(NH3)2(H2O)]+, which preferentially binds to guanine, forming crosslinks between guanine bases, interfering with cell division during mitosis, leading cell to apoptosis (Xie 2012). The drug is excreted in the urine and accounts for 13–17% of the dose. The biodegradation of cisplatin is close to zero, easily soluble in water, and polar with a very low log Kow equal to −2.2 (Table 15.2). According to Isidori et al. (2016a), CDDP was found in both Spanish and Slovenian municipal and hospital wastewaters. In Spanish wastewaters, the concentrations were below detection limit (LOD = 0.7–0.8 ng/L), whereas in Slovenian hospital effluents, total platinum compounds were found in the order of magnitude of hundreds of ng/L. Furthermore, Vyas et al., in Vyas et al. 2014, measured platinum-based anticancer drugs as total Pt compounds in the wastewaters of a major UK hospital in concentrations ranging from 0.02 to 140 μg/L in the oncology effluent. The median acute toxicity observed for CDDP in crustaceans and rotifers (Table 15.3) ranged from 0.94 to 8.44 mg/L (Zounková et al. 2007; Parrella et al. 2014), while in the fish *D. rerio*, LC50 was found at higher concentrations equal to 64.5 mg/L (Kovács et al. 2016). In addition, the median growth inhibition in *P. putida*, in *P. subcapitata*, and in *S. leopoliensis* was observed from few tenths to units of mg/L (Zounková et al. 2007; Brezovšek et al. 2014). Furthermore, CDDP was causing 50% of the offspring reduction in crustaceans at concentrations ranging from 1.63 μg/L (EC50 – *D. magna*) to 16.8 μg/L (EC50 – *C. dubia*) and in rotifers at concentrations in the order of hundreds of μg/L. Platinum compounds are, generally, found among the most toxic antineoplastic drugs most likely for the direct DNA strand breaks and ROS production. Furthermore, due to CDDP physico-chemical properties, it is rapidly hydrolyzed in different residues forming new, stable, and more toxic mixtures (Parrella et al. 2014).

15.2.5.2 Subgroup L01XE: *Imatinib*

Imatinib mesylate (IM, L01XE01) is a tyrosine kinase inhibitor mainly used in the treatment of Philadelphia chromosome-positive chronic myeloid leukemia. IM, binding an intracellular pocket located within tyrosine kinases, inhibits the ATP binding and prevents phosphorylation and the subsequent activation of growth receptors, resulting in decreased proliferation and enhancement of cell apoptosis (An et al. 2010). IM is excreted in 81% in unchanged form, and the fecal to urinary excretion ratio is approximately 5:1. This chemical is soluble in water ($pH < 5.5$) and its log Kow is 2.89 (Table 15.2). In addition, Nageswari and collaborators (Nageswari et al. 2012) did not observe degradation when IM was subjected to base hydrolysis (NaOH for 30 h), water hydrolysis, and light and heat conditions. Scientific literature regarding the environmental characteristics, the occurrence, and the toxicity of this chemical in the aquatic system is scarce. Nevertheless, according to Isidori et al. (2016a), IM was found in Slovenian wastewaters at concentrations lower than LOQ value (180 ng/L), but higher than LOD (54 ng/L). The PEC values for IM refined by excretion rates ranged from 0.18 to 5.64 ng/L (Besse et al. 2012; Booker et al. 2014; Santos et al. 2017). Acute toxicity was observed at L(E)C50 values ranging from 11.97 to 43.27 mg/L in the crustaceans *D. magna*, *C. dubia*, and *T. platyurus* (Parrella et al. 2014), while in *D. rerio*, IM median lethal effect was observed at 70.8 mg/L (Kovács et al. 2016).

As regards growth inhibition, 50% of the effect was found in *P. subcapitata* and in *S. leopoliensis* after 72-h exposure at concentrations in the order of few mg/L (Brezovšek et al. 2014; Białk-Bielińska et al. 2017). Furthermore, the concentrations causing 50% inhibition of the reproduction were observed both in crustaceans and in rotifers in the order of hundreds of μg/L (Parrella et al. 2014). As regards fish embryo toxicity and the inhibition of reproduction in the plant *L. minor*, the lowest observed effective concentration was observed at dozens of mg/L (Kovács et al. 2016; Białk-Bielińska et al. 2017).

15.3 Are Acute and Chronic Toxicity Data Sufficient to Identify Antineoplastic Drugs of Environmental Concern?

On investigation of acute toxicity data toward non-target organisms, the selected drugs, described herein, would not seem to be of environmental concern since acute toxicity was found at the lowest value (EC50) of 0.044 mg/L for 5-FU in *P. putida*, which are far from those of environmental concern, but are still relevant regarding environmental risk assessment when chronic data are lacking. The effects observed in chronic toxicity studies are shown in Table 15.4. 5-Fluorouracil and CDDP were found to be the most hazardous antineoplastic drugs causing the inhibition of reproduction in the range from 1 to 10 μg/L. Potential risk is not foreseeable when comparing chronic toxicity data of all antineoplastic drugs to the respective MEC values, as the effective concentrations are at least one or two orders of magnitude higher than the environmental drug occurrence and crustaceans represented the most susceptible species. In Chap. 18, the risk quotient (RQ), specifically the ratio between the PEC or the MEC value and the Predicted No Effect Concentrations (PNEC), will be calculated as suggested by EMA (2006), to assess the risk posed by antineoplastic drug residues in the environment. The PNEC value is given by the ratio between the NOEC from the long-term toxicity tests corrected by an assessment factor depending on the number of trophic levels, taxonomic groups, and feeding strategies. Invertebrates, along the trophic chain, hold a strategic role in the determination of the reproductive success because they constitute >90% of extant species (Jha 2008). In line with

Table 15.4 Chromatic table of the Long-term effects observed in producers and consumers for the antineoplastic drugs with EC50 values (mg/L) in different orders of magnitude concentration ranges in grayscale

		CP	IFO	MTX	5-FU	CAP	ET	DOX	CDDP	IM
Bacteria	*V. fischeri*		-	-	-	-	-	-	-	-
Cyanobacteria	*S. leopoliensis*									
Algae	*P. subcapitata*					-				
	D. subspicatus	-	-	-		-	-	-	-	-
Plants	*L. minor*					-	-	-	-	
Rotifers	*B. calyciflorus*			-						
Crustaceans	*D. magna*		-	-				-		
	C. dubia			-				-		

NOT EVALUATED -	> 100	10-100	1-10	0.1-1	0.01-0.1	0.001-0.01

the same author, the reproductive success is the most relevant and important ecotoxicological parameter since reproduction impairment at the whole organism level depends on the genotoxic, mutagenic, and carcinogenic potentialities of antineoplastic drugs, closely linked to the alterations in DNA.

Thus, for chemicals that challenge biological systems in various ways as do antineoplastic drugs, a concurrent evaluation of toxicity and genotoxicity in the same bioindicators may be of relevant environmental interest. Isidori et al. (2016b) observed an increase of developmental malformations in the South African clawed frog, *Xenopus laevis*, at concentrations of dozens of mg/L, testing 5-FU, ET, CAP, and IM. Moreover, in Gačić et al. 2014, Gačić et al. found DNA damage in mussels at dozens of μg/L. Unexpectedly, Parrella et al., in Parrella et al. 2015, identified DNA strand breaks in crustaceans at concentrations in the order of dozens and/or hundreds of ng/L for 5-FU, CDDP, DOX, and ET in *D. magna* and also for IM in *C. dubia* confirming that daphnids are extremely sensitive primary consumers in the aquatic food chain and in the detection of genotoxic alterations. These alarming findings indicate that EMA and FDA guideline trigger values for environmental risk assessment of drugs (0.01 μg/L and 1 μg/L, respectively) could underestimate possible risks related to the presence in aquatic systems of such hazardous active compounds as antineoplastic drugs. The consumption of antineoplastic drugs is expected to increase in the near future, and considering their resistance to WWTP removal, their concentration levels within the environment are most likely to rise. Therefore, a safe level for non-target organisms and humans is difficult to establish since genetic modification and cancer risk may exist at any level of exposure. Furthermore, a trigger limit may not be set for DNA-damaging drugs, since drugs are found in association with other parent drug residues, metabolites, biotic and abiotic transformation products, and other pollutants in authentic samples causing potential additive/synergistic effects. This is undoubtedly a harmful combination for the environment and for human health.

References

An X, Tiwari AK, Sun Y, Ding PR, Ashby CR Jr, Chen ZS (2010) BCR-ABL tyrosine kinase inhibitors in the treatment of Philadelphia chromosome positive chronic myeloid leukemia: a review. Leuk Res 34:1255–1268

ASTM E 1440-1491 (2004) Standard guide for acute toxicity with the rotifer Brachionus. American Society for Testing and Materials, Philadelphia. USA reapproved

Baldwin EL, Osheroff N (2005) Etoposide, topoisomerase II and cancer. Curr Med Chem Anti-cancer Agents 5:363–372

Baldwin WS, Marko PB, Nelson DR (2009) The cytochrome P450 (CYP) gene superfamily in Daphnia pulex. BMC Genomics 10:169

Besse JP, Latour JF, Garric J (2012) Anticancer drugs in surface waters. What can we say about the occurrence and environmental significance of cytotoxic, cytostatic and endocrine therapy drugs? Environ Int 39:73–86

Białk-Bielińska A, Mulkiewicz E, Stokowski M, Stolte S, Stepnowski P (2017) Acute aquatic toxicity assessment of six anti-cancer drugs and one metabolite using biotest battery – biological effects and stability under test conditions. Chemosphere 189:689–698

Booker V, Halsall C, Llewellyn N, Johnson A, Williams R (2014) Prioritising anticancer drugs for environmental monitoring and risk assessment purposes. Sci Total Environ 473–474:159–170

Brezovšek P, Eleršek T, Filipič M (2014) Toxicities of four anti-neoplastic drugs and their binary mixtures tested on the green alga Pseudokirchneriella subcapitata and the cyanobacterium Synechococcus leopoliensis. Water Res 52:168–177

Brooker V, Halsall C, Llewellyn N, Johnson A, Williams R (2014) Prioritising anticancer drugs for environmental monitoring and risk assessment purposes. Sci Total Environ 473–474:159–170

Burns EE, Thomas-Oates J, Kolpin DW, Furlong ET, Boxall ABA (2017) Are exposure predictions, used for the prioritization of pharmaceuticals in the environment, fit for purpose? Environ Toxicol Chem 36(10):2823–2832

Catastini C, Mullot JU, Boukari S, Mazellier P, Levi Y, Cervantes P, Ormsby JN (2008) Assessment of antineoplastic drugs in effluents of two hospitals [Identification de molecules anticancereuses dans les effluents hospitaliers]. J Eur Hydrol 39:171–180

Česen M, Kosjek T, Laimou-Geraniou M, Kompare B, Širok B, Lambropolou D, Heath E (2015) Occurrence of cyclophosphamide and ifosfamide in aqueous environment and their removal by biological and abiotic wastewater treatment processes. Sci Total Environ 527–528:465–473

Česen M, Kosjek T, Busetti F, Kompare B, Heath E (2016a) Human metabolites and transformation products of cyclophosphamide and ifosfamide: analysis, occurrence and formation during abiotic treatments. Environ Sci Pollut Res 23:11209–11223

Česen M, Eleršek T, Novak M, Žegura B, Kosjek T, Filipič M, Heath E (2016b) Ecotoxicity and genotoxicity of cyclophosphamide, ifosfamide, their metabolites/transformation products and their mixtures. Environ Pollut 210:192–201

Delgado LF, Faucet-Marquis V, Pfohl-Leszkowicz A, Dorandeu C, Marion B, Schetrite S, Albasi C (2011) Cytotoxicity micropollutant removal in a crossflow membrane bioreactor. Bioresour Technol 102:4395–4401

DIN 38412-37 (1991) German standard methods for the examination of water, wastewater and sludge – Bio-assays (group L) – Part 37: determination of the inhibitory effect of water on bacteria (Photobacterium phosphoreum) (L 37)

DIN 38412-L34 (1991) German standard methods for the examination of water, waste water and sludge - Bio-assays (group L) – Part 34: determination of the inhibitory effect of waste water on the light emission of photobacterium phosphoreum; luminescent bacteria waste water test using conserved bacteria (L 34)

EMA (2006) Guideline on the environmental risk assessment of medicinal products for human use. London, 01 June 2006, Ref. EMEA/CHMP/SWP/4447/00

Ferrando-Climent L, Rodriguez-Mozaz S, Barceló D (2014) Incidence of anticancer drugs in an aquatic urban system: from hospital effluents through urban wastewater to natural environment. Environ Pollut 193:216–223

Gačić Z, Kolarević S, Sunjog K, Kračun-Kolarević M, Paunović M, Knežević-Vukčević J, Vuković-Gačić B (2014) The impact of in vivo and in vitro exposure to base analogue 5-FU on the level of DNA damage in haemocytes of freshwater mussels Unio pictorum and Unio tumidus. Environ Pollut 191:145–150

Grung M, Källqvist T, Sakshaug S, Skurtveit S, Thomas KV (2008) Environmental assessment of Norwegian priority pharmaceuticals based on the EMEA guideline. Ecotoxicol Environ Saf 71 (2):328–340

Han Y, Zhang JP, Qian JQ, Hu CQ (2015) Cardiotoxicity evaluation of anthracyclines in zebrafish (Danio rerio). J Appl Toxicol 35:241–252

Isidori M, Lavorgna M, Russo C, Kundi M, Žegura B, Novak M, Filipič M, Mišík M, Knasmueller S, Lopez de Alda M, Barceló D, Žonja B, Česen M, Ščančar J, Kosjek T, Heath E (2016a) Chemical and toxicological characterisation of anticancer drugs in hospital and municipal wastewaters from Slovenia and Spain. Environ Pollut 219:275–287

Isidori M, Piscitelli C, Russo C, Smutná M, Bláha L (2016b) Teratogenic effects of five anticancer drugs on Xenopus laevis embryos. Ecotoxicol Environ Saf 133:90–96

ISO 10706 (2000) Water quality-determination of long-term toxicity of substances to Daphnia magna Straus (Cladocera, Crustacea). International Organization for Standardization, Geneva

ISO 10712 (1995) Water quality – Pseudomonas putida growth inhibition test (Pseudomonas cell multiplication inhibition test) International Organization for Standardization. Switzerland, Geneva

ISO 14380 (2011) Water quality-determination of the acute toxicity to Thamnocephalus Platyurus (Crustacea, Anostraca). International Organization for Standardization, Geneva

ISO 20665 (2008) Water quality-determination of chronic toxicity to Ceriodaphnia Dubia in 7 days-population growth inhibition test. International Organization for Standardization, Geneva

ISO 20666 (2008) Water quality-determination of chronic toxicity to Brachionus Calyciflorus in 48 H-population growth inhibition test. International Organization for Standardization, Geneva

ISO 6341 (1996) Water quality-determination of the inhibition of the mobility of Daphnia magna Straus (Cladocera, Crustacea)-Acute toxicity test. International Organization for Standardization, Geneva

Jha AN (2008) Ecotoxicological applications and significance of the comet assay. Mutagenesis 23:207–221

Köhler C, Venditti S, Igos E, Klepiszewski K, Benetto E, Cornelissen A (2012) Elimination of pharmaceutical residues in biologically pretreated hospital wastewater using advanced UV irradiation technology: a comparative assessment. J Hazard Mater 239–240:70–77

Kosjek T, Heath E (2011) Occurrence, fate and determination of cytostatic pharmaceuticals in the environment. Trends Anal Chem 30:1065–1087

Kosjek T, Negreira N, Heath E, de Alda ML, Barceló D (2016) Biodegradability of the anticancer drug etoposide and identification of the transformation products. Environ Sci Pollut Res Int 23:14706–14717

Kovács R, Bakos K, Urbányi B, Kövesi J, Gazsi G, Csepeli A, Appl AJ, Bencsik D, Csenki Z, Horvát A (2016) Acute and sub-chronic toxicity of four cytostatic drugs in zebrafish. Environ Sci Pollut Res 23:14718–14729

Kovalova L, Siegrist H, Singer H, Wittmer A, McArdell CS (2012) Hospital wastewater treatment by membrane bioreactor: performance and efficiency for organic micropollutant elimination. Environ Sci Technol 46:1536–1545

Kümmerer K, Haiß A, Schuster A, Hein A, Ebert I (2016) Antineoplastic compounds in the environment-substances of special concern. Environ Sci Pollut Res Int 23:14791–14804

Lenz K, Mahnik SN, Weissenbacher N, Mader RM, Krenn P, Hann S, Koellensperger G, Uhl M, Knasmueller S, Ferk F, Bursh W, Fuerhacker M (2007) Monitoring, removal and risk assessment of cytostatic drugs in hospital wastewater. Water Sci Technol 56:141–149

Lutterbeck CA, Kern DI, Machado EL, Kümmerer K (2014) Ecotoxicological evaluation of selected anticancer drugs. Toxicol Lett 229(S):71–72

Lutterbeck CA, Machado ÊL, Kümmerer K (2015a) Photodegradation of the antineoplastic cyclophosphamide: a comparative study of the efficiencies of UV/H2O2, UV/Fe2+/H2O2 and UV/TiO2 processes. Chemosphere 120:538–546

Lutterbeck CA, Baginska E, Machado ÊL, Kümmerer K (2015b) Removal of the anti-cancer drug methotrexate from water by advanced oxidation processes: aerobic biodegradation and toxicity studies after treatment. Chemosphere 141:290–296

Mahnik SN, Lenz K, Weissenbacher N, Mader RM, Fuerhacker M (2007) Fate of 5-fluorouracil, doxorubicin, epirubicin, and daunorubicin in hospital wastewater and their elimination by activated sludge and treatment in a membrane-bio-reactor system. Chemosphere 65:30–37

Nageswari A, Reddy KV, Mukkanti K (2012) Stability-indicating UPLC method for determination of Imatinib Mesylate and their degradation products in active pharmaceutical ingredient and pharmaceutical dosage forms. J Pharm Biomed Anal 66:109–115

Negreira N, de Alda LM, Barceló D (2014) Cytostatic drugs and metabolites in municipal and hospital wastewaters in Spain: filtration, occurrence, and environmental risk. Sci Total Environ 1:68–77

OECD (2006) Fish embryo toxicity (FET) test, OECD guideline for the testing of chemicals, draft proposal for a new guideline. OECD Publishing, Paris

OECD 201 (2011) Guideline for the testing of chemicals 201: freshwater alga and cyanobacteria, growth inhibition test. Organization for Economic Cooperation and Development, Paris

OECD 202 (2004) Daphnia sp., acute immobilization test. Organization for Economic Co-operation and Development, Paris

OECD 203 (1992) Fish, acute toxicity test, OECD guidelines for the testing of chemicals, section 2. OECD Publishing, Paris. https://doi.org/10.1787/9789264069961-en

OECD 211 (2008) Daphnia magna reproduction test. Organization for Economic Cooperation and Development, Paris

OECD 221 (2006) Guideline for the testing of chemicals 221: Lemna sp. Growth inhibition test. Organization for Economic Cooperation and Development, Paris

Olalla A, Negreira N, López de Alda M, Barceló D, Valcárcel Y (2018) A case study to identify priority cytostatic contaminants in hospital effluents. Chemosphere 190:417–430

Parrella A, Lavorgna M, Criscuolo E, Russo C, Fiumano V, Isidori M (2014) Acute and chronic toxicity of six anticancer drugs on rotifers and crustaceans. Chemosphere 115:59–66

Parrella A, Lavorgna M, Criscuolo E, Russo C, Isidori M (2015) Eco-genotoxicity of six anticancer drugs using comet assay in daphnids. J Hazard Mater 286:573–580

Russo C, Lavorgna M, Česen M, Kosjek T, Heath E, Isidori M (2018) Evaluation of acute and chronic ecotoxicity of cyclophosphamide, ifosfamide, their metabolites/transformation products and UV treated samples. Environ Pollut 233:356–363

Santos MSF, Franquet-Griell H, Lacorte S, Madeira LM, Alves A (2017) Anticancer drugs in Portuguese surface waters – estimation of concentrations and identification of potentially priority drugs. Chemosphere 184:1250–1260

Straub JO (2010) Combined environmental risk assessment for 5-fluorouracil and Capecitabine in Europe. Integr Environ Assess Manag 6:540–566

Suhail N, Bilal N, Khan HY, Hasan S, Sharma S, Khan F, Mansoor T, Banu N (2012) Effect of vitamins C and E on antioxidant status of breast-cancer patients undergoing chemotherapy. J Clin Pharm Ther 37:22–26

US EPA (1993) Methods for measuring the acute toxicity of effluents and receiving waters to freshwater and marine organisms, 4th edn. EPA-600-4-90. Washington, DC, US Environmental Protection Agency

Vyas N, Turner A, Sewell G (2014) Platinum-based anticancer drugs in waste waters of a major UK hospital and predicted concentrations in recipient surface waters. Sci Total Environ 493:324–329

Xie H (2012) Occurrence, ecotoxicology, and treatment of anticancer agents as water contaminants. J Environ Anal Toxicol S2:002. https://doi.org/10.4172/2161-0525

Yin J, Shao B, Zhang J, Li K (2010) A preliminary study on the occurrence of cytostatic drugs in hospital effluents in Beijing, China. Bull Environ Contam Toxicol 84:39–45
Załeska-Radziwiłł M, Łebkowska M, Affek K, Chrzanowska N (2011) Environmental risk assessment of selected pharmaceuticals present in surface waters in relation to cyanobacteria and plants. Ochr Środowiska i Zasobów Nat 48:372–382
Zhang J, Chang VWC, Giannis A, Wang JY (2013) Removal of cytostatic drugs from aquatic environment: a review. Sci Total Environ 445–446:281–298
Zounková R, Odráska P, Dolezalová L, Hilscherová K, Marsálek B, Bláha L (2007) Ecotoxicity and genotoxicity assessment of cytostatic pharmaceuticals. Environ Toxicol Chem 26:2208–2214
Zounková R, Kovalova L, Blaha L, Dott W (2010) Ecotoxicity and genotoxicity assessment of cytotoxic antineoplastic drugs and their metabolites. Chemosphere 81:253–260

Chapter 16
Genotoxicity of the Residues of Anticancer Drugs: A Hazard for Aquatic Environment

Metka Filipič, Matjaž Novak, and Bojana Žegura

Abstract Anticancer drugs are a group of pharmaceuticals that are used in cancer treatment. These drugs have high pharmacological potency and are designed to kill tumour cells or to prevent and disrupt tumour cell division by interfering with genetic material or processes that govern their replication. However, anticancer drugs do not affect only cancer cells but also dividing normal cells. After human consumption, anticancer drug residues are released into the environment as parent compounds and their metabolites, where they might affect non-target environmental organisms even at the level of sub- to few ng/L in particular during chronic exposure. Recent ecotoxicological studies that included also detection of genotoxic effects of selected anticancer drugs with different mechanisms of chemotherapeutic action demonstrated high differences in the sensitivity of different aquatic organisms in regard to lethal and reproductive effects. However, in all organisms, the concentrations at which mortality and reproductive effects were observed were higher than the concentrations that were detected or expected in the environmental samples. On the contrary, the genotoxic effects of certain anticancer drugs were in crustacean and fish detected at concentrations that may occur in the aquatic environment. Thus, potential ecological risks for invertebrates and vertebrates cannot be ruled out. The results clearly demonstrated that residues of certain anticancer drugs are hazardous for aquatic environment; thus, further research and activities are needed that will enable reliable environmental risk assessment and introduction of measures to reduce their release into the environment.

Keywords Aquatic organisms · Anticancer drugs · Genotoxic · Ecotoxic · Environmental risk

---

M. Filipič (✉) · M. Novak · B. Žegura
Department of Genetic Toxicology and Cancer Biology, National Institute of Biology, Ljubljana, Slovenia
e-mail: metka.filipic@nib.si

E. Heath et al. (eds.), *Fate and Effects of Anticancer Drugs in the Environment*,
https://doi.org/10.1007/978-3-030-21048-9_16

16.1 Introduction

Pharmaceuticals are a broad group of chemicals with different mechanisms of action used to diagnose, treat, cure or even prevent various diseases in humans and animals. These biologically active substances have brought great benefits to humanity in terms of healthier and longer lives and improved the quality of life for human beings and animals. Their overall consumption is substantial, and it is increasing globally mainly due to population growth, aging and expiration of patents, leading to less expensive and more available generics (Daughton 2002; Kummerer 2009). Scientific as well as public concern regarding the occurrence of residues of pharmaceuticals as environmental micropollutants has been increasing during the last two decades, as their presence has been demonstrated across the aquatic and terrestrial environment, where they have been consistently detected (Fent et al. 2006; Hughes et al. 2013; Rodriguez-Mozaz and Weinberg 2010). After the consumption, pharmaceuticals are excreted through faeces and urine as mixtures of metabolites and the unchanged parent compounds. Predominantly, these compounds then enter the aquatic environment via effluents from hospital and municipal wastewater treatment plants and landfill leakages and, to a minor extent, in the discharge from the pharmaceutical industry. Due to their ubiquitous presence in the environment that arises from their continual input into the aquatic compartment, they are considered as 'pseudo'-- persistent pollutants (Hernando et al. 2006).

According to the purpose of use and biological activity, different groups of pharmaceuticals can be classified such as antibiotics, analgetics, anticancer drugs, anti-inflammatory drugs, oestrogens and hormonal drugs, antiepileptics, cardiovascular pharmaceuticals, etc. (Kummerer 2009). Since 3000–10,000 different pharmaceutical preparations are currently in use (Boxall et al. 2012; Dong et al. 2013), it is impossible to experimentally assess the hazards and risks of all of these compounds in a timely manner. Therefore, different prioritization approaches have been used to identify those groups that are likely to pose risk for the environment and need special attention when dealing with environmental risk (Boxall et al. 2012; Kuster and Adler 2014). One such group are anticancer drugs (ACDs), which have been until recently overlooked in the environmental risk assessments, mainly due to their low consumption amounts, low predicted environmental concentrations and scarce information regarding their occurrence in the environment (Besse et al. 2012; Kosjek and Heath 2011; Kummerer et al. 2014). However, many of these drugs are classified as mutagenic, genotoxic, carcinogenic, embryotoxic and/or teratogenic, and it has been postulated that they can cause adverse effects in aquatic ecosystem also at low concentrations (Besse et al. 2012; Kosjek and Heath 2011; Kummerer et al. 2014; Toolaram et al. 2014).

The concentrations of the residues of anticancer drugs in the aquatic environment are compared to the concentrations of the residues of many other pharmaceuticals very low, at the level of pg/L to ng/L. Higher concentrations up to μg/L are detected in hospital and municipal wastewaters (Kosjek and Heath 2011; Kummerer et al. 2014). In a recent paper, Toolaram et al. (2014) gave a comprehensive review on genotoxicity characterization of ACDs and their transformation products focusing on their potential risk for the environment. The authors concluded that from an environmental perspective, there is a lack of studies assessing genotoxicity of these drugs in the aquatic organisms. In this chapter, we present the new findings on potential adverse effects of the residues of ACDs to aquatic organisms and discuss the relevance of these findings for the environmental risk assessment of ACDs.

16.2 Mechanisms of Therapeutic Action of Anticancer Drugs

Anticancer drugs (ACDs) are mainly used in chemotherapy to fight cancer, a disease involving uncontrolled multiplication of the body's own cells and spreading abnormal forms within the body. The Anatomical Therapeutic Chemical (ATC) Classification System (http://www.whocc.no/atc_ddd_index/) classifies anticancer drugs according to their chemical structures and therapeutic properties as Class L – Antineoplastic and Immunomodulating Agents – with more than 260 drugs. According to their mechanism of chemotherapeutic action, the antineoplastic agents can be further classified into alkylating agents, antimetabolites, topoisomerase inhibitors, mitotic spindle inhibitors and protein kinase inhibitors (Table 16.1). Although the mechanisms of chemotherapeutic action of the different classes are different, most of ACDs are designed to interact directly or indirectly with DNA causing DNA damage and/or inhibit DNA synthesis resulting in inhibition of cancer cell proliferation and cancer cell death. However, most of these mechanisms of action are not very specific, and ACDs affect also normal cells, in particular those rapidly dividing. Therefore, once the residues of ACDs are present in the environment, they might affect non-target environmental organisms.

16.3 Ecotoxicity of Anticancer Drugs

Ecotoxicity of ACDs is poorly investigated. The data are available mainly for those ACDs that are consumed in high amounts or have been present on the market for many years (Booker et al. 2014; Kummerer et al. 2014). Moreover, in most cases, only acute ecotoxicity data are available, which however do not allow for prediction of the adverse effects of life-cycle exposure to these compounds in aquatic organisms.

Within the recently completed FP7 project, CytoThreat, the ecotoxicity of an antimetabolite 5-fluorouracil (5-FU), an alkylating agent cisplatin (CDDP), a topoisomerase inhibitor etoposide (ET) and a tyrosine kinase inhibitor imatinib mesylate (IM) has been evaluated. The four compounds with different mechanisms of chemotherapeutic action also differ in their inherent genotoxic properties (Table 16.2).

Table 16.1 Classification of anticancer drugs according to their mechanisms of chemotherapeutic action

Class	Representatives	Mechanism of therapeutic action	Major clinical uses
Alkylating agents	Nitrogen mustard analogues: Cyclophosphamide, ifosfamide Platinum based: Cisplatin, carboplatin	Formation of DNA adducts and DNA alkylation of nucleic bases within the same or complementary DNA strands forming intrastrand and interstrand cross-links that interfere with DNA replication and transcription (Fleming 1997; Kelland 2007)	Lymphoma, breast and ovarian cancer, small cell lung cancer, testicular cancer, head and neck cancer, ovarian, multiple myeloma
Antimetabolites	Folic acid analogues: Methotrexate	Inhibition key folate-dependent enzymes resulting in nucleotide biosynthesis inhibition and consequently inhibition of cell proliferation and further damage that lead to cell death (Hagner and Joerger 2010)	Breast cancer, head and neck cancer, leukaemia, Burkitt lymphoma, lung carcinoma, osteosarcoma, bladder cancer
	Purine and pyrimidine analogues: Thioguanine, fludarabine, 5-fluorouracil, capecitabine, gemcitabine, cytarabine	Inhibition of thymidylate synthase that leads to imbalance of deoxynucleotides that disrupt DNA repair and synthesis, resulting in lethal DNA damage Misincorporation of analogues into DNA and RNA during the DNA repair eventually leads to DNA strand breaks and cell death and disruption of protein synthesis, respectively (Longley et al. 2003)	Breast, ovarian, cervical, bladder and prostate cancer; gastrointestinal adenocarcinomas; metastatic pancreatic cancer; small cell lung cancer
Topoisomerase inhibitors	Topoisomerase I inhibitors: Topotecan and irinotecan Topoisomerase II inhibitors: Etoposide, amascarine, teniposide	Interference with normal functions of DNA topoisomerases I and II that are involved in controlling, modifying and maintaining structures and topology of DNA, leading to an increase in DNA and chromosomal breakage and cell death (Pommier 2013)	Ovarian germ cell cancers, testicular cancer, small cell lung cancer, acute leukaemia

(continued)

Table 16.1 (continued)

Class	Representatives	Mechanism of therapeutic action	Major clinical uses
Mitotic spindle inhibitors	Vincristine, vinblastine, paclitaxel	Bind to tubulin dimers causing dissolution, disassembly or polymerization of the microtubules, which leads to the destruction of the mitotic spindle and arrest of the cell cycle in metaphase (Jackson et al. 1996)	Breast, ovarian, lung, head and neck tumours; leukaemia; paediatric tumours; neuroblastoma
Tyrosine kinase inhibitors	Imatinib, nilotinib, lapatinib	Designed to inhibit catalytic activity of the aberrantly expressed tyrosine kinase by interfering with the ATP, resulting in the inhibition of proliferation and induction of apoptosis (Arora and Scholar 2005)	Targeted cancer treatment: Chronic myeloid leukaemia, colorectal cancer

Table 16.2 Inherent genotoxic properties of 5-fluorouracil, cisplatin, etoposide and imatinib mesylate

	Genotoxicity			
Genotoxicity assay	5-Fluorouracil	Cisplatin	Etoposide	Imatinib mesylate
Bacterial reverse mutation assay	Positive	Positive	Negative	Negative
In vitro mammalian cell mutation assay	Positive	Positive	Positive	Negative
In vitro mammalian cell clastogenicity assay	Positive	Positive	Positive	Positive/negative
In vivo rodent clastogenicity assay	Positive	Positive	Positive	Negative
Carcinogenicity (IARC)	Group 3	Group 2A	Group 1	Not classified

In the standard genotoxicity assays, 5-FU, CDDP and ET are genotoxic in vitro and in vivo. IM is not genotoxic in vivo, while the in vitro results are inconclusive.

The acute and chronic ecotoxicity of the four drugs was studied in algae, crustacean and zebrafish using the standardized assays where available (Table 16.3). In addition to standard lethal and sublethal toxic effects, the specific genotoxicity endpoints such as DNA strand breaks and micronuclei induction were evaluated as indicators of adverse long-term effects associated with damage of the genetic material. DNA damage was determined with the comet assay, which is a sensitive method for the detection of single-strand (and double-strand) breaks, alkali-labile sites, DNA-DNA/DNA-protein cross-linking and single-strand breaks associated with incomplete excision repair (Tice et al. 2000). The micronucleus assay was applied to determine chromosomal damage. Micronuclei are formed during cell

Table 16.3 Ecotoxicity test systems used for the determination of acute and chronic toxicity and genotoxicity of ACDs towards algae, crustacean and fish

	Alga: *P. subcapitata* Cyanobacteria: *S. leopoliensis*	Crustacea: *D. magna and C. dubia*	Fish: *Danio rerio* (zebrafish)
Acute toxicity	Freshwater alga and cyanobacteria, growth inhibition test (OECD TG 201 2011)	*D. magna*: Acute immobilisation test (OECD TG 202 2004):	Fish, acute toxicity test: Limit test (OECD TG 203 1992)
		C. dubia: Acute mortality assay (EPA-600-4-90/027F 1993)	Fish embryo acute toxicity (FET) test (OECD TG 236 2013)
Chronic toxicity		Reproduction inhibition	Fish, early-life stage toxicity test (OECD TG 210 2013)
		D. magna: (OECD TG 211 2008);	
		C. dubia (ISO 2008)	Fish two-generation toxicity test
Genotoxicity		Comet assay	Comet assay, micronucleus assay
Gene expression analyses			Gene expression analyses: microarrays, QRT-PCR

division and can occur at different times after DNA damage, depending on the cell-cycle kinetics and the mechanisms of induction (Bolognesi and Hayashi 2011). Both of these assays have broad applicability in aquatic animals, and they have been applied in numerous laboratory exposure and environmental monitoring studies (Bolognesi and Hayashi 2011; Frenzilli et al. 2009).

In primary producers, alga *Pseudokirchneriella subcapitata* and cyanobacterium *Synechococcus leopoliensis*, the four drugs showed different toxic potential, and the two species examined showed differences in their susceptibilities towards the tested drugs (Table 16.4). In *P. subcapitata*, the most toxic of these drugs was 5-FU, followed by CDDP, IM and ET. In *S. leopoliensis*, the most toxic was CDDP, followed by 5-FU and IM, while ET was not toxic at concentrations up to 351 mg/L (Brezovšek et al. 2014). Although these assays are generally considered as acute toxicity assays, they are in principle multigenerational assays and can be considered as chronic toxicity tests. According to the EU Technical Guidance Document (European Commission 2003), when the exposure is 72 h (or longer), the EC_{50} values can be considered as parameter of acute toxicity, and the no observed effect concentration (NOEC) values can be considered as chronic exposure parameter.

The concentrations at which growth inhibition was observed are rather high and not relevant for environmental contamination except for 5-FU, which caused growth inhibition at concentrations that have been detected in hospital wastewaters (Mahnik et al. 2004) and in surface water in Taiwan (Lin et al. 2014) (Table 16.7).

Acute and chronic toxicity of the 5-FU, CDDP, ET, IM and capecitabine (CAP) and doxorubicin (DOX) has been tested in primary consumers of the aquatic chain

Table 16.4 Acute toxicity of 5-fluorouracil, etoposide, cisplatin and imatinib mesylate towards algae, crustacean and fish

Test species	Toxicological endpoint	Exposure duration	5-FU	ET	CDDP	IM	References
			EC_{50} (mg/L)[a]	EC_{50} (mg/L)[a]	EC_{50} (mg/L)[a]	EC_{50} (mg/L)[a]	
P. subcapitata	Growth inhibition	72 h	0.13 (0.07–0.22)	30.4 (25.4–36.4)	1.52 (1.26–1.83)	2.29 (1.88–2.77)	Brezovšek et al. (2014)
S. leopoliensis	Growth inhibition	72 h	1.20 (0.72–2.02)	Nd	0.67 (0.34–1.34)	5.36 (4.29–6.70)	Brezovšek et al. (2014)
C. dubia	Mortality	24 h	501 (351–854)	Nd	2.50 (2.13–2.97)	31.9 (27.6–36.9)	Parrella et al. (2014)
D. magna	Immobilization	48 h	20.8 (18.1–24.0)	Nd	0.94 (0.90–0.97)	11.9 (9.3–15.4)	Parrella et al. (2014)
D. rerio FET test	Mortality	120 h	2222 (2006–2462)	>300	81.3 (72.9–90.9)	65.9 (59.9–71.6)	Kovacs et al. (2016)
D. rerio adult fish	Mortality	96 h	>100	>100	64.5 (59.1–69.2)	70.8 (63.7–81.1)	Kovacs et al. (2016)

[a]EC_{50} half maximal effective concentration; nd not determinable

Daphnia magna, *Ceriodaphnia dubia*, *Brachionus calyciflorus* and *Thamnocephalus platyurus*. Acute ecotoxicological effects occurred at concentrations in the range of mg/L (Table 16.4), whereas the growth inhibition after chronic exposure was observed in the range of μg/L (Parrella et al. 2014) (Table 16.5). These concentrations are for 5-FU, CDDP and ET in the range of concentrations that were detected in hospital and municipal effluents (Table 16.7). The genotoxicity of the six antineoplastic drugs was studied, applying the in vivo comet assay on cells from whole organisms of *Daphnia magna* and *Ceriodaphnia dubia* (Parrella et al. 2015). The results showed that all tested drugs induced DNA damage, in both organisms (Table 16.6). It is notable that DNA damage determined after 24 h of exposure was observed at concentrations that were orders of magnitude lower than the concentrations at which inhibition of reproduction was observed. Parrella et al. (2015) concluded that the evaluation of DNA damage after 24-h exposure in the whole organisms of crustacean could be considered an early biomarker of the effect on survival and/or reproductive toxicity representing a useful tool for environmental monitoring and risk assessment of anticancer drugs.

In zebrafish (*Danio rerio*), the acute and sub-chronic toxicity of the four drugs was low (Tables 16.4 and 16.5). In adult fish, 5-FU and ET were not toxic, while the EC_{50} values for CDDP and IM were 64.5 and 70.8 mg/L, respectively. Also in the zebrafish embryo toxicity test, the EC_{50} values were in the range of tens to hundreds mg/L. The sub-chronic toxicity of 5-FU and IM was determined in the fish early life stage toxicity test. Significant increase in mortality by both drugs was observed at concentrations $\geq$10 mg/L (Table 16.4). 5-FU also affected body weight and length that was significantly reduced at concentrations $\geq$1 mg/L (Kovacs et al. 2016).

However, none of these assays with zebrafish is appropriate to detect long-term and delayed effects that may be associated with the genotoxic activity of ACDs. Thus, 5-FU was tested in a two-generation study, which has not been performed before with any ACD (Kovacs et al. 2015). The concentration ranges used were 0.01, 1 and 100 μg/L. In addition to standard toxicological endpoints, such as the survival, growth and reproduction, the induction of DNA damage and micronucleus formation were determined as genotoxicity endpoints. The DNA damage was determined in gills, liver, kidneys, gonads and blood cells using the comet assay. Complementary to the comet assay, the micronucleus assay was applied to determine chromosomal damage in erythrocytes. Both of these assays have broad applicability in aquatic animals, and they have been applied in numerous laboratory exposure and environmental monitoring studies. Furthermore, whole genome gene expression profiling was performed on liver samples from F1 generation. Changes in gene expression can either be related to adaptive processes or can be used as indicators of toxic effects; they can indicate harmful impacts of chemicals in cases where classical toxicological endpoints show no obvious adverse effects.

The study showed that the exposure to 5-FU did not affect survival, growth and reproduction of the zebrafish; however, histopathological changes were observed in the liver and kidney, along with genotoxic effects, at all 5-FU concentrations (Table 16.5). Increases in DNA damage determined with the comet assay were significant in the liver and blood cells, but not in the gills and gonads. In

Table 16.5 Chronic toxicity of 5-fluorouracil, etoposide, cisplatin and imatinib mesylate towards algae, crustacean and fish

Test	Toxicological endpoint	Exposure duration	5-FU LOEC (μg/L)[a]	ET LOEC (μg/L)[a]	CDDP LOEC (μg/L)[a]	IM LOEC (μg/L)[a]	References
P. subcapitata	Growth inhibition	72 h	20	34,260	980	1190	Brezovšek et al. (2014)
S. leopoliensis	Growth inhibition	72 h	390	Nd.	310	4000	Brezovšek et al. (2014)
C. dubia	Reproduction inhibition	7 days	6.67	312.5	14,7	0.87	Parrella et al. (2014)
D. magna	Reproduction inhibition	21 days	6.17	333.3	3	9.54	Parrella et al. (2014)
D. rerio early life stage	Mortality	33 days	10,000	–	–	10,000	Kovacs et al. (2016)
	Body parameters		1000	–	–	Not affected	
D. rerio 2 generation exposure	Mortality, reproduction	7 months	> 100	–	–	> 100	Kovacs et al. (2015)
	Histopathological changes (liver, kidney)	7 months	0.01	–	–	Not affected	

[a]LOEC lowest observed effect concentration; n.d. not determinable; – not tested

Table 16.6 Genotoxicity of 5-fluorouracil, etoposide, cisplatin and imatinib mesylate towards algae, crustacean and fish

Test species	Toxicological endpoint	Exposure duration	5-FU	ET	CDDP	IM	References
			LOAEC (μg/L)[a]	LOAEC (μg/L)[a]	LOAEC (μg/L)[a]	LOAEC (μg/L)[a]	
C. dubia	DNA damage (comet assay)	24 h	0.06	0.1	0.3	0.3	Parrella et al. (2015)
D. magna	DNA damage (comet assay)	24 h	0.5	0.3	0.01	2	Parrella et al. (2015)
D. rerio	DNA damage (comet assay)	7 months	100	–	–	–	Kovacs et al. (2015)
	Micronucleus assay	7 months	0.01	–	–	–	Kovacs et al. (2015)

[a]LOAEC lowest observed adverse effect concentration; – not tested

Table 16.7 Predicted and measured environmental concentrations of 5-fluorouracil, etoposide, cisplatin and imatinib mesylate

Anticancer drug	PEC (ng/L)[a]	Matrix	Concentration (ng/L)	References
5-fluorouracil	7.91	Surface water	5–160	Lin et al. (2014)
	–	Hospital effluent	< 5000–124,000	Mahnik et al. (2004)
		Municipal water effluent	< 1.6–17	Isidori et al. (2016)
Cisplatin	0.54	Surface water	< LOQ	Isidori et al. (2016)
		Hospital effluent	10–50	Fuerhacker et al. (2006)
		Municipal water effluent	3–25	Negreria et al. (2014)
Etoposide	0.87	Surface water	–	
		Hospital effluent	110–600	Catastini et al. (2008)
		Municipal plant effluent	3–15	Martin et al. (2011)
Imatinib	4.99	Surface water	–	
		Hospital effluent	75–577	Olalla et al. (2018)
		Municipal plant effluent	< 36	Negreira et al. (2013)

[a]PEC predicted environmental concentration refined by excretion rates estimated by Besse et al. (2012)

erythrocytes, a significant, dose-dependent increase in the frequency of micronuclei was observed at all 5-FU concentrations (Table 16.6). Whole genome transcriptomic analysis of liver samples of F1 generation zebrafish exposed to 0.01 μg/L and 1 μg/L 5-FU revealed dose-dependent increases in the number of differentially expressed genes, including up-regulation of several DNA-damage-responsive genes and oncogenes (i.e. jun, myca) (Kovacs et al. 2015). Although chronic exposure to environmentally relevant concentrations of 5-FU did not affect the reproduction of the exposed zebrafish, it cannot be excluded that 5-FU can lead to degenerative changes, including cancers, which over long-term exposure of several generations might affect fish populations.

16.4 Occurrence of the Residues of Anticancer Drugs in the Environment

Since ACDs are not completely removed in conventional wastewater treatment plants, they may remain in the dissolved phase in the aqueous effluents and leave wastewater treatment plants only in partly reduced or even in unchanged concentrations (Kosjek and Heath 2011; Zhang et al. 2013). Thus, it is considered that effluents from wastewater treatment plants are the most important direct source of ACDs into the aquatic environment. However, for the majority of ACDs, the environmental burden is unknown, mainly because they have been ignored due to their low concentrations in the environment and the lack of analytical techniques sensitive enough to detect such low concentrations in complex environmental matrices.

During the last years, new analytical methods are being developed; however, still most of the studies are focused on the detection of the residues of ACD in wastewaters, while the data on the concentrations of ACDs in surface waters is rather limited. The most frequently detected ACDs in surface waters are CP and IF, ranging from sub-ng/L up to 20 ng/L (Buerge et al. 2006; Ferrando-Climent et al. 2014; Franquet-Griell et al. 2017; Zuccato et al. 2000). Also tamoxifen (up to 38 ng/L), cytarabine (13 ng/L), ciprofloxacin (up to 102 ng/L) and bleomycin (17 ng/L) have been found in aquatic environment at relatively high concentrations (> 10 ng/L) (Aherne et al. 1990; Ferrando-Climent et al. 2014; Martin et al. 2011). In Thailand, in Chao Phraya River, CP was detected at 1907 ng/L, hydroxycarbamide at 788 ng/L and 5-FU at 578 ng/L (Usawanuwat et al. 2014), whereas in rivers in Taiwan, Lin et al. (2014) reported concentrations of CP up to 96 ng/L and 5-FU up to 160 ng/L.

Since the data regarding the occurrence of ACDs in surface waters is limited, for the purposes of environmental risk assessment, their predicted environmental concentrations (PECs) are often used. PEC values are calculated based on the consumption data for ADCs in a given population and present the expected fraction of the drug in surface waters once discharged and diluted. However, PECs usually do not take into account local specificities and possible hot spots and thus provide only rough estimation on national or regional scale (Zhang et al. 2013). Table 16.7 summarizes PEC values for 5-FU, CDDP, ET and IM published by Besse et al. (2012) and recently published data on their detected concentrations in hospital and municipal wastewater and in surface waters. For all compounds, it can be seen that higher concentrations are present in hospital wastewaters than in the municipal ones. In surface waters, only 5-FU and cisplatin were detected, while no data on the eventual presence of ET and IM is reported.

16.5 Environmental Risk Assessment

For the environment risk assessment of pharmaceuticals, the European Medicines Agency (EMA 2006) suggests to use the risk quotient (RQ) that is defined as the ratio between predicted environmental concentration (PEC) and predicted no effect concentrations (PNEC). The PNEC value is determined based on the lowest no effect concentration (NOEC) result from the base set of long-term toxicity tests corrected by the assessment factor (AF) 10 that takes into account uncertainties due to possible interspecies and intraspecies differences in susceptibility. The RQ above 1 indicate potential risk from the compound towards the environment. In the presented studies, the most sensitive organisms were crustacean; thus, NOEC values for reproduction

Table 16.8 RQ for 5-fluorouracil, etoposide, cisplatin and imatinib mesylate calculated on the basis of the NOEC/NOAEC results obtained in the most sensitive species in chronic exposure toxicity and genotoxicity studies

Compound	PEC[a] (ng/L)	Test system with the lowest NOEC[b]/NOAEC[c]	NOEC/ NOAEC (ng/L)	PNEC[d] (ng/L)	RQ
5-FU	7.91	Reproduction inhibition *D. magna*	2060	206	0.04
		Comet assay *C. dubia*	6	0.6	**13.2**
		Micronucleus *D. rerio*	10	1	**7.9**
ET	0.87	Reproduction inhibition *C. dubia*	97,600	9760	0.00009
		Comet assay *C. dubia*	10	1	0.87
CDDP	0.54	Reproduction inhibition *D. magna*	1000	100	0.005
		Comet assay *D. magna*	1	0.1	**5.4**
IM	5.00	Reproduction inhibition *C. dubia*	270	27	0.19
		Comet assay *C. dubia*	30	3	**1.7**

[a]PEC (predicted environmental concentration) values are obtained from Besse et al. (2012)
[b]NOEC (no effect concentration) values for reproduction inhibition are from Parrella et al. (2014)
[c]NOAEC (no adverse effect concentration) values for genotoxicity for crustacean are from Parrella et al. (2015) and for zebrafish from Kovacs et al. (2015)
[d]PNEC (predicted no effect concentration): NOEC or NOAEC/AF (assessment factor (AF) is 10) (EMA 2006)

inhibition were used for calculating PNEC. It can be seen in Table 16.8 that the RQ for all four drugs is <1. According to EMA (2006), it can be concluded that the residues of these drugs are unlikely to represent the risk for the environment. However, the amounts of FU, CDDP and ET in the hospital waste waters may be 2–3 orders of magnitude higher than the PEC. In some cases (i.e. 5-FU), they reach the PNEC and even LOEC levels. Thus, locally near hospital effluents, it cannot be excluded that these residues may affect reproduction of aquatic organisms.

It is notable that DNA damage (comet assay) in crustacean caused by short-term treatment for 24 hrs and micronucleus assay in peripheral blood cells of zebrafish chronically exposed to 5-FU through F0 and F1 generation were highly sensitive endpoints. Therefore, we calculated the RQ for these endpoints. In Table 16.8, it can be seen that the RQ values are in this case $\geq$1 for 5-FU, CDDP and IM, but not for ET.

The data indicate that when considering reproductive effects as the most sensitive toxicological endpoint, the four ACDs do not represent risk for the environment. However, when genotoxicity is considered as the most sensitive toxicological endpoint, 5-FU, CDDP and IM represent potential environmental risks. It is notable that 5-FU at 0.01 μg/L in the two-generation study induced in addition to micronuclei also histopathological changes in liver and kidney of exposed fish as well as substantial changes in gene expression, including up-regulation of DNA damage responsive genes and several oncogenes (Kovacs et al. 2015). This result indicates that 5-FU can upon chronic exposure affect fish at environmental concentrations. However, the reproductive parameters were not affected, which raises the question of the significance of these effects in the context of ecological risk assessment which is, on the contrary to human risk assessment that is aimed to protect individuals, aimed to protect natural populations. Although there is a clear evidence linking exposure of aquatic organisms to genotoxic contaminants and adverse effects such as DNA damage, chromosome aberrations, and cancer, it is still not clear how these effects are expressed as adverse effect at the level of populations.

16.6 Conclusions

The EMA guideline (2006) describes a stepwise-tiered procedure for environmental risk assessment for pharmaceuticals. The Phase I is a pre-screening assessment aiming to estimate exposure and has an action limit for predicted environmental concentration (PEC) at 0.01 μg/L. If the PEC in surface water is below this limit, it is assumed that the compound is unlikely to represent a risk for the environment. If the PEC value is equal to or above 0.01 μg/L, a Phase II environmental effect analysis and fate has to be performed. The battery of ecotoxicological tests required in Phase II is generally based on organisms from three different trophic levels of the aquatic chain: algae (growth inhibition test), crustacean (inhibition of population growth), and fish (early-life stage toxicity assay). The PEC values for ACDs are generally below the action limit (0.01 μg/L) that would require Phase II environmental fate and effect analysis, which may also explain why so far ACDs were not considered to be of environmental relevance. However, the mechanism of action of these drugs is via direct or indirect interference with genetic material. For genotoxic substances that directly interact with DNA a linear dose–response relationship is assumed. In the linear model, DNA damage induction is believed to be directly proportional to dose; leading to the implication that direct acting genotoxic agents induce DNA damage even at the lowest concentration and that no 'safe' dose range exists (Jenkins et al. 2010). This means that even exposure to very low concentrations of genotoxic ACDs may represent a threat for exposed organisms and thus the limit PEC value at 0.01 μg/L may not apply for antineoplastic drugs.

Concurrent acute and chronic toxicity studies of 5-FU, CDDP, ET and IM that included also detection of genotoxic effects indicated high differences in the sensitivity of different aquatic organisms in regard to lethal and reproductive effects. The most sensitive species were crustacean, whereas the least sensitive was zebrafish. However, in all organisms the concentrations at which mortality and reproductive effects were observed were higher than the concentrations that were detected or expected in the environmental samples ($RQ < 1$). Based on these data it can be concluded that residues of ACDs do not pose threat to the aquatic organisms. On the contrary, the genotoxic effects in crustacean exposed to 5-FU, CDDP or IM were detected at concentrations that may occur in the aquatic environment (RQ >1). Thus potential ecological risks of aquatic exposure of invertebrates to 5-FU, CDDP and IM cannot be ruled out. Notably only ET did not induce genotoxic effects at concentrations below the PEC values.

Of particular relevance for the future environmental risk assessment related to genotoxicity of ACDs is the two-generation toxicity study of 5-FU in zebrafish, which revealed histopathological changes in the liver and kidney, as well as genotoxic effects and changes in gene expression at exposure concentration of 0.01 μg/L. It is known that DNA damaging agents have a significant ecological relevance since they cause genetic alterations in somatic and germ cells that are associated with serious adverse effects, which may occur even at low exposure levels (Bolognesi and Cirillo 2014). This indicates that the trigger PEC (0.01 μg/L) for Phase II studies proposed by EMA (2006) may not be relevant for genotoxic ACDs.

Due to lack of data on the occurrence of the residues of ACD in aquatic environment, the environmental risk assessment for the four ACDs could be performed only on the basis PEC values that may overestimate the real situation. However, the consumption of ACDs is constantly increasing in developed countries due to the population aging, earlier cancer diagnosis and better health care; thus, also, increase in the residues of these drugs can be expected.

The results of the presented studies not only clearly demonstrated that residues of ACDs are hazardous for aquatic environment but also revealed many knowledge gaps that prevent more accurate environmental risk assessment as well as risk management of ACDs. We recommend:

- The most consumed ACDs should be included into systematic monitoring across EU countries to obtain reliable data on the occurrence of their residues in aquatic environment.
- Further genotoxicity studies in aquatic organisms are needed with more ACDs.
- Studies are needed to explore the relevance of the observed genotoxic effects for the effects at the level of environmental populations.
- In the guidelines for environmental risk assessment of pharmaceuticals (EMA 2006), the trigger PEC value (0.01 μg/L) for Phase II environmental effect and fate analysis is not suitable for ACDs. Phase II environmental effect and fate analysis should be for genotoxic ACDs required irrespective to the PEC values.
- The ecotoxicity tests proposed for Phase II environmental effect analysis of ACDs should, in addition to conventional toxicological parameters, include also genotoxicity endpoints in aquatic organisms.

References

Aherne GW, Hardcastle A, Nield AH (1990) Cytotoxic drugs and the aquatic environment: estimation of bleomycin in river and water samples. J Pharm Pharmacol 42:741–742

Arora A, Scholar EM (2005) Role of tyrosine kinase inhibitors in cancer therapy. J Pharmacol Exp Ther 315:971–979

Besse JP, Latour JF, Garric J (2012) Anticancer drugs in surface waters. What can we say about the occurrence and environmental significance of cytotoxic, cytostatic and endocrine therapy drugs? Environ Int 39:73–86

Bolognesi C, Cirillo S (2014) Genotoxicity biomarkers in aquatic bioindicators. Curr Zool 60:273–284

Bolognesi C, Hayashi M (2011) Micronucleus assay in aquatic animals. Mutagenesis 26:205–213

Booker V, Halsall C, Llewellyn N, Johnson A, Williams R (2014) Prioritising anticancer drugs for environmental monitoring and risk assessment purposes. Sci Total Environ 473-474:159–170

Boxall AB, Rudd MA, Brooks BW, Caldwell DJ, Choi K, Hickmann S et al (2012) Pharmaceuticals and personal care products in the environment: what are the big questions? Environ Health Perspect 120:1221–1229

Brezovšek P, Elersek T, Filipic M (2014) Toxicities of four anti-neoplastic drugs and their binary mixtures tested on the green alga Pseudokirchneriella subcapitata and the cyanobacterium Synechococcus leopoliensis. Water Res 52:168–177

Buerge IJ, Buser HR, Poiger T, Muller MD (2006) Occurrence and fate of the cytostatic drugs cyclophosphamide and ifosfamide in wastewater and surface waters. Environ Sci Technol 40:7242–7250

Catastini C, Mullot J-U, Boukari S, Mazellier P, Levi Y, Cervantes P, Ormsby J-N (2008) Assessment of antineoplastic drugs in effluents of two hospitals lIdentification de molecules anticancéreuses dans les effluents hospitaliers. J Eur Hydrol 39(2):171–180

Daughton CG (2002) Environmental stewardship and drugs as pollutants. Lancet 360:1035–1036

Dong Z, Senn DB, Moran RE, Shine JP (2013) Prioritizing environmental risk of prescription pharmaceuticals. Regul Toxicol Pharmacol 65:60–67

EMA (European Medicines Agency) (2006) Guidelines on the environmental risk assessment of medicinal products for human use. European Chemical Agency. Doc ref. EMEA/CHMP/SWP/4447/00

European Commission (2003) Technical guidance document on risk assessment in support of commission Directive 93/67/EEC on Risk assessment for new notified substances commission regulation (EC) No 1488/94, on Risk assessment for existing substances directive 98/8/EC of the European parliament and of the council concerning the placing of biocidal products on the market. Institute for Health and Consumer Protection, European Chemicals Bureau

Fent K, Weston AA, Caminada D (2006) Ecotoxicology of human pharmaceuticals. Aquat Toxicol 76:122–159

Ferrando-Climent L, Rodriguez-Mozaz S, Barcelo D (2014) Incidence of anticancer drugs in an aquatic urban system: from hospital effluents through urban wastewater to natural environment. Environ Pollut 193:216–223

Fleming RA (1997) An overview of cyclophosphamide and ifosfamide pharmacology. Pharmacotherapy 17:146S–154S

Franquet-Griell H, Cornadó D, Caixach J, Ventura F, Lacorte S (2017) Determination of cytostatic drugs in Besòs River (NE Spain) and comparison with predicted environmental concentrations. Environ Sci Pollut Res 24:6492

Frenzilli G, Nigro M, Lyons BP (2009) The comet assay for the evaluation of genotoxic impact in aquatic environments. Mutat Res Rev Mutat Res 681:80–92

Fuerhacker M et al (2006) Chemische Analyse, Risikobewertungund Entfernung von ausgewählten Zytostatika aus Abwasserströmen aus Krankenhäusern Risikoabschätzung und Risikomanagement, Teil II “Risikoabschätzung und Risikomanagement”. Fördervertrag der ÖKK:GZ A201860, Lebensministerium: 1–168, Wien

Hagner N, Joerger M (2010) Cancer chemotherapy: targeting folic acid synthesis. Cancer Manag Res 2:293–301

Hernando MD, Mezcua M, Fernndez-Alba AR, Barcelo D (2006) Environmental risk assessment of pharmaceutical residues in wastewater effluents, surface waters and sediments. Talanta 69:334–342

Hughes SR, Kay P, Brown LE (2013) Global synthesis and critical evaluation of pharmaceutical datasets collected from river systems. Environ Sci Technol 47:661–677

Isidori M, Lavorgna M, Russo C, Kundi M, Zegura B, Novak M, Filipic M, Misik M, Knasmueller S, de Alda ML et al (2016) Chemical and toxicological characterisation of anticancer drugs in hospital and municipal wastewaters from Slovenia and Spain. Environ Pollut 219:275–287

ISO 20665 (2008) Water quality-determination of chronic toxicity to Ceriodaphnia dubia in 7 days-population growth inhibition test. International Organization for Standardization, Geneva

Jackson MA, Stack HF, Waters MD (1996) Genetic activity profiles of anticancer drugs. Mutat Res 355:171–208

Jenkins GJS, Zair Z, Johnson GE, Doak SH (2010) Genotoxic thresholds, DNA repair, and susceptibility in human populations. Toxicology 278:305–310

Kelland L (2007) The resurgence of platinum-based cancer chemotherapy. Nat Rev Cancer 7:573–584

Kosjek T, Heath E (2011) Occurrence, fate and determination of cytostatic pharmaceuticals in the environment. TrAC Trends Anal Chem 30:1065–1087

Kovacs R, Csenki Z, Bakos K, Urbanyi B, Horvath Ă, Garaj-Vrhovac V, Gajski G, Geric M, Negreira N, Lopez de Alda M et al (2015) Assessment of toxicity and genotoxicity of low doses of 5-fluorouracil in zebrafish (Danio rerio) two-generation study. Water Res 77:201–212

Kovacs R, Bakos K, Urbanyi B, Kovesi J, Gazsi G, Csepeli A et al (2016) Acute and sub-chronic toxicity of four cytostatic drugs in zebrafish. Environ Sci Pollut Res 23:14715–14729

Kummerer K (2009) The presence of pharmaceuticals in the environment due to human use-present knowledge and future challenges. J Environ Manag 90:2354–2366

Kummerer K, Haiss A, Schuster A, Hein A, Ebert I (2014) Antineoplastic compounds in the environment-substances of special concern. Environ Sci Pollut Res Int 23:14791–14804

Kuster A, Adler N (2014) Pharmaceuticals in the environment: scientific evidence of risks and its regulation. Philos Trans R Soc Lond Ser B Biol Sci 369:20130571

Lin AY, Lin YC, Lee WN (2014) Prevalence and sunlight photolysis of controlled and chemotherapeutic drugs in aqueous environments. Environ Pollut 187:170–181

Longley DB, Harkin DP, Johnston PG (2003) 5-fluorouracil: mechanisms of action and clinical strategies. Nat Rev Cancer 3:330–338

Mahnik SN, Rizovski B, Fuerhacker M, Mader RM (2004) Determination of 5-fluorouracil in hospital effluents. Anal Bioanal Chem 380:31–35

Martin J, Camacho-Munoz D, Santos JL, Aparicio I, Alonso E (2011) Simultaneous determination of a selected group of cytostatic drugs in water using high-performance liquid chromatography-triple-quadrupole mass spectrometry. J Sep Sci 34:3166–3177

Negreira N, Lopez de Alda M, Barcelo D (2013) On-line solid phase extraction-liquid chromatography-tandem mass spectrometry for the determination of 17 cytostatics and metabolites in waste, surface and ground water samples. J Chromatogr A 1280:64–74

OECD TG 201 (2011) Freshwater alga and cyanobacteria, growth inhibition test. Organization for Economic Cooperation and Development, Paris

OECD TG 202 (2004) Daphnia sp., acute immobilization test. Organization for Economic Co-operation and Development, Paris

OECD TG 203 (1992) Fish, acute toxicity test. Organization for Economic Cooperation and Development, Paris

OECD TG 210 (2013) Fish, early-life stage toxicity test. Organization for Economic Cooperation and Development, Paris

OECD TG 211 (2008) Daphnia magna reproduction test. Organization for Economic Cooperation and Development, Paris

OECD TG 236 (2013) Fish embryo acute toxicity (FET) test. Organization for Economic Cooperation and Development, Paris

Olalla A, Negreira N, Lopez de Alda M, Barcelo D, Valcarcel Y (2018) A case study to identify priority cytostatic contaminants in hospital effluents. Chemosphere 190:417–430

Parrella A, Lavorgna M, Criscuolo E, Russo C, Fiumano V, Isidori M (2014) Acute and chronic toxicity of six anticancer drugs on rotifers and crustaceans. Chemosphere 115:59–66

Parrella A, Lavorgna M, Criscuolo E, Russo C, Isidori M (2015) Eco-genotoxicity of six anticancer drugs using comet assay in daphnids. J Hazard Mater 286:573–580

Pommier Y (2013) Drugging topoisomerases: lessons and challenges. ACS Chem Biol 8:82–95

Rodriguez-Mozaz S, Weinberg HS (2010) Meeting report: pharmaceuticals in water e an interdisciplinary approach to a public health challenge. Environ Health Perspect 118:1016–1020

Tice RR, Agurell E, Anderson D, Burlinson B, Hartmann A, Kobayashi H, Miyamae Y, Rojas E, Ryu JC, Sasaki YF (2000) Single cell gel/comet assay: guidelines for in-vitro and in-vivo genetic toxicology testing. Environ Mol Mutagen 35:206–221

Toolaram AP, Kummerer K, Schneider M (2014) Environmental risk assessment of anti-cancer drugs and their transformation products: a focus on their genotoxicity characterization-state of knowledge and short comings. Mutat Res Rev Mutat Res 760:18–35

US EPA (1993) Methods for measuring the acute toxicity of effluents and receiving waters to freshwater and marine organisms, 4th edn. EPA-600-4-90. US Environmental Protection Agency, Washington, DC

Usawanuwat J, Boontanon N, Boontanon SK (2014) Analysis of three anticancer drugs (5-fluorouracil, cyclophosphamide and hydroxyurea) in water samples by HPLC-MS/MS. Int J Adv Agric Environ Eng 1:2349–1523

Zhang J, Chang VW, Giannis A, Wang JY (2013) Removal of cytostatic drugs from aquatic environment: a review. Sci Total Environ 445–446:281–298

Zuccato E, Calamari D, Natangelo M, Fanelli R (2000) Presence of therapeutic drugs in the environment. Lancet 355:1789–1790

Chapter 17
Toxicity of Antineoplastic Drug Mixtures

Marko Gerić, Goran Gajski, and Verica Garaj Vrhovac

Abstract Securing potable water and preserving aquatic ecosystems are growing problems that will become even more important in the near future. Well-defined environmental pollutants such as pesticides, heavy metals, or organic matter have been thoroughly examined, but over the recent decade, a spotlight was put on a new and emerging group of pollutants – pharmaceuticals, antineoplastic drugs being considered as the most hazardous ones among these. Patients' excretion is the most important routes by which antineoplastic residues end up in the effluents from hospitals and households and reach freshwater where aquatic organisms can be affected. Low concentrations of antineoplastic drug can be expected in occupational settings, and medical staff could be chronically exposed to these highly active agents, representing a potential threat to their health.

The development of more sensitive analytical methods facilitated the detection of lower concentrations of antineoplastic drugs in the environment, which triggered the research of their toxicological effects on different organisms. Most studies usually focus on single compound toxicity, but due to the complexity of environmental exposure, the era of mixture testing has begun.

In this chapter, we will focus on the toxicity of mixtures of the most commonly used antineoplastic drugs with different modes of action (5-fluorouracil, cisplatin, etoposide, imatinib mesylate, cyclophosphamide, etc.) and their transformation products toward different experimental models (bacteria, algae, animals, plants, and human cells).

Keywords Anticancer drugs · Environment · Occupational exposure · Mixtures

M. Gerić · G. Gajski · V. Garaj Vrhovac (✉)
Mutagenesis Unit, Institute for Medical Research and Occupational Health, Zagreb, Croatia
e-mail: vgaraj@imi.hr

E. Heath et al. (eds.), *Fate and Effects of Anticancer Drugs in the Environment*,
https://doi.org/10.1007/978-3-030-21048-9_17

17.1 Introduction

It is impossible to imagine life without water, yet pure water is often taken for granted (Hrudey et al. 2006; Rush 2013). Recent UN Water Reports highlight the need for urgent measures in order to sustain pure freshwater and ecosystems. Food and energy production are the most common sectors where high volumes of water are needed. If population growth is taken into account, by the middle of the twenty-first century, the need for water will have increased dramatically. It is estimated that in 2050 there will be 9 billion inhabitants, which will result in 70% more food demand, more than 50% of energy demand, and higher freshwater demand. Entire ecosystems are also threatened by an excessive water withdrawal, and they are responsible for sustaining of water cycles and providing precious resources for humans (UN WWAP 2012, 2015). Therefore, it is of top priority to promote sustainable water systems and protection of ecosystems, especially aquatic ones.

Since cancer is one of the leading causes of human mortality responsible for more than 8 million deaths annually, appropriate strategies to fight cancer have been developed during the last few decades (Ferlay et al. 2013). The most commonly used therapy, besides radiation and surgery, is still chemotherapy. Antineoplastic drugs (usually named cytostatic drugs) are intended to inhibit division of cancer cells and to promote cell death. Therefore, these pharmaceuticals are highly active compounds aimed at different biomolecules and are usually not specific, which is their greatest disadvantage (Arora and Scholar 2005; Heath et al. 2016).

So, where is the connection between antineoplastic drugs and the aquatic environment? Large amounts of chemotherapy are given every day in many hospitals worldwide, which is the first step toward the release of these pharmaceuticals into the environment. Drug preparation and patients' excretion are common routes by which antineoplastic residues end up in the effluents from hospitals and households. To some extent, releases from the pharmaceutical industry might also be expected. Wastewater treatment plants (WWTPs) represent the sites that should lower the amounts of pollutants discharged into the environment. There are several drawbacks with WWTPs: they are usually not implemented in hospitals, several chemicals are poorly degraded, and in some world regions wastewaters are released without previous treatment (Besse et al. 2012; UN WWAP 2012; Zhang et al. 2013).

With the development of more sensitive and reliable methods, the residues of antineoplastic drugs can be detected in the aquatic environment, usually in ng/L scale, but their concentrations can reach hundreds of μg/L in hospital effluents. It has to be taken into account that due to a constant release and presence in the environment, antineoplastic drugs are considered as pseudo-persistent pollutants (Hernando et al. 2006; Kosjek and Heath 2011; Xie 2012; Zhang et al. 2013; Negreira et al. 2013a, b, 2014, 2015; Česen et al. 2015).

Although toxicity of antineoplastic drugs has been elucidated in many studies using different study models, generally those studies dealt with single compound toxicity (Gačić et al. 2014; Parrella et al. 2014b, 2015; Gerić et al. 2014; Kovács et al. 2015; Kračun-Kolarević et al. 2015; Novak et al. 2016, 2017b; Isidori et al.

2016b; Kovacs et al. 2016; Gajski et al. 2016b; Mišik et al. 2016b). However, in the environment, the residues of antineoplastic drugs occur in complex mixtures; thus, for an appropriate risk assessment of aquatic ecosystems, studies determining the effects of antineoplastic drugs mixtures were needed. Two most common mathematical models applied for predicting the effects of more than one compound on certain target molecule or tissue are concentration addition (CA) and independent action (IA). The CA is a model applied when assuming that compounds in a mixture act by the same mode of action, whereas the IA model is applied when assuming that compounds in a mixture do not act by the same mode of action. There are many tools in mixture toxicity assessment including the relative potency factor (RPF) in which each compound is scaled relative to a well-characterised index chemical, deriving mixture to single chemical; the toxic equivalency factor (TEF) where concentrations of each compound and appropriate factors are used to derive the impact of a mixture; the hazard index (HI) that is based on known reference doses and concentrations of each compound in the mixture; and the maximum cumulative ratio (MCR) which highlights the impact of the most contributing chemical in the mixture (Monosson 2005; Ermler et al. 2014; Kienzler et al. 2016). However, many controversies remain in mixture toxicity assessment since all of the approaches have some strengths and limitations, hence not providing reliable assumptions when compared to experimental data.

The aim of this chapter is to sum up up-to-date knowledge on the toxicity of mixtures from the most commonly used antineoplastic drugs with different modes of action (5-fluorouracil (5-FU), cisplatin (cDDP), etoposide (ET), imatinib mesylate (IM), cyclophosphamide (CP), etc.) and their transformation products (TPs) toward different experimental models.

17.2 Search Strategy and Inclusion/Exclusion Criteria

The extensive literature search was done using PubMed (US National Library of Medicine, National Institutes of Health, Bethesda, MD, USA – http://www.ncbi.nlm.nih.gov/PubMed) limiting search to papers published in English during recent 15 years (January 1, 2002–August 23, 2017). The search terms included "antineoplastic" or "cytostatic" or "anticancer", "drug", "mixture∗" or "combined", and "toxic∗", excluding terms "therapy", "radio∗", or "pesticide∗". Articles excluded from the review (a) did not cover the toxicological assessment of antineoplastic drug mixture, (b) had no chemical characterisation of mixture, (c) did not assess mixture toxicity, but single compounds only, (d) provided a toxicological assessment with non-cytostatic compound (such as natural products, other classes of pharmaceuticals, etc.), (e) aimed at providing better therapy outcomes, or (f) examined the toxicity of antineoplastics and radiation. Hence, a total of 12 papers were included in this literature review.

17.3 Environmental Toxicity of Antineoplastic Drug Mixtures

The idea of investigating the toxicological potential of highly active chemicals occurring in the environment at low concentrations has only recently been put in the spotlight of research. Over the past 15 years, several studies assessed the occurrence, fate, and effects of antineoplastic drugs mostly in aquatic environment in respect to various organisms represented at different trophic levels from bacteria to human cells. Moreover, ecotoxicity studies turned from single compound testing to mixture testing in order to eliminate underestimations of the threat represented by mixtures.

The study by Brezovšek et al. (2014) showed that binary mixtures formed by antineoplastic drugs with different modes of action (5-FU, cDDP, ET, IM) could cause both synergistic and antagonistic growth inhibition in green algae (*Pseudokirchneriella subcapitata*) and cyanobacteria (*Synechococcus leopoliensis*), depending on the mixture (Table 17.1). More importantly, neither CA nor IA model correctly predicted the actual effects retrieved experimentally. From the environmental standpoint, the concentrations tested were somewhat higher than those found or predicted in freshwater. However, due to the possibility of reaching such concentrations in wastewater and synergistic effects of some mixtures, growth inhibition of primary producers cannot be ruled out. In fact, when three compounds (5-FU, ET, and IM) were combined, the same decline in CA and IA models was observed, showing an additive and synergistic response at lower effective concentrations (EC) and an antagonistic response at high EC (Table 17.1). More importantly, such mixture caused an abnormal algal cell aggregation as a consequence of changes in cell surface properties (Eleršek et al. 2016). Furthermore, the aquatic environment consists not only of parent compounds but also of metabolites and/or TPs, adding up to this already complex puzzle. At the level of primary producers, cytotoxic and genotoxic effects of ifosfamide (IF) and CP and their metabolites/TPs were observed in *S. leopoliensis* and *Salmonella typhimurium*, while the effect was again underestimated in the CA model (Česen et al. 2016). Just like algae are the most contributing taxa for primary production in aquatic ecosystems, vascular plants are major producers on land (Sigman and Hain 2012). The study on *Tradescantia* by Mišik et al. (2016a) revealed that antineoplastic drugs in certain mixtures have synergistic effects in the formation of micronuclei (MNi) in plant tetrads. This trend was observed when IM (kinase inhibitor) was part of a binary mixture. Interestingly, when studying the effects of single compounds on the induction of pollen aberrations, the concentrations needed to show the effects were considerably higher than those that could be found in the environment (Mišik et al. 2016b). Since upon exposure to low concentrations of cytostatic drug/TPs mixtures the changes in plant tetrads DNA integrity, growth, and positioning of algal cells were detected, it is possible that the changes in producer biodiversity could lead to reduction in O_2 production and CO_2 uptake and to a decrease in inorganic nutrients removal. In other words, this could lead to changes in the processes vital for sustaining ecosystems (Cardinale et al. 2011).

Table 17.1 The review of antineoplastic drug mixtures on different model organisms

Reference	Treating agents	Study model	Biomarker	Concentration		Results	
Borgatta et al., 2016	TAM, OH-TAM	*Daphnia pulex*	mortality, survival, size, reproduction, intrinsic rate	TAM/OH-TAM (0.12/0.16 μg/L)		mortality Ø, survival Ø, size Ø, reproduction ↓, intrinsic rate Ø	
Brezovšek et al., 2014	5-FU, cDDP, ET, IM	*Pseudokirchneriella subcapitata* and *Synechococcus leopoliensis*	growth inhibition	*P. subcapitat*[b] 5-FU/cDDP $EC_5/2$, $EC_{10}/2$, $EC_{20}/2$, $EC_{50}/2$, $EC_{90}/2$ cDDP/ET $EC_5/2$, $EC_{10}/2$, $EC_{20}/2$, $EC_{50}/2$, $EC_{90}/2$ 5-FU/IM $EC_5/2$, $EC_{10}/2$, $EC_{20}/2$, $EC_{50}/2$, $EC_{90}/2$	*S. leopoliensis*[b] 5-FU/cDDP $EC_5/2$, $EC_{10}/2$, $EC_{20}/2$, $EC_{50}/2$, $EC_{90}/2$ cDDP/ET $EC_5/2$, $EC_{10}/2$, $EC_{20}/2$, $EC_{50}/2$, $EC_{90}/2$ 5-FU/IM $EC_5/2$, $EC_{10}/2$, $EC_{20}/2$, $EC_{50}/2$, $EC_{90}/2$	*P. subcapitata* 5-FU/cDDP S++ cDDP/ET A++ 5-FU/IM S++	*S. leopoliensis* 5-FU/cDDP S+ cDDP/ET A++ 5-FU/IM A++
Česen et al., 2016	CP, IF, ketoCP, NdCP, CPCOOH	*Synechococcus leopoliensis* and *Salmonella typhimurium TA1535/pSK1002*	growth inhibition, SOS/umuC	*S. leopoliensis*[c] CP/IF/ketoCP/NdCP/CPCOOH 6, 7.9, 10.5, 17.1, and 37.2 mg/L	*S. typhimurium*[c] CP/IF/ketoCP/NdCP/CPCOOH 6, 7.9, 10.5, 17.1, and 37.2 mg/L	*S. leopoliensis* S+	*S. typhimurium* cytotoxicity: Ø for all concentrations tested genotoxicity: IR significantly higher at 37.2 mg/L
Danesi et al., 2010	5-FU, cDDP, PAC	*Drosophila melanogaster*	somatic mutation and recombination test	cDDP/PAC/5-FU (0.003/0.0025/0.025 mM and 0.006/0.005/0.05 mM)		significant results: - mwh/flr^3 genotype: ↑small single spots, ↑large single spots and ↑total spots for both concentrations and twin spots at lower concentration - mwh/TM3 genotype: ↑small single spots and ↑total spots for both concentrations - no synergism detected	

(continued)

Table 17.1 (continued)

Reference	Treating agents	Study model	Biomarker	Concentration		Results	
				P. subcapitata 5-FU/IM/ET	*S. leopoliensis* 5-FU/IM/ET	*P. subcapitata* 5-FU/IM/ET	*S. leopoliensis* 5-FU/IM/ET
Eleršek et al., 2016	5-FU, ET, IM	*Pseudokirchneriella subcapitata* and *Synechococcus leopoliensis*	growth inhibition	2.9/184.2/4279.2	19.8/866.3/36578.9	S++	S++
				5.8/264.1/5621.6	42.5/1041.0/36578.9	S++	Ad
				12.1/390.3/7567.1	97.3/1270.7/36578.9	S++	Ad
				42.7/761.7/12569.2	400.7/1786.3/36578.9	Ad	Ad
				315.0/2197.3/28087.5 μg/mL	3776.7/3065.0/36578.9 μg/mL	A++	A++
Gajski et al., 2016a	5-FU, ET, IM	human blood, plasma and mononuclear cells	lymphocyte micronuclei, comet assay tail intensity and sister chromatid exchanges, fpg-comet, oxidised proteins, MDA	5-FU/ET/IM (0.000001 – 10 μg/mL of each compound)		significant results after 4 and 24 h of exposure, respectively cytotoxicity: - cell viability↓ at 0.01 and 0.00001 μg/mL - NDI↓ at 1 and 1 μg/mL - PRI↓ 0.01 and 0.01 μg/mL genotoxicity: - TI↑ at Ø and 10 μg/mL - AST↑ at Ø and 0.000001 μg/mL - MNed↑ at 0.000001 μg/mL and 0.1 μg/mL - SCEs↑ at 0.000001 μg/mL and 0.000001 μg/mL oxidative stress: - fpg↑ at 1 and 0.1 μg/mL - MDA↑ at Ø and 0.000001 μg/mL - OXP↑ at 0.00001 μg/mL and 0.00001 μg/mL	

Table 17.1 (continued)

Reference	Treating agents	Study model	Biomarker	Concentration	Results
Hirose et al., 2005	CP, MET, CYT, 5-FU, MMC, BLE, EPI, ET, VCR, IRI, PAC, cDDP, L-ASP	Molt-4 human lymphoblastoid cells	WST-8 cytotoxicity test	CP/MET/CYT/5-FU/MMC/BLE/EPI/ET/VCR/IRI/PAC/cDDP/L-ASP (4464.3/1785.7/1785.7/2232.1/17.9/44.6/17.9/714.3/8.9/357.1/267.9/89.3 μg/mL/44.6 K.U./mL	CC50↓ as the removal of cytostatic mixture progressed
Isidori et al., 2016a	wastewater tested for: CP, IF, ketoCP, NdCP, CPCOOH, 5-FU, GEM, MET, OH-MET, IRI, ERL, CAP, OH-D-TAM, OH-TAM, TAM, TMZ, DOX, PAC, OH-PAC, IM, ET, Pt	*Ceriodaphnia dubia, Tradescantia sp., Allium cepa,* zebrafish liver cell line	acute and chronic toxicity, tetrads' micronuclei, root micronuclei, comet assay	10 wastewater samples[d]: 1. 10 of 22 agents quantified (total 20 210.1 ng/L)	1. *C. dubia* acute toxicity LC50 64.9%, chronic toxicity EC50 3.7%, *Tradescantia* MN NA, *A. cepa* MN Ø, ZFL TI at 20%
				2. 11 of 22 agents quantified (total 1 006.2 ng/L)	2. *C. dubia* acute toxicity LC50 68.6%, chronic toxicity EC50 3.0%, *Tradescantia* MN NA, *A. cepa* MN Ø, ZFL TI at 20%
				3. 11 of 22 agents quantified (total 94 450 ng/L)	3. *C. dubia* acute toxicity LC50 Ø, chronic toxicity EC50 7.4%, *Tradescantia* MN↑ at 100%, *A. cepa* MN Ø, ZFL TI↑ at 10%
				4. 6 of 22 agents quantified (total 243.5 ng/L)	4. *C. dubia* acute toxicity LC50 44.3%, chronic toxicity EC50 3.6%, *Tradescantia* MN↑ at 100%, *A. cepa* MN Ø, ZFL TI↑ at 20%
				5. 3 of 22 agents quantified (total 28.4 ng/L)	5. *C. dubia* acute toxicity LC50 Ø, chronic toxicity EC50 7.0%, *Tradescantia* MN Ø, *A. cepa* MN Ø, ZFL TI Ø

(continued)

Table 17.1 (continued)

Reference	Treating agents	Study model	Biomarker	Concentration		Results	
				6. 5 of 22 agents quantified (total 79.6 ng/L)		6. *C. dubia* acute toxicity LC50 Ø, chronic toxicity EC50 5.6%, *Tradescantia* MN NA, *A. cepa* MN Ø, ZFL TI↑ at 20%	
				7. 6 of 22 agents quantified (total 139.9 ng/L)		7. *C. dubia* acute toxicity LC50 64.0%, chronic toxicity EC50 3.5%, *Tradescantia* MN NA, *A. cepa* MN Ø, ZFL TI↑ at 20%	
				8. 2 of 22 agents quantified (total 9.8 ng/L)		8. *C. dubia* acute toxicity LC50 77.5%, chronic toxicity EC50 1.4%, *Tradescantia* MN Ø, *A. cepa* MN Ø, ZFL TI↑ at 30%	
				9. 4 of 22 agents quantified (total 24.6 ng/L)		9. *C. dubia* acute toxicity LC50 28.9%, chronic toxicity EC50 15.9%, *Tradescantia* MN↑ at 100%, *A. cepa* MN Ø, ZFL TI↑ at 30%	
				10. 2 of 22 agents quantified (total 17.3 ng/L)		10. *C. dubia* acute toxicity LC50 51.9%, chronic toxicity EC50 11.7%, *Tradescantia* MN Ø, *A. cepa* MN Ø, ZFL TI↑ at 30%	
Kundi et al., 2016	5-FU, cDDP, ET, IM	*Daphnia magna* and *Ceriodaphnia dubia*	whole organism comet assay	*D. magna*	*C. dubia*	*D. magna*[a]	*C. dubia*[a]
				IM/ET (0.1/13 and 2.2/247 μg/L)	IM/ET (0.08/0.32 and 2.28/8.80 μg/L)	IM/ET (Ø and Ø)	IM/ET (A++ and A++)
				IM/cDDP (0.1/0.1 and 1.8/3.5 μg/L)	IM/cDDP (0.14/0.01 and 2.80/0.10 μg/L)	IM/cDDP (Ø and Ø)	IM/cDDP (Ø and Ø)
				IM/5-FU (0.05/14 and 1.6/424 μg/L)	IM/5-FU (0.03/0.20 and 1.56/10.74 μg/L)	IM/5-FU (A++ and A++)	IM/5-FU (Ø and Ø)
				ET/cDDP	ET/cDDP	ET/cDDP	ET/cDDP

Table 17.1 (continued)

Reference	Treating agents	Study model	Biomarker	Concentration		Results	
				(9/0.15 and 212/3.6 μg/L)	(0.06/0.0006 and 3.39/0.03 μg/L)	(Ø and Ø)	(A++ and A++)
				ET/5-FU (6.5/15 and 186/440 μg/L)	ET/5-FU (0.01/0.02 and 1.36/2.43 μg/L)	ET/5-FU (Ø and Ø)	ET/5-FU (A++ and A++)
				cDDP/5-FU (0.07/10 and 2.5/354 μg/L)	cDDP/5-FU (0.0002/0.04 and 0.02/3.53 μg/L)	cDDP/5-FU (Ø and Ø)	cDDP/5-FU (A++ and A++)
Mišík et al., 2016a	5-FU, cDDP, ET, IM	*Tradescantia sp.*	tetrads' micronuclei	IM/ET (16.93/0.62 and 42.02/1.54 μM) IM/cDDP (11.00/0.04 and 30.51/0.11 μM) IM/5-FU (3.75/19.48 and 14.80/76.85 μM) ET/cDDP (0.68/0.07 and 1.63/0.16 μM) ET/5-FU (0.24/34.26 and 0.87/123.13 μM) cDDP/5-FU (0.01/21.59 and 0.06/83.43 μM)		IM/ET[a] (S++ and Ø) IM/cDDP (Ø and S++) IM/5-FU (S++ and S++) ET/cDDP (A+ and Ø) ET/5-FU (Ø and A++) cDDP/5-FU (Ø and Ø)	
Novak et al., 2017a	IF, CP, 5-FU, cDDP	zebrafish liver cell line	cell viability, comet assay, micronucleus test	CP/IF/5-FU/cDDP (12/10/0.09/0.6 and 120/100/0.9/6 μg/L)		cytotoxicity: - Ø and ↓viability genotoxicity: - ↑ TI and ↑ TI - Ø MNed, Ø NDI and Ø MNed, Ø NDI	
Parrella et al., 2014a	5-FU, cDDP, ET, IM	*Daphnia magna* and *Ceriodaphnia dubia*	offspring reduction	*D. magna* IM/ET (35.32/20.42 and 223.16/129.01 μg/L) IM/cDDP (20.83/0.08 and	*C. dubia* IM/ET (18.3/97.3 μg/L) IM/cDDP (7.0/2.1 μg/L)	*D. magna*[a] IM/ET (Ø and A++) IM/cDDP (Ø and A++)	*C. dubia*[a] IM/ET (Ø) IM/cDDP (S+)

(continued)

Table 17.1 (continued)

Reference	Treating agents	Study model	Biomarker	Concentration		Results	
				127.90/0.49 µg/L)			
				IM/5-FU (26.58/1.88 and 153.67/10.87 µg/L)	IM/5-FU (8.1/0.6 µg/L)	IM/5-FU (Ø and Ø)	IM/5-FU (Ø)
				ET/cDDP (61.43/0.40 and 155.00/1.02 µg/L)	ET/cDDP (41.97/11.10 and 189.80/50.20 µg/L)	ET/cDDP (Ø and Ø)	ET/cDDP (Ø and A++)
				ET/5-FU (77.87/9.53 and 165.38/20.23 µg/L)	ET/5-FU (76.10/2.92 and 218.05/8.37 µg/L)	ET/5-FU (Ø and Ø)	ET/5-FU (Ø and A++)
				cDDP/5-FU (0.27/5.00 and 0.79/14.66 µg/L)	cDDP/5-FU (8.55/1.24 and 40.19/5.83 µg/L)	cDDP/5-FU (Ø and Ø)	cDDP/5-FU (Ø and Ø)

Abbreviations: *5-FU* 5-fluorouracil, *AST* atypically sized tails, *BLE* bleomycin, *CA* chromosome aberrations, *CAP* capecitabine, *cDDP* cisplatin, *CP* cyclophosphamide, *CPCOOH* carboxy-cyclophosphamide, *CYT* cytarabine, *DOX* doxorubicin, *EPI* epirubicin, *ERL* erlotinib, *ET* etoposide, *fpg* formamidopyrimidine DNA glycosylase, *GEM* gemcitabine, *IF* ifosfamide, *IM* imatinib mesylate, *IR* induction ratio, *IRI* irinotecan, *ketoCP* keto-cyclophosphamide, *L-ASP* L-asparaginase, *MDA* malondialdehyde, *MET* methotrexate, *MMC* mitomycin C, *MN* micronucleus, *MNed* micronucleited cells, *NA* not applicable, *NdCP* N-dechloroethyl-cyclophosphamide, *NDI* nuclear division index, *OH-MET* hydroxymethotrexate, *OH-PAC* 6(a)-hydroxypaclitaxel, *OH-TAM* (Z)-4-hydroxytamoxifen, *OH-D-TAM* 4-hydroxy-Ndesmethyl-tamoxifen, *OXP* oxidised proteins, *PAC* paclitaxel, *PRI* proliferation rate index, *Pt* cisplatin as total platinum, *SCEs* sister chromatid exchanges, *VCR* vincristine, *TAM* tamoxifen, *TI* tail intensity, *TMZ* temozolomide, *ZFL* zebrafish liver cell lines

Results: *Ø* no change/Bliss independence, A+ trend for antagonism, A++ antagonism, Ad additive, S+ trend for synergism, S++ synergism

[a]relation to Bliss independence reference model

[b]exact concentrations not described; EC_x determined in single compound toxicity

[c]mixture mass ratio 1:1:1:1:1

[d]sum of concentration averages of analysed agents

Moving up the food web and entering the first level of heterotrophs, the effects of antineoplastic drug mixtures were assessed in planktonic crustaceans *Daphnia magna* and *Ceriodaphnia dubia* as primary consumers. When exposed to a binary mixture of commonly used antineoplastics with different modes of action (5-FU, cDDP, ET, IM), cladocerans responded in offspring reduction and increase in DNA damage. The application of the IA and CA models in these studies provided good predictive power, providing under- and overestimation of mixtures only occasionally. An independent action of active compounds can be expected at a concentration lower than needed to achieve the same effect when acting as a single compound (Parrella et al. 2014a; Kundi et al. 2016). Similar pattern was also observed in *Daphnia pulex* treated with tamoxifen (TAM) and its TP 4-hydroxy-tamoxifen mixture. A decrease in the reproductive performance was detected, whereas exposure to single compounds in much higher concentration did not affect reproduction (Borgatta et al. 2016). Another primary heterotroph that served as a model for a toxicological study of an antineoplastic drug mixture was *Drosophila melanogaster*. During their larval development, they were fed with 5-FU, cDDP, and paclitaxel at two different concentrations. Such mixture was responsible for the induction of recombination (~30%) and mutations (~70%), which were detected using different parameters of the SMART assay (Danesi et al. 2010). From the ecological point of view, so far we have reviewed the effects of antineoplastic residues on primary producers and consumers that are dominant in biomass and demonstrated their vulnerability toward these pseudo-persistent pollutants. Since they represent the basis of every ecosystem providing nutrients for higher trophic levels, ignoring the changes at these levels of food web could disturb the ecosystem's development, productivity, stability, resilience, and other key processes (Van Der Putten et al. 2004).

When exposing test models in in vitro study, the concentration range used should be high enough to detect certain response (if possible), yet justified with the data of the occurrence of particular compounds in the environment to get relevant results. Another approach is to expose laboratory model organisms to pollutants, in this case antineoplastic drug mixtures, sampled from the environment. Toxicity of a synthetic mixture derived from hospital ward effluents was tested on zebrafish liver cell line (ZFL) (Novak et al. 2017a). The cytotoxic effects of IF, CP, 5-FU, and cDDP mixture on ZFL were observed only when the concentrations of antineoplastics were ten-fold higher than those observed in the environment. However, the induction of DNA strand breaks was detected using the comet assay, indicating possible genotoxicity of the studied mixture. The concept of exposing models to environmental samples was applied in a study where 10 different wastewater samples from Spain and Slovenia were toxicologically assessed using *Tradescantia*, *Allium cepa*, *C. dubia*, and zebrafish liver cell lines (Isidori et al. 2016a). Among various antineoplastic drugs, their metabolites, and transformation products analysed, kinase inhibitor erlotinib (ERL) was detected in all wastewater samples in the range 2.0–7.2 ng/L, while carboxy-cyclophosphamide was found at only two sites but in the range

17,700–60,600 ng/L. At one sampling site, the cumulative concentrations of antineoplastic drugs reached 100,000 ng/L. Several different hospital wastewaters and an influent to WWTP induced an increase in the frequency of MNi in *Tradescantia* tetrads but did not induce genotoxic effects in the test system with *Allium*. The worthy information is that after the water treatment, WWTP effluent did not induce DNA damage, highlighting the need of wastewater treatment before its release into the environment. At the level of primary consumers, *C. dubia* was less sensitive in the acute test compared to the reproduction test after chronic exposure. While LD50 in acute tests ranged from 28.9% to 100%, the EC50 for reproduction toxicity ranged from 1.4% to 15.9% dilution of the original sample. Regarding the toxicity studies on ZFL cell lines, 20–30% diluted samples induced significant DNA damage (Isidori et al. 2016a). After performing correlation and regression analysis, the compounds detected in the wastewaters explain up to 50% of the observed effects. Such estimation left many questions unanswered and provided challenge for further research. According to literature, many pollutants have recently been investigated for their occurrence, fate, and toxicological potential toward aquatic ecosystems including other pharmaceuticals, heavy metals, pesticides, etc., representing mixture interactions that are even more complex (Pal et al. 2010; Schriks et al. 2010; Tchounwou et al. 2012; Meffe and de Bustamante 2014; Albuquerque et al. 2016; Balakrishna et al. 2017; Elzwayie et al. 2017; Madikizela et al. 2017).

We have elucidated the negative effects of antineoplastic drug mixtures toward organisms represented at different levels of the food web. The changes in the ecosystem will indirectly influence human life, but the following question remains: Can such mixtures have a direct impact on human health? The in vitro studies on human cells, although they cannot be directly extrapolated to humans, are widely used in the estimation of possible effects of certain compound. A toxicological study on a mixture of 13 antineoplastic drugs (epirubicin, CP, methotrexate [MET], cytarabine, 5-FU, irinotecan hydrochloride, vincristine sulfate, mitomycin C, etc.) in mg/mL and μg/mL scale induced cytotoxic effects in human Molt-4 lymphoblastoid cells (Hirose et al. 2005). This study also provided practical solution of electrolytic treatment of hospital wastewaters, thus reducing the concentrations of antineoplastic drugs and their toxicity. Gajski et al. (2016a) examined antineoplastic drug mixtures at environmentally relevant concentrations where human blood cells were exposed to 5-FU, ET, and IM at concentrations from 1 ng/L upward. Mixture of these drugs induced genotoxic effects at environmentally relevant concentrations (formation of micronuclei, nucleoplasmic bridges, nuclear buds, and induction of sister chromatid exchanges and abnormally sized tails in the comet assay), what was not observed when testing single compounds (Novak et al. 2017b; Gajski et al. 2016b). Moreover, 5-FU, ET, and IM mixture induced oxidative stress at DNA (fpg-sensitive sites), protein (protein carbonyls), and lipid level (malondialdehyde). These results confirmed the potential threat these mixtures represent to the environment and also to human health, opening an interesting research field of how these compounds influence occupationally exposed populations.

17.4 Occupational Toxicity of Antineoplastic Drug Mixtures

Biomonitoring of occupationally exposed populations aims at lowering the incidence of occupationally related diseases. A specific subpopulation of people occupationally exposed to antineoplastic drugs includes primarily nursing, pharmacy, medical, and veterinarian personnel, physicians, and their assistants, and also waste handlers, laundry workers, and other staff that come into contact with contaminated surfaces (Connor 2006; Agbonifo et al. 2017; Walton and Rogers 2017). The process of chemotherapy has several critical points during drug preparation, transport, and administration at which unwanted exposure to anticancer drugs might occur. Patients' excretion after receiving therapy is a known route of environmental exposure, but some body fluids might contaminate surrounding surfaces, as well (Connor 2006; Zhang et al. 2013; Rioufol et al. 2014; Marie et al. 2017).

There are several protective measures used to reduce such unwanted exposure. The use of personal protective equipment is gold standard in every guideline when handling anticancer drugs. The use of protective clothing (gowns, masks) and wearing thick double nitrile gloves which are frequently changed are the recommendation for reducing dermal exposure during drug preparation (Landeck et al. 2015). More advanced systems include a centralised preparation of each therapy in aseptic pharmaceutical isolators that reduce surface and air contamination to a limited closed volume, while closed system drug transfer devices and robotic help reduce surface contamination inside such isolators. Another advantage of using these devices is the reduction of contamination of bags that are transported to patients and avoidance of therapy spilling during administration, thus significantly reducing unwanted anticancer drugs exposure (Wakui et al. 2013a, b; Simon et al. 2016; Schierl et al. 2016).

Another important factor in keeping occupational safety at high levels is staff education. Although there are guidelines for handling drugs and using protective equipment that are designed to minimise health risks, the lack of appropriate personnel trainings decreases the awareness of anticancer drugs hazard, leading to lower adherence to standardised guidelines (Jeong et al. 2015; Boiano et al. 2016). However, despite all the mentioned mechanisms of preventing exposure to anticancer drugs, workspace monitoring and staff biomonitoring reveal the presence and adverse effects of these drugs in occupational settings (Dranitsaris et al. 2005; Hall et al. 2013; Keat et al. 2013; Nersesyan et al. 2016).

There are several levels at which biomonitoring can be performed (Suspiro and Prista 2011; Vyas et al. 2014). Some studies aimed at monitoring the workplace by determining the traces of anticancer drug residues on working surfaces and equipment. The percentage of samples in which drug residues were detected varied from 1% to 95% per study, while the concentration of anticancer drugs found ranged from 0.005 ng/cm^2 on countertops of patients' checking area to 2110 ng/cm^2 on top of the biological safety cabinets. Additionally, the concentration on the internal side of gloves from the staff working on cytostatic preparation and administration ranged 70–3770 ng (Sottani et al. 2012; Crul and Simmons-Sanders 2017; Marie et al. 2017; Graeve et al. 2017). Another strategy is to monitor the effects in the staff exposed to these drugs. The biomarkers of exposure, in this case anticancer drugs and their metabolites, are usually determined in urine of the working personnel. Positive urine samples from the staff handling anticancer drugs ranged from 0% to 100% per certain study, while the concentration of drugs and TPs ranged from 0.008 ng/mL to 462.8 ng/24 h urine (Suspiro and Prista 2011; Yoshida et al. 2013; Hon et al. 2015; Poupeau et al. 2016). On the other hand, the biomarkers of effect serve as the indicators of changes in the body that are associated with unwanted exposure. In fact, many studies indicate that occupational exposure to anticancer drugs has an impact on human genome integrity (Dranitsaris et al. 2005; Ursini et al. 2006; Cavallo et al. 2007; Kopjar et al. 2010; Suspiro and Prista 2011; Sottani et al. 2012; Ladeira et al. 2014; Moretti et al. 2015; Villarini et al. 2016). Irrespectively of the biomarker chosen, a lot of effort was invested in order to describe and evaluate the risk of occupationally exposed population to anticancer drug mixtures. Further studies are aimed at assessing potential risks using multibiomarker and -omics approach, as well as at minimising possible exposure routes, thus preventing threats to human health.

17.5 Conclusion and Future Prospective

The studies reviewed in this chapter suggest that anticancer drug mixtures have a potential to cause unwanted changes in the aquatic environment and to human health. Although present at low concentrations, these pharmaceuticals are constantly released into the environment having pseudo-persistent properties. When observing the toxicological studies investigating such mixtures using different model organisms, high species differences were detected. Another interesting piece of information is the deviations from commonly used CA and IA models for predicting mixture toxicities. Therefore, it is difficult to highlight the ideal model organism for conducting further studies. It is definitely a good approach to cover different trophic levels to explore the impact of mixtures on several compartments of the ecosystem.

Occupational exposure to anticancer mixtures is still a major issue in protecting professionally exposed populations that include much more job positions than just medical staff. It has to be underlined that staff education and appropriate modern tools improve occupational health.

The complexity of mixtures is another issue in toxicology. As the literature suggests, freshwater is polluted by many different organic and inorganic pollutants, and it is difficult to assess the toxicological potential of all possible combinations. Therefore, there is a need for further development of mathematical and computational models that will be supported by experimental results in order to improve mixtures risk assessment.

The application of appropriate wastewater treatment is crucial for reducing the unwanted residues release into the environment. There is still a lot of space for improvement of wastewater treatment, particularly in developing and third world countries. Finally, there is an exciting era of new discoveries in mixture testing ahead of us.

References

Agbonifo N, Hittle B, Suarez RKD (2017) Occupational exposures of home healthcare workers. Home Heakthcare Now 35:150–159. https://doi.org/10.1097/NHH.0000000000000509

Albuquerque AF, Ribeiro JS, Kummrow F et al (2016) Pesticides in Brazilian freshwaters: a critical review. Environ Sci Process Impacts 18:779–787. https://doi.org/10.1039/C6EM00268D

Arora A, Scholar EM (2005) Role of tyrosine kinase inhibitors in cancer therapy. Perspect Pharmacol 315:971–979. https://doi.org/10.1124/jpet.105.084145.have

Balakrishna K, Rath A, Praveenkumarreddy Y et al (2017) A review of the occurrence of pharmaceuticals and personal care products in Indian water bodies. Ecotoxicol Environ Saf 137:113–120. https://doi.org/10.1016/j.ecoenv.2016.11.014

Besse J, Latour J, Garric J (2012) Anticancer drugs in surface waters What can we say about the occurrence and environmental signi fi cance of cytotoxic, cytostatic and endocrine therapy drugs? Environ Int 39:73–86. https://doi.org/10.1016/j.envint.2011.10.002

Boiano J, Steege A, Sweeney M (2016) Adherence to safe handling guidelines by health care workers who administer antineoplastic drugs. J Occup Environ Hyg 11:728–740. https://doi.org/10.1016/j.bbamem.2015.02.010.Cationic

Borgatta M, Waridel P, Decosterd LA et al (2016) Multigenerational effects of the anticancer drug tamoxifen and its metabolite 4-hydroxy-tamoxifen on Daphnia pulex. Sci Total Environ 545–546:21–29. https://doi.org/10.1016/j.scitotenv.2015.11.155

Brezovšek P, Eleršek T, Filipič M (2014) Toxicities of four anti-neoplastic drugs and their binary mixtures tested on the green alga Pseudokirchneriella subcapitata and the cyanobacterium Synechococcus leopoliensis. Water Res 52:168–177. https://doi.org/10.1016/j.watres.2014.01.007

Cardinale BJ, Matulich KL, Hooper DU et al (2011) The functional role of producer diversity in ecosystems. Am J Bot 98:572–592. https://doi.org/10.3732/ajb.1000364

Cavallo D, Ursini CL, Omodeo-Salè E, Iavicoli S (2007) Micronucleus induction and FISH analysis in buccal cells and lymphocytes of nurses administering antineoplastic drugs. Mutat Res Genet Toxicol Environ Mutagen 628:11–18. https://doi.org/10.1016/j.mrgentox.2006.10.014

Česen M, Kosjek T, Laimou-Geraniou M et al (2015) Occurrence of cyclophosphamide and ifosfamide in aqueous environment and their removal by biological and abiotic wastewater treatment processes. Sci Total Environ 527–528:465–473. https://doi.org/10.1016/j.scitotenv.2015.04.109

Česen M, Eleršek T, Novak M et al (2016) Ecotoxicity and genotoxicity of cyclophospamide, ifosfamide, their metabolites/transformation products and their mixtures. Environ Pollut 210:192–201. https://doi.org/10.1016/j.envpol.2015.12.017

Connor TH (2006) Hazardous anticancer drugs in health care: environmental exposure assessment. Ann N Y Acad Sci 1076:615–623. https://doi.org/10.1196/annals.1371.021

Crul M, Simmons-Sanders K (2017) Carry-over of antineoplastic drug contamination in Dutch hospital pharmacies. J Oncol Pharm Pract 24:1–7. https://doi.org/10.1177/1078155217704990

Danesi CC, Bellagamba BC, Dihl RR et al (2010) Evaluation of the genotoxicity of cisplatin, paclitaxel and 5-fluorouracil combined treatment in the Drosophila wing-spot test. Food Chem Toxicol 48:3120–3124. https://doi.org/10.1016/j.fct.2010.08.005

Dranitsaris G, Johnston M, Poirier S et al (2005) Are health care providers who work with cancer drugs at an increased risk for toxic events? A systematic review and meta-analysis of the literature. J Oncol Pharm Pract 11:69–78. https://doi.org/10.1191/1078155205jp155oa

Eleršek T, Milavec S, Korošec M et al (2016) Toxicity of the mixture of selected antineoplastic drugs against aquatic primary producers. Environ Sci Pollut Res 23:14780–14790. https://doi.org/10.1007/s11356-015-6005-2

Elzwayie A, Afan HA, Allawi MF, El-Shafie A (2017) Heavy metal monitoring, analysis and prediction in lakes and rivers: state of the art. Environ Sci Pollut Res 24(13):12104–12117. https://doi.org/10.1007/s11356-017-8715-0

Ermler S, Scholze M, Kortenkamp A (2014) Genotoxic mixtures and dissimilar action: concepts for prediction and assessment. Arch Toxicol 88:799–814. https://doi.org/10.1007/s00204-013-1170-x

Ferlay J, Soerjomataram I, Ervik M, et al (2013) GLOBOCAN 2012 v1.0, cancer incidence and mortality worldwide. IARC Cancer Base No. 11. Lyon, France, International Agency Research Cancer. http://globocan.iarc.fr

Gačić Z, Kolarević S, Sunjog K et al (2014) The impact of in vivo and in vitro exposure to base analogue 5-FU on the level of DNA damage in haemocytes of freshwater mussels Unio pictorum and Unio tumidus. Environ Pollut 191:145–150. https://doi.org/10.1016/j.envpol.2014.04.024

Gajski G, Gerić M, Domijan AM, Garaj-Vrhovac V (2016a) Combined cyto/genotoxic activity of a selected antineoplastic drug mixture in human circulating blood cells. Chemosphere 165:529–538. https://doi.org/10.1016/j.chemosphere.2016.09.058

Gajski G, Gerić M, Žegura B et al (2016b) Genotoxic potential of selected cytostatic drugs in human and zebrafish cells. Environ Sci Pollut Res Int 23:14739–14750. https://doi.org/10.1007/s11356-015-4592-6

Gerić M, Gajski G, Garaj-Vrhovac V (2014) γ-H2AX as a biomarker for DNA double-strand breaks in ecotoxicology. Ecotoxicol Environ Saf 105:13–21. https://doi.org/10.1016/j.ecoenv.2014.03.035

Graeve CU, McGovern PM, Alexander B et al (2017) Occupational exposure to antineoplastic agents. Workplace Health Saf 65:9–20. https://doi.org/10.1177/2165079916662660

Hall A, Davies H, Demers P et al (2013) Occupational exposure to antineoplastic drugs and ionizing radiation in Canadian veterinary settings: findings from a national surveillance project. Can J Public Health 104:e460–e465

Heath E, Filipič M, Kosjek T, Isidori M (2016) Fate and effects of the residues of anticancer drugs in the environment. Environ Sci Pollut Res 23:14687–14691. https://doi.org/10.1007/s11356-016-7069-3

Hernando M, Mezuca M, Fernandez-Alba A, Barcelo D (2006) Environmental risk assessment of pharmaceutical residues in wastewater effluents, surface waters and sediments. Talanta 69:334–342. https://doi.org/10.1016/j.talanta.2005.09.037

Hirose J, Kondo F, Nakano T et al (2005) Inactivation of antineoplastics in clinical wastewater by electrolysis. Chemosphere 60:1018–1024. https://doi.org/10.1016/j.chemosphere.2005.01.024

Hon CY, Teschke K, Shen H et al (2015) Antineoplastic drug contamination in the urine of Canadian healthcare workers. Int Arch Occup Environ Health 88:933–941. https://doi.org/10.1007/s00420-015-1026-1

Hrudey SE, Hrudey EJ, Pollard SJT (2006) Risk management for assuring safe drinking water. Environ Int 32:948–957. https://doi.org/10.1016/j.envint.2006.06.004

Isidori M, Lavorgna M, Russo C et al (2016a) Chemical and toxicological characterisation of anticancer drugs in hospital and municipal wastewaters from Slovenia and Spain. Environ Pollut 219:275–287. https://doi.org/10.1016/j.envpol.2016.10.039

Isidori M, Piscitelli C, Russo C et al (2016b) Teratogenic effects of five anticancer drugs on Xenopus laevis embryos. Ecotoxicol Environ Saf 133:90–96. https://doi.org/10.1016/j.ecoenv.2016.06.044

Jeong KW, Lee BY, Kwon MS, Jang JH (2015) Safety management status among nurses handling anticancer drugs: nurse awareness and performance following safety regulations. Asian Pac J Cancer Prev 16:3203–3211. https://doi.org/10.7314/APJCP.2015.16.8.3203

Keat CH, Sooaid NS, Yun CY, Sriraman M (2013) Improving safety-related knowledge, attitude and practices of nurses handling cytotoxic anticancer drug: pharmacists' Experience in a General Hospital, Malaysia. Asian Pac J Cancer Prev 14:69–73. https://doi.org/10.7314/APJCP.2013.14.1.69

Kienzler A, Bopp SK, van der Linden S et al (2016) Regulatory assessment of chemical mixtures: requirements, current approaches and future perspectives. Regul Toxicol Pharmacol 80:321–334. https://doi.org/10.1016/j.yrtph.2016.05.020

Kopjar N, Želježić D, Kasuba V, Rozgaj R (2010) Antineoplastic drugs as a potential risk factor in occupational settings: mechanisms of action at the cell level, genotoxic effects, and their detection using different biomarkers. Arh Hig Rada Toksikol 61:121–146. https://doi.org/10.2478/10004-1254-61-2010-2025

Kosjek T, Heath E (2011) Occurrence, fate and determination of cytostatic pharmaceuticals in the environment. TrAC Trends Anal Chem 30:1065–1087. https://doi.org/10.1016/j.trac.2011.04.007

Kovács R, Csenki Z, Bakos K et al (2015) Assessment of toxicity and genotoxicity of low doses of 5-fluorouracil in zebrafish (Danio rerio) two-generation study. Water Res 77:201–212. https://doi.org/10.1016/j.watres.2015.03.025

Kovacs R, Bakos K, Urbanyi B et al (2016) Acute and sub-chronic toxicity of four cytostatic drugs in zebrafish. Environ Sci Pollut Res 23:14718–14729. https://doi.org/10.1007/s11356-015-5036-z

Kračun-Kolarević M, Kolarević S, Atanacković A et al (2015) Effects of 5-fluorouracil, etoposide and CdCl2 in aquatic oligochaeta Limnodrilus udekemianus Claparede (Tubificidae) measured by comet assay. Water Air Soil Pollut 226:242. https://doi.org/10.1007/s11270-015-2511-6

Kundi M, Parrella A, Lavorgna M et al (2016) Prediction and assessment of ecogenotoxicity of antineoplastic drugs in binary mixtures. Environ Sci Pollut Res 23:14771–14779. https://doi.org/10.1007/s11356-015-4884-x

Ladeira C, Viegas S, Pádua M et al (2014) Assessment of genotoxic effects in nurses handling cytostatic drugs. J Toxicol Environ Health Part A 77:879–887. https://doi.org/10.1080/15287394.2014.910158

Landeck L, Gonzalez E, Koch OM (2015) Handling chemotherapy drugs – do medical gloves really protect? Int J Cancer 137:1800–1805. https://doi.org/10.1002/ijc.29058

Madikizela LM, Tavengwa NT, Chimuka L (2017) Status of pharmaceuticals in African water bodies: occurrence, removal and analytical methods. J Environ Manag 193:211–220. https://doi.org/10.1016/j.jenvman.2017.02.022

Marie P, Christophe C, Manon R et al (2017) Environmental monitoring by surface sampling for cytotoxics: a review. Environ Monit Assess 189:52. https://doi.org/10.1007/s10661-016-5762-9

Meffe R, de Bustamante I (2014) Emerging organic contaminants in surface water and groundwater: a first overview of the situation in Italy. Sci Total Environ 481:280–295. https://doi.org/10.1016/j.scitotenv.2014.02.053

Mišik M, Filipič M, Nersesyan A et al (2016a) Analyses of combined effects of cytostatic drugs on micronucleus formation in the Tradescantia. Environ Sci Pollut Res 23:14762–14770. https://doi.org/10.1007/s11356-015-5837-0

Mišik M, Kundi M, Pichler C et al (2016b) Impact of common cytostatic drugs on pollen fertility in higher plants. Environ Sci Pollut Res 23:14730–14738. https://doi.org/10.1007/s11356-015-4301-5

Monosson E (2005) Chemical mixtures: considering the evolution of toxicology and chemical assessment. Environ Health Perspect 113:383–390. https://doi.org/10.1289/ehp.6987

Moretti M, Grollino MG, Pavanello S et al (2015) Micronuclei and chromosome aberrations in subjects occupationally exposed to antineoplastic drugs: a multicentric approach. Int Arch Occup Environ Health 88:683–695. https://doi.org/10.1007/s00420-014-0993-y

Negreira N, Lopez de Alda M, Barcelo D (2013a) On-line solid phase extraction liquid chromatography tandem mass spectrometry for the determination of 17 cytostatics and metabolites in waste, surface and ground water samples. J Chromatogr A 1280:64–74. https://doi.org/10.1016/j.chroma.2013.01.031

Negreira N, Mastroianni N, López de Alda M, Barceló D (2013b) Multianalyte determination of 24 cytostatics and metabolites by liquid chromatography–electrospray–tandem mass spectrometry and study of their stability and optimum storage conditions in aqueous solution. Talanta 116:290–299. https://doi.org/10.1016/j.talanta.2013.04.070

Negreira N, Lopez de Alda M, Barcelo D (2014) Study of the stability of 26 cytostatic drugs and metabolites in wastewater under different conditions. Sci Total Environ 482–483:389–398. https://doi.org/10.1016/j.scitotenv.2014.02.131

Negreira N, Lopez de Alda M, Barcelo D (2015) Degradation of the cytostatic etoposide in chlorinated water by liquid chromatography coupled to quadrupole-Orbitrap mass spectrometry: identification and quantification of by-products in real water samples. Sci Total Environ 506–507:36–45. https://doi.org/10.1016/j.scitotenv.2014.10.097

Nersesyan A, Fenech M, Bolognesi C et al (2016) Use of the lymphocyte cytokinesis-block micronucleus assay in occupational biomonitoring of genome damage caused by in vivo exposure to chemical genotoxins: past, present and future. Mutat Res Mutat Res 770:1–11. https://doi.org/10.1016/j.mrrev.2016.05.003

Novak M, Žegura B, Baebler Š et al (2016) Influence of selected anti-cancer drugs on the induction of DNA double-strand breaks and changes in gene expression in human hepatoma HepG2 cells. Environ Sci Pollut Res 23:14751. https://doi.org/10.1007/s11356-015-5420-8

Novak M, Bojana Ž, Modič B et al (2017a) Cytotoxicity and genotoxicity of anticancer drug residues and their mixtures in experimental model with zebra fish liver cells. Sci Total Environ 601–602:293–300. https://doi.org/10.1016/j.scitotenv.2017.05.115

Novak M, Žegura B, Nunić J et al (2017b) Assessment of the genotoxicity of the tyrosine kinase inhibitor imatinib mesylate in cultured fish and human cells. Mutat Res Genet Toxicol Environ Mutagen 814:14–21. https://doi.org/10.1016/j.mrgentox.2016.12.002

Pal A, Gin KYH, Lin AYC, Reinhard M (2010) Impacts of emerging organic contaminants on freshwater resources: review of recent occurrences, sources, fate and effects. Sci Total Environ 408:6062–6069. https://doi.org/10.1016/j.scitotenv.2010.09.026

Parrella A, Kundi M, Lavorgna M et al (2014a) Toxicity of exposure to binary mixtures of four antineoplastic drugs in Daphnia magna and Ceriodaphnia dubia. Aquat Toxicol 157:41–46. https://doi.org/10.1016/j.aquatox.2014.09.012

Parrella A, Lavorgna M, Criscuolo E et al (2014b) Acute and chronic toxicity of six anticancer drugs on rotifers and crustaceans. Chemosphere 115:59–66. https://doi.org/10.1016/j.chemosphere.2014.01.013

Parrella A, Lavorgna M, Criscuolo E et al (2015) Eco-genotoxicity of six anticancer drugs using comet assay in daphnids. J Hazard Mater 286:573–580. https://doi.org/10.1016/j.jhazmat.2015.01.012

Poupeau C, Roland C, Bussieres J-F (2016) Surveillance urinaire des professionnels de la santé exposés aux antinéoplasiques dans le cadre de leur travail : revue de la littérature de 2010 à 2015. Can J Hosp Pharm 69:376–387

Rioufol C, Ranchon F, Schwiertz V et al (2014) Administration of anticancer drugs: exposure in hospital nurses. Clin Ther 36:401–407. https://doi.org/10.1016/j.clinthera.2014.01.016

Rush EC (2013) Water: neglected, unappreciated and under researched. Eur J Clin Nutr 67:492–495. https://doi.org/10.1038/ejcn.2013.11

Schierl R, Masini C, Groeneveld S et al (2016) Environmental contamination by cyclophosphamide preparation: comparison of conventional manual production in biological safety cabinet and robot-assisted production by APOTECAchemo. J Oncol Pharm Pract 22:37–45. https://doi.org/10.1177/1078155214551316

Schriks M, Heringa MB, van der Kooi MME et al (2010) Toxicological relevance of emerging contaminants for drinking water quality. Water Res 44:461–476. https://doi.org/10.1016/j.watres.2009.08.023

Sigman DM, Hain MP (2012) The biological productivity of the Ocean. Nat Educ Knowl 3:1–16

Simon N, Vasseur M, Pinturaud M et al (2016) Effectiveness of a closed-system transfer device in reducing surface contamination in a new antineoplastic drug-compounding unit: a prospective, controlled, parallel study. PLoS One 11:1–17. https://doi.org/10.1371/journal.pone.0159052

Sottani C, Porro B, Imbriani M, Minoia C (2012) Occupational exposure to antineoplastic drugs in four Italian health care settings. Toxicol Lett 213:107–115. https://doi.org/10.1016/j.toxlet.2011.03.028

Suspiro A, Prista J (2011) Biomarkers of occupational exposure do anticancer agents: a minireview. Toxicol Lett 207:42–52. https://doi.org/10.1016/j.toxlet.2011.08.022

Tchounwou PB, Yedjou CG, Patlolla AK, Sutton DJ (2012) Heavy metal toxicity and the environment. EXS Suppl 101:133–164

UN WWAP (2012) The United Nations World water development report 4: managing water under uncertainty and risk. UNESCO, Paris

UN WWAP (2015) The United Nations World water development report 2015: water for a sustainable world. UNESCO, Paris

Ursini CL, Cavallo D, Colombi A et al (2006) Evaluation of early DNA damage in healthcare workers handling antineoplastic drugs. Int Arch Occup Environ Health 80:134–140. https://doi.org/10.1007/s00420-006-0111-x

Van Der Putten WH, De Ruiter PC, Bezemer TM et al (2004) Trophic interactions in a changing world. Basic Appl Ecol 5:487–494. https://doi.org/10.1016/j.baae.2004.09.003

Villarini M, Gianfredi V, Levorato S et al (2016) Occupational exposure to cytostatic/antineoplastic drugs and cytogenetic damage measured using the lymphocyte cytokinesis-block micronucleus assay: a systematic review of the literature and meta-analysis. Mutat Res Rev Mutat Res 770:35–45. https://doi.org/10.1016/j.mrrev.2016.05.001

Vyas N, Yiannakis D, Turner A, Sewell GJ (2014) Occupational exposure to anti-cancer drugs: a review of effects of new technology. J Oncol Pharm Pract 20:278–287. https://doi.org/10.1177/1078155213498630

Wakui N, Ookubo T, Iwasaki Y et al (2013a) Determination of exposure of dispensary drug preparers to cyclophosphamide by passive sampling and liquid chromatography with tandem mass spectrometry. J Oncol Pharm Pract 19:31–37. https://doi.org/10.1177/1078155212451196

Wakui N, Ookubo T, Iwasaki Y et al (2013b) Development of a closed drug preparation method for oral anticancer drugs. J Oncol Pharm Pract 19:315–320. https://doi.org/10.1177/1078155212467554

Walton A, Rogers B (2017) Workplace hazards faced by nursing assistants in the United States: a focused literature review. Int J Environ Res Public Health 14:544. https://doi.org/10.3390/ijerph14050544

Xie H (2012) Occurrence, ecotoxicology, and treatment of anticancer agents as water contaminants. J Environ Anal Toxicol 02. https://doi.org/10.4172/2161-0525.S2-002

Yoshida J, Koda S, Nishida S et al (2013) Association between occupational exposure and control measures for antineoplastic drugs in a pharmacy of a hospital. Ann Occup Hyg 57:251–260. https://doi.org/10.1093/annhyg/mes061

Zhang J, Chang VWC, Giannis A, Wang J (2013) Removal of cytostatic drugs from aquatic environment : a review. Sci Total Environ 445–446:281–298. https://doi.org/10.1016/j.scitotenv.2012.12.061

Chapter 18
Environmental Metabolomics: A Powerful Tool to Investigate Biochemical Responses to Drugs in Nontarget Organisms

Giovanni Mastroianni, Monica Scognamiglio, Chiara Russo, Antonio Fiorentino, and Margherita Lavorgna

Abstract Metabolomics is the analysis of endogenous and exogenous low molecular mass metabolites within cells, tissues, or biofluids of an organism in response to an external stressor. In this chapter, we highlight the importance of the subdiscipline of environmental metabolomics, which investigates the interactions of organisms with environmental stressors such as biotic and abiotic factors, xenobiotics, temperature shifts, and chemical contaminants. Over the past decade, there has been increasing scientific interest in environmental metabolomics, most likely attributable to the comprehensive nature of nontargeted metabolomics. Hypotheses have therefore been developed on complex environmental stressor effects, especially those with unknown modes of action. The availability of a wide variety of model organisms such as freshwater organisms of the food chain has promoted the potential of metabolomics to detect stress from an extensive range of external factors. Furthermore, these dynamics may shift from individuals to populations, contemplating the traditional fields of the ecophysiology and ecology from instantaneous effects to those over evolutionary timescales. In this chapter, we provide an overview of analytical instrumentation, extraction methods, general experimental design, and the statistical methods generally used in environmental metabolomics. Despite the difficulty in understanding the consequences of environmental exposure due to inter- and intra-individual variability, we believe that environmental metabolomics may enrich our understanding of the responses of organisms to the numerous types of environmental stressors.

Keywords Metabolomics · Aquatic organisms · NMR · Xenobiotics · Environment

---

Giovanni Mastroianni and Monica Scognamiglio equally contributed to the manuscript.

G. Mastroianni · M. Scognamiglio · C. Russo · A. Fiorentino (✉) · M. Lavorgna
Department of Environmental, Biological and Pharmaceutical Sciences and Technologies, University of Campania "L.Vanvitelli", Caserta, Italy
e-mail: antonio.fiorentino@unicampania.it

E. Heath et al. (eds.), *Fate and Effects of Anticancer Drugs in the Environment*,
https://doi.org/10.1007/978-3-030-21048-9_18

Abbreviations

CE-MS	Capillary electrophoresis mass spectrometry
DIMS	Desorption ionization mass spectrometry
FT-ICR-MS	Fourier transform ion cyclotron resonance mass spectrometry
FT-IR	Fourier transform infrared spectroscopy
GC/ToF-MS	Gas chromatography–time-of-flight mass spectrometry
GC-MS	Gas chromatography mass spectroscopy
H-NMR	Proton nuclear magnetic resonance
LC-HRMS	Liquid chromatography high resolution mass spectrometry
LC-MS	Liquid chromatography mass spectrometry
MAS-NMR	Magic angle spinning nuclear magnetic resonance
NMR	Nuclear magnetic resonance
Q-TOF LC/MS	Quadrupole time-of-flight liquid chromatography mass spectrometry

18.1 Introduction

Great scientific change in molecular biology and biochemistry approaches and techniques has begun since the sequencing of human genome in the 1990s. Automated micro-array methods to detect changes in gene expression and the ability to assay and identify proteins by mass spectroscopy methods led respectively to two new revolutionary disciplines: transcriptomics and proteomics which are useful for the comprehension of the complex interactions between genetic make-up and environmental factors. It is crucial to underline that the small molecules involved in biochemical processes provide a wide range of information on the status of living systems when studying changes in genes expression due to every variation in life conditions and external perturbations. The process of monitoring and evaluating such changes is termed “metabonomics” or “metabolomics” when mostly model organism or plant system are studied (Lindon et al. 2006).

The use of NMR spectroscopy combined with multivariate statistic investigation of the complex data sets led to the following definition: “the quantitative measurement of the dynamic multiparametric metabolic response of living systems to pathophysiological stimuli or genetic modification” by Jeremy Nicholson and colleagues (1999).

The metabolites, or small molecules, within a cell, tissue, organ, biological fluid, or entire organism constitute the metabolome (Miller 2007). Metabolomic analyses can be categorized as either nontargeted or targeted (Issaq et al. 2008; Verpoorte et al. 2008). Nontargeted metabolomics is a nonbiased quantitative analysis of all—or a large number of—metabolites found in a biological sample (Issaq et al. 2008). By contrast, targeted metabolomics analyzes a specific group of metabolites (Issaq et al. 2008; Verpoorte et al. 2008).

The complementary role of metabolomics in regard to other omic techniques may actually provide a potential solution to the many weak points that are encountered with other omic methods (Griffin and Bollard 2004; Bilello 2005; van Ravenzwaay et al. 2007). Indeed, despite the development of methods to detect changes in genomic, transcriptomic, and proteomic profiles, key information needed to make significant interpretations based on these data are usually not promptly available (Ankley et al. 2006). Changes in gene expressions and protein synthesis due to external stressor exposure of an organism usually cause the activation of homeostatic controls and feedback mechanisms; these changes could be intensified at the metabolome level (Nicholson et al. 2008; Ankley et al. 2006; van Ravenzwaay et al. 2007). As a result, metabolomics could be considered a more sensitive and reliable indicator of the external stress than other omic technologies (Nicholson et al. 2008; Ankley et al. 2006; van Ravenzwaay et al. 2007).

The popularity of metabolomics in many fields of scientific research like nutrition, medicine, clinical pharmacology, and toxicology has grown considerably thanks to the ability to detect subtle molecular changes and the comprehensive nature of metabolite measurements (Lin et al. 2006; Chen et al. 2008; Fialho et al. 2008). As a result, metabolomics is nowadays considered a rapid and sensitive technique that would be able of clear up relationships between metabolite levels and an external stressor, be it contaminant exposure, nutritional deficit, or disease (Miller 2007). For example, cancer cells' metabolic profiles have been used to monitor and comprehend tumor progression (Griffiths et al. 2002; Griffiths and Stubbs 2003; Griffin and Shockcor 2004). Metabolomic investigations in non-model organisms will be particularly exciting in the characterization of organism metabolic responses to anthropogenic (or manmade) stressors, such as pollutants and climate changes, highlighting another important application of the environmental metabolomics. This approach would utilize "sentinel" (or representative) organisms of a particular ecosystem to reveal the condition of the environment (Viant 2008b).

In the last decades, in order to understand the toxic effects of several kinds of xenobiotic residues (pharmaceuticals, pesticides, nanoparticles, heavy metals) in the environment and the related biological changes, various nontarget organisms of the aquatic food chain have been selected as bioindicators for their suitable characteristics. In fact, invertebrates such as bivalve molluscs *Mytilus edulis* and *Mytilus galloprovincialis* (Tuffnail et al. 2009; Fasulo et al. 2012; Bonnefille et al. 2018), crustaceans *Daphnia magna* and *Gammarus pulex* (Taylor et al. 2010; Gómez-Canela et al. 2016; Kovacevic et al. 2016; Nagato et al. 2016; Wagner et al. 2017) and vertebrates like fish such as *Danio rerio* and *Oncorhynchus mykiss* (Samuelsson et al. 2006; De Sotto et al. 2017) represent a valid tool to study the ecotoxicogenomics and the metabolomics. These bioindicators are chosen for these kinds of studies because: they are important components of the aquatic ecosystem, are easy to recognize, are sensitive to a wide range of stressors, are abundant and widely distributed, and are suitable for laboratory experiments. In addition, these aquatic nontarget organisms have been utilized by different scientific environmental researchers (Brezovšek et al. 2014; Parrella et al. 2014, 2015; Isidori et al. 2016a; Kundi et al. 2016; Russo et al. 2018; Kovács et al. 2015, 2016; Gačić et al. 2014) through specific tests according to international standard guidelines not only for highlighting physiological alterations (mortality, offspring reduction, inhibition of growth, malformations) but also for evaluating genotoxic, mutagenic and teratogenic damages. Since there is not a clear relationship between biochemical mode of action of xenobiotics and defined endpoints such as mortality and reproduction, a detection of the metabolic profile in aquatic sentinel organisms may be of great scientific interest. In fact, significant variations in amino acids, in glucose and other metabolite concentrations in cells, tissues, or biofluids (Kovacevic et al. 2016; Wagner et al. 2017) may occur in organisms after sub-lethal exposure to hazardous and pseudo-persistent contaminants like pharmaceuticals. Indeed, among all contaminants, pharmaceuticals are frequently detected in aquatic ecosystems because of their high consumption and scarce removal efficiencies by wastewater treatment plants (Negreira et al. 2014; Lenz et al. 2007; Isidori et al. 2016b).

Among pharmaceuticals, antineoplastic drugs are suspected to be hazardous for aquatic nontarget species (Parrella et al. 2014, 2015; Isidori et al. 2016a; Kundi et al. 2016; Russo et al. 2018). To the best of our knowledge, there are no studies in literature regarding the overall metabolomic responses in different nontarget organisms exposed to antineoplastic drugs, while only few studies (Table 18.1) have utilized metabolomics as a robust tool to evaluate the biological environmental responses in the aquatic sentinel species exposed to other pharmaceutical classes. On the other hand, in recent studies (Laith et al. 2017; Mumtaz et al. 2017; Ruiz-Torres et al. 2017), metabolomic approach has been considered a good strategy to identify, select, and provide secondary metabolites from natural promising sources, such as plants and marine invertebrates and vertebrates, as new drugs for cancer therapy, rather than using metabolomics for studying environmental biological effects.

In this chapter, we provide an overview of the experimental design, analytical techniques, and statistical methods used in environmental metabolomics as well as an overview of recent studies using aquatic nontarget organisms in metabolomics to demonstrate the potential of this technique to detect and understand the mechanisms of exposure to some pharmaceuticals.

18.2 Experimental Design and Analytical Methods Used in Environmental Metabolomics

The basic procedure used in an environmental metabolomics study is outlined in Fig. 18.1. Furthermore, an accurate outline of the entire metabolomic experimental scheme from the experimental design, to data mining and biological interpretation is described by Wolfender et al. (2013).

The first step in any metabolomics experiment is the experimental design, which, in case of environmental metabolomics, involves the selection of a model organism or microorganism, a type of external stressor (e.g., exposure to contaminants, heat/cold, starvation, or disease), and the mode/route of exposure. A good experimental design mainly depends on the starting point and goals of the research. Focusing on the biological question underpinning the research is the most crucial step, as

Table 18.1 Summary of metabolomic responses to pharmaceuticals in aquatic organisms

Organism	Pharmaceutical	Metabolic response	Method	References
Mytilus galloprovincialis (Mussel)	Diclofenac	Tyrosine and tryptophan metabolism	LC-HRMS	Bonnefille et al. (2018)
Daphnia magna (Crustacean)	Propranolol	−Growth; +amino acid metabolites; −glucose; +Fatty acid and oxylipid metabolites	^{1}H-NMR	Wagner et al. (2017)
			FT-ICR MS	Taylor et al. (2010)
	Carbamazepine	+Growth; −amino acid metabolites; +glucose	^{1}H-NMR	Kovacevic et al. (2016)
	Ibuprofen	−Growth; −amino acid metabolites; −glucose	^{1}H-NMR	Kovacevic et al. (2016)
	Triclosan	−Growth; +amino acid metabolites; −glucose	^{1}H-NMR	Kovacevic et al. (2016)
Gammarus pulex (Crustacean)	Nimesulide	Protein synthesis, oxidative stress and signaling cascades	LC-HRMS	Gómez-Canela et al. (2016)
	Triclosan			
	Propranolol			
Tautogolabrus adspersus (Fish)	Tamoxifen	Estrogen metabolism	GC/ToF-MS	Mills et al. (2016)
Oncorhynchus mykiss (Fish)	Ethinylestradiol (EE2)	Lipid metabolism	^{1}H-NMR	Samuelsson et al. (2006)
Danio rerio (Fish)	Clarithromycin	Purines metabolism	Q-TOF LC/MS	De Sotto et al. (2017)
	Florfenicol			
	Sulfamethazine			

+ increase, − decrease

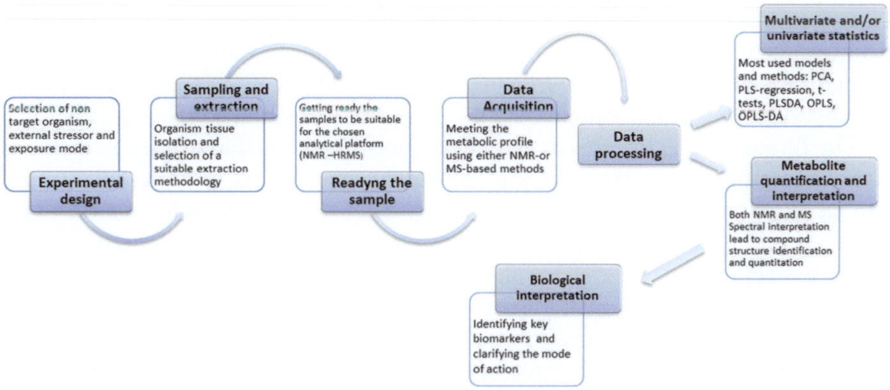

Fig. 18.1 Environmental metabolomics experiment workflow (from Simpson and McKelvie 2009; Lankadurai et al. 2013). NMR, nuclear magnetic resonance spectroscopy; MS, mass spectrometry; PCA, principal component analysis; PLS-regression, partial least-squares regression analysis. In some cases, a further step might be present between the design of experiment and the sampling, consisting in the experimental performance when the experiment has to be performed in laboratory controlled conditions

metabolomics usually takes into account a large number of samples (Brown et al. 2008). Taking into consideration the biological variability and choosing a suitable number or replicates from an early stage of the research is also compelling, for organisms grown in either natural or controlled conditions.

When planning and carrying out the experiment, it is extremely important to consider that metabolism is highly dynamic and changes occur at different timescales, depending on the organism, and on the metabolic pathway considered. For example, some changes might occur according to the developmental stage or phenology (Scognamiglio et al. 2014) while others to circadian clock (Eckel-Mahan et al. 2012), with the results that the first changes can only be observed on a longer timescale, while the second ones are responsible for cyclic changes in metabolites' concentration during the light/darkness alternation. It has to be specified that there are many other physiological reasons for metabolic alterations that also depend on the organism (and on the biological medium) taken into consideration. Therefore, this dynamic feature has to be borne in mind when planning timing for treatment, sampling, and so on, in order to make sure to detect and discriminate the changes caused by the external perturbation from the ones due to physiological changes in the metabolism. Taking into consideration the two parameters previously mentioned (developmental stage and circadian clock), for example, it is crucial to collect samples at the same moment of the day and at the same organism growth stage. Monitoring environmental variations (e.g., photoperiod, relative humidity, temperature) throughout the progress of the experiment is important as well.

Once the organism has been exposed to the external stressor, the biological medium to study will be harvested and may include blood, urine, or other biological fluids, and (or) tissue/organ extracts (Simpson and McKelvie 2009). Quenching metabolism immediately after exposure is essential to minimize the influence of puzzling variables in the analysis of the metabolic response (Lin et al. 2006). This is commonly done by flash freezing with liquid nitrogen and storing at −80°C. However, in order to improve precipitation and inactivation of soluble enzymes, acid treatment, or extraction with cold mixtures of organic solvents such as methanol, ethanol, acetone, or acetonitrile might be used (Lin et al. 2006). When suitable, also freeze drying of the samples is a good practice (Kim and Verpoorte 2010).

The further step is the choice and development of a suitable extraction method. Considering the high variability and complexity of matrixes to be tested, sample extraction and preparation methods can vary a lot depending on the matrix to be analyzed and on the analytical platform used (Kim and Verpoorte 2010). As no single extraction method can isolate every metabolite within a sample, a proper extraction procedure will need to be selected and tested, based on the goals of the experiment. In general, an aqueous buffer extraction is sufficient to obtain a polar metabolite profile, but a more rigorous extraction involving a mixture of polar and nonpolar solvents is required to extract both polar and nonpolar metabolites (Wu et al. 2008). Various extraction methods involving a mixture for organic solvents (methanol, chloroform, and acetonitrile) and water in varying ratios have been examined by Lin et al. (2006). The methanol/chloroform/water (final solvent ratio of 2/2/1.8, respectively) extraction method, which was first described by Bligh and Dyer (1959), has been considered one of the most reproducible and with the highest recovery of both polar and nonpolar metabolites. Wu et al. (2008) then went a step further, and examined three different strategies to add the methanol, chloroform, and water to the tissue samples for extraction: (i) stepwise addition—the original Bligh and Dyer (1959) method of adding each solvent one by one; (ii) two-step addition—methanol and water are added in step one and chloroform and water are added in step two; (iii) all-in-one addition—all three solvents are added together. They stated that the two-step addition was the best out of the three: this assessment is based on metabolite yield, extraction reproducibility, and sample throughput. Recently, Liebeke and Bundy (2012) compared four different solvent systems for the extraction of metabolites from the tissue of the earthworm *Lumbricus rubellus*: (i) chloroform: methanol: water, 2:1:1 (CMW); (ii) 75% hot ethanol (hEt); (iii) acetonitrile:methanol:water, 2:2:1 (AMW); (iv) isopropanol:methanol: water, 2:5:2 (IMW). Extracts were analyzed using NMR, gas chromatography (GC)–MS and Fourier transform ion cyclotron resonance (FT-ICR) MS. They determined that the AMW extraction gave the best results in terms of reproducibility and good yield for metabolite extraction (Liebeke and Bundy 2012). It has furthermore been shown that in case of plant material, a mixture of phosphate buffer and MeOH (1:1) is usually able to extract big part of the metabolome (Kim and Verpoorte 2010).

Also, the presence of proteins binding the typical used NMR internal standards 2, 2 dimethyl-2-silapentane-5-sulfonate sodium salt (DSS) and sodium 3-trimethylsilyltetradeuteriopropionate (TSP) in aqueous buffer extracts has led to large differences in the quantification of metabolite concentrations (Nowick et al. 2003). Consequently, the development of extraction methods that not only capture both polar and nonpolar metabolites but also include the precipitation of proteins is essential. Nevertheless, one of the main problems of metabolomics is still the lack of a standardized and reproducible extraction protocol, so a great effort should be made in this direction.

Once the extractions are completed, the samples need to be prepared for the analytical platform of choice. A list of the available techniques and of the viewpoints on their advantages and disadvantages in metabolomics applications have been extensively discussed in the literature (Table 18.2) (Scognamiglio et al. 2015). It is important to emphasize that the choice of the analytical platform and strategy heavily depends on the object of the research, and it is commonly acknowledged that the better results can be achieved combining different extractions and analytical measurements (Allwood et al. 2008; Kim and Verpoorte 2010). However, nuclear magnetic resonance (NMR) spectroscopy or mass spectrometry (MS) are doubtlessly considered the most powerful tools and the ones that will be here described more in detail. NMR and MS use in metabolomics is so widespread that in the past there was a common misconception that "metabonomics" dealt with NMR-derived metabolic profiling studies, while "metabolomics" dealt with MS-derived metabolic profiling studies (Robertson 2005).

Table 18.2 Reviews and recent environmental metabolomics papers with corresponding analytical techniques (highlighted box indicates the reviewed\used methodology)

Paper	NMR	MAS-NMR	LC-MS	GC-MS	CE-MS	DIMS	FT-ICR-MS	FT-IR	
Allwood et al. 2008	■	■	■	■	■	■	■	■	Reviews on analytical techniques used in metabolomics
Alonso et al. 2015	■		■	■					
Dunn and Ellis 2005	■		■	■	■	■		■	
Dunn et al. 2012			■	■	■	■			
Fukusaki and Kobayashi 2005	■		■	■	■	■		■	
Goodacre et al. 2004	■		■	■			■		
Hall 2006	■		■	■	■	■	■		
Kim et al. 2011	■	■							
McGhie and Rowan 2012	■		■	■					
Schripsema 2009	■		■	■	■		■		
Sumner et al. 2003	■		■	■					
Verpoorte et al. 2007	■		■	■					
Wishart 2008	■								
Wolfender et al. 2013	■	■	■	■	■	■	■		
Xiao et al. 2012			■						
Bird et al. 2011			■						Recent environmental metabolomics papers
Cappello et al 2016	■								
Cappello et al 2017	■								
Yingrong Chen et al 2015	■								
Chen et al. 2011			■						
Chiu et al 2017	■								
Creek et al. 2012			■						
De Sotto et al 2017			■						
Garreta-Lara et al 2016				■					
Ghaste et al 2016							■		
Gillis et al 2017				■					
Kokushi et al 2017				■					
Kovacevic et al 2016	■								
Lloyd et al 2017			■						
Nagato et al 2017	■								
Gil-Solsona et al 2017			■						
Van der Hooft et al 2016			■	■					
Wang et al 2017	■								
Wu et al 2017				■					
Zhao et al 2016	■			■					

In MS, the analysis is done after the fragmentation of the molecules, which this leads to the generation of ions that are later separated by their mass-to-charge ratio and finally analyzed by a detector. A number of ion sources and of analyzers is available (Xiao et al. 2012). Direct injection MS enables the injection of a crude extract directly into an electrospray mass spectrometer, resulting in one spectrum per sample, but this method is not particularly quantitative and not often used (Lin et al. 2006).

MS use in metabolomics is usually coupled to liquid chromatography (LC; Wu et al. 2005; Bajad and Shulaev 2007; Hughes et al. 2009), gas chromatography (GC; McKelvie et al. 2009; Flores-Valverde et al. 2010; Warren et al. 2012), or capillary electrophoresis (CE; Sato et al. 2004; Ramautar et al. 2012; Yamamotoya et al. 2012). These chromatographic techniques separate the complex sample mixtures so that they can be analyzed by MS, but this can make the overall analysis time consuming (Robertson 2005; Pan and Raftery 2007). Also, GC methods usually entail elaborate derivatization steps that are very lengthy and thus inconvenient for high-throughput analysis (Lin et al. 2006), besides introducing bias in the analysis due to the used chemical reactions. However, the combination of MS with chromatography, when there is availability of pure certified chemical standards, is a useful approach for identification and quantification. Indeed, retention time can be considered as an additional hint of metabolite identity, while the chemical standards are used also to set up proper external calibration curves for compounds quantification (Allwood et al. 2008).

Mass fragment databases are usually employed for easy preliminary identification of compounds (Noctor et al. 2007; Pan and Raftery 2007). Only the use of tandem MS (MS-MS), preferably HR $(MS)^n$ instruments allows for structural definition of compound identities. This usually excludes the countless, to date unknown metabolites, which are reported as "unknowns" or as putatively identified metabolites. In this case, isolation and characterization by NMR is fundamental to definite structural elucidation (Sumner et al. 2007).

The newest LC-MS/MS approaches are a useful tool for metabolite identification and quantification (Xiao et al. 2012), although it is always recommended to follow published guidelines and to refer to minimum reporting standards for the level of confidence of chemical structural elucidation (Fernie et al. 2011; Goodacre et al. 2007; Sumner et al. 2007).

The biggest advantage of MS-based techniques is sensitivity (typically picogram level), making the technique very suitable in studies that are targeting novel biomarkers (Robertson 2005; Pan and Raftery 2007) but, on the other hand, the identification of unknown metabolites is problematic. FT-ICR MS in particular provides high resolution and mass accuracy, but the instrument is costly and is thus not widely used (Pan and Raftery 2007).

The principal downside of MS-based approaches is their difficult standardization, as a consequence of a number of combinations of chromatographic systems (the separation step also includes bias in the analysis), ion sources, and different analyzers that highly impact the analysis output.

Beyond the potentially long and not always reproducible chromatographic separations, the difficult standardization, and the structural elucidation power limitation, other drawbacks of MS include: the presence of matrix effects, the destructive nature, its selectivity for certain analytes, ion suppression causing extensive variations of signal intensities, and the lack of more robust methods for chromatographic separations (Robertson 2005; Pan and Raftery 2007).

Often, due to its selectivity, MS has been used in targeted metabolomics studies (Edwards et al. 2006; Lutz et al. 2006; Issaq et al. 2008; Southam et al. 2011; García-Cañaveras et al. 2012) or in the elucidation or confirmation of metabolites first observed by NMR (Bundy et al. 2002a; Crockford et al. 2006).

On the other hand, NMR is nondestructive, nonselective, possesses cross-laboratory reproducibility, and lacks sample bias (Robertson 2005; Pan and Raftery 2007; Viant et al. 2009). As a result, NMR has been used extensively in the nontargeted or comprehensive studies of all or most of all the metabolites in a biological sample (Verpoorte et al. 2008; Simpson and McKelvie 2009; Whitfield Åslund et al. 2011a, b; Li et al. 2012; Ritota et al. 2012).

NMR is an instrumental analytical technique that allows obtaining detailed information on the structure of molecules by observing the behavior of atomic nuclei in a magnetic field. The frequency at which each nucleus resonates depends on its chemical surrounding environment so each compound has a highly specific spectrum, which is an information-rich graph. Rapid identification of all of the compounds present in a mixture can be performed thanks to the combination of one-dimensional (1D) and two-dimensional (2D) NMR techniques. Recent advances in the identification of unknown compounds in the analyzed mixtures allow NMR to obtain important structural information without requiring further purification (Forseth and Schroeder 2011). Indeed, one of the main strengths of this technique is the unique set of structural information furnished that in most cases guarantees the definitive structural elucidation of the compounds, sometimes including stereochemistry.

Besides its power in structural elucidation, ^{1}H NMR is commonly used in metabolomics thanks to several other advantages: easy sample preparation, ease of standardization, and high reproducibility, and the solvent used and the magnetic field strength being the only variables (Verpoorte et al. 2007). Indeed, in NMR-based metabolomics the only bias is introduced by the solvent choice (Allwood et al. 2008) and the reproducibility appears very good and allows a comprehensive identification and quantification of a large number of compounds with short analytical times (including the extraction procedures). The need of a standardization procedure for the metabolome extraction previously mentioned makes NMR-based metabolomics convenient thanks to minimum sample preparation. In fact, extraction can be carried out directly in deuterated solvents, often with a mixture of phosphate buffer in D_2O and MeOD (1:1). Furthermore, NMR is fully quantitative (Kim et al. 2011; Wishart 2008). The calculated precision and accuracy of a 500 MHz NMR spectrometer in a quantitative ^{1}H NMR analysis of external standards has been demonstrated to be around 1% (Burton et al. 2005).

The main disadvantage of NMR, compared to MS, is its low sensitivity. This is an issue in the analyses of endogenous metabolites and in particular detecting novel biomarkers. Indeed, these metabolites are usually present at very low levels that cannot be reached with NMR. The sensitivity of ^{1}H NMR is also dependent on the number of protons in the molecule, the structure and size of the molecule. Nowadays, the problem of low sensitivity can be attenuated using ultrahigh magnetic field strength NMR spectrometers and probes that are cryogenically cooled to 4.5 K; these probes may result in a four-times raise in sensitivity (Logan et al. 1999; Griffin 2003; Pan and Raftery 2007; Grimes and O'Connell 2011) reaching the stage at which structures can be solved using very small quantities (in the microgram range) (Harvey et al. 2015). The other advantage of microcoil probe use is also the lower sample mass requirements, which is a big benefit for small organisms (Lacey et al. 1999; Pan and Raftery 2007; Grimes and O'Connell 2011; Poynton et al. 2011). The only disadvantage with these methods is the affordability of such high-end instrumentation and although sensitivity has been drastically increased, NMR is still surely less sensitive when compared to mass spectrometry (Forseth and Schroeder 2011; Kim et al. 2011). Nonetheless, compared to MS, the sensitivity of NMR is independent from metabolite pKa or hydrophobicity, making it a very adaptable choice for representative analyses (Pan and Raftery 2007). An additional downside of NMR is that some classes of lipids can only be identified as total groups and not as individual compounds by means of 1D NMR.

A large part of environmental metabolomics studies still use NMR because of the comprehensive nature of nontargeted metabolomics and the ability to generate hypotheses involving complex environmental stressors for which there are no known modes of action (Bundy et al. 2001; Tjeerdema 2008; Viant 2008a; Viant et al. 2009; Brown et al. 2009; Simpson and McKelvie 2009). For all these reasons, NMR is an ideal environmental metabolomics discovery tool. It should be noted that once metabolites of interest are discovered using NMR, targeted MS-based methods can be subsequently developed for the routine monitoring of these metabolites.

The majority of NMR-based metabolomics studies still use one-dimensional (1D) ^{1}H NMR experiments (Bundy et al. 2002b, c; Samuelsson et al. 2006; Brown et al. 2010; McKelvie et al. 2010, 2011). 1D ^{1}H NMR experiments are advantageous for metabolomic studies, which usually have hundreds of samples, because of their short acquisition times (10–15 min per sample), allowing for high-throughput analyses and a high number of sample replicates (Pan and Raftery 2007; Yuk et al. 2010).

Analyzing aqueous samples using ^{1}H NMR requires the application of water suppression techniques (Nicholson and Wilson 1989; McKay 2009). Water concentration in samples is much higher (50 mol/L) compared to the millimolar metabolite concentrations, leading to the suppression of signal intensities in the peaks of other compounds due to saturation of the NMR receiver by the H_2O signal (Bothwell and Griffin 2011). While deuterated solvents are mostly used (deuterium resonates at a different frequency than ^{1}H in the NMR), there is always residual H_2O present in samples. The best and most used water suppression methods for metabolomics are presaturation, nuclear overhauser effect spectroscopy (NOESY) presaturation, and presaturation utilizing relaxation gradients and echoes (PURGE; Bundy et al. 2002a; Viant et al. 2003; Simpson and Brown 2005; Wishart 2008; McKelvie et al. 2010; Poynton et al. 2011). Several studies have used PURGE water suppression for NMR-based metabolomic analyses (McKelvie et al. 2011, 2013; Yuk et al. 2011, 2013). Among the several water suppression techniques, PURGE provided superior water suppression with the least amount of parameter optimization and the fewest number of spectral regions that need to be excluded because of variations in the suppression of the solvent peak. Furthermore, comparing various 1D and 2D NMR techniques, PURGE ^{1}H NMR has demonstrated to be the most rapid, informative, and economic method for analyzing aqueous metabolomics samples (McKay 2009; Yuk et al. 2010).

The complexity of any biological sample due to the large number of molecules that they possess, results in a large number of peaks within the small chemical shift range of a ^{1}H NMR spectrum (0–14 ppm). This leads to difficulty in identifying compounds that are present at low concentrations, considering the common chance of peak overlapping generated by different metabolites. This usually means that some peaks are masked by larger peaks from compounds present in higher concentrations (Pan and Raftery 2007). Some of the best techniques to alleviate the spectral overlap and improve resolution between peaks are: Carr–Purcell–Meiboom–Gill (CPMG), J-resolved spectroscopy (J-RES), and other various 2D NMR techniques.

Carr–Purcell–Meiboom–Gill (CPMG) is used to remove broad resonances associated with molecules of high molecular mass or molecules with constrained motion and hereby offers better resolution of low molecular mass metabolites (Weckwerth 2007; Wishart 2008).

J-resolved spectroscopy (J-RES) projection may improve spectral resolution. J-RES is a two-dimensional (2D) NMR technique, in which the chemical shift information is on one axis and the spin–spin coupling information is on another. Projecting only the chemical shift axis, an equivalent to a 1D proton decoupled spectrum is obtained, which has less spectral overlap and enables better detection of specific metabolites (Viant et al. 2003; Pan and Raftery 2007; Yuk et al. 2010), although it also results in the loss of some information.

Other 2D NMR techniques have also been used to increase spectral resolution because they have an additional dimension into which the signals can be dispersed. Besides J-RES, some of the other common 2D NMR techniques in metabolomics are ^{1}H correlation spectroscopy (COSY) and ^{1}H-^{13}C heteronuclear single quantum coherence (HSQC) spectroscopy (Xi et al. 2008; Ludwig and Viant 2010; Yuk et al. 2010; Flores-Sanchez et al. 2012). The main benefit of using 2D NMR techniques, such as HSQC, is that the ^{13}C axis has a large chemical shift range (200 ppm), which allows for greater spectral dispersion and enhanced resolution (Xi et al. 2008; Chylla et al. 2011; Hu et al. 2011a, b). However, a drawback of most 2D NMR techniques is the lower sensitivity, which results in longer acquisition times—sometimes many times more than 1D experiments (Jacobsen 2007). In fact, 2D experiments such as HSQC may require long acquisition times for adequate signal-to-noise (S/N) ratios. For this reason, the use of 2D NMR techniques in metabolomic studies is limited to complementing compound identification from 1D ^{1}H NMR experiments (Yuk et al. 2010).

After the samples are analyzed, the data are processed and statistical analyses are performed using multivariate and univariate analyses. These are done in conjunction with the quantification and identification of the metabolites.

The final step then involves biological interpretation of the data to make a connection between the external stressor and the metabolic response of the organism.

18.3 Statistical Methods Used in Metabolomics

The interest in metabolomics is due to its ability to generate large volumes of data in a high-throughput way, so one of the biggest challenges is to find a way to visually analyze all of the collected data (i.e., NMR or MS data) to identify differences between samples in a reasonable time frame (Robertson 2005). Both, multivariate statistical and pattern recognition methods are employed to smooth the analysis of metabolomics data sets and to obtain meaningful relationships between the external stressor and the metabolic response (Trygg et al. 2007; Coen et al. 2008). Pattern recognition methods are able to reduce the dimensionality of metabolomics data from hundreds of variables into two or three components that are orthogonal to each other (Trygg et al. 2007).

PCA is an unsupervised method, meaning that the model is not provided with any prior information concerning the identity of the samples (Holmes and Antti 2002), and is the most widely used multivariate statistical approach in metabolomics (Bundy et al. 2002a, 2004; Trygg et al. 2007; Wishart 2008; Simpson and McKelvie 2009). The association of the samples in a PCA scores plot is based on the similarities in their metabolic profiles. PCA figures out the comprehensive variability in a data set, which is explained by a set of uncorrelated variables called principal components (PCs); these are linear combinations of the original variables (Trygg et al. 2007). The first PC (PC1) explains most of the variation in the data and PC2, which is orthogonal to PC1, explains the second most variation in the data and so on. PCA allows for dimensional reduction of the data into a low-dimensional plane, such as PC1 *versus* PC2. The scores plot (e.g., PC1 *vs.* PC2) allows for a visual examination of the relationship between the samples based on their metabolic profiles.

PLS regression analysis and PLS discriminant analysis (PLS-DA) are also used often as multivariate statistical tools in metabolomics (Barker and Rayens 2003; van Ravenzwaay et al. 2007; Ekman et al. 2008; Jones et al. 2008; Whitfield Åslund et al. 2011b). PLS-regression and PLS-DA are supervised methods. In this case, the classification of the samples as either control or experimental is known to the model. Predictive models are built adding predefined variables to maximize the separation between the sample classes. These variables are measurable quantities such as the contaminant exposure concentration. In PLS-DA these are dummy variables: for example, we can assign all the controls a value of zero and the experimental group a value of one to distinguish the sample classes (Trygg et al. 2007).

In order to reduce models' components, make it easier to interpret and more relevant, supervised methods such as orthogonal projections to latent structures (OPLS) and OPLS-DA have been increasingly used in metabolomics studies (Trygg and Wold 2002). These are basically extensions of PLS and PLS-DA where the orthogonal variation to the predefined variables is removed from the model (Trygg and Wold 2002; Bylesjö et al. 2006), but could also be analyzed together with the identification of the uncorrelated variables sources.

Cross-validation methods such as the leave-one-out cross validation (LOOCV) are required to evaluate the robustness of the supervised methods such as PLS-regression, PLSDA, OPLS, and OPLS-DA (Westerhuis et al. 2008; Varmuza and Filzmoser 2009; Whitfield Åslund et al. 2011b). LOOCV is performed by first randomly eliminating one of the samples from the original data set, which is called the test set, then the model (PLS/PLS-DA or OPLS/OPLS-DA) is built on the remaining samples (the training set). This process is repeated until all of the samples have been left out of the model at least once. The training set models created are eventually used to predict the test set. The Q2Y, which is known as the goodness of prediction (Westerhuis et al. 2008), represents the ability of the model to predict the test set. This value can be used to evaluate the robustness of a model: typically, a Q2Y > 0.4 is considered a strong model (Jones et al. 2008; Westerhuis et al. 2008). The significance of PLS/OPLS models needs to be evaluated, and this can be done using permutation testing (Eriksson et al. 2006; Alam et al. 2010; Whitfield Åslund et al. 2011b). Permutation testing consists of maintaining the data set constant, while randomly permuting the order of the predefined variables a set number of times. For each permutation a new PLS/OPLS model is fitted, and the Q2Y is calculated providing a reference distribution of the Q2Y statistic. The significance of the original PLS/OPLS model and the confidence in its validity is increased if its Q2Y value is higher than the values obtained for all the PLS/OPLS models built during the permutation tests (Eriksson et al. 2006).

Although, metabolomics studies mostly use multivariate statistics, complementary univariate statistical analyses are also attended to further increase the amount of information obtained from the research. T tests are commonly used to assess the significance of the separation between the controls and stressed organisms in PCA and PLS-DA scores plots, and to define which metabolites in the ^{1}H NMR spectra of the treatment class increased or decreased significantly relative to the controls. A T test filtered difference ^{1}H NMR spectra can also be created by subtracting the buckets of the average controls from that of each average exposure class. Not statistically significant (= 0.05) bucket values metabolite peaks can be replaced with a zero, resulting in a T test filtered ^{1}H NMR difference spectrum (Ekman et al. 2008, 2009). T test filtered difference ^{1}H NMR spectra and loading plots can be used together to determine which metabolites are potential biomarkers of exposure to a particular contaminant.

18.4 Metabolomic Responses Observed in Aquatic Nontarget Organisms Exposed to Pharmaceuticals

Despite the immense potential of metabolomic research for assessing environmental pollutants, only a small number of studies have been conducted till now to evaluate the metabolomic responses observed in various aquatic nontarget organisms exposed to pharmaceuticals. In fact, Bonefille et al. in 2018, evaluated the effects of the nonselective, nonsteroidal anti-inflammatory drug diclofenac against the marine

Mytilus galloprovincialis, chosen for its ease in being handled, for its capability in accumulating toxins, and for its sedentary nature. In this mussel, these authors studied metabolomic perturbations caused by 100 μg/L diclofenac, concentration not affecting organisms' viability; then, the metabolomic analysis was performed by liquid chromatography-hyphenated to high-resolution mass spectrometry (LC-HRMS) in extracts of digestive gland, and alterations in the tyrosine and tryptophan metabolisms were observed at concentrations only few orders of magnitude higher than those found in seawater (1 μg/L, Gaw et al. 2014). In particular, after a 7-day exposure, tyrosine pathways were down-modulated, while steroid hormone biosynthesis and tryptophan pathways were positively modulated.

In addition to mussels, other nontarget invertebrates such as crustaceans were suitable tools for metabolomic analysis. In fact, Taylor et al. (2010), Kovacevic et al. (2016), and Wagner et al. (2017) studied metabolomic responses in the cladoceran crustacean *Daphnia magna* after exposure to various pharmaceuticals. In particular, Taylor and coauthors, in 2010, explored *D. magna* metabolic changings after 24 h exposure to 1.4 mg/L of the nonselective β-adrenergic receptor blocker propranolol by direct infusion Fourier transform ion cyclotron resonance mass spectrometry (FT-ICR MS). High-quality metabolite profiles were detected both in hemolymph and in the whole-organism extracts from 14-day-old daphnids and metabolic perturbations were found in the multiple fatty acid and oxylipid metabolites. Wagner et al. (2017) performed a similar study testing 0.67 mg/L of propranolol in both neonates (<24 h old) and adult daphnids (18 days old) by nuclear magnet resonance spectroscopy ^{1}H-NMR observing in both populations an increase in amino acid metabolites and a reduction in glucose levels when compared to control. Furthermore, Kovacevic et al. (2016) studied the metabolic profile in the same freshwater crustacean testing three different pharmaceuticals: triclosan (μg/L), carbamazepine, and ibuprofen (mg/L), at sublethal concentrations after 48 h exposure. Triclosan is used for impeding bacterial growth by inhibiting enzymes involved in fatty acid synthesis, carbamazepine is a sodium channel blocker used for the treatment of epilepsy and neuropathic pain for its effects on serotonin systems, while ibuprofen is a nonsteroidal anti-inflammatory drug inhibiting cyclooxygenase enzymes involved in prostaglandins synthesis. Kovacevic and coauthors analyzed adult daphnid metab olites by ^{1}H-NMR, and alterations in amino acid content as well as in sugar glucose levels were observed according to a concentration-dependent relationship between daphnids' metabolic responses and drug exposure concentrations, reflecting changes at organ, organismal, and population levels. In light of the foregoing, the freshwater consumer *D. magna* seems to be a very sensitive aquatic bioindicator for the evaluation of the metabolomic responses to many environmental pollutants also considering the advances used in analytical metabolomic techniques.

Other nontarget sentinel species belonging to a higher level of the food aquatic chain are represented by marine and freshwater fish. These organisms were among the earliest organisms used in environmental metabolomics thanks to their similar biochemical mechanisms in comparison to humans, in response to pharmaceuticals. In fact, in 2006, Samuelsson et al., using ^{1}H-NMR, studied the effects of the synthetic contraceptive estrogen ethinylestradiol on the rainbow trout *Oncorhynchus mykiss* (11 months old) and observed alterations in the plasma metabolite profile at

10 ng/L with a strong induction of the lipoprotein vitellogenin synthesis. Furthermore, Mills et al. (2016) explored physiological responses to the endocrine-active pharmaceutical Tamoxifen in adult fish *Tautogolabrus adspersus* using the gas chromatograph coupled to a time-of-flight mass spectrometer (GC/ToF-MS). Thus, Mills et al. observed high levels of proline, threonine, alanine, lysine, tyrosine, and tryptophan, and found down-regulated metabolites involved in amino acid synthesis and metabolism, phospholipid synthesis, glucoronidation, and glycolysis, proving that *T. adspersus* could represent a sensitive nontarget organism, useful for studying metabolomics perturbations after exposure to pharmaceuticals. As reported in scientific literature, fish have been used not only to observe the metabolomics responses of estrogen-like molecules, but also to study metabolomics alterations caused by the exposure to antibiotics. In fact, De Sotto et al. (2017) studied environmental effects of 0.1 mg/L of clarithromycin, florfenicol and sulfamethazine, individually and in mixtures, on adults of *Danio rerio* after 72 h exposure using high-performance liquid chromatography with quadrupole time-of-flight (QTOF) mass spectrometer. When clarithromycin and florfenicol were tested individually, they were able to yield more metabolites than those found for sulfamethazine, and the most affected pathway was the metabolism of purines, especially guanosine involved in protecting neurons against excitotoxic damages. The similarity between clarithromycin and florfenicol could be explained thanks to the same mode of action of these two antibiotics, which inhibit protein biosynthesis interacting with 50 S subunit in nontarget organisms. Surprisingly, when De Sotto et al. (2017) tested antibiotics in mixtures, a small amount of metabolites was observed, probably due to antagonistic interactions. In line with the scientific literature taken into account here, *Danio rerio* is a good model in environmental metabolomics to identify the effects of pharmaceuticals, due to its similarity to human metabolism and its ease in absorption of small molecules through skin and gills.

In conclusion, scientific interest is constantly increasing in the wide field of environmental metabolomics, a very useful approach to understand the impact of various environmental xenobiotics in nontarget organisms, through different analytical platforms. In the last years, it has been applied to evaluate metabolic changes in different aquatic organisms of the trophic chain after pharmaceutical exposure (only few scientific papers to date); to the best of our knowledge, no studies on anticancer drugs exist at the moment. Since the consumption and the administration of this class of pseudo-persistent pharmaceuticals are increasing as also their occurrence in the aquatic systems, it would be advisable to use metabolomic strategies to understand anticancer drugs' environmental toxic effects.

References

Alam TM, Alam MK, Neerathilingam M, Volk DE, Sarkar S, Ansari GAS, Luxon BA (2010) ^{1}H NMR metabonomic study of rat response to triphenyl phosphate and tri-butyl phosphate exposure. Metabolomics 6(3):386–394. https://doi.org/10.1007/s11306-010-0205-z

Allwood JW, Ellis DI, Goodacre R (2008) Metabolomic technologies and their application to the study of plants and plant–host interactions. Physiol Plant 132:117–135. https://doi.org/10.1111/j.1399-3054.2007.01001.x

Alonso A, Marsal S, JuliÃ A (2015) Analytical methods in untargeted metabolomics: state of the art in 2015. Front Bioeng Biotechnol 5(3):23

Ankley GT, Daston GP, Degitz SJ, Denslow ND, Hoke RA, Kennedy SW, Miracle AL, Perkins EJ, Snape J, Tillitt DE, Tyler CR, Versteeg D (2006) Toxicogenomics in regulatory ecotoxicology. Environ Sci Technol 40(13):4055–4065

Bajad S, Shulaev V (2007) Highly-parallel metabolomics approaches using LC–MS2 for pharmaceutical and environmental analysis. Trends Anal Chem 26(6):625–636. https://doi.org/10.1016/j.trac.2007.02.009. PMID:19756204

Barker M, Rayens W (2003) Partial least squares for discrimination. J Chemom 17(3):166–173. https://doi.org/10.1002/cem.785

Bilello JA (2005) The agony and ecstasy of "OMIC" technologies in drug development. Curr Mol Med 5(1):39–52. https://doi.org/10.2174/1566524053152898. PMID:15720269

Bird SS, Marur VR, Sniatynski MJ, Greenberg HK, Kristal BS (2011) Serum Lipidomics profiling using LC–MS and high-energy collisional dissociation fragmentation: focus on triglyceride detection and characterization. Anal Chem 83(17):6648–6657

Bligh EG, Dyer WJ (1959) A rapid method of total lipid extraction and purification. Can J Biochem Physiol 37(8):911–917

Bonnefille B, Gomez E, Alali M, Rosain D, Fenet H, Courant F (2018) Metabolomics assessment of the effects of diclofenac exposure on Mytilus galloprovincialis: potential effects on osmoregulation and reproduction. Sci Total Environ 613–614:611–618

Bothwell JHF, Griffin JL (2011) An introduction to biological nuclear magnetic resonance spectroscopy. Biol Rev 86(2):493–510. https://doi.org/10.1111/j.1469-185X.2010.00157.x. PMID:20969720

Brezovšek P, Eleršek T, Filipič M (2014) Toxicities of four anti-neoplastic drugs and their binary mixtures tested on the green alga Pseudokirchneriella subcapitata and thecyanobacterium Synechococcus leopoliensis. Water Res 52:168–177

Brown SAE, Simpson AJ, Simpson MJ (2008) Evaluation of sample preparation methods for nuclear magnetic resonance metabolic profiling studies with Eisenia Fetida. Environ Toxicol Chem 27(4):828

Brown SAE, Simpson AJ, Simpson MJ (2009) ^{1}H NMR metabolomics of earthworm responses to sub-lethal PAH exposure. Environ Chem 6:432–440. https://doi.org/10.1071/EN09054

Brown SAE, McKelvie JR, Simpson AJ, Simpson MJ (2010) ^{1}H NMR metabolomics of earthworm exposure to sub-lethal concentrations of phenanthrene in soil. Environ Pollut 158(6):2117–2123. https://doi.org/10.1016/j.envpol.2010.02.023. PMID:20338676

Bundy JG, Osborn D, Weeks JM, Lindon JC, Nicholson JK (2001) An NMR-based metabonomic approach to the investigation of coelomic fluid biochemistry in earthworms under toxic stress. FEBS Lett 500(1–2):31–35. https://doi.org/10.1016/S0014-5793(01)02582-0. PMID:11434921

Bundy JG, Lenz EM, Bailey NJ, Gavaghan CL, Svendsen C, Spurgeon D, Hankard PK, Osborn D, Weeks JA, Trauger SA, Speir P, Sanders I, Lindon JC, Nicholson JK, Tang H (2002a) Metabonomic assessment of toxicity of 4-fluoroaniline, 3,5-difluoroaniline and 2 fluoro-4-methylaniline to the earthworm Eisenia veneta (Rosa): identification of new endogenous biomarkers. Environ Toxicol Chem 21(9):1966–1972. https://doi.org/10.1002/etc.5620210926. PMID:12206438

Bundy JG, Lenz EM, Osborn D, Weeks JM, Lindon JC, Nicholson JK (2002b) Metabolism of 4-fluoroaniline and 4-fluorobiphenyl in the earthworm Eisenia veneta characterized by high-resolution NMR spectroscopy with directly coupled HPLC–NMR and HPLC–MS. Xenobiotica 32(6):479–490. https://doi.org/10.1080/00498250210124156. PMID:12160481

Bundy JG, Spurgeon DJ, Svendsen C, Hankard PK, Osborn D, Lindon JC, Nicholson JK (2002c) Earthworm species of the genus Eisenia can be phenotypically differentiated by metabolic profiling. FEBS Lett 521(1–3):115–120. https://doi.org/10.1016/S0014-5793(02)02854-5. PMID:12067738

Bundy JG, Spurgeon DJ, Svendsen C, Hankard PK, Weeks JM, Osborn D, Lindon JC, Nicholson JK (2004) Environmental Metabonomics: applying combination biomarker analysis in earthworms at a metal contaminated site. Ecotoxicology 13(8):797–806

Burton IW, Quilliam MA, Walter JA (2005) Quantitative ^{1}H NMR with external standards: use in preparation of calibration solutions for algal toxins and other natural products. Anal Chem 77(10):3123–3131. https://doi.org/10.1021/ac048385h. PMID:15889900

Bylesjö M, Rantalainen M, Cloarec O, Nicholson JK, Holmes E, Trygg J (2006) OPLS discriminant analysis: combining the strengths of PLS-DA and SIMCA classification. J Chemom 20(8–10):341–351. https://doi.org/10.1002/cem.1006

Cappello T, Pereira P, Maisano M, Mauceri A, Pacheco M, Fasulo S (2016) Advances in understanding the mechanisms of mercury toxicity in wild golden grey mullet (*Liza aurata*) by 1H NMR-based metabolomics. Environ Pollut 219:139–148

Cappello T, Maisano M, Mauceri A, Fasulo S (2017) 1 H NMR-based metabolomics investigation on the effects of petrochemical contamination in posterior adductor muscles of caged mussel *Mytilus galloprovincialis*. Ecotoxicol Environ Saf 142:417–422

Chen C, Gonzalez FJ, Idle JR (2008) LC-MS-based metabolomics in drug metabolism. Drug Metab Rev 39(2–3):581–5971

Chen L, Zhou L, Chan ECY, Neo J, Beuerman RW (2011) Characterization of the human tear metabolome by LC–MS/MS. J Proteome Res 10(10):4876–4882

Chiu KH, Dong CD, Chen CF, Tsai ML, Ju YR, Chen TM, Chen CW (2017) NMR-based metabolomics for the environmental assessment of Kaohsiung Harbor sediments exemplified by a marine amphipod (*Hyalella azteca*). Mar Pollut Bull 124(2):714–724

Chylla RA, Hu K, Ellinger JJ, Markley JL (2011) Deconvolution of twodimensional NMR spectra by fast maximum likelihood reconstruction: application to quantitative metabolomics. Anal Chem 83(12):4871–4880. https://doi.org/10.1021/ac200536b. PMID:21526800

Coen M, Holmes E, Lindon JC, Nicholson JK (2008) NMR-based metabolic profiling and metabonomic approaches to problems in molecular toxicology. Chem Res Toxicol 21(1):9–27. https://doi.org/10.1021/tx700335d. PMID:18171018

Creek DJ, Anderson J, McConville MJ, Barrett MP (2012) Metabolomic analysis of trypanosomatid protozoa. Mol Biochem Parasitol 181(2):73–84

Crockford DJ, Holmes E, Lindon JC, Plumb RS, Zirah S, Bruce SJ, Rainville P, Stumpf CL, Nicholson JK (2006) Statistical heterospectroscopy, an approach to the integrated analysis of NMR and UPLC–MS data sets: application in metabonomic toxicology studies. Anal Chem 78(2):363–371. https://doi.org/10.1021/ac051444m. PMID:16408915

De Sotto RB, Medriano CD, Choc Y, Kimd H, Chungd IC, Seokd SKG, Honge SW, Park Y, Kima S (2017) Sub-lethal pharmaceutical hazard tracking in adult zebrafish using untargeted LC–MS environmental metabolomics. J Hazard Mater 339:63–72

Dunn WB, Ellis DI (2005) Metabolomics: current analytical platforms and methodologies. TrAC Trends Anal Chem 24(4):285–294

Dunn WB, Wilson ID, Nicholls AW, Broadhurst D (2012) The importance of experimental design and QC samples in large-scale and MS-driven untargeted metabolomic studies of humans. Bioanalysis 4(18):2249–2264

Eckel-Mahan KL, Patel VR, Mohney RP, Vignola KS, Baldi P, Sassone-Corsi P (2012) Coordination of the transcriptome and metabolome by the circadian clock. Proc Natl Acad Sci 109(14):5541–5546

Edwards JL, Chisolm CN, Shackman JG, Kennedy RT (2006) Negative mode sheathless capillary electrophoresis electrospray ionization-mass spectrometry for metabolite analysis of prokaryotes. J Chromatogr A 1106(1–2):80–88. https://doi.org/10.1016/j.chroma.2005.08.082

Ekman DR, Teng Q, Villeneuve DL, Kahl MD, Jensen KM, Durhan EJ, Ankley GT, Collette TW (2008) Investigating compensation and recovery of fathead minnow (Pimephales promelas) exposed to 17-ethynylestradiol with metabolite profiling. Environ Sci Technol 42(11):4188–4194. https://doi.org/10.1021/es8000618. PMID:18589986

Ekman DR, Teng Q, Villeneuve DL, Kahl MD, Jensen KM, Durhan EJ, Ankley GT, Collette TW (2009) Profiling lipid metabolites yields unique information on sex- and time-dependent responses of fathead minnows (Pimephales promelas) exposed to 17-ethynylestradiol. Metabolomics 5(1):22–32. https://doi.org/10.1007/s11306-008-0138-y

Eriksson L, Johansson E, Kettaneh-Wold N, Trygg J, Wikstrom C, Wold S (2006) Multi- and megavariate data analysis. Part 1: Basic principles and applications, 2nd edn. Umea, Umetrics, AB

Fasulo S, Iacono F, Cappello T, Corsaro C, Maisano M, D'Agata A, Giannetto A, De Domenico E, Parrino V, Lo Paro G, Mauceri A (2012) Metabolomic investigation of Mytilus galloprovincialis (Lamarck1819) caged in aquatic environments. Ecotoxicol Environ Saf 84:139–146

Fernie AR, Aharoni A, Willmitzer L, Stitt M, Tohge T, Kopka J, Carroll AJ, Saito K, Fraser PD, DeLuca V (2011) Recommendations for reporting metabolite data. Plant Cell 23(7):2477–2482

Fialho E, Moreno FS, Ong TP (2008) Nutrição no pós-genoma: fundamentos e aplicações de ferramentas ômicas. Rev Nutr 21(6):757–766

Flores-Sanchez IJ, Choi YH, Verpoorte R (2012) Metabolite analysis of Cannabis sativa L. by NMR spectroscopy. Methods Mol Biol 815:363–375. https://doi.org/10.1007/978-1-61779-424-7

Flores-Valverde AM, Horwood J, Hill EM (2010) Disruption of the steroid metabolome in fish caused by exposure to the environmental estrogen 17 - ethinylestradiol. Environ Sci Technol 44(9):3552–3558. https://doi.org/10.1021/es9039049. PMID:20356033

Forseth RR, Schroeder FC (2011) NMR-spectroscopic analysis of mixtures: from structure to function. Curr Opin Chem Biol 15(1):38–47

Fukusaki E, Kobayashi A (2005) Plant metabolomics: potential for practical operation. J Biosci Bioeng 100(4):347–354

Gačić Z, Kolarević S, Sunjog K, Kračun-Kolarević M, Paunović M, Knežević-Vukčević J, Vuković-Gačić B (2014) The impact of in vivo and in vitro exposure to base analogue 5-FU on the level of DNA damage in haemocytes of freshwater mussels Unio pictorum and Unio tumidus. Environ Pollut 191:145–150

García-Cañaveras JC, Donato MT, Castell JV, Lahoz A (2012) Targeted profiling of circulating and hepatic bile acids in human, mouse, and rat using a UPLC-MRM-MS-validated method. J Lipid Res 53(10):2231–2241. https://doi.org/10.1194/jlr.D028803. PMID:22822028

Garreta-Lara E, Campos B, Barata C, Lacorte S, Tauler R (2016) Metabolic profiling of *Daphnia magna* exposed to environmental stressors by GC–MS and chemometric tools. Metabolomics 12:86. https://doi.org/10.1007/s11306-016-1021-x

Gaw S, Thomas KV, Hutchinson TH (2014) Sources, impacts and trends of pharmaceuticals in the marine and coastal environment. Philos Trans R Soc B 369:20130572. https://doi.org/10.1098/rstb.2013.0572

Ghaste M, Mistrik R, Shulaev V (2016) Applications of Fourier transform ion cyclotron resonance (FT-ICR) and Orbitrap based high resolution mass spectrometry in metabolomics and Lipidomics. Int J Mol Sci 17(6):816

Gil-Solsona R, Nácher-Mestre J, Lacalle-Bergeron L, Sancho JV, Alvar Calduch-Giner J, Hernández F, Pérez-Sánchez J (2017) Untargeted metabolomics approach for unraveling robust biomarkers of nutritional status in fasted gilthead sea bream (*Sparus aurata*). Peer J 5:e2920

Gillis JD, Price GW, Prasher S (2017) Lethal and sub-lethal effects of triclosan toxicity to the earthworm Eisenia fetida assessed through GC–MS metabolomics. J Hazard Mater 323:203–211

Gómez-Canela C, Miller TH, Bury NR, Tauler R, Barron LP (2016) Targeted metabolomics of *Gammarus pulex* following controlled exposures to selected pharmaceuticals in water. Sci Total Environ 562:777–788

Goodacre R, Vaidyanathan S, Dunn WB, Harrigan GG, Kell DB (2004) Metabolomics by numbers: acquiring and understanding global metabolite data. Trends Biotechnol 22(5):245–252

Goodacre R, Broadhurst D, Smilde AK, Kristal BS, Baker JD, Beger R, Bessant C, Connor S, Capuani G, Craig A, Ebbels T, Kell DB, Manetti C, Newton J, Paternostro G, Somorjai R, Sjöström M, Trygg J, Wulfert F (2007) Proposed minimum reporting standards for data analysis in metabolomics. Metabolomics 3(3):231–241

Griffin JL (2003) Metabonomics: NMR spectroscopy and pattern recognition analysis of body fluids and tissues for characterisation of xenobiotic toxicity and disease diagnosis. Curr Opin Chem Biol 7(5):648–654. https://doi.org/10.1016/j.cbpa.2003.08.008. PMID:14580571

Griffin JL, Bollard ME (2004) Metabonomics: its potential as a tool in toxicology for safety assessment and data integration. Curr Drug Metab 5(5):389–398. https://doi.org/10.2174/1389200043335432. PMID:15544433

Griffin JL, Shockcor JP (2004) Metabolic profiles of cancer cells. Nat Rev Cancer 4(7):551–561. https://doi.org/10.1038/Nrc1390

Griffiths JR, Stubbs M (2003) Opportunities for studying cancer by metabolomics: preliminary observations on tumors deficient in hypoxia-inducible factor 1. Adv Enzym Regul 43(1):67–76

Griffiths JR, McSheehy PMJ, Robinson SP, Troy H, Chung YL, Leek RD, Williams KJ, Stratford IJ, Harris AL, Stubbs M (2002) Metabolic changes detected by in vivo magnetic resonance studies of HEPA-1 wild-type tumors and tumors deficient in hypoxia-inducible factor-1β (HIF-1β): evidence of an anabolic role for the HIF-1 pathway. Cancer Res 62(3):688–695. PMID:11830521

Grimes JH, O'Connell TM (2011) The application of micro-coil NMR probe technology to metabolomics of urine and serum. J Biomol NMR 49(3–4):297–305. https://doi.org/10.1007/s10858-011-9488-2

Hall RD (2006) Plant metabolomics: from holistic hope, to hype, to hot topic. New Phytol 169(3):453–468

Harvey AL, Edrada-Ebel R, Quinn RJ (2015) The re-emergence of natural products for drug discovery in the genomics era. Nat Rev Drug Discov 14(2):111

Holmes E, Antti H (2002) Chemometric contributions to the evolution of metabonomics: mathematical solutions to characterising and interpreting complex biological NMR spectra. Analyst 127(12):1549–1557. https://doi.org/10.1039/b208254n. PMID:12537357

Hu K, Ellinger JJ, Chylla RA, Markley JL (2011a) Measurement of absolute concentrations of individual compounds in metabolite mixtures by gradient-selective time-zero 1H-13C HSQC with two concentration references and fast maximum likelihood reconstruction analysis. Anal Chem 83(24):9352–9360. https://doi.org/10.1021/ac201948f. PMID:22029275

Hu K, Westler WM, Markley JL (2011b) Simultaneous quantification and identification of individual chemicals in metabolite mixtures by twodimensional extrapolated time-zero 1H-13C HSQC (HSQC0). J Am Chem Soc 133(6):1662–1665. https://doi.org/10.1021/ja1095304. PMID:21247157

Hughes SL, Bundy JG, Want EJ, Kille P, Sturzenbaum SR (2009) The metabolomic responses of Caenorhabditis elegans to cadmium are largely independent of metallothionein status, but dominated by changes in cystathionine and phytochelatins. J Proteome Res 8(7):3512–3519. https://doi.org/10.1021/pr9001806. PMID:19466807

Isidori M, Piscitelli C, Russo C, Smutná M, Bláha L (2016a) Teratogenic effects of five anticancer drugs on Xenopus laevis embryos. Ecotoxicol Environ Saf 133:90–96

Isidori M, Lavorgna M, Russo C, Kundi M, Zegura B, Novak M, Filipic M, Misík M, Knasmueller S, Lopez de Alda M, Barcelo D, Zonja B, Cesen M, Scancar J, Kosjek T, Heath E (2016b) Chemical and toxicological characterisation of anticancer drugs in hospital and municipal wastewaters from Slovenia and Spain. Environ Pollut 219:275–287

Issaq HJ, Abbott E, Veenstra TD (2008) Utility of separation science in metabolomicstudies. J Sep Sci 31(11):1936–1947. https://doi.org/10.1002/jssc.200700601. PMID:18348322

Jacobsen NE (2007) NMR spectroscopy explained: simplified theory, applications and examples for organic chemistry and structural biology. Wiley, Hoboken, NJ

Jones OAH, Spurgeon DJ, Svendsen C, Griffin JL (2008) A metabolomics based approach to assessing the toxicity of the polyaromatic hydrocarbon pyrene to the earthworm Lumbricus rubellus. Chemosphere 71(3):601–609. https://doi.org/10.1016/j.chemosphere.2007.08.056. PMID:17928029

Kim HK, Verpoorte R (2010) Sample preparation for plant metabolomics. Phytochem Anal 21(1):4–13

Kim HK, Choi YH, Verpoorte R (2011) NMR-based metabolomics: where do we stand, where do we go? Trends Biotechnol 29:267–275

Kokushi E, Shintoyo A, Koyama J, Uno S (2017) Evaluation of 2,4-dichlorophenol exposure of Japanese medaka, *Oryzias latipes*, using a metabolomics approach. Environ Sci Pollut Res 24(36):27678–27686

Kovacevic V, Simpson AJ, Simpson MJ (2016) ^{1}H NMR-based metabolomics of *Daphnia magna* responses after sub-lethal exposure to triclosan, carbamazepine and ibuprofen. Comp Biochem Physiol Part D 19:199–210

Kovács R, Csenki Z, Bakos K, Urbányi B, Horváth Á, Garaj-Vrhovac V, Gajski G, Gerić M, Negreira N, López de Alda M, Barceló D, Heath E, Kosjek T, Žegura B, Novak M, Zajc I, Baebler Š, Rotter A, Ramšak Ž, Filipič M (2015) Assessment of toxicity and genotoxicity of low doses of 5-fluorouracil in zebrafish (Danio rerio) two-generation study. Water Res 77:201–212

Kovács R, Bakos K, Urbányi B, Kövesi J, Gazsi G, Csepeli A, Appl ÁJ, Bencsik D, Csenki Z, Horváth Á (2016) Acute and sub-chronic toxicity of four cytostatic drugs in zebrafish. Environ Sci Pollut Res Int 23(15):14718–14729

Kundi M, Parrella A, Lavorgna M, Criscuolo E, Russo C, Isidori M (2016) Prediction and assessment of ecogenotoxicity of antineoplastic drugs in binary mixtures. Environ Sci Pollut Res 23:14771–14779

Lacey ME, Subramanian R, Olson DL, Webb AG, Sweedler JV (1999) High-resolution NMR spectroscopy of sample volumes from 1 nL to 10 _L. Chem Rev 99(10):3133–3152. https://doi.org/10.1021/cr980140f. PMID:11749512

Laith AA, Ambaka M, Abol-Munafi AB, Nurhafizah WWI, Najiaha M (2017) Metabolomic analysis of marine and mud crabs based on antibacterial activity. Aquac Rep 7:7–15

Lankadurai BP, Nagato EG, Simpson MJ (2013) Environmental metabolomics: an emerging approach to study organism responses to environmental stressors. Environ Rev 21(3):180–205

Lenz K, Mahnik SN, Weissenbacher N, Mader RM, Krenn P, Hann S, Koellensperger G, Uhl M, Knasmueller S, Ferk F, Bursh W, Fuerhacker M (2007) Monitoring, removal and risk assessment of cytostatic drugs in hospital wastewater. Water Sci Technol 56(12):141–149

Li W, Ruan CJ, Teixeira da Silva JA, Guo H, Zhao CE (2012) NMR metabolomics of berry quality in sea buckthorn (Hippophae L.). Mol Breed 31:57–67. https://doi.org/10.1007/s11032-012-9768-x

Liebeke M, Bundy JG (2012) Tissue disruption and extraction methods for metabolic profiling of an invertebrate sentinel species. Metabolomics 8(5):819–830. https://doi.org/10.1007/s11306-011-0377-1

Lin CY, Viant MR, Tjeerdema RS (2006) Metabolomics: methodologies and applications in the environmental sciences. J Pestic Sci 31(3):245–251. https://doi.org/10.1584/jpestics.31.245

Lindon JC, Nicholson JK, Holmes E (2006) The handbook of metabonomics and metabolomics. Elsevier Science, Amsterdam. ISBN 9780444528414.

Lloyd AJ, Beckmann M, Wilson T, Tailliart K, Allaway D, Draper J (2017) Ultra high performance liquid chromatography–high resolution mass spectrometry plasma lipidomics can distinguish between canine breeds despite uncontrolled environmental variability and non-standardized diets. Metabolomics 13(2):1–15

Logan TM, Murali M, Wang GS, Jolivet C (1999) Application of a highresolution superconducting NMR probe in natural product structure determination. Magn Reson Chem 37(7):512–515. https://doi.org/10.1002/(SICI)1097-458X(199907)37:7<512::AID-MRC502>3.0.CO;2-B

Ludwig C, Viant MR (2010) Two-dimensional J-resolved NMR spectroscopy: review of a key methodology in the metabolomics toolbox. Phytochem Anal 21(1):22–32. https://doi.org/10.1002/pca.1186. PMID:19904730

Lutz U, Lutz RW, Lutz WK (2006) Metabolic profiling of glucuronides in human urine by LC–MS/MS and partial least-squares discriminant analysis for classification and prediction of gender. Anal Chem 78(13):4564–4571. https://doi.org/10.1021/ac0522299. PMID:16808466

McGhie TK, Rowan DD (2012) Metabolomics for measuring phytochemicals, and assessing human and animal responses to phytochemicals, in food science. Mol Nutr Food Res 56(1):147–158

McKay RT (2009) Chapter 2 Recent advances in solvent suppression for solution NMR: a practical reference. Annu Rep NMR Spectrosc 66:33–76. https://doi.org/10.1016/S0066-4103(08)00402-X

McKelvie JR, Yuk J, Xu YP, Simpson AJ, Simpson MJ (2009) ^{1}H NMR and GC/MS metabolomics of earthworm responses to sub-lethal DDT and endosulfan exposure. Metabolomics 5(1):84–94. https://doi.org/10.1007/s11306-008-0122-6

McKelvie JR, Wolfe DM, Celejewski M, Simpson AJ, Simpson MJ (2010) Correlations of Eisenia fetida metabolic responses to extractable phenanthrene concentrations through time. Environ Pollut 158(6):2150–2157. https://doi.org/10.1016/j.envpol.2010.02.022. PMID:20338675

McKelvie JR, Wolfe DM, Celejewski MA, Alaee M, Simpson AJ, Simpson MJ (2011) Metabolic responses of Eisenia fetida after sub-lethal exposure to organic contaminants with different toxic modes of action. Environ Pollut 159(12):3620–3626. https://doi.org/10.1016/j.envpol.2011.08.002. PMID:21856054

McKelvie JR, Whitfield Åslund M, Celejewski M, Simpson AJ, Simpson MJ (2013) Reduction in the earthworm metabolomic response after phenanthrene exposure in soils with high soil organic carbon content. Environ Pollut 175:75–81. https://doi.org/10.1016/j.envpol.2012.12.018. PMID:23337355

Miller MG (2007) Environmental metabolomics: a SWOT analysis (strengths, weaknesses, opportunities, and threats). J Proteome Res 6(2):540–545. https://doi.org/10.1021/pr060623x. PMID:17269710

Mills LJ, Henderson WM, Jayaraman S, Gutjahr-Gobell RE, Zaroogian GE, Horowitz DB, Laws SC (2016) Approaches for predicting effects of unintended environmental exposure to an endocrine active pharmaceutical, tamoxifen. Environ Toxicol 31(12):1834–1850

Mumtaz MW, Hamid AA, Akhtar MT, Anwar F, Rashid U, AL-Zuaidy MH (2017) An overview of recent developments in metabolomics and proteomics – phytotherapic research perspectives. Front Life Sci 10(1):1–37

Nagato EG, Simpson AJ, Simpson MJ (2016) Metabolomics reveals energetic impairments in Daphnia magna exposed to diazinon, malathion and bisphenol-A. Aquat Toxicol 170:175–186

Negreira N, Lopez de Alda M, Barcelo D (2014) Cytostatic drugs and metabolites in municipal and hospital wastewaters in Spain: filtration, occurrence, and environmental risk. Sci Total Environ 497-498:68–77

Nicholson JK, Wilson ID (1989) High resolution proton magnetic resonance spectroscopy of biological fluids. Prog Nucl Magn Reson Spectrosc 21:449–501. https://doi.org/10.1016/0079-6565(89)80008-1

Nicholson JK, Lindon JC, Holmes E (2008) 'Metabonomics': understanding the metabolic responses of living systems to pathophysiological stimuli via multivariate statistical analysis of biological NMR spectroscopic data. Xenobiotica 29(11):1181–1189

Noctor G, Bergot GL, Mauve C, Thominet D, Lelarge-Trouverie C, Prioul JL (2007) A comparative study of amino acid measurement in leaf extracts by gas chromatography-time of flight-mass spectrometry and high performance liquid chromatography with fluorescence detection. Metabolomics 3(2):161–174. https://doi.org/10.1007/s11306-007-0057-3

Nowick JS, Khakshoor O, Hashemzadeh M, Brower JO (2003) DSA: a new internal standard for NMR studies in aqueous solution. Org Lett 5(19):3511–3513. https://doi.org/10.1021/ol035347w. PMID:12967312

Pan ZZ, Raftery D (2007) Comparing and combining NMR spectroscopy and mass spectrometry in metabolomics. Anal Bioanal Chem 387(2):525–527. https://doi.org/10.1007/s00216-006-0687-8. PMID:16955259

Parrella A, Lavorgna M, Criscuolo E, Russo C, Fiumano V, Isidori M (2014) Acute and chronic toxicity of six anticancer drugs on rotifers and crustaceans. Chemosphere 115:59–66

Parrella A, Lavorgna M, Criscuolo E, Russo C, Isidori M (2015) Eco-genotoxicity of six anticancer drugs using comet assayin daphnids. J Hazard Mater 286:573–580

Poynton HC, Taylor NS, Hicks J, Colson K, Chan SR, Clark C, Scanlan L, Loguinov AV, Vulpe C, Viant MR (2011) Metabolomics of microliter hemolymph samples enables an improved understanding of the combined metabolic and transcriptional responses of Daphnia magna to cadmium. Environ Sci Technol 45(8):3710–3717. https://doi.org/10.1021/es1037222. PMID:21417318

Ramautar R, Busnel JM, Deelder AM, Mayboroda OA (2012) Enhancing the coverage of the urinary metabolome by sheathless capillary electrophoresis–mass spectrometry. Anal Chem 84(2):885–892. https://doi.org/10.1021/ac202407v. PMID:22148170

Ritota M, Casciani L, Han BZ, Cozzolino S, Leita L, Sequi P, Valentini M (2012) Traceability of Italian garlic (Allium sativum L.) by means of HRMAS–NMR spectroscopy and multivariate data analysis. Food Chem 135(2):684–693. https://doi.org/10.1016/j.foodchem.2012.05.032. PMID:22868146

Robertson DG (2005) Metabonomics in toxicology: a review. Toxicol Sci 85(2):809–822. https://doi.org/10.1093/toxsci/kfi102. PMID:15689416

Ruiz-Torres V, Encinar JA, Herranz-López M, Pérez-Sánchez A, Galiano V, Barrajón-Catalán E, Micol V (2017) An updated review on marine anticancer compounds: the use of virtual screening for the discovery of small-molecule cancer drugs. Molecules 22:1037. https://doi.org/10.3390/molecules22071037

Russo C, Lavorgna M, Cesen M, Kosjek T, Heath T, Isidori M (2018) Evaluation of acute and chronic ecotoxicity of cyclophosphamide, ifosfamide, their metabolites/transformation products and UV treated samples. Environ Pollut 233:356–363

Samuelsson LM, Forlin L, Karlsson G, Adolfsson-Erici M, Larsson DGJ (2006) Using NMR metabolomics to identify responses of an environmental estrogen in blood plasma of fish. Aquat Toxicol 78:341–349

Sato S, Soga T, Nishioka T, Tomita M (2004) Simultaneous determination of the main metabolites in rice leaves using capillary electrophoresis mass spectrometry and capillary electrophoresis diode array detection. Plant J 40(1):151–163. https://doi.org/10.1111/j.1365-313X.2004.02187.x. PMID:15361149

Schripsema J (2009) Application on NMR in plant metabolomics: techniques, problems and prospects. Phytochem Anal 21:14–21

Scognamiglio M, D'Abrosca B, Fiumano V, Golino M, Esposito A, Fiorentino A (2014) Seasonal phytochemical changes in Phillyrea angustifolia L.: metabolomic analysis and phytotoxicity assessment. Phytochem Lett 8:163–170

Scognamiglio M, D'Abrosca B, Esposito A, Fiorentino A (2015) Chemical Composition and Seasonality of Aromatic Mediterranean Plant Species by NMR-Based Metabolomics. J Anal Methods Chem 2015(2015):258570. https://doi.org/10.1155/2015/258570

Simpson AJ, Brown SA (2005) Purge NMR: effective and easy solvent suppression. J Magn Reson 175(2):340–346. https://doi.org/10.1016/j.jmr.2005.05.008. PMID:15964227

Simpson MJ, McKelvie JR (2009) Environmental metabolomics: new insights into earthworm ecotoxicity and contaminant bioavailability in soil. Anal Bioanal Chem 394(1):137–149

Southam AD, Lange A, Hines A, Hill EM, Katsu Y, Iguchi T, Tyler CR, Viant MR (2011) Metabolomics reveals target and off-target toxicities of a model organophosphate pesticide to roach (Rutilus rutilus): implications for biomonitoring. Environ Sci Technol 45(8):3759–3767. https://doi.org/10.1021/es103814d. PMID:21410251

Sumner LW, Mendes P, Dixon RA (2003) Plant metabolomics: large-scale phytochemistry in the functional genomics era. Phytochemistry 62(6):817–836

Sumner LW, Amberg A, Barrett D, Beale MH, Beger R, Daykin CA, Fan TW-M, Fiehn O, Goodacre R, Griffin JL, Hankemeier T, Hardy N, Harnly J, Higashi R, Kopka J, Lane AN, Lindon JC, Marriott P, Nicholls AW, Reily MD, Thaden JJ, Viant MR (2007) Proposed minimum reporting standards for chemical analysis. Metabolomics 3(3):211–221

Taylor NS, Weber RJM, White TA, Viant MR (2010) Discriminating between different acute chemical toxicities via changes in the Daphnid Metabolome. Toxicol Sci 118(1):307–317

Tjeerdema RS (2008) Application of NMR-based techniques in aquatic toxicology: brief examples. Mar Pollut Bull 57(6–12):275–279. https://doi.org/10.1016/j.marpolbul.2008.01.002. PMID:18282584

Trygg J, Wold S (2002) Orthogonal projections to latent structures (O-PLS). J Chemom 16(3):119–128. https://doi.org/10.1002/cem.695

Trygg J, Holmes E, Lundstedt T (2007) Chemometrics in metabonomics. J Proteome Res 6(2):469–479. https://doi.org/10.1021/pr060594q. PMID:17269704

Tuffnail W, Mills GA, Cary P, Greenwood R (2009) An environmental ^{1}H NMR metabolomic study of the exposure of the marine mussel Mytilus edulis to atrazine, lindane, hypoxia and starvation. Metabolomics 5:33–43

van der Hooft JJJ, Padmanabhan S, Burgess KEV, Barrett MP (2016) Urinary antihypertensive drug metabolite screening using molecular networking coupled to high-resolution mass spectrometry fragmentation. Metabolomics 12(7):1–15

van Ravenzwaay B, Cunha GCP, Leibold E, Looser R, Mellert W, Prokoudine A, Walk T, Wiemer J (2007) The use of metabolomics for the discovery of new biomarkers of effect. Toxicol Lett 172(1–2):21–28. https://doi.org/10.1016/j.toxlet.2007.05.021. PMID:17614222

Varmuza K, Filzmoser P (2009) Introduction to multivariate statistical analysis in chemometrics. CRC Press, Boca Raton, FL

Verpoorte R, Choi YH, Kim HK (2007) NMR-based metabolomics at work in phytochemistry. Phytochem Rev 6(1):3–14

Verpoorte R, Choi YH, Mustafa NR, Kim HK (2008) Metabolomics: back to basics. Phytochem Rev 7(3):525–537. https://doi.org/10.1007/s11101-008-9091-7

Viant MR (2008a) Environmental metabolomics using ^{1}H-NMR spectroscopy. Methods Mol Biol 410:137–150. PMID:18642599

Viant MR (2008b) Recent developments in environmental metabolomics. Mol Bio Syst 4 (10):980–986. https://doi.org/10.1039/b805354e

Viant MR, Rosenblum ES, Tjeerdema RS (2003) NMR-based metabolomics: a powerful approach for characterizing the effects of environmental stressors on organism health. Environ Sci Technol 37(21):4982–4989. https://doi.org/10.1021/es034281x. PMID:14620827

Viant MR, Bearden DW, Bundy JG, Burton IW, Collette TW, Ekman DR, Ezernieks V, Karakach TK, Lin CY, Rochfort S, de Ropp JS, Teng Q, Tjeerdema RS, Walter JA, Wu H (2009) International NMR-based environmental metabolomics intercomparison exercise. Environ Sci Technol 43(1):219–225. https://doi.org/10.1021/es802198z. PMID:19209610

Wagner ND, Simpson AJ, Simpson MJ (2017) Metabolomic responses to sublethal contaminant exposure in neonate and adult daphnia magna. Environ Toxicol Chem 36:938–946

Wang M, Chen L, Liu D, Chen H, Tang D-D, Zhao Y-Y (2017) Metabolomics highlights pharmacological bioactivity and biochemical mechanism of traditional Chinese medicine. Chem Biol Interact 273:133–141

Warren CR, Aranda I, Cano FJ (2012) Metabolomics demonstrates divergent responses of two Eucalyptus species to water stress. Metabolomics 8(2):186–200. https://doi.org/10.1007/s11306-011-0299-y

Weckwerth W (2007) Metabolomics: methods and protocols. Humana Press Inc., Totowa, NJ

Westerhuis JA, Hoefsloot HCJ, Smit S, Vis DJ, Smilde AK, van Velzen EJJ, van Duijnhoven JPM, van Dorsten FA (2008) Assessment of PLSDA cross validation. Metabolomics 4(1):81–89. https://doi.org/10.1007/s11306-007-0099-6

Whitfield Åslund M, Celejewski M, Lankadurai BP, Simpson AJ, Simpson MJ (2011a) Natural variability and correlations in the metabolic profile of healthy Eisenia fetida earthworms observed using ^{1}H NMR metabolomics. Chemosphere 83(8):1096–1101. https://doi.org/10.1016/j.chemosphere.2011.01.036. PMID:21316730

Whitfield Åslund M, Simpson AJ, Simpson MJ (2011b) ^{1}H NMR metabolomics of earthworm responses to polychlorinated biphenyl (PCB) exposure in soil. Ecotoxicology 20(4):836–846. https://doi.org/10.1007/s10646-011-0638-9. PMID:21424327

Wishart DS (2008) Quantitative metabolomics using NMR. TrAC Trends Anal Chem 27(3):228–237

Wolfender J-L, Rudaz S, Hae Choi Y, Kyong Kim H (2013) Plant metabolomics: from holistic data to relevant biomarkers. Curr Med Chem 20(8):1056–1090

Wu B, Zhu L, Le XC (2017) Metabolomics analysis of TiO 2 nanoparticles induced toxicological effects on rice (*Oryza sativa* L.). Environ Pollut 230:302–310

Wu L, Mashego MR, van Dam JC, Proell AM, Vinke JL, Ras C, van Winden WA, van Gulik WM, Heijnen JJ (2005) Quantitative analysis of the microbial metabolome by isotope dilution mass spectrometry using uniformly 13C-labeled cell extracts as internal standards. Anal Biochem 336(2):164–171. https://doi.org/10.1016/j.ab.2004.09.001. PMID:15620880

Wu H, Southam AD, Hines A, Viant MR (2008) High-throughput tissue extraction protocol for NMR- and MS-based metabolomics. Anal Biochem 372(2):204–212. https://doi.org/10.1016/j.ab.2007.10.002. PMID:17963684

Xi Y, de Ropp JS, Viant MR, Woodruff DL, Yu P (2008) Improved identification of metabolites in complex mixtures using HSQC NMR spectroscopy. Anal Chim Acta 614(2):127–133. https://doi.org/10.1016/J.Aca.2008.03.024

Xiao JF, Zhou B, Ressom HW (2012) Metabolite identification and quantitation in LC-MS/MS-based metabolomics. TrAC Trends Anal Chem 32:1–14

Yamamotoya T, Dose H, Tian Z, Faure A, Toya Y, Honma M, Igarashi K, Nakahigashi K, Soga T, Mori H, Matsuno H (2012) Glycogen is the primary source of glucose during the lag phase of E. coli proliferation. Biochim Biophys Acta 1824(12):1442–1448. https://doi.org/10.1016/j.bbapap.2012.06.010

Yingrong Chen, Zhihong Ma, Lishan Min, Hongwei Li, Bin Wang, Jing Zhong, Licheng Dai, (2015) Biomarker identification and pathway analysis by serum metabolomics of lung Cancer. Biomed Res Int 1–9

Yuk J, McKelvie JR, Simpson MJ, Spraul M, Simpson AJ (2010) Comparison of 1-D and 2-D NMR techniques for screening earthworm responses to sub-lethal endosulfan exposure. Environ Chem 7(6):524–536. https://doi.org/10.1071/EN10084

Yuk J, Simpson AJ, Simpson MJ (2011) 1-D and 2-D NMR metabolomics of earthworm responses to sub-lethal trifluralin and endosulfan exposure. Environ Chem 8:281–294. https://doi.org/10.1071/EN11033

Yuk J, Simpson MJ, Simpson AJ (2013) 1-D and 2-D NMR-based metabolomics of earthworms exposed to endosulfan and endosulfan sulfate in soil. Environ Pollut 175:35–44. https://doi.org/10.1016/j.envpol.2012.12.007. PMID:23333485

Zhao L, Huang Y, Hu J, Zhou H, Adeleye AS, Keller AA (2016) Environ Sci Technol 50(4):2000–2010

Index

E. Heath et al. (eds.), *Fate and Effects of Anticancer Drugs in the Environment*,
https://doi.org/10.1007/978-3-030-21048-9

MIX
Papier aus verantwortungsvollen Quellen
Paper from responsible sources
FSC® C105338

If you have any concerns about our products,
you can contact us on
ProductSafety@springernature.com

In case Publisher is established outside the EU,
the EU authorized representative is:
Springer Nature Customer Service Center GmbH
Europaplatz 3, 69115 Heidelberg, Germany

Printed by Libri Plureos GmbH
in Hamburg, Germany